MEMBRANE DISTILLATION

MEMBRANE DISTILLATION
PRINCIPLES AND APPLICATIONS

by

MOHAMED KHAYET

Dept of Applied Physics I, Univ Complutense of Madrid, Spain

and

TAKESHI MATSUURA

*Industrial Membrane Res Laboratory, Dept of Chem Biol Eng,
University of Ottawa, Ottawa, Canada*

ELSEVIER

AMSTERDAM • BOSTON • HEIDELBERG • LONDON
NEW YORK • OXFORD • PARIS • SAN DIEGO
SAN FRANCISCO • SINGAPORE • SYDNEY • TOKYO

Elsevier
Radarweg 29, PO Box 211, 1000 AE Amsterdam, The Netherlands
The Boulevard, Langford Lane, Kidlington, Oxford OX5 1GB, UK

First Published 2011
Reprinted 2011

Copyright © 2011 Elsevier B.V. All rights reserved.

No part of this publication may be reproduced, stored in a retrieval system, or transmitted in any form or by any means, electronic, mechanical, photocopying, recording, or otherwise, without the prior written permission of the publisher.

Permissions may be sought directly from Elsevier's Science & Technology Rights Department in Oxford, UK: phone: (+44) 1865 843830, fax: (+44) 1865 853333, E-mail: permissions@elsevier.com. You may also complete your request online via the Elsevier homepage (http://elsevier.com), by selecting "Support & Contact"then "Copyright and Permission" and then "Obtaining Permissions."

Library of Congress Cataloging-in-Publication Data
A catalog record for this book is available from the Library of Congress

British Library Cataloguing-in-Publication Data
A catalogue record for this book is available from the British Library.

978-0-444-53126-1

For information on all Elsevier Publications
visit our Web site at www.elsevierdirect.com

11 12 13 14 15   10 9 8 7 6 5 4 3 2

Printed and bound in Great Britain

Working together to grow
libraries in developing countries

www.elsevier.com | www.bookaid.org | www.sabre.org

ELSEVIER   BOOK AID International   Sabre Foundation

This book is dedicated to our families in appreciation for their support.

Contents

Preface ix
Acknowledgements xi
Author Biographies xiii

1. Introduction to Membrane Distillation 1
2. Membranes Used in MD and Design 17
3. Formation of Flat Sheet Phase Inversion MD Membranes 41
4. Formation of Hollow Fibre MD Membranes 59
5. Thermally Induced Phase Separation for MD Membrane Formation 89
6. Membrane Modification for MD Membrane Formation 121
7. Formation of Nano-Fibre MD Membranes 163
8. MD Membrane Characterization 189
9. MD Membrane Modules 227
10. Direct Contact Membrane Distillation 249
11. Sweeping Gas Membrane Distillation 295
12. Vacuum Membrane Distillation 323
13. Air Gap Membrane Distillation 361
14. Membrane Distillation Hybrid Systems 399
15. Economics, Energy Analysis and Costs Evaluation in MD 429
16. Future Directions in Membrane Distillation 453

Index 461

Preface

As human population grows, serious problems are imposed upon the current production systems such as carbon dioxide emission into the atmosphere, shortage in fresh water supply and large energy consumption. Sustainable growth of human activities is thus becoming increasingly more difficult. Under these circumstances, development of novel industrial processes requiring less energy is of vital importance. It is hence natural that the membrane separation process has continued to replace conventional separation processes from the time of its inception almost a half century ago due to its inherently less energy requirement. As a result, pressure-driven processes such as reverse osmosis, nano-filtration, ultrafiltration, membrane gas and vapour separation are nowadays considered as well established and reliable separation processes. As they grow mature, membrane separation processes of second generation are searched for to further enhance the productivity and to alleviate the stress to our environment. Membrane distillation (*MD*) is one of such emerging membrane separation processes and is now on the verge of industrial-scale applications.

MD is one of the *non-isothermal* membrane separation processes known for about 47 years and is still being developed. It refers to a thermally driven transport of vapour through non-wetted porous hydrophobic membranes by vapour pressure difference between two sides of the membrane pores as a driving force. Heat and mass transfer through the membrane take place simultaneously in the process. Furthermore, different *MD* configurations (direct contact membrane distillation, sweeping gas membrane distillation, vacuum membrane distillation and air gap membrane distillation) can be used for various applications (desalination, environmental/waste cleanup, water-reuse, food, medical, etc.). These characteristics make *MD* attractive within the academic community as well as industrial sectors. Further, temperatures lower than in the conventional distillation, the operating hydrostatic pressures lower than in the pressure-driven processes, less demanding membrane mechanical properties and high rejection factors of non-volatile solutes achievable make *MD* look more attractive than any other separation processes. The possibility of using waste heat and renewable energy sources also enables *MD* technique to be used in conjunction with other processes in an industrial scale.

Various books have already been published on synthetic membranes and membrane separation processes but all of them are on the *isothermal* membrane processes. On the other hand, the *non-isothermal* membrane processes such as *MD* are only briefly mentioned in the final few chapters of the books as "other membrane processes" or "emerging membrane applications". This urges the authors to write a monograph in which experimental and theoretical aspects of *MD* are detailed. In this book clear discussions are made for each chosen topic in an adequate depth. Further, the book provides a broad reference base that covers a wide range of information, such as *MD* membrane preparation and characterization, transport theory in *MD*, *MD* configurations, *MD* systems and *MD* economics. The authors believe that the book is unique as the first book exclusively dedicated to the thermally driven *MD* process.

The book consists of the following chapters:

An introduction to the terminology and fundamental concepts associated with *MD* as well as a historical review of the *MD* development are presented in Chapter 1.

Chapter 2 covers both commercial and laboratory made *MD* membranes. Materials selection for membrane preparation and membrane design for *MD* application are also provided.

The technique to fabricate flat sheet *MD* membranes by the phase inversion method is described in Chapter 3.

Chapter 4 is dedicated for hollow fibre spinning and the effects of spinning parameters on hollow fibre properties.

Chapter 5 describes thermally induced phase separation (*TIPS*) method for *MD* membrane fabrication.

Chapter 6 deals with the modification of *MD* membranes.

In the above last four chapters the principles involved in *MD* membrane fabrication and parameters affecting the *MD* membrane properties are thoroughly discussed.

Chapter 7 is for the novel electro-spun nanofibre membranes used for *MD* process.

Chapter 8 deals with physical and chemical methods for characterization of *MD* membranes.

Chapter 9 covers the *MD* module types designed for each individual *MD* configuration. Different preparation processes and industrial manufacturing techniques are described in detail and illustrated by a number of clear and instructive schematic drawings and photographs.

Chapters 10–13 are for direct contact membrane distillation (*DCMD*), sweeping gas membrane distillation (*SGMD*), vacuum membrane distillation (*VMD*) and air gap membrane distillation (*AGMD*), respectively. For each configuration, mathematical model for process simulation is presented with a thorough discussion on heat and mass transfer. Temperature and concentration polarization effects and applications of each *MD* configuration are highlighted showing their advantages and disadvantages.

One of the benefits of *MD* is that it is workable in conjunction with other conventional processes such as distillation and other membrane processes. This is outlined in Chapter 14 with examples of various *MD* hybrid systems. The advantages and the drawbacks of each system are also discussed.

Economics and energy analysis as well as cost evaluations are presented in Chapter 15. Desalination cost of *MD* is compared with other membrane processes such as reverse osmosis.

Finally, future directions in *MD* are shown in Chapter 16.

The book is written for engineers, scientists, professors, graduate students as well as general readers in universities, research institutions and industries, who are engaged in membrane materials, membrane design and membrane processes. All will benefit from reading the book as many engineering aspects are included. In general, the book will be useful for all levels in the universities and for readers of the following scientific disciplines: Chemical Engineering, Applied Physics, Polymer Chemistry, Materials Science, Environmental Science and Engineering and Civil Engineering.

The authors believe that the book will have a strong impact on membrane science and technology of the future and thus will contribute significantly to the well-being of the human society.

The authors are deeply indebted to many of their colleagues and students with whom they performed theoretical and experimental studies learning together membrane technology and research.

January 2011
Mohamed Khayet and Takeshi Matsuura

Acknowledgements

Prof. Khayet wishes to express his gratitude to his Spanish family López Escorial for their unceasing encouragements, invaluable assistance and for sharing the joys during the writing of this book. Needless to say, he is also indebted to his Moroccan family.

Author Biographies

Prof. Mohamed Khayet: was born in Marrakech, Morocco, in 1966 and received his Degree in Sciences Physics (1990) from the Faculty of Sciences, University Cadi Ayyad of Marrakech (Morocco). Granted by the Spanish Agency of International Cooperation (AECI) he pursued his doctoral studies at the Faculty of Sciences Physics, Department of Atomic Molecular & Nuclear Physics, University Complutense of Madrid (UCM, Spain) and received his PhD degree in Sciences Physics from UCM in 1997. He joined the Department of Applied Physics I (UCM) in 1997 serving as an assistant and then as an associate Professor.

Granted by UCM, he realized a postdoctoral stay at the Industrial Membrane Research Institute (IMRI) in Ottawa (Canada) during the period 2000/2001. He got other grants and worked as a visiting researcher at universities and research institutions. Currently, he is professor of Thermodynamics, statistical Physics and Renewable Energy Applications in the Department of Applied Physics I (UCM) and director of the UCM research group "Membranes and Renewable Energy".

He published over 100 scientific papers in various international refereed journals, various book chapters and 3 books. He has given over 70 presentations at scientific conferences, workshops and congresses. He is author of 2 patents in the field of Membrane Science and Technology and supervised various national and international projects as well as research and academic studies. He is referee of various international journals and member of the editorial board of the international journals: Desalination; Applied Membrane Science & Technology, Membrane Water Treatment, Polymers and Membranes. His current research interests include among others future desalination technologies, renewable energy, membrane fabrication and characterization, membrane processes, modelling and transport phenomena.

Prof. Takeshi Matsuura: was born in Shizuoka, Japan, in 1936. He received his B.Sc. (1961) and M.Sc. (1963) degrees from the Department of Applied Chemistry at the Faculty of Engineering, University of Tokyo. He went to Germany to pursue his doctoral studies at the Institute of Chemical Technology of the Technical University of Berlin and received Doktor-Ingenieur in 1965.

After working at the Department of Synthetic Chemistry of the University of Tokyo as an assistant and at the Department of Chemical Engineering of the University of California, Davis, as a postdoctoral research associate, he joined the National Research Council of Canada in 1969. He came to the University of Ottawa in 1992 as a professor and the chairholder of the British Gas/NSERC Industrial Research. He served as a professor of the Department of Chemical Engineering and the director of the Industrial Membrane Research Institute (IMRI) until he retired in 2002. He was appointed to professor emeritus in 2003. As well, he served as a visiting professor at numerous overseas universities.

CHAPTER 1

Introduction to Membrane Distillation

OUTLINE

Concept of MD	2	Mechanism of MD Transport	9
Nomenclature in MD	4	Engineering Aspects: MD Applications	12
A Historical Survey of MD	5		

Membrane Science and Technology has made a tremendous progress during the last decades and membrane processes have become competitive to the conventional separation methods in a wide variety of applications. Different membrane separation processes have been developed during the past half century and new membrane applications are constantly emerging from industries or from academic and government laboratories. Membrane Distillation (MD) is one of the emerging *non-isothermal* membrane separation processes known for about 47 years and still needs to be developed for its adequate industrial implementation. It refers to a thermally driven transport of vapour through non-wetted porous hydrophobic membranes, the driving force being the vapour pressure difference between the two sides of the membrane pores. As in other membrane separation processes, the driving force is the chemical potential difference through the membrane thickness. Simultaneous heat and mass transfer occur in this process and, as will be explained later, different MD configurations such as (i) direct contact membrane distillation, (ii) sweeping gas membrane distillation, (iii) vacuum membrane distillation and (iv) air gap membrane distillation, can be used for various applications (desalination, environmental/waste cleanup, water-reuse, food, medical, etc.)

The involved simultaneous heat and mass transfer phenomena through the membrane, the different MD configurations and the various MD applications make MD attractive within the academic community as a kind of didactic application. Additionally, the possibility of using waste heat and/or alternative energy sources, such as solar and geothermal energy, enables MD to be combined with other processes in integrated systems, making it a more promising separation technique for an industrial scale. Furthermore, the lower temperatures than in the conventional distillation, the lower operating hydrostatic pressures than in the pressure-driven processes (i.e., reverse osmosis (RO), nanofiltration (NF), ultrafiltration (UF) and microfiltration (MF)), the less demanding membrane mechanical properties and the high rejection factors achievable especially during water treatment containing non-volatile solutes make MD more attractive than any other popular separation processes.

Benefiting from the low temperature and transmembrane hydrostatic pressure required to perform MD operations, several approaches to make the MD a viable separation technique were proposed. These approaches ranged from finding new areas of MD applicability and the cooperation of MD with other processes as a pre-treatment or post-treatment step, to researches devoted to preparation of membranes together with MD modules and studies of factors affecting MD production associated with the application of some enhancement techniques.

MD has been the subject of worldwide academic studies by many experimentalists and theoreticians. Unfortunately, from the commercial stand point, MD has gained only little acceptance and is yet to be implemented in the industry. The major barriers include MD membrane and module design, membrane pore wetting, low permeate flow rate and flux decay as well as uncertain energetic and economic costs.

CONCEPT OF MD

MD is a process mainly suited for applications in which water is the major component present in the feed solution. As stated earlier, MD is a thermally driven process, in which only vapour molecules are transported through porous hydrophobic membranes. The liquid feed to be treated by MD must be maintained in direct contact with one side of the membrane without penetrating its dry pores unless a trans-membrane pressure higher than the membrane liquid entry pressure (i.e., breakthrough pressure, LEP, explained in Chapter 8) is applied. The hydrophobic nature of the membrane prevents liquid solutions from entering its pores due to the surface tension forces. As a result, liquid/vapour interfaces are formed at the entrances of the membrane pores. Various MD modes differing in the technology applied to establish the driving force can be used. The differences between them are localized only in the permeate side as can be seen in Fig. 1.1.

The MD driving force may be maintained with one of the four following possibilities applied in the permeate side:

i) An aqueous solution colder than the feed solution is maintained in direct contact with the permeate side of the membrane giving rise to the configuration known as Direct Contact Membrane Distillation (DCMD). Both the feed and permeate aqueous solutions are circulated tangentially to the membrane surfaces by means of circulating pumps or are stirred inside the membrane cell by means of a magnetic stirrer. In this case the transmembrane temperature difference induces a vapour pressure difference. Consequently, volatile molecules evaporate at the hot liquid/vapour interface, cross the membrane pores in vapour phase and condense in the cold liquid/vapour interface inside the membrane module. A DCMD with liquid gap is another DCMD variant, in which a stagnant cold liquid, frequently distilled water, is kept in direct contact with the permeate side of the membrane (Fig. 1.1).

ii) Vacuum is applied in the permeate side of the membrane module by means of a vacuum pump. The applied vacuum pressure is lower than the saturation pressure of volatile molecules to be separated from the feed solution. In this case, condensation takes place outside of the membrane module. This MD configuration is termed Vacuum Membrane Distillation (VMD).

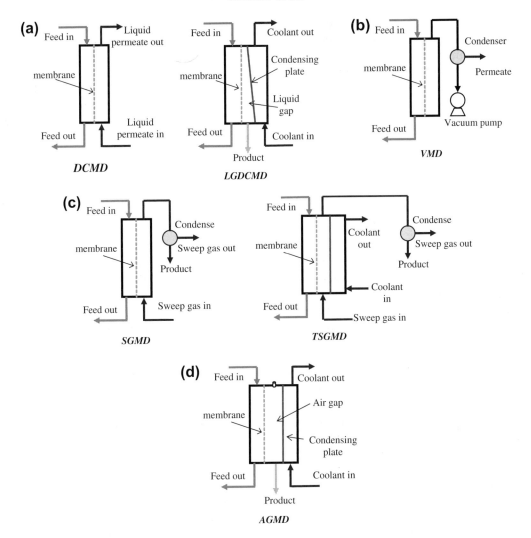

FIGURE 1.1 MD process configurations: (a) DCMD and DCMD with liquid gap; (b) VMD; (c) SGMD and thermostatic SGMD; (d) AGMD.

iii) A stagnant air gap is interposed between the membrane and a condensation surface. In this case, the evaporated volatile molecules cross both the membrane pores and the air gap to finally condense over a cold surface inside the membrane module. This MD configuration is called Air Gap Membrane Distillation (AGMD).

iv) A cold inert gas sweeps the permeate side of the membrane carrying the vapour molecules and condensation takes place outside the membrane module. This type of configuration is termed Sweeping Gas Membrane Distillation (SGMD). In this configuration, due to the heat transferred from the feed side through the membrane, the sweeping gas temperature in the permeate side increases considerably along the membrane module length. A SGMD variant termed Thermostatic Sweeping Gas Membrane Distillation (TSGMD) has been proposed recently. In this mode of SGMD the increase in the gas temperature is minimized by using a cold wall in the permeate side.

Each one of the above MD configurations has its advantages and inconveniencies for a given application as will be explained later in this book. Two examples are given below:

i) To solve the problem of heat loss by conduction through the membrane, which leads to relatively low efficiency of the MD process, an air gap was placed inside the membrane module between the permeate side of the membrane and the condensing surface. This reduces considerably both the heat loss by conduction and temperature polarization, thereby improving the separation effect. However, the permeate flux has to overcome the air barrier and therefore it is drastically reduced depending on the effective air gap width. On the other hand, because permeate is condensed on a cold surface rather than directly on membrane surface, AGMD can be applied in fields where the DCMD is limited such as the removal of organic compounds from aqueous solutions.

ii) Generally, in VMD, membranes of smaller pore size (i.e., less than 0.45 µm) than in the other MD configurations are used because in this case vacuum is applied and the risk of pore wetting is high.

NOMENCLATURE IN MD

Various authors involved in MD investigations often abbreviate DCMD term to MD and SGMD is named Membrane Air Stripping (MAS). To avoid misconceptions the adequate term should be used as explained below.

Following the 1996 International Union of Pure and Applied Chemistry (IUPAC) recommendations [1], the 21st term 'Membrane Distillation' was defined as 'distillation process in which the liquid and gas phases are separated by a porous membrane, the pores of which are not wetted by the liquid phase'. In fact, the term MD comes from the similarity of the MD to conventional distillation as both processes are based on the vapour/liquid equilibrium (VLE) for separation and both processes require heat to be supplied to the feed solution in order to achieve the required latent heat of vapourization. Before the Workshop on Membrane Distillation held in Rome on 5 May 1986, various terms had been used to identify MD, such as transmembrane distillation, thermo-pervaporation, pervaporation, membrane evaporation and capillary distillation.

The terminology for MD was first discussed by the committee formed by six members during the Workshop on Membrane Distillation held in Rome on 5 May 1986: V. Calabro (University della Calabria, Calabria, Italy), A.C.M. Franken (Twente University, Enschede, Netherlands), S. Kimura (University of Tokyo, Tokyo, Japan), S. Ripperger (Enka Membrana, Wuppertal, Germany), G. Sarti (Universita di Bologna, Bologna, Italy) and R. Schofield (University of New South Wales, Kensington, Australia). Terms, definitions and symbols related with MD have been discussed, standardized and lately reported in [2,3].

Mainly, MD should be applied for nonisothermal membrane operations in which the driving force is the partial pressure gradient across the membrane that complies with the following characteristics:

i) Porous.
ii) Not wetted by the process liquids.
iii) Does not alter the VLE of the involved species.

iv) Does not permit condensation to occur inside its pores.
v) Is maintained in direct contact at least with the hot feed liquid solution to be treated.

A HISTORICAL SURVEY OF MD

On 3 June 1963, Bodell filed the first MD patent [4] and four years later Findley published the first MD paper in the international journal *Industrial & Engineering Chemistry Process Design Development* [5]. Within the 16 cited references throughout the paper, Findley did not mention the first Patent made by Bodell [4]. Findley used the DCMD configuration using various types of membrane materials (paper hot cup, gum wood, aluminum foil, cellophane, glass fibres, paper plate, diatomaceous earth mat and nylon). Silicone and Teflon have been used as coating materials to achieve the required membrane hydrophobicity. Based on the obtained MD experimental results, Findley outlined the most suitable membrane characteristics needed for a MD membrane discussed in Chapter 2 and stated throughout the paper the following: 'calculations indicate possible economical performance, especially at high temperatures, if high temperature, long life and low cost membrane are obtainable'. Findley also suggested the possibility of using infinite-stage flash evaporation through porous membranes.

In 1967 another U.S. patent was filed by Weyl on 14 May 1964 [6] claiming an improved method and apparatus for the recovery of demineralized water from saline waters also using DCMD. It was stated that the two bodies of water may be stationary or moving, that is passing with respect to the membrane, and the process may be effected in a single stage or may be multi-staged. The membrane used was a polytetrafluoroetylene (PTFE) membrane having a thickness of 3175 µm, an average pore size of 9 µm and a porosity of 42%. In this patent, the use of other suitable hydrophobic membranes made of polyethylene (PE), polypropylene (PP) and polyvinyl chloride (PVC) was suggested. As it was considered recently and discussed in the present book in Chapter 7, Weyl [6] also stated that membranes may be constructed of non-hydrophobic material coated with a hydrophobic substance, for example, by liquid or vapour impregnation. In this patent, an alternate geometrical form for multi-stage operation was presented by having the membrane coiled up into a cylinder giving rise to the actual known DCMD with liquid gap in the spiral wound module.

A year later, in 1968, a second U.S. patent was made by Bodell [7], partly as a continuation of his first U.S. patent [4]. Bodell described a system and a method to convert impotable aqueous solutions to potable water using a parallel array of tubular silicone membranes having a 0.3 mm inner and 0.64 mm outer diameter. No membrane characteristics such as pore size and porosity were presented. Air was circulated through the lumen side of the tubular membranes and condensation was carried out in an external condenser, giving rise to the actual known Sweeping Gas Membrane Distillation (SGMD) configuration. The patent provided novel apparatus and methods for desalting seawater in an economical manner. The improved apparatus was also provided for extracting potable water from brine, sewage, urine, wastewater, bacteria-containing water and other impotable water sources. Bodell recommended the water vapour pressure in the air side of the SGMD system to be at least 4 kPa below that of the aqueous medium. Moreover, Bodell suggested, for the first time, an alternative means of providing low water vapour pressure in the tubes by applying vacuum leading to the actual known VMD configuration [4,7].

A second MD paper has been published by Findley and co-authors [8] without mentioning any of the previous cited U.S. patents [4,6,7]. The study concerns heat and mass transfer of water vapour from a hot salt solution through a hydrophobic porous membrane to a cooled water condensate. Their experimental studies indicated that the major factor influencing the rates of transfer was the diffusion through the stagnant gas (i.e., air) in the membrane pores. First theoretical calculations have been reported taking into account the membrane thermal conductivity and the film heat transfer coefficients. An empirical correction related to the possible internal condensation and diffusion along the surfaces has been considered to perform their calculations [8].

In Europe, a seawater desalination 'SGMD' process using dry air was also proposed by Van Haute and Henderyckx [9,10] in the 2nd European Symposium on Fresh Water from the Sea held in Athens (Greece) in May of 1967. The authors stated that the proposed apparatus can utilize waste hot water and it should be possible to use it for solar distillation; but it has not been further developed.

After this short period of time, interest in the MD process has faded quickly losing its brightness due partly to the observed lower MD production compared to that of reverse osmosis (RO) process. Rodgers [11,12], in his successive patents related to distillation, presented a system and a method of desalination using a stack of flat sheet membranes separated by non-permeable corrugated heat transfer films and working under DCMD configuration. A temperature gradient is applied over the membrane stack, and the latent heat passing from feed to distillate is recovered by heat transfer to a lower temperature feed. Thus the latent heat of vapourization may be used several times, as with multiple effect evaporation. The feed liquid supplied was subjected to treatment including heating and deaeration. The main object was to provide an improved economical desalination system and method of desalination. It was disclosed in the aforementioned patents that the distillation unit comprises a multiplicity of sheet-like elements all substantially rectangular having the same dimensions and arranged with their edges in alignment so that the distillation unit takes the form of a relatively thin parallelepiped. It was also indicated that the suitable materials for the membranes are those which permit the formation of microporous membranes having high porosity, that is 70–80%, uniform pore size distribution and must be either poorly wettable or non-wettable (i.e., hydrophobic) by the used liquids or can be treated to render them non-wettable. The cited polymers were polycarbonates, polyesters, polyethylene, polypropylene and the halogenated polyethylenes, particularly the fluorocarbons. Particular mention has been made of polyvinylidene fluoride (PVDF) as a preferred membrane material and the so called 'solvent–non-solvent' casting process as the preferred method of forming the membranes. The use of cellulose nitrate, cellulose acetate and cellulose triacetate microscopic porous filter media coated with silicone water repellant to provide a non-wetting porous membrane was also mentioned.

MD process has recovered much interest within the academic communities in the early 1980s when novel membranes with better characteristics and modules became available [13–23].

A Gore-Tex membrane, which is an expanded PTFE membrane having a thickness of 50 μm and 0.5 μm pore size, has been used first by Esato and Eiseman [13] as a biologically inert membrane oxygenator, and later was proposed by Gore & Associated Co. under the name 'Gore-Tex Membrane Distillation' for MD application in a spiral-type module using the liquid gap DCMD configuration (Fig. 1.1) [14]. The proposed Gore-Tex membranes are made of PTFE having a thickness as low as 25 μm, porosity up to 80% and a pore size of 0.2–0.45 μm.

Other types of MD membranes, method and apparatuses have been proposed by Cheng and Wiersma in a series of U.S. patents [15–18]. The object of the first patent [15], filed by Cheng on 14 February 1979, was to provide an improved thermal membrane distillation process with continuous distillate production over a prolonged period of time. A desalination system with three cell stages has been presented. Multiple-layered (i.e., composite) membranes have been proposed comprising a thin hydrophobic microporous layer or membrane and a thin hydrophilic layer or membrane. The hydrophilic layer was maintained adjacent to the distilland (i.e., salt water) whereas the hydrophobic layer was kept adjacent to the distillate (i.e., fresh water). The proposed composite porous membrane was formed by clamping the hydrophobic/hydrophilic layers closely together to form a cell with a suitable support backing to maintain the integrity of the composite membrane. The evaporation and condensation phenomena took place within the micropores of the hydrophobic layer while the hydrophilic layer prevents intrusion of distilland into the pores of the hydrophobic layer. It was reported that, in the case of salt water distillation, generally higher distillate production rates have been observed in composite membranes with the smaller pore sizes in the hydrophilic layer than in the hydrophobic layer. The best results have been obtained with the hydrophobic layer having a mean pore size smaller than 0.5 µm. It was also reported that the hydrophilic layer can be non-porous. The proposed hydrophobic materials for the composite membrane include PTFE and PVDF, whereas the proposed hydrophilic materials included cellulose acetate, cellulose nitrate, mixed esters of cellulose and polysulfone.

The following patents [16–18] filed by Cheng and Wiersma claimed the use of composite membranes having a thin microporous hydrophobic layer and one or two thin hydrophilic layers with the hydrophobic layer sandwiched between the two hydrophilic layers. It was stated that the two hydrophilic layers may be of different materials (cellulose acetate, mixed esters of cellulose, polysulfone and polyallylamine) and the composite membrane could be formed by coating the hydrophilic layers on the hydrophobic layer. It was found that fresh water production rate for a distilland bulk temperature 62.8 °C and distillate temperature 56.7 °C was 75.2 kg/m^2·day. For the same distilland and distillate temperature conditions a composite membrane of similar structure except with non-hydrophilic layer on the distillate side of the hydrophobic membrane yielded a fresh water production rate of 51.3 kg/m^2·day. In other words, it was reported that the addition of the hydrophilic layer of the distillate side of the hydrophobic membrane increased the fresh water production rate by almost 50%.

In the fourth patent, [18] filed by Cheng and Wiersma on 4 March 1982, an improved apparatus and method for MD was proposed using a composite membrane comprising a microporous hydrophobic layer having deposited thereon an essentially non-porous hydrophilic coating. The hydrophobic layer of the membrane had either asymmetrical or symmetrical shaped micropores. Fluoro-substituted vinyl polymers, which are suitably hydrophobic, were proposed as ideal materials for the microporous hydrophobic layer of the composite membrane. The coating material and method were selected taking into account the adequate adhesion to the hydrophobic substrate, its resistance to both mechanical abrasion and chemical damage from the distilland, its ability to be coatable as a thin continuous layer on the surface of the porous substrate and to allow certain liquids to pass through to the hydrophobic layer of the membrane. Examples were plasma polymerized allylamine, dehydrated polyvinyl alcohol and polyacrylic acid.

At the same time, the Swedish National Development Co. (Svenska Utvecklings AB) developed the plate and frame membrane module applying the AGMD configuration [19,20]. The German company Enka AG presented polypropylene (PP) hollow fibre membranes in tubular modules at the Europe–Japan Joint Congress on Membranes

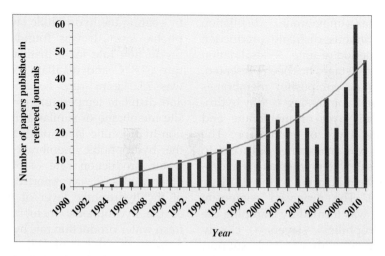

FIGURE 1.2 Growth of research MD activity up to the year 2010 represented as a plot of number of papers published in refereed journals for each year.

and Membrane Processes, held at Stresa (Italy) in 1984 [21]. Their experiments using DCMD process with heat recovery took more academic orientation by publication of their results in international journals [22,23]. In the same congress, other papers on MD have also been communicated [24,25]. More MD papers were presented at the Second World Congress on Desalination and Water Reuse in 1985, held in Bermuda. This renewed interest is a result of the development of various types of porous hydrophobic membranes used in different MD configurations [26–30]. Since then, numerous studies have been carried out but taking more academic interest rather than industrial. Most of those studies are published in national and/or international journals such as *Journal of Membrane Science and Desalination* as will be shown later in this book. Recently, interest in MD has increased significantly as can be seen in Fig. 1.2, presenting the number of published papers in journals per year in the MD field.

It must be stated that the number of MD papers referenced in the 1997 MD review by Lawson and Lloyd [31] was below 87 and that mentioned in the recent MD review (2006) by El-Bourawi et al. [32] is 168; however, actually the number of MD papers published in international journals is more than 500.

It is worth quoting that within the published papers in international journals, DCMD is the most studied MD configuration, as it is illustrated in Fig. 1.3, although the heat loss by conduction through the membrane matrix is higher than in the other MD configurations. 63.3% of the MD studies are focused on DCMD as in this configuration condensation step is carried out inside the membrane module leading to a simple operation mode. In contrast, SGMD is the least studied configuration, only 4.5% of the MD published papers. This is due to the fact that SGMD requires external condensers to collect the permeate and a source for gas circulation.

Most of the considered publications in Figs 1.2 and 1.3 are concerned with theoretical models of MD and experimental studies on the effects of the operating conditions. 50.4% of the MD publications dealt with theoretical models (i.e., 40.6% for DCMD, 48.4% for AGMD, 40.3% for VMD and 72.2% for SGMD) whereas only about 16.7% of the MD papers

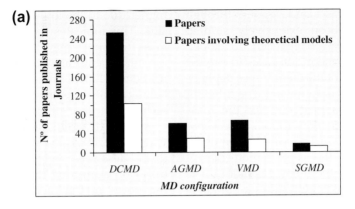

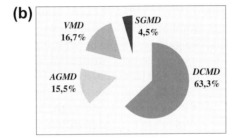

FIGURE 1.3 Number of MD papers published in refereed journals for each configuration including the papers presenting theoretical models (a) and percentages of each MD configuration (b).

reported in journals are focused on the preparation of MD membranes. Few authors have considered the possibility of manufacturing novel membranes and membrane module designs specifically for MD applications [33]. As a matter of fact, commercial microporous hydrophobic membranes available in capillary or flat sheet forms have been used in MD experiments although these membranes were prepared initially for other purposes, for example MF. Discussions on this subject are given in the next chapter.

MECHANISM OF MD TRANSPORT

In MD process, both heat and mass transfer through porous hydrophobic membranes are involved simultaneously. The mass transfer occurs through the pores of the membrane whereas heat is transferred through both the membrane matrix and its pores. The heat transfer within the membrane is due to the latent heat accompanying vapour or gas flux and the heat transferred by conduction across both the membrane material and the gas-filled membrane pores. One must pay attention that only water vapour or volatile compounds are transported through the membrane pores from the feed side to the permeate side as the membrane is hydrophobic. In addition, there is a presence of fluid boundary layers adjoining both the feed and permeate membrane sides giving rise to the phenomena called *temperature polarization* and *concentration polarization*. These phenomena are explained in this book for each MD configuration.

The transport of gases and vapours through porous media has been extensively studied and theoretical models have been developed based on the kinetic theory of gases to predict the MD performance of the membranes depending on the MD configuration used. The different types of mechanisms proposed for the mass transport are (Knudsen flow model, viscous or Poiseuille flow model, ordinary molecular diffusion model and/or the combination thereof often summarized as the dusty gas model.) The governing quantity which provides a guideline in determining the operating mass transport mechanism in a given pore under given experimental conditions is Knudsen number (Kn) defined as the ratio of the mean free path, λ, of the transported molecules to the pore size of the membrane. The mean free path (λ_i) for a species i can be calculated using the following expression:

$$\lambda_i = \frac{k_B T}{\sqrt{2}\pi \, \bar{p} \, \sigma_i^2} \quad (1.1)$$

where σ_i is the collision diameter (2.641 Å for water vapour), k_B is the Boltzmann constant, $\bar{p}$ the mean pressure within the membrane pores and T the absolute temperature.

For the binary mixture (i and j) in air, the mean free path can be evaluated by the following equation:

$$\lambda_{i/j} = \frac{k_B T}{\pi \, \bar{p}((\sigma_i + \sigma_j)/2)^2} \frac{1}{\sqrt{1 + M_j/M_i}} \quad (1.2)$$

where σ_i and σ_j are the collision diameters and M_i and M_j the molecular weight of the components i and j, respectively.

In DCMD, the mean free path for water vapour at 50 °C under atmospheric pressure is approximately 0.14 µm, which is around the pore sizes of the membranes used in MD. However, in VMD, the mean free path value is higher due to the low pressure in the permeate side. This indicates that the physical nature of mass transport may be different when using the same membrane under different MD configurations. Furthermore, for the membrane having a pore size distribution, different mechanisms may occur simultaneously.

In AGMD configuration, the transport of vapours through the membrane was assumed to be described by the theory of molecular diffusion admitting the presence of air inside the pores of the membrane and in the gap width as a stagnant film. Stefan diffusion and binary type relations (i.e., Fick's equation of molecular diffusion) as well as Stefan–Maxwell equations were used to describe the multicomponent mass transfer in AGMD systems [34–39]. In all these theoretical models, the pore size was not considered although experimental studies proved the dependence of the AGMD flux on this parameter [40,41]. Recently, attempts were made to predict the AGMD performance process using the dusty gas model that takes into account all membrane parameters to describe the simultaneous Knudsen diffusion, molecular diffusion and viscous flow models [42,43].

In MD theoretical studies, generally a membrane of uniform and non-interconnected cylindrical pores is assumed. The pore size distribution of MD membranes rather than their uniform pore size has been considered to a lesser extent in DCMD configuration [44–47], in VMD configuration [48] and in SGMD configuration [49]. A three-dimensional network of interconnected cylindrical pores with a pore size distribution was considered for Monte Carlo simulation of DCMD [50,51] and VMD [52]. The agreements between the predicted MD permeate fluxes and the experimental ones were found to be good.

It must be pointed out that the presence of air within the membrane pores between the feed and permeate liquid/vapour interfaces hinders the mass transfer resulting in a reduction of the DCMD flux. Deaerated DCMD systems were proposed [22,53–55]. This can be carried out by lowering the pressure of the liquid streams hence controlling the maximum pressure of gas within the membrane pores. For membranes having small pore sizes, Knudsen flow is predominant and the removal of air results only in a small increase in the DCMD flux; however, for membranes having larger pores a substantial increase in the DCMD flux can be achieved by deaeration.

In MD processes, the transport of molecules through the membrane matrix (i.e., surface diffusion) is neglected due to the fact that the diffusion area of the membrane matrix is small compared to the pore area. For hydrophobic MD membranes, the 'affinity' between water and the membrane material is very low and it may be allowed to neglect the contribution of transport through the membrane matrix especially for porous membranes with large pore sizes and high porosities. Nevertheless, when other compounds are present in the aqueous feed solution especially for compounds having strong 'affinity' with the membrane material, the transport mechanism through the matrix of the membrane may have a significant effect. Systematic studies are needed to clarify this point in MD. It was reported that surface diffusion may affect MD performance in membranes with small pore sizes (<0.02 μm) [56,57]. A theoretical model considering mass transport through the membrane matrix by solution–diffusion mechanism has been proposed for VMD and an extensive comparative study between pervaporation separation process (PV) and VMD employing the same membrane material (PVDF) has been carried out [48]. This is explained in detail in Chapter 12.

As stated above, the heat transfer within the membrane is due to the latent heat accompanying vapour flux and the heat transferred by conduction across both the membrane material and the gas-filled membrane pores. The following equation was applied in various studies.

$$Q_m = \frac{k_m}{\delta}(T_{m,f} - T_{m,p}) + \sum_{i=1}^{s} J_i^t \Delta H_{v,i} \quad (1.3)$$

where k_m is the thermal conductivity of the membrane, δ is the membrane thickness, $\Delta H_{v,i}$ is the evaporation enthalpy of the species i of the transmembrane flux J_i^t, s is the number of permeated components, $T_{m,f}$ is the temperature of the feed aqueous solution at the membrane surface and $T_{m,p}$ is the temperature of the permeate at the membrane surface.

It is worth quoting that of the total heat flux transferred through the membrane, typically 50–80% is consumed as latent heat for permeate production, while the remainder is lost by thermal conduction. In fact, the heat loss by conduction through the membrane matrix becomes less significant when the MD system works under high operating temperatures, which are lower than the boiling point of the feed aqueous solution. This may be considered one method to minimize heat loss through the membrane, which is one of the inconveniences of MD process, in general, and DCMD, in particular.

In VMD, the boundary layer resistance in the permeate side and the contribution of the heat transported by conduction through the membrane are frequently neglected [31,32,58,59]. This makes VMD of pure water useful to determine the temperature of the feed solution at the membrane surface ($T_{m,f}$) as it cannot be measured directly and therefore the boundary layer heat transfer coefficients in the membrane module can be evaluated [60]. This procedure has been used for selecting the adequate empirical heat transfer correlation of a given MD system, which is a complex task when developing theoretical models to determine the temperature polarization coefficients. In fact, the use of empirical heat transfer correlations in MD was questioned and even criticized as these correlations were developed originally for only heat exchangers.

ENGINEERING ASPECTS: MD APPLICATIONS

The MD process is currently applied mostly at the laboratory scale and the MD applications are very appropriate for environmental, chemical, petrochemical, food, pharmaceutical and biotechnology industries. Recently, some pilot plant applications have been proposed for desalination and nuclear desalination but are still under experimental tests and their use is not fully extended [61–67].

The major MD application has been in desalination for production of high purity water. Near 100% rejection of non-volatile elyctrolytes (i.e., sodium chloride, NaCl; potassium chloride, KCl; lithium bromide, LiBr; etc.) and non-electrolytes (i.e., glucose, sucrose, fructose, etc.) solutes present in aqueous solutions was achieved. A quality water as low as 0.8 μS/cm electrical conductivity with 0.6 ppm TDS (total dissolved solids) was produced [68]. As the permeate product is very pure it is suitable for use in medical and pharmaceutical sectors. In fact, in the case of a solution with non-volatile components only water molecules flow through the membrane pores. It must be mentioned here that Weyl [6] was the first in conducting desalination by DCMD. However, the obtained permeate fluxes were up to $1 \text{ kg/m}^2 \cdot \text{h}$, which were lower than the RO permeate fluxes ($20-75 \text{ kg/m}^2 \cdot \text{h}$). Actually, due to MD membrane module improvement, the MD production begins to be competitive to RO process in the field of desalination with nearly total rejection factors, which can not be accomplished by RO at high permeate fluxes.

MD has been applied successfully to wastewater treatment at a laboratory scale, either to produce a permeate less hazardous to the environment or to recover valuable compounds. MD has been tested for the treatment of pharmaceutical wastewater containing taurine, textile wastewater contaminated with dyes such as methylene blue, aqueous solutions contaminated with boron, arsenic, heavy metals, ammonia (NH_3), coolant liquid (i.e., glycols), humic acid and acid solutions rich in specific compounds, oil-water emulsions, olive oil mill wastewater for polyphenols recovery and radioactive wastewater solutions. It was proved that DCMD is feasible to process low and medium-level radioactive wastes giving high decontamination factor in only one stage and can be applied for nuclear desalination [62,69,70]. Recently, DCMD was proposed for wastewater reclamation in space in a combined direct osmosis system [71].

Due to the fact that MD can be conducted at relatively low feed temperatures, it was successfully tested in many areas where high temperature applications lead to degradation of the process fluids especially in food processing. It was demonstrated that MD can be used for the concentration of milk [30], for the recovery of volatile aroma compounds from black currant juice [72], for the concentration of must (i.e., the juice obtained from grape pressing containing sugars and a wide variety of aroma compounds) [73] and for the concentration of many other types of juices including orange juice, mandarin juice, apple juice, sugarcane juice, etc. It was concluded that the utilization of either osmotic distillation (OD) and/or MD in the food industry for concentration or separation is promising especially at high feed concentration degrees. This will be discussed in Chapter 10.

MD also has potential applications in biotechnology. As an example, MD has been used for the removal of toxic products from culture broths. The application of DCMD unit connected to a laboratory bioreactor for the selective recovery of ethanol from the culture medium has been reported [74]. The experiments were run at a constant temperature of 38 °C on anaerobic cultures of *fragilis*. MD was also applied for the concentration of biological solutions such as bovine plasma and bovine blood [75,76]. It was demonstrated that MD was suitable for stable removal of solute free water from blood with a haematocrit of 45%. DCMD was applied to the direct concentration of protein (0.4% and 1% bovine serum albumin at pH 7.4) aqueous solutions at low temperatures and found that fouling effects were practically absent, while the limiting factor of the process was the temperature polarization [77].

It is known that azeotropic mixtures are impossible to be separated by simple distillation. Thus, the application of MD for breaking azeotropic mixtures was proposed and tested for the separation of hydrochloric acid/water, propionic acid/water and formic acid/water azeotrope mixtures [78,79]. It was demonstrated that MD is of potential interest in breaking azeotropic mixtures. The effect of the inert gases, helium, air and sulfur hexafluoride, in breaking the formic acid/water azeotropic mixtures was studied [80]. The selectivity was found to be larger and near unity when using helium (around 0.96), followed by that in air (about 0.9) and then in sulfur hexafluoride (0.85–0.86). The results were related with the different diffusivities of the components in the inert gas.

MD has been proposed for the extraction of volatile organic compounds (VOCs) from dilute aqueous solutions. Various types of dilute binary mixtures containing VOCs at different concentrations were tested by different MD configurations and membrane modules. Values of the selectivity different from those calculated on the basis of the corresponding VLE data were found. Removal from water of alcohols such as methanol, ethanol, isopropanol and n-butanol; halogenated VOCs such as chloroform, trichloroethylene and tetrachloroethylene, benzene, acetone, acetonitrile, ethylacetate, methylacetate and methyltertbutyl ether among others were studied. The potential advantage of MD for ethanol recovery from fermentation broth was also reported [34]. It must be mentioned here that the addition of salt such as magnesium chloride ($MgCl_2$) during the treatment of aqueous alcohol feed solutions was found to increase the alcohol selectivity significantly with only a slight decrease in the total permeate flux. This was attributed to the reduction in water vapour pressure leading to a decrease in the water mass transfer through the membrane [36,78].

The concentration of aqueous solutions containing sodium hydroxide (NaOH) and the strong mineral acid, sulfuric acid (H_2SO_4), at different pH values has been investigated [30]. Comparable MD permeate flux and electrical conductivity to those obtained using sodium chloride (NaCl) aqueous solutions was noticed. MD separation of aqueous solutions containing volatile solutes such as nitric acid (HNO_3) and hydrochloric acid (HCl) have been conducted and similar trends for both components were found, different from that of the aqueous solutions containing non-volatile solutes [30]. Attempts were made for the concentration of hydrogen iodide (HI) and sulphuric acid aqueous solutions in relation to hydrogen energy production from water using DCMD and AGMD [81].

Details are given in the following chapters of the present book.

References

[1] W.J. Koros, Y.H. Ma, T. Shimadzu, Terminology for membranes and membrane processes (IUPAC recommendations 1996), J. Membr. Sci. 120 (1996) 149–159.

[2] A.C.M. Franken, S. Ripperger, Terminology for membrane distillation, European Society of Membrane Science and Technology, Issued January (1988).

[3] C.A. Smolders, A.C.M. Franken, Terminology for membrane distillation, Desalination 72 (1989) 249–262.

[4] B.R. Bodell, Silicone rubber vapor diffusion in saline water distillation, United States Patent Serial No. 285,032 (1963).

[5] M.E. Findley, Vaporization through porous membranes, Ind. & Eng. Chem. Process Des. Dev. 6 (1967) 226−237.
[6] P.K. Weyl, Recovery of demineralized water from saline waters, United States Patent 3, 340, 186 (1967).
[7] B.R. Bodell, Distillation of saline water using silicone rubber membrane, United States Patent Serial No. 3,361,645 (1968).
[8] M.E. Findley, V.V. Tanna, Y.B. Rao, C.L. Yeh, Mass and heat transfer relations in evaporation through porous membranes, AIChE J. 15 (1969) 483−489.
[9] A. Van Haute, Y. Henderyckx, The permeability of membranes to water vapor, Desalination 3 (1967) 169−173.
[10] Y. Henderyckx, Diffusion doublet research, Desalination 3 (1967) 237−242.
[11] F.A. Rodgers, Stacked microporous vapor permeable membrane distillation system, United States Patent Serial No. 3,650,905 (1972).
[12] F.A. Rodgers, Compact multiple effect still having stacked impervious and previous membranes, United States Patent Serial No. Re.27,982 (1974), original No. 3,497,423.
[13] K. Esato, B. Eiseman, Experimental evaluation of Gore-Tex membrane oxygenator, The Journal of Thoracic and Cardiovascular Surgery 69 (1975) 690−697.
[14] D.W. Gore, Gore-Tex membrane distillation, Proc. of the 10th Ann. Convention of the Water Supply Improvement Assoc., Honolulu, USA, July 25−29, (1982).
[15] D.Y. Cheng, Method and apparatus for distillation, United States Patent Serial No. 4,265,713 (1981).
[16] D.Y. Cheng, S.J. Wiersma, Composite membrane for a membrane distillation system, United States Patent Serial No. 4,316,772 (1982).
[17] D.Y. Cheng, S.J. Wiersma, Composite membrane for a membrane distillation system, United States Patent Serial No. 4,419,242 (1983).
[18] D.Y. Cheng, S.J. Wiersma, Apparatus and method for thermal membrane distillation, United States Patent Serial No. 4,419,187 (1983).
[19] L. Carlsson, The new generation in sea water desalination: SU membrane distillation system, Desalination 45 (1983) 221−222.
[20] S.I. Andersson, N. Kjellander, B. Rodesjo, Design and field tests of a new membrane distillation desalination process, Desalination 56 (1985) 345−354.
[21] Catalogue of Enka AG presented at Europe−Japan Joint Congress on Membranes and Membrane Processes, Stresa, Italy, June (1984).
[22] K. Schneider, T.J. van Gassel, Membrandestillation, Chem. Ing. Tech. 56 (1984) 514−521 (in German).
[23] T.J. van Gassel, K. Schneider, An energy-efficient membrane distillation process, in: E. Drioli, M. Nagaki (Eds.), Membranes and Membrane Processes, Plenum Press, New York, 1986, pp. 343−348.
[24] G.C. Sarti, C. Gostoli, Use of hydrophobic membranes in thermal separation of liquid mixture, Proc. Europe−Japan Meeting on Membranes and Membrane Processes, Stresa (June 1984).
[25] K. Schneider, T.J. van Gassel, Membrane distillation, Proc. Europe−Japan Meeting on Membranes and Membrane Processes, Stresa (June 1984).
[26] E. Drioli, Y. Wu, Membrane distillation: An experimental study, Desalination 53 (1985) 339−346.
[27] G.C. Sarti, C. Gostoli, S. Matulli, Low energy cost desalination processes using hydrophobic membranes, Desalination 56 (1985) 277−286.
[28] A.S. Jonsson, R. Wimmerstedt, A.C. Harrysson, Membrane distillation: A theoretical study of evaporation through microporous membranes, Desalination 56 (1985) 237−249.
[29] W.T. Hanbury, T. Hodgkiess, Membrane distillation: An assessment, Desalination 56 (1985) 287−297.
[30] S. Kimura, S. Nakao, Transport phenomena in membrane distillation, J. Membr. Sci. 33 (1987) 285−298.
[31] K.W. Lawson, D.R. Lloyd, Review: membrane distillation, J. Membr. Sci. 124 (1997) 1−25.
[32] M.S. El-Bourawi, Z. Ding, R. Ma, M. Khayet, A framework for better understanding membrane distillation separation process: Review, J. Membr. Sci. 285 (2006) 4−29.
[33] M. Khayet, T. Matsuura, J.I. Mengual, M. Qtaishat, Design of novel direct contact membrane distillation membranes, Desalination 192 (2006) 105−111.
[34] C. Gostoli, G.C. Sarti, Separation of liquid mixtures by membrane distillation, J. Membr. Sci. 41 (1989) 211−224.
[35] F.A. Banat, J. Simandl, Desalination by membrane distillation: A parametric study, Sep. Sci. & Tech. 33 (1998) 201−226.
[36] F.A. Banat, J. Simandl, Membrane distillation for dilute methanol: Separation from aqueous streams, J. Membr. Sci. 163 (1999) 333−348.
[37] F.A. Banat, F. Abu Al-Rub, R. Jumah, M. Al-Shannag, Application of Stefan-Maxwell approach to azeotropic separation by membrane distillation, Chem. Eng. J. 73 (1999) 71−75.
[38] F.A. Banat, F. Abu Al-Rub, M. Shannag, Modeling of dilute ethanol-water mixture separation by membrane distillation, Sep. & Pur. Tech. 16 (1999) 119−131.

REFERENCES

[39] F.A. Banat, F. Abu Al-Rub, R. Jumah, M. Shannag, Theoretical investigation of membrane distillation role in breaking the formic acid-water azeotropic point: Comparison between Fickian and Stefan-Maxwell-based models, Int. Comm. Heat Mass Transfer 26 (1999) 879–888.

[40] M.A. Izquierdo-Gil, M.C. García-Payo, C. Fernández-Pineda, Air gap membrane distillation of sucrose aqueous solutions, J. Membr. Sci. 155 (1999) 291–307.

[41] M.A. Izquierdo-Gil, M.C. García-Payo, C. Fernández-Pineda, Air gap membrane distillation of sucrose aqueous solutions, J. Membr. Sci. 155 (1999) 291–307.

[42] C.M. Guijt, G.W. Meindersma, T. Reith, A.B. Haan, Air gap membrane distillation: 1. Modelling and mass transport properties for hollow fiber membranes, Sep. & Pur. Tech. 43 (2005) 233–244.

[43] M.A. Izquierdo-Gil, M.C. García-Payo, C. Fernández-Pineda, Direct contact membrane distillation of sugar aqueous solutions, Sep. Sci. & Tech. 34 (1999) 1773–1801.

[44] L. Martínez, F.J. Florido-Díaz, A. Hernández, P. Prádanos, Characterization of three hydrophobic porous membranes used in membrane distillation: Modelling and evaluation of their water vapor permeabilities, J. Membr. Sci. 203 (2002) 15–27.

[45] L. Martínez, F.J. Florido-Díaz, A. Hernández, P. Prádanos, Estimation of vapor transfer coefficient of hydrophobic porous membranes for applications in membrane distillation, Sep. & Pur. Tech. 33 (2003) 45–55.

[46] J. Phattaranawik, R. Jiraratananon, A.G. Fane, Effect of pore size distribution and air flux on mass transport in direct contact membrane distillation, J. Membr. Sci. 215 (2003) 75–85.

[47] F. Laganà, G. Barbieri, E. Drioli, Direct contact membrane distillation: Modelling and concentration experiments, J. Membr. Sci. 166 (2000) 1–11.

[48] M. Khayet, T. Matsuura, Pervaporation and vacuum membrane distillation processes: Modeling and experiments, AIChE J. 50 (2004) 1697–1712.

[49] M. Khayet, J. Rodríguez-López, Efecto de la distribución de tamaño de poro en la predicción de la permeabiliadad de membranas hidrófobas: Destilación en membrana con gas de barrido, in: J.M. Ortiz de Zárate, M. Khayet Souhaimi, La investigación del Grupo Especializado de Termodinámica de las Reales Sociedades Españolas de Física y de Química, Madrid (Spain) (2006) 43–56.

[50] A.O. Imdakm, T. Matsuura, A Monte Carlo simulation model for membrane distillation processes: Direct contact (MD), J. Membr. Sci. 237 (2004) 51–59.

[51] A.O. Imdakm, T. Matsuura, Simulation of heat and mass transfer in direct contact membrane distillation (MD): The effect of membrane physical properties, J. Membr. Sci. 262 (2005) 117–128.

[52] A.O. Imdakm, M. Khayet, T. Matsuura, A Monte Carlo simulation model for vacuum membrane distillation process, J. Membr. Sci. 306 (2007) 341–348.

[53] R.W. Schofield, A.G. Fane, C.J.D. Fell, R. Macoun, Factors affecting flux in membrane distillation, Desalination 77 (1990) 279–294.

[54] R.W. Schofield, A.G. Fane, C.J.D. Fell, Gas and vapor transport through microporous membranes: I. Knudsen-Poiseuille transition, J. Membr. Sci. 53 (1990) 159–171.

[55] A.G. Fane, R.W. Schofield, C.J.D. Fell, The efficient use of energy in membrane distillation, Desalination 64 (1987) 231–243.

[56] Y. Fujii, S. Kigoshi, H. Iwatani, M. Aoyama, Selectivity and characteristics of direct contact membrane distillation type experiment: I. Permeability and selectivity through dried hydrophobic fine porous membranes, J. Membr. Sci. 72 (1992) 53–72.

[57] Y. Fujii, S. Kigoshi, H. Iwatani, M. Aoyama, Y. Fusaoka, Selectivity and characteristics of direct contact membrane distillation type experiment: II. Membrane treatment and selectivity increase, J. Membr. Sci. 72 (1992) 73–89.

[58] K.W. Lawson, D.R. Lloyd, Membrane Distillation. I. Module design and performance evaluation using vacuum membrane distillation, J. Membr. Sci. 120 (1996) 111–121.

[59] S. Bandini, A. Saavedra, G.C. Sarti, Vacuum membrane distillation: Experiments and modeling, AIChE J. 43-2 (1997) 398–408.

[60] J.I. Mengual, M. Khayet, M.P. Godino, Heat and mass transfer in vacuum membrane distillation, Int. J. Heat & Mass Transfer 47 (2004) 865–875.

[61] J. Koschikowski, M. Wieghaus, M. Rommel, Solar thermal-driven desalination plants based on membrane distillation, Desalination 156 (2003) 295–304.

[62] M. Khayet, J.I. Mengual, G. Zakrzewska-Trznadel, Direct contact membrane distillation for nuclear desalination. Part II. Experiments with radioactive solutions, Int. J. Nuclear Desalination 2 (2006) 56–73.

[63] F. Banat, N. Jwaied, Economic evaluation of desalination by small-scale autonomous solar-powered membrane distillation units, Desalination 220 (2008) 566–573.

[64] J. Gilron, L. Song, K.K. Sirkar, Design for cascade of crossflow direct contact membrane distillation, Industrial & Engineering Chemistry Research 46 (2007) 2324–2334.

[65] Y. Xu, B.K. Zhu, Y.Y. Xu, Pilot test of vacuum membrane distillation for seawater desalination on a ship, Desalination 189 (2006) 165–169.

[66] J.H. Hanemaaijer, Memstill® — low cost membrane distillation technology for seawater desalination, Desalination 168 (2004) 355.

[67] J.H. Hanemaaijer, J. van Medevoort, A.E. Jansen, C. Dotremont, E. van Sonsbeek, T. Yuan, L.D. Ryck, Memstill membrane distillation — a future desalination technology, Desalination 199 (2006) 175–176.

[68] K. Karakulski, M. Gryta, A. Morawski, Membrane processes used for potable water quality improvement, Desalination 145 (2002) 315–319.

[69] G. Zakrewska-Trznadel, M. Harasimowicz, A.G. Chmielewski, Membrane processes in nuclear technology-application for liquid radioactive waste treatment, Sep. Pur. Tech. 22-23 (2001) 617–625.

[70] G. Zakrewska-Trznadel, M. Harasimowicz, A.G. Chmielewski, Concentration of radioactive components in liquid low-level radioactive waste by membrane distillation, J. Membr. Sci. 163 (1999) 257–264.

[71] T.Y. Cath, D. Adams, A.E. Childress, Membrane contactor processes for wastewater reclamation in space, II. Combined direct osmosis, osmotic distillation, and membrane distillation for treatment of metabolic wastewater, J. Membr. Sci. 257 (2005) 111–119.

[72] R. Bagger-Jorgensen, A.S. Meyer, C. Varming, G. Jonsson, Recovery of volatile aroma compounds from black currant juice by vacuum membrane distillation, J. Food Engineering 64 (2004) 23–31.

[73] S. Bandini, G.C. Sarti, Concentration of must through vacuum membrane distillation, Desalination 149 (2002) 253–259.

[74] H. Udriot, S. Ampuero, I.W. Marison, U. von Stokar, Extractive fermentation of ethanol using membrane distillation, Biotechnology letters 11 (1989) 509–514.

[75] K. Sakai, T. Muroi, K. Ozawa, S. Takesawa, M. Tamura, T. Nakane, Extraction of solute-free water from blood by MD, Am. Soc. Artif. Intern. Organs 32 (1986) 397–400.

[76] K. Sakai, T. Koyano, T. Muroi, M. Tamura, Effects of temperature and concentration polarization on water vapor permeability for blood in membrane distillation, The Chemical Engineering Journal 38 (1988) B33–B39.

[77] J.M. Ortiz de Zárate, C. Rincón, J.I. Mengual, Concentration of bovine serum albumin aqueous solutions by membrane distillation, Sep. Sci. & Tech. 33 (1998) 283–296.

[78] H. Udriot, A. Araque, U. von Stockar, Azeotropic mixtures may be broken by membrane distillation, Cheng. Eng. J. 54 (1994) 87–93.

[79] M.C. García-Payo, C.A. Rivier, I.W. Marison, U. von Stokar, Separation of binary mixtures by thermostatic sweeping gas membrane distillation: II. Experimental results with aqueous formic acid solutions, J. Membr. Sci. 198 (2002) 197–210.

[80] F.A. Banat, F. Abu Al-Rub, R. Jumah, M. Shannag, On the effect of inert gases in breaking the formic acid-water azeotrope by gas-gap membrane distillation, Chem. Eng. J. 73 (1999) 37–42.

[81] G. Caputo, C. Felici, P. Tarquini, A. Giaconia, S. Sau, Membrane distillation of HI/H_2O and H_2SO_4/H_2O mixtures for the sulphur-iodine thermochemical process, Int. J. of Hydrogen Energy 32 (2007) 4736–4743.

CHAPTER 2

Membranes Used in MD and Design

OUTLINE

Introduction	17	Copolymer Flat Sheet and Hollow Fibre Membranes	27
Commercial Membranes	19	Composite Bi- and Multi-Layered Membranes with Different Hydrophobicity Levels	28
Laboratory-Made MD Membranes: Porous Hydrophobic Membrane Preparation Techniques	21	Nanofibre Membranes	33
		Track-Etched Membranes	34
Flat Sheet Single Hydrophobic Layer Membranes	22	MD Membrane Engineering and Membrane Material Selection for MD	35
Hollow Fibre Single Hydrophobic Layer Membranes	24		

INTRODUCTION

As stated in Chapter 1, the membranes to be used in membrane distillation (MD) must be porous and hydrophobic. It can be a single hydrophobic layer (i.e. conventional and most used membrane), a composite porous bilayer hydrophobic/hydrophilic membrane or a composite trilayer hydrophilic/hydrophobic/hydrophilic or hydrophobic/hydrophilic/hydrophobic porous membrane. A hydrophilic/hydrophobic membrane type was also proposed for MD and tested in desalination [1–3]. In fact, both supported and unsupported membranes can be used in this process. The pore size of the membranes frequently used in MD lies between 10 nm and 1 μm and the porosity should be as high as possible. It is generally admitted that the MD permeate flux increases with the increase of the pore size and/or porosity. The choice of a membrane for MD applications is a compromise between a low heat transfer flux by conduction achieved using thicker membranes and a high permeate flux achieved using thin membranes having large pore size, low pore tortuosity, and high porosity.

It is agreed upon that in MD the membrane itself acts only as a physical barrier sustaining the liquid–vapour interfaces formed at the entrances of the membrane pores. Volatile compounds present in feed solution are transported across the membrane pores according to the vapour/liquid equilibrium (VLE) principle and both heat and mass transfer occur simultaneously through the membrane. Many MD researchers believe that selective properties in MD process are not governed by the membrane transport phenomena. Since the hydrophobic character of the membrane is a crucial requirement in MD, membranes have to be made by materials with low values of surface energy. Hydrophobicity can be achieved by either using hydrophobic materials or making the hydrophilic membrane surface energy as low as possible applying different surface modification techniques.

It must be remembered that more than 35 years ago Rodgers [4,5] presented successive U.S. patents related to distillation, by which a system and a method of desalination were claimed. A stack of flat sheet membranes separated by non-permeable corrugated heat transfer films were used in direct contact membrane distillation (DCMD) configuration. It was indicated that the suitable materials for the membranes were those which permitted the formation of microporous membranes having high porosity (i.e. 70–80%) and uniform pore size distribution. They were also either poorly wettable or non-wettable (i.e. hydrophobic or treated to render them non-wettable) by the used liquids. The cited polymers were polycarbonates (PCs), polyesters (PSTs), polyethylene (PE), polypropylene (PP) and the halogenated PEs, particularly the fluorocarbons. Special attention was made to polyvinylidene fluoride (PVDF) as a preferred membrane material and to the 'solvent–non-solvent' casting process as the preferred method for preparation of membranes. Moreover, it was mentioned the use of cellulose nitrate (CT), cellulose acetate (CA) and cellulose triacetate (CTA) microscopic hydrophilic porous filters for MD when they were coated with a silicone water repellent material to render them non-wettable porous membranes.

As reported in Chapter 1, MD was introduced in the late 1960s but did not attain a fully commercial status as a water desalination process like reverse osmosis (RO) technology, partly because membranes with the characteristics most suitable for the MD process were not available, especially at reasonable prices. These characteristics include negligible permeability to the liquids and non-volatile components (i.e. high hydrophobicity, low surface energy, small maximum pore size, narrow pore size distribution), high porosity (i.e. high permeate flux and low thermal conductivity), high resistance to heat flow by conduction through the membrane matrix (i.e. heat transfer by conduction through the membrane matrix is considered heat loss in MD as no corresponding mass transfer is produced), sufficient but not excessive thickness (i.e. permeate flux is inversely proportional to the membrane thickness and mechanical strength is proportional to the membrane thickness), low moisture absorptivity inside membrane pores (i.e. no condensation should occur inside membrane pores) and long life. In other words, a good MD membrane should exhibit a low membrane resistance to mass transfer, a high liquid entry pressure (LEP) of distilled water or feed solutions to be treated in order to guarantee dryness of the membrane pores, low thermal conductivity of the membrane material, good thermal stability and excellent chemical resistance to most of the feed solutions. One of the important controlling parameters to be considered in MD process is the thermal conductivity of the whole membrane (i.e. membrane material), which must be as low as possible.

Conventionally, the membranes used in most of the MD studies were fabricated for other separation processes such as microfiltration (MF) rather than MD. In fact, most of the required characteristics for MD membranes are met with commercially available MF membranes. On the other hand, the need for novel membranes fabricated especially for MD purposes has been widely accepted and recommended by many MD investigators. Only few researchers, however, have worked on the design, preparation and testing of MD membranes as will be shown later on in this chapter. Details of commercial membranes used in different MD studies will also be presented in this chapter.

It is worth quoting that developments in the MD process were made mainly in the early 1980s when newer, more suitable membranes became available, for instance hydrophobic polytetrafluoroethylene (PTFE) membranes. In fact, PTFE represents an ideal material for membrane manufacturing for MD since among polymers it exhibits a high hydrophobic character, good chemical resistance and high thermal stability. The basic disadvantage of PTFE lies in its difficult processability. Moreover, at present, commercial PTFE membranes are usually produced through complicated extrusion, rolling and stretching or sintering procedures [6]. Other polymers such as PP and PVDF were employed in the preparation of membranes for filtration purposes, in which the driving force was the hydrostatic pressure. But these membranes were also used for different MD applications. For instance, PP membranes are prepared either by molten extrusion technique followed by stretching or by the so-called thermal phase separation process that needs polymer dissolution at high temperature in less common solvents [6]. PVDF dissolves at room temperature in a variety of solvents and therefore porous membranes can be easily produced by phase separation (i.e. phase inversion) process, simply immersing the cast solution film in a coagulant bath (i.e. non-solvent, frequently water). In this case, membrane porosity is controlled by the additives (i.e. pore-forming agents) in the casting solution or by replacing water in the coagulation bath with a different non-solvent media [6]. Recently, copolymers such as poly(vinylidene fluoride-hexafluoropropylene) (PVDF-HFP) and poly(vinylidene fluoride-co-tetrafluoroethylene) (PVDF-TFE) are used to prepare MD membranes in flat sheet or hollow fibre using the phase inversion technique [7–9]. Membrane surface modification using different technologies such as grafting, coating or blending fluorinated surface-modifying macromolecules (SMMs) with hydrophilic polymers were also tested for different MD systems and configurations [10–13]. Lately, attempts were made to use nanofibre membranes prepared by electrospinning method in MD desalination [14]. It should especially be noted that significant results were obtained recently in the preparation and modification of polymeric membranes. Also, the improvement of the MD permeate flux has increased the reliability of MD process.

COMMERCIAL MEMBRANES

Commercial micro-porous hydrophobic membranes made of PP, PVDF or PTFE (Teflon), available in tubular, capillary or flat sheet forms have been used in MD experiments. Tables 2.1 and 2.2 summarize the commercial membranes commonly used in MD studies together with their principal characteristics as specified by the manufacturers. However, these membranes were developed for MF applications and the manufactured data (rejection and hydraulic permeability) are characteristics for MF. The morphological architecture of some synthetic membranes comes close to fulfilling the requirements of membranes for MD.

Measurements of the hydraulic permeability are not relevant for membranes to be applied in MD process, because hydraulic permeation involves mass transfer mechanisms other than vapour transport, e.g. liquid viscous flow. Chapters 10–13 are dedicated for mass transport mechanisms in different MD configurations.

Detailed explanations of the characterization of some commercial membranes are given in Chapter 8 together with relevant figures.

Membranes were studied in commercial modules formed in shell-and-tube, plate-and-frame and spiral-wound configurations and were tested in laboratory and pilot plant scale experiments. Related information is given in Chapter 9.

TABLE 2.1 Flat Sheet Commercial Membranes Commonly Used in MD

Membrane type	Membrane trade name	Manufacturer	Material	δ (μm)	d_p (μm)	ε (%)	LEP_w (kPa)
Flat sheet membranes	TF200	Gelman	PTFE/PP[a]	178	0.20	80	282
	TF450				0.45		138
	TF1000				1.00		48
	Taflen		PTFE	60	0.8	50	–
	GVHP	Millipore	PVDF[b]	110	0.22	75	204
	HVHP			140	0.45		105[c]
	FGLP		PTFE/PE[a]	130	0.2	70	280
	FHLP			175	0.5	85	124
	FALP			150	1.0	85	48.3
	Gore		PTFE	64	0.2	90	368[d]
				77	0.45	89	288[d]
			PTFE/PP[a]	184	0.2	44	463[d]
	Enka		PP	100	0.1	75	
				140	0.2		
	Celgard 2500	Hoechst Celanese Co.	PP	28	0.05[e]	45	–
	Celgard 2400			25	0.02	38	
	Metricel[f]	Gelman		90	0.1	55	
	Vladipore[g]		–	120	0.25	70	–
	3MA	3M Corporation	PP	91	0.29[h]	66	
	3MB			81	0.40[h]	76	
	3MC			76	0.51[h]	79	–
	3MD			86	0.58[h]	80	
	3ME			79	0.73[h]	85	
	Teknokrama[i]		PTFE	–	0.2	80	–
					0.5		
					1.0		
	G-4.0-6-7[j]	GoreTex Sep GmbH	PTFE	100	0.20	80	463[c]

δ, membrane thickness; d_p, mean pore size; ε, porosity; LEP_w, liquid entry pressure of water.

[a] Flat sheet polytetrafluoroethylene (PTFE) membranes supported by polypropylene (PP) or polyethylene (PE).
[b] Flat sheet polyvinylidene fluoride (PVDF) membranes.
[c] Measured value [21].
[d] Measured value [22].
[e] Maximum pore size (0.07 μm).
[f] Reported in [18].
[g] Membrane used in [23].
[h] Maximum pore size [24].
[i] Membrane used in [25].
[j] Spiral-wound module, SEP Gesellschaft für Technische Studien, Entwicklung, Planung mbH, filtration area: 4 m².

TABLE 2.2 Capillary Commercial Membranes Commonly Used in MD

Membrane type	Membrane trade name	Manufacturer	Material	δ (μm)	d_p (μm)	ε (%)	LEP_w (kPa)
Capillary membranes	Accurel® S6/2 MD020CP2N[a]	AkzoNobel Microdyn	PP	450	0.2	70	140
	MD020TP2N	Enka Microdyn		1550	0.2	75	
	Accurel® BFMF 06-30-33[b]	Enka A.G. Euro-Sep		200	0.2	70	
	Celgard X-20	Hoechst Celanese Co.		25	0.03	35	–
	Sartocon®-Mini SM 3031 750701W[c]	Sartorius	Polyolefine	–	0.22	–	
	PTFE[d]	Sumitomo Electric	POREFLON	550	0.8	62	
	PTFE[e]	Gore-tex	TA001	400	2 (maximum pore size)	50	

PP, polypropylene; δ, membrane thickness; d_p, mean pore size; ε, porosity; LEP_w, liquid entry pressure of water.

[a] Shell-and-tube capillary membrane module: Filtration area, 0.1 m²; inner capillary diameter, 1.8 mm; length of capillaries, 470 mm.
[b] Shell-and-tube capillary membrane module: Filtration area, 0.3 m²; inner capillary diameter, 0.33 mm; length of capillaries, 200 mm.
[c] Plate-and-frame module; dimensions, 138/117/7 mm; filtration area, 0.1 m².
[d] polytetrafluoroethylene (PTFE) hollow fibre: inner/outer diameters, 0.9/2 mm [26].
[e] PTFE hollow fibre: inner diameters, 1 mm [27].

It is worth mentioning that the role of some commercial membranes in the MD process transport was discussed in several articles published in scientific journals rather than by the manufacturers [15–20]. For example, the membranes TF200, TF450 and TF1000 have a heterogeneous structure with a thin PTFE layer supported by a PP net, which is wetted by aqueous solutions when DCMD is considered. The pores of the PTFE layer are maintained dry and constitute the resistance for vapour transport through the membrane, while the PP net contributes to the mass transfer resistance in the liquid phase [18].

The choice of a membrane for MD process from those cited in Tables 2.1 and 2.2 is a compromise between a high permeate flux, a high separation factor and a low thermal conductivity under given MD operating conditions and for a given feed aqueous solution to be treated. These characteristics should be given by the manufacturers to help customers select their appropriate membrane.

Other than the commercial membranes that were patented, reported in journals or communicated in meetings and congresses [4,5,7,28–36], very few more studies have been performed on the preparation and modification of membranes for MD process as will be shown in the next section.

LABORATORY-MADE MD MEMBRANES: POROUS HYDROPHOBIC MEMBRANE PREPARATION TECHNIQUES

Other than the commercial membranes shown in the previous section, which were prepared for other membrane processes, very

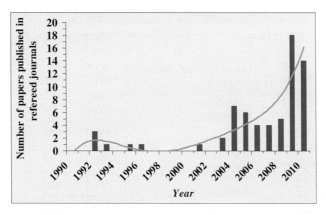

FIGURE 2.1 MD membrane preparation and testing activities up to the year 2010 represented as a plot of number of papers published in refereed journals per year.

few laboratory researches have been performed on the preparation and modification of membranes designed specifically for MD process. It is despite the fact that one can find easily in the membrane literature different types of porous and hydrophobic membranes prepared and characterized for other purposes rather than MD. As will be shown later, the MD membrane architecture must exhibit specific characteristics and fulfil specific requirements. Therefore, the study on the characteristics of MD membranes is very necessary. Fortunately, some significant results were obtained recently in the preparation and modification of polymeric membranes and their testing in MD, providing increase reliability for the MD process. Fig. 2.1 shows the number of scientific papers on membrane preparation and testing in different MD applications. An abrupt increase in the number of papers on MD membrane engineering (i.e. design, preparation and testing in MD) is seen since only 6 years ago and it is hoped that this trend will continue in the future. Improved MD membranes with specific morphology and microstructure are highly demanded. Membranes with different pore sizes, porosities, thicknesses and materials are required in order to carry out systematic MD studies for better understanding mass transport in different MD configurations, thereby improving the permeate flux.

MD is usually carried out with commercial MF membranes, but a number of research laboratories worldwide have started to produce their own membranes in order to improve the MD flux and selectivity. In what follows some of the reported studies in the field of MD membrane design are reviewed.

Flat Sheet Single Hydrophobic Layer Membranes

Ortiz de Zárate et al. [37] presented asymmetric PVDF flat sheet membranes prepared by the phase inversion technique from binary solutions of PVDF/dimethyl acetamide (DMAC) or PVDF/dimethyl formamide (DMF) of different polymer concentrations (10–25 wt%). The membranes were tested in DCMD configuration. It was observed that both the pore size and the porosity increased with the decrease of the PVDF concentration in the polymer casting solution, although there was no improvement in DCMD flux compared to the commercial membranes (GVHP in Table 2.1). A maximum in transmembrane mass transfer coefficient appeared at different polymer concentrations, depending on the solvent (i.e. 13.2 wt% PVDF for DMAC and 15 wt% PVDF for DMF). This result was attributed to the membrane top layer, which became denser at concentrations higher than the optimum value and therefore the DCMD fluxes were reduced. Also, it was mentioned that for PVDF concentrations lower than 15 wt%, the formed membrane matrix was defective with big holes, which were wetted easily by the feed solution, and therefore could not contribute to the DCMD process.

Another study was carried out by Tomaszewska [38] on the preparation and characterization of PVDF membranes for MD using DMF and DMAC, the same solvents employed by Ortiz de Zárate et al. [37]. But lithium chloride (LiCl) salt was introduced in the casting solution as a non-solvent additive. The effects of LiCl concentration, casting solution composition, solvent evaporation time prior to coagulation and temperature of the coagulation bath on the structural properties and permeability of the PVDF flat sheet membranes prepared by the phase inversion method were investigated. The PVDF concentration in the casting solutions was varied from 8 to 15 wt%. It was observed that an increase in both porosity (> 79%) and pore size (0.0698–0.349 μm) occurred with the enhancement of the LiCl concentration, which was varied from 0 to 3 wt%. The LEP_w decreased from 290 to 45 kPa. The addition of only 0.5 g of LiCl per 8 g of PVDF in the DMAC solution increased the permeate flux very strongly from 90 to about 180 l/m²·day. Further addition of LiCl in the PVDF casting solution caused a gradual increase of the permeate flux from 180 to 240 l/m²·day when the feed solution containing 1–2% sodium chloride (NaCl) was treated at feed and permeate temperatures of 60 and 20 °C, respectively. Solute rejection factors higher than 99% have been achieved. The drawback of using LiCl as an additive is the observed drastic decrease of the mechanical resistance. This was due to the presence of big cavities in the membrane structure. It was further concluded that the characteristics and properties of the membrane were affected by the composition of the casting solution as well as by the temperature of the coagulation bath. It is worth noting that the same type of membranes were prepared previously by Bottino et al. [39] for ultrafiltration (UF) applications, in which an increase of the porosity and UF permeation flux of the PVDF membranes was observed with an increase in the content of LiCl in the casting solution.

Liquid water was selected by Khayet and Matsuura [40] and Khayet et al. [41] as a non-solvent additive to improve the VMD permeability and to reduce the cost of the MD membranes, which must also be taken into account as an important parameter for the MD membrane design. Both supported and unsupported PVDF flat sheet membranes were prepared by 15 wt% PVDF in the solvent DMAC. The support was a non-woven PST backing material (Osmonics, Inc.). The concentration of water in the casting solution was varied from 0 to 6.8 wt%. The prepared membranes were tested for the removal of volatile organic compounds (e.g. chloroform) from water by VMD configuration. Various membrane characterization techniques were applied to determine the porosity, pore size, pore size distribution and LEP_w. An increase of the porosity (26.8–79.6%) and pore size (0.02–0.7 μm) with an increase of water content in the casting solution was observed for both the supported and unsupported membranes, whereas the LEP_w decreased. The VMD permeate flux increased exponentially with the water content in the PVDF casting solution for both the supported and unsupported membranes (i.e. 1–14 kg/m²·h for the supported membranes and 0.6–16 kg/m²·h for the unsupported membranes). Both the VMD permeate flux and the mass transfer coefficients of the supported membranes were found to be higher than those of the unsupported membranes when the amount of water in the casting solution was lower than 4.25 wt%. However, at the 4.3 and 5.1 wt% water concentrations, the overall mass transfer

TABLE 2.3 Tarflen Membranes Prepared in Szczecin University of Technology (Poland)

Membrane type	Membrane trade name	Material	δ (μm)	d_{min} (μm)	d_p (μm)	d_{max} (μm)	ε (%)	τ^a
Flat sheet Membranes	Tarflen1	PTFE	62.0	0.16	0.21	0.27	81	2.1 (2.7)
	Tarflen2		60.0	–	–	0.34	65	(2.2)
	Tarflen5		84.5	–	–	0.25	59	(2.7)
	Tarflen6		72.7	–	–	019	21	(3.2)

PTFE, polytetrafluoroethylene; δ, membrane thickness; d_p, mean pore size; d_{min}, minimum pore size; d_{max}, maximum pore size; ε, porosity; τ, pore tortuosity determined by gas permeation measurements.
^{a} Values in brackets were estimated using d_{max}.
Source: Data were published in [17] (www.inderscience.com).

coefficients were lower for the supported membranes. This was attributed to the resistance of the backing material. Furthermore, it was found that the separation factor decreased with increasing the concentration of water in the PVDF casting solution for both the supported and unsupported membranes and was generally lower for the supported membranes. The decrease of the separation factor was explained by the increase of the permeation rate of water through the membrane pores.

Microporous hydrophobic flat sheet membranes (Tarflen, Poland) were prepared from the commercial PTFE tape by extraction of the lubricating compounds, followed by controlled uniaxial stretching and heating [42–45]. The measured contact angle of water at the surface of this type of membranes was found to be 106° and their characteristics are shown in Table 2.3. These membranes were applied successfully in DCMD by different Polish laboratories in desalination and for the concentration of acids and separation of aqueous isotopic solutions.

Hollow Fibre Single Hydrophobic Layer Membranes

Other than flat sheet membranes, hollow fibre-type membranes were also prepared using different techniques and tested in MD. Fujii et al. [46,47] span porous hollow fibre membranes for DCMD from different polymers including PVDF, polysulfone (PS), poly(phenylene oxide) (PPO), polyacrylonitrile (PAN) and CTA using the dry/wet spinning technique. Dimethyl sulfoxide (DMSO) was used as solvent for PVDF and CTA, while N-methyl-2-pyrrolidone was used as solvent for PS and PPO. It must be pointed out that some of the used polymers such as CTA are relatively hydrophilic, but according to the authors the membrane process is still called DCMD (see section 1.2 in Chapter 1). No differences were given for the transport mechanism between the hydrophobic membrane such as PVDF and the hydrophilic membrane such as CTA, and no details were given why the authors thought that the process used could be termed DCMD when hydrophilic porous membranes were used. In any case, the prepared hollow fibre membranes exhibited pore sizes smaller than those of MF membranes. For example, PVDF hollow fibres were prepared with 4.0–24.8 nm mean pore size, 56–73% porosity, 0.675–0.844 mm internal diameter and 0.982–1.071 mm external diameter. When aqueous ethanol solutions were subjected to DCMD, the permeate fluxes of ethanol and water were found to be directly proportional to their partial vapour pressure difference across the membrane and the separation factors varied according to the membrane polymers and the DCMD operation conditions. Ethanol permeability in the DCMD experiments was fairly constant but that of water varied widely for different operation conditions and membranes. PPO membrane exhibited an ethanol selectivity of 7.3 with an ethanol flux of 0.168 kg/m^2·h and a water flux of 0.422 kg/m^2·h. The ethanol selectivity of PVDF membranes was found to be lower (i.e. 3.7−5.2) with an ethanol flux of 0.239−0.640 kg/m^2·h and a water flux of 0.878−2.314 kg/m^2·h. PS membrane exhibited very low ethanol selectivity (1.6−3.1) with an ethanol flux of 0.272−1.21 kg/m^2·h and a water flux of 1.637−10.34 kg/m^2·h. The ethanol selectivity of PAN membrane was 1.3−3.2 with an ethanol flux of 0.045−0.199 kg/m^2·h and a water flux of 0.307−2.737 kg/m^2·h. CTA membrane exhibited the lowest ethanol selectivity (i.e. 1.1) with an ethanol flux of 0.177 kg/m^2·h and a water flux of 2.962 kg/m^2·h.

Using the wet spinning technique, various asymmetric microporous PVDF hollow fibre membranes with different pore sizes (i.e. 0.031−0.068 μm), effective porosities (71−1516 m^{-1}) and morphologies were prepared for VMD by the phase inversion method using the solvent DMAC and the nonsolvent additives LiCl and water [48,49]. Techniques and procedures to determine the mean pore size and effective porosity of MD membranes are given in detail in Chapter 8. The different pore sizes and porosities were achieved by varying the dope composition and spinning conditions. It was found that the polymer dope composition was the most significant parameter controlling the morphology and permeation characteristics of the PVDF hollow fibre membranes. The increase of the PVDF polymer yields a reduction of the effective surface porosity and mean pore size of both the internal and external surfaces of the hollow fibres. The PVDF hollow fibre membranes were employed to remove 1,1,1-trichloroethane (TCA) from aqueous solutions of different TCA concentrations and for toluene and benzene removal from water. It was observed that the PVDF hollow fibres post-treated by the solvent exchange method using ethanol exhibited higher porosity and higher permeability since the applied post-treatment prevented pore collapse and closure due to the membrane shrinkage during air drying step. Under optimum VMD operating parameters, particularly a downstream pressure of 8−10.7 kPa, a feed temperature of 50 °C and a feed flow rate of 10^{-3} m^3/h, up to 97% TCA removal efficiency has been achieved. For benzene and toluene removal from water by VMD, a separation factor over 99% was obtained under the following optimal operating conditions (downstream pressure: 8 kPa, feed temperature: 50 °C and feed flow rate: 10^{-3} m^3/h).

It must be mentioned here that first attempts have been made to cast PVDF dopes on the outer surface of porous tubular supports in order to prepare PVDF porous hollow fibres and capillaries for MD [50]. In this case, the porous tubular support acts as a pore-forming agent by simply stretching the nascent membrane during immersion in the coagulation water bath. In other words, it can be explained that during immersion the free shrinkage of the cast solution was hindered by the support and as a consequence the nascent membrane was stretched. This stretching action of the support on the nascent membrane during coagulation in water bath was responsible for the formation of pores with sizes in the range of a few hundred nanometers. These sizes along with the hydrophobic character of the polymer make the developed membranes potentially useful for MD application. An additional advantage of this type of membranes lies in their improved mechanical properties due, obviously, to the support. Various PVDF membranes with different structures, porosities and permeation properties were fabricated and applied in SGMD desalination [50]. It was concluded that among the different casting parameters studied, the PVDF solution concentration was the most relevant parameter affecting PVDF membrane properties. In this case, diluted and less viscous solutions yielded a more homogeneous distribution of the PVDF coating on the braid support and permitted the formation of membranes with higher permeability. When using 3 w/v% NaCl aqueous solution, high salt rejection factors were observed (i.e. 99.5–99.9%) and the permeate flux was slightly higher than that of some commercial membranes but lower than 2.6 l/m^2·h. It was also observed that the permeate flux was practically independent from the immersion time of the braid in the polymer solution. The variation of the casting parameters did not practically affect the structure of the cross-section of the PVDF membrane since it was largely controlled by the polymer solution and in particular by the exchange rate between the solvent and the water during the final immersion step of the cast solution. Dilute and less viscous casting solutions penetrated into the support and formed thinner and more fragile coatings that were more affected by the stretching action of the support and consequently more porous and highly permeable PVDF membranes were formed.

Recently, Wang et al. [51] prepared a PVDF hollow fibre membrane for DCMD using the solvent N-methyl-1-pyrrolidone (NMP) and ethylene glycol as a non-solvent additive. The dry/jet wet spinning method was employed for hollow fibre fabrication. The PVDF concentration was 12 wt% and that of ethylene glycol was 8 wt%. The fabricated PVDF hollow fibre exhibited 0.16 μm mean pore size, a very narrow pore size distribution and an external ultra-skin layer over a porous support layer. It was mentioned that the fully porous membrane structure had the advantage of decreasing the vapour transport resistance, leading to an enhancement of the permeation flux. When using an aqueous salt solution of 3.5 wt% as a feed, a feed temperature of 79.3 °C and a permeate temperature of 17.5 °C, the PVDF hollow fibre membrane produced 41.5 kg/m^2·h with a rejection as high as 99.99%. The authors claimed developing a membrane with a DCMD performance comparable or superior to most of commercially available PVDF hollow fibre membranes.

It is worth quoting that a series of studies have been carried out in an effort to improve the properties of PVDF membranes by introducing non-solvent additives in the PVDF polymer dope, although these membranes were not prepared for MD purposes. Deshmuck and Li [52] and Wang et al. [53] introduced poly(vinylpyrrolidone) (PVP) in the PVDF casting solution as an additive in order to obtain highly porous PVDF hollow fibre membranes. However, trace quantities of PVP in the membrane affect the hydrophobicity of the PVDF membrane. These membranes cannot be used in MD. Uragami et al. [54] tested the effect of polystyrene sulfonic acid in the PVDF dope as well as the influence of poly(ethylene glycol) (PEG). The permeability of the prepared flat sheet membranes was improved, as it was expected. By contrast, the strength of the membranes was reduced. Khayet et al. [55,56] also used the additive PEG for preparation of PVDF hollow fibre membranes for UF. Glycerol and phosphoric acid were employed by Benzinger and Robinson [57] as pore-forming agents in PVDF membranes to increase the membrane permeability. Shih et al. [58] used ethanol to increase the gas permeability through dried PVDF membranes. Small molecular additives such as water, ethanol and i-propanol were used by Wang et al. [59] for the preparation of PVDF asymmetric hollow fibre membranes. It was found that the wet fibres exhibited high water permeability, while the dry fibres had a high gas permeability, good mechanical strength and excellent hydrophobicity.

Melt-extruded/cold-stretching method was used by Li et al. [60] to prepare PE and PP hollow fibre membranes for desalination by DCMD and VMD. Compared to PP hollow fibre membranes, higher water fluxes have been obtained for PE membranes in both DCMD and VMD. This was attributed to the larger pore size of the PE membranes compared to PP ones. The highest permeate flux reported was 0.8 l/m^2·h in DCMD and about 4 l/m^2·h in VMD.

Copolymer Flat Sheet and Hollow Fibre Membranes

Other flat sheet and hollow fibre MD membranes have been prepared by the phase inversion method using copolymers of PVDF, PTFE and PP such as PVDF-TFE and PVDF-HFP. The membrane characteristics and MD performances were compared to those of the PVDF membranes prepared under similar conditions. Flat sheet PVDF-TFE asymmetric microporous membranes were fabricated by Feng et al. [7,8] and it was found that the PVDF-TFE membranes prepared from the copolymer casting solutions containing LiCl exhibited better mechanical performances, higher contact angles (i.e. higher hydrophobicity) and lower DCMD permeate fluxes than those of the PVDF membranes (i.e. the DCMD flux was 6.5 and 3.5 kg/m^2·h for the PVDF and PVDF-TFE membranes, respectively, for a feed temperature of 55 °C and a permeate temperature of 20 °C). The low permeate flux observed for the PVDF-TFE membrane was attributed to its smaller pore size ($< 2.4 \times 10^{-2}$ μm) and porosity ($<80\%$) compared to those of PVDF membranes [7]. However, the PVDF-TFE membranes exhibited excellent mechanical properties (i.e. stretching strain and extension ratios at break of PVDF-TFE membranes were approximately 6–8 times greater than those of PVDF membranes). For all tested membranes, close to 100% separation factors were observed when salt aqueous solutions were employed as feed. Moreover, the same authors [8] used lithium perchlorate trihydrate (LiClO$_4$·3H$_2$O)/trimethyl phosphate (TMP) as a pore-forming additive. The pore size, porosity and DCMD permeate fluxes of the prepared PVDF-TFE membranes were found to be higher than those prepared using LiCl.

García-Payo et al. [9] proposed the design, preparation, characterization and application in DCMD of hollow fibre membranes prepared by the dry/wet spinning technique employing the copolymer PVDF-HFP at different concentrations from 17 to 24 wt%. The solvent was DMAC and the non-solvent additive was PEG. All the spinning parameters were kept constant except the copolymer concentration. The temperature of both the internal and external coagulants was maintained at 40 °C. The effects of the copolymer concentration on the morphological properties of the PVDF-HFP hollow fibres were studied by atomic force microscopy (AFM), scanning electron microscopy (SEM) and DCMD experiments. Changes of the cross-section structure of the PVDF-HFP hollow fibres with the increase of the copolymer concentration were detected as can be seen in Fig. 2.2. The cross-section of the hollow fibre prepared with the lowest PVDF-HFP concentration exhibited a finger-like structure at both the external and internal layers, which disappeared first from the internal layer as the PVDF-HFP concentration was increased and, finally, a sponge-like structure was formed throughout the cross-section of the hollow fibre at the highest PVDF-HFP concentration. These results were explained based on the decrease of the coagulation rate with the increase of the copolymer concentration in the dope solution. The proposed copolymer hollow fibre membranes, especially those prepared with the lowest PVDF-HFP concentrations, were proved to be promising for MD applications.

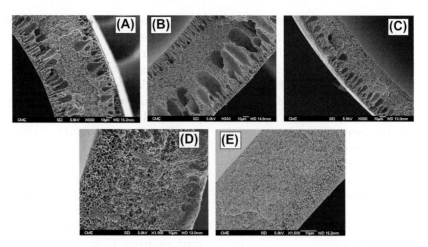

FIGURE 2.2 Cross-sectional morphology of PVDF-HFP hollow fibre membranes prepared by the dry/wet spinning technique using different copolymer concentrations in DMAC and 3 wt% PEG as non-solvent additive: (a) 17 wt%, (b) 19 wt%, (c) 20 wt%, (d) 22 wt% and (e) 24 wt%. *Source: Reprinted from [9]. Copyright 2009, with kind permission from Elsevier.*

Composite Bi- and Multi-Layered Membranes with Different Hydrophobicity Levels

Composite membranes were prepared by other authors with either two or three layers of different hydrophobicity levels (i.e. different water contact angles). As stated in Chapter 1, the use of composite membranes in MD has been reported by Cheng and Wiersma for the first time in late 1980s [32–35] as a series of patents. Since then only few papers have been published using the patented concept.

Ohta et al. [61] prepared a partially hydrophilic dense fluoro-carbon composite membrane and tested it for seawater desalination. The authors used the term MD for both porous and dense membranes. DCMD configuration was used and the obtained permeate fluxes ($< 6 \, \text{kg/m}^2 \cdot \text{h}$) were of magnitude similar to those achieved with porous hydrophobic membranes. The effects of the DCMD operation parameters were similar to those observed for a single porous hydrophobic layer. It was also found that the permeability and thermal efficiency of fluoro-carbon membrane were superior to those of silicon membrane. However, no details were given on the transport mechanism through this type of membranes.

Kong et al. [62] employed a hydrophilic microporous CN membrane surface modified via plasma polymerization using two monomer systems: octafluorocyclobutane (OFCB) and vinyltrimethylsilicon/carbon tetrafluoride (VTMS/CF_4). A tri-layer membrane with a hydrophilic layer sandwiched between two hydrophobic layers was prepared and tested in DCMD employing 0.3–0.5 M NaCl feed aqueous solutions. The effects of various polymerization conditions on the DCMD performance and structure of the membrane were investigated. The tri-layer membrane exhibited similar DCMD behaviour to the typical DCMD behaviour frequently observed when using single-layer porous hydrophobic membranes. Moreover, it was found that the DCMD performance of the plasma-modified CN membrane by VTMS/CF_4 was better than that prepared using OFCB, and the DCMD permeate flux of both membranes decreased while the salt rejection factor increased when the discharge time was getting longer. This was attributed to the gradual decrease of the pore size with increasing discharge time until a dense layer was formed on the membrane surfaces. The same group, Wu et al. [63], extended their study by presenting other composite membranes for MD. These membranes were prepared via radiation polystyrene grafting of CA to achieve the required hydrophobicity. The morphological structure of the membranes was investigated by various techniques including SEM, contact angle measurements and X-ray photoelectron spectroscopy (XPS). Compared to the asymmetric PVDF membrane having a permeability of 0.68×10^{-3} kg/m$^2 \cdot$h$\cdot$Pa and a rejection factor of 99.9%, it was found that the permeability of the polystyrene-grafted CA membrane ranged from 0.14×10^{-3} to 0.96×10^{-3} kg/m$^2 \cdot$h$\cdot$Pa with a rejection factor decreasing from 99.1 to 66.7%; the OFCB plasma-modified CN membrane exhibited higher permeability (0.82×10^{-3} to 1.32×10^{-3} kg/m$^2 \cdot$h$\cdot$Pa) and lower rejection factor (99.5–92.1%); the permeability of the VTMS/CF$_4$ plasma-modified CN membrane was higher than that of the PVDF membrane (1.81–2.22×10^{-3} kg/m$^2 \cdot$h$\cdot$Pa) with a rejection factor varying from 99.9 to 96.4%.

The effects of coating polymers and heat treatment for PVDF hollow fibre membranes were investigated later on by Fujii et al. [47]. Silicone rubber (Si-LTV, Dow Corning Toray Silicone Co. Ltd.), poly(1-trimethylsilyl-1-propyne) (PMSP) and polyketone (Honshu Kagaku Co. Ltd.) were used as coating materials of the inner of the PVDF hollow fibre membranes. It was observed that ethanol permeability through both silicone-coated and silicone-uncoated PVDF hollow fibre membranes was similar but that of water decreased remarkably after coating treatment. Thus the ethanol selectivity for the coated membranes was higher and in many cases exceeded the corresponding VLE. The PMSP-coated PVDF membranes yielded lower ethanol and water permeability; however, the decrease in the water permeability was much greater and thus the selectivity was increased.

Similar results were obtained with the polyketone-coated PVDF membranes. Moreover, when using acetone, acetonitrile and n-butanol aqueous solutions, the selectivity of the same coated PVDF membranes was found to be higher than their relative volatility. This was attributed to the decrease of the partial water permeate flux. On the other hand, the heat treatment of the PVDF membrane had the effect of increasing the hydrophobicity of the membrane. The contact angle increased from 94.3° to 102° after heat treatment at 80 °C for 89 h. Therefore, the water permeate flux was decreased, while the ethanol permeate flux was maintained stable and thereby the selectivity was increased to become higher than the relative volatility. These results may be explained by the fact that ethanol molecule has a hydrophobic ethyl group and a hydrophilic hydroxyl group (—OH), and a change of hydrophobicity has little effect on the diffusion of this molecule as the effects on both groups may be compensated. By contrast, water molecule has only the hydrophilic hydroxyl groups and a change of hydrophobicity has stronger effects on the sorption parameter, while the structural changes have less effect because of its small molecular size.

During the past few years, Khayet et al. [64–66] proposed a new type of porous composite hydrophobic/hydrophilic flat sheet membranes for DCMD application. The membranes were prepared by the simple phase inversion method using SMMs. As presented in Fig. 2.3, the hydrophobic side of the membrane was brought into contact with the hot feed aqueous solution, while the hydrophilic layer of the membrane was maintained in contact with cold water, which penetrates into the pores of the hydrophilic layer. The composite porous hydrophobic/hydrophilic membranes were found to be promising for desalination by DCMD as they combine the low resistance to mass flux, achieved by the diminution of the water vapour transport path through the hydrophobic thin top layer, and a low

FIGURE 2.3 Schema of the DCMD mechanism of transport through porous membranes: (a) homogenous hydrophobic and (b) composite hydrophobic/hydrophilic

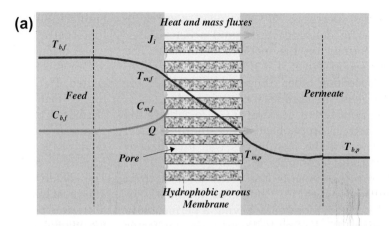

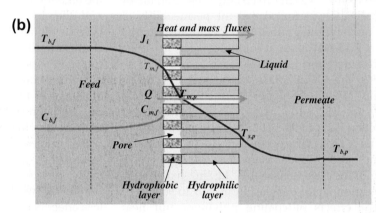

conductive heat loss through the membrane, obtained by using a thicker hydrophilic sublayer. This type of membranes were prepared by the phase inversion technique in a single casting step from polyetherimide (PEI) polymer solutions containing SMMs, the solvent DMAC and the non-solvent γ-butyrolactone (GBL). Only 2 wt% of SMMs was added to the casting solution. These SMMs are oligomeric fluoropolymers synthesized by polyurethane chemistry and tailored with fluorinated end groups. SMMs preparation and characterization were reviewed more in detail by Khayet et al. [67]. During membrane formation, these SMMs migrate towards the top air/polymer interface rendering the membrane hydrophobic. This was confirmed by contact angle measurements and XPS analysis, which indicated the gradient in fluorine across the membrane as a result of the migration of fluorinated end groups to the air-side surface during membrane formation.

The effects of PEI concentration in the casting solution on the DCMD permeate flux and on the membrane characteristics were investigated. For each PEI concentration, it was found that the LEP_w of the SMM-modified membrane was higher than that of the unmodified membrane, while the pore sizes were smaller and decreased as the concentration of PEI was increased. The pore sizes of the SMM-modified PEI membranes determined by the gas permeation test were 12–23 nm, smaller than the pore sizes of the commercial membranes shown in Table 2.1. However, their DCMD permeabilities were found to be of the same order of magnitude to those of the commercial PTFE membranes (TF200 and TF450 in Table 2.1) that have higher porosity and an order of magnitude greater pore sizes. Furthermore, the DCMD permeate flux of the SMM/PEI membrane prepared with 12 wt% PEI was found to be higher than those of the commercial PTFE membranes. Very high salt (NaCl) separation factors (>99.7%) were observed for the SMM-modified PEI membranes. A theoretical model was developed to estimate the thickness of the two layers of these composite membranes using the DCMD experiments together with other membrane parameters and the heat and mass transfer equations (see Chapter 10). The calculated thickness of the hydrophobic layer was lower than 8 μm, which is an order of magnitude smaller than that of the PTFE layer of the commercial membranes [66]. During the past few years, polyethersulfone (PES) and PS flat sheet membranes were further modified by the research group of Khayet and Matsuura (Department of Applied Physics I, Faculty of Physics, University Complutense of Madrid, Spain; Industrial Membrane Research Laboratory, University of Ottawa, Canada) using different types of SMMs, different solvents, different additives and different membrane preparation conditions in order to optimize the MD performance of the composite hydrophobic/hydrophilic type of membranes.

Larbot et al. [68] used ceramic tubular fibre membranes grafted by fluoroalkylsilanes for water desalination by DCMD configuration. The ceramic membranes were alumina (Al_2O_3) and zirconia (ZrO_2) of pore sizes 200 and 50 nm. The measured water contact angles of the membranes were higher after grafting, indicating that the grafted membranes were more hydrophobic. It was stated that when using the grafted Al_2O_3 membrane having pore size of 200 nm, the rejection factor ranged from 90% to 96% at low NaCl concentrations (10^{-3}–10^{-2} M), whereas for higher NaCl concentrations higher rejection factors up to 99% were achieved. DCMD permeate flux of 163.2 $l/m^2 \cdot day$ was reached when the feed temperature was 95 °C. No indications were given on the cause of the lower rejection factor observed when using lower NaCl concentrations. When the grafted ZrO_2 membrane having smaller pore size (i.e. 50 nm) was used, lower DCMD permeate fluxes ($\approx 95\ l/m^2 \cdot day$) and near 100% rejection factors were obtained, whereas higher permeate fluxes ($\approx 202\ l/m^2 \cdot day$) with lower rejection factors (99.5%) were obtained for the grafted ZrO_2 membrane having higher pore size (i.e. 200 nm). Finally, it was concluded that the use of ceramic tubular fibres would be of great interest in MD. Similar type of membranes were used by Belleville et al. [69] and Brodard et al. [70] in osmotic evaporation process. Tubular macroporous alumina membranes, provided by Pall Corporation (Exekia Division) with mean pore sizes of 0.2 and 0.8 μm, were grafted by siloxane compounds. It was reported that the permeate flux was independent of the membrane pore size, concluding that a limiting water vapour diffusion occurred through the membrane pores.

One of the limitations of the MD process is the risk of hydrophobic membrane pore wetting which reduces the MD performance. Composite hydrophilic/hydrophobic membrane type was proposed to prevent the pore wetting of the hydrophobic layer by the presence of a hydrophilic layer. For example, Xu et al. [71] proposed coating hydrophobic microporous membranes for MD and osmotic distillation (OD) with hydrophilic sodium alginate hydrogel for protection against pore wetting by surface-active agents such as oils, fats and detergents. A microporous PTFE membrane (Poreflon 020-40, Sumitomo Electric Fine Polymer, Osaka, Japan) of pore size 0.2 μm was alginate coated and then cross-linked by a water-soluble carbodiimide. The observed reduction of the overall mass transfer coefficient due to coating was less than 5% and the coated membranes were resistant to wetting for at least 300 min when exposed to orange—oil—water mixture.

Peng et al. [3] cast a dense hydrophilic polymer solution on porous PVDF (GVHP, Table 2.1) membrane. The polymer solution was a blend of polyvinyl alcohol (PVA) and polyethylene glycol (PEG) cross-linked by aldehydes and sodium acetate. The composite hydrophilic/hydrophobic membranes were tested for desalination by DCMD configuration using 3.5 wt% NaCl aqueous solution. The effects of the feed temperature and salt concentration were investigated. The behaviour of the coated membranes was found similar to that of the uncoated PVDF membrane. Separation factor of more than 99% was achieved and the DCMD permeate flux of the coated membrane was only 9% lower than that of the uncoated membrane, which was 23.7 kg/m^2·h at 70 °C feed temperature and 22 °C permeate temperature. It was speculated that the feed liquid was first dissolved in the hydrophilic layer of the membrane, whose molecular network exhibited a suitable swelling, followed by diffusion through a continuous pathway along the wriggling polymer chain and finally evaporated at the interface between the hydrophobic and hydrophilic layers. It was concluded that the hydrophilic layer could prevent wetting of the hydrophobic membrane pores. However, the authors [3] did not measure the *LEP* of water or the *LEP* of their feed solutions for the tested hydrophilic/hydrophobic membrane type.

Li and Sirkar [1,2] reported on novel hollow fibre membrane and module for use in both DCMD and VMD configurations. Their novel membranes were commercial porous PP hollow fibres (Accurel Membrana, Wuppertal, Germany) of different dimensions and thicknesses coated with a variety of ultrathin microporous silicone-fluoropolymer (fluorosilicone) layer on their external surface by plasma polymerization. The reason for applying the coating layer was to provide an additional porous layer having higher hydrophobicity than PP, which itself is one of the polymeric materials with very low surface energy. The coated fibres were arranged in a rectangular cross-flow module design, permitting the hot feed solution to flow over the outside surface of the fibres and thus reducing the temperature polarization effect. Both the DCMD and VMD experiments were carried out at feed temperatures ranging from 60 to 90 °C with 1% NaCl aqueous solution. Higher permeate fluxes (41—79 kg/m^2·h), complete absence of membrane pore wetting, higher temperature polarization coefficients (93—99% in VMD) than in other MD systems have been reported.

Novel hydrophobic/hydrophilic poly(phthalazinone ether sulfone ketone) (PPESK) hollow fibre composite membranes were prepared by coating their inner surface with silicone rubber and sol-gel polytrifluoropropylsiloxane to be used in desalination by VMD [72]. First, the PPESK hollow fibre UF membranes were prepared by the dry/jet wet spinning technique. About 5—20 g/l of silicone rubber (middle-temperature vulcanizing addition-type silicone rubber) was dissolved in petroleum ether and the obtained solution was forced to flow through the lumen side of the PPESK hollow fibre membranes. This last procedure was repeated several times. The coated membranes were then heated at different temperatures (40, 60 and 100 °C) for different heating periods of time. The effects of coating conditions (i.e. coating time, coating temperature and the concentration of silicone rubber) on the VMD performance were studied and the membrane stability was evaluated based on long-term experiments (up to 14 days). The higher VMD permeate flux was obtained at the lower coating temperature and lower concentration of silicone rubber in the coating solution. A permeate VMD flux as high as 3.5 l/h·m^2 with a salt rejection factor of 99% was reached using a feed NaCl aqueous solution having a concentration of 5 g/l, a feed temperature of 40 °C and a downstream pressure of 0.078 MPa and employing the coated hollow fibre with 60 °C coating temperature, 9 h coating time and 5 g/l silicone rubber. Moreover, the studied composite hollow fibre membrane showed stable VMD performance in a long-term experiment. In the case of polytrifluoropropylsiloxane-coated hollow fibre membranes, the highest permeate flux of 3.7 l/h.m^2 with NaCl rejection factor of 94.6% was obtained for 30 min prepolymerization time. It was proved that the prepolymerization time was an important factor, affecting the VMD performance of this type of membranes. Higher prepolymerization time decreased the permeability of the membrane. It must be mentioned here that PPESK is an ideal membrane material for high temperature gas separation, UF and nanofiltration (NF) separation processes.

Co-extrusion dry/jet wet spinning method was employed for the first time by Bonyadi and Chung [73] to prepare composite hydrophilic/hydrophobic hollow fibre membranes for DCMD. For the outer hydrophobic layer, a 12.5/87.5 wt% PVDF/NMP solution containing 30 wt% hydrophobic cloisite 15A was employed, while a low polymer solution 8.5/4/87.5 wt% PVDF/PAN/NMP containing 50 wt% hydrophilic cloisite NA$^+$ was used for the formation of the inner layer. Methanol/water (i.e. 80/20 wt%) was used as internal and external coagulant because it is a weak coagulant and can induce a high porous structure on PVDF membranes with high surface and bulk porosities, large pore size and sharp pore size distribution but weak mechanical properties. Details of fibre spinning are given in Chapter 4. The clay particles were incorporated in order to compensate for the weak mechanical properties and modify the hydrophobicity of the PVDF hollow fibre membrane matrix forming composite hydrophobic/hydrophilic membranes. A permeate flux of 55.2 kg/m^2·h based on the fibre outer diameter (i.e. 115 kg/m^2·h based on the inner diameter) was reached with a salt separation factor of 99.8% using a 3.5 wt% NaCl aqueous solution, a feed temperature of 90.3 °C and a permeate temperature of 16.5 °C. It was claimed that the proposed composite hydrophilic/hydrophobic hollow fibre membranes exhibited much higher DCMD permeation flux than most of the previous reported membranes.

Nanofibre Membranes

Recently, a novel type of membrane was proposed for MD, with which the first attempt was made in desalination by AGMD [14]. These are nanofibre membranes prepared by electrospinning or electrostatic spinning method. Detailed description of the electrospinning method, controlling parameters and preparation of these membranes is given in Chapter 5. Feng et al. [14] electro-span PVDF membrane whose SEM and AFM images are presented in Fig. 2.4. No support was used. About 18 wt% PVDF solution in the solvent DMF was employed and the voltage applied was 18 kV. The contact angle of the PVDF nanofibre membrane was found to be much higher (130°) than that of the conventional phase inversion membrane (83°). The enhanced hydrophobic property of the MD membrane surface is particularly important, since it prevents the penetration of the feed and/or permeate liquid into the membrane. The nanofibre membrane was applied for desalination of saline water by AGMD at different NaCl feed concentrations and feed and permeate temperatures. Similar AGMD trends as those of the other types of membranes were observed. A permeate flux as high as 11.5 kg/m^2·h with NaCl rejection factor higher than 98.5% was obtained and after many days of operation, the nanofibre membrane was intact and unplugged. It was claimed that the MD permeate flux is comparable to the

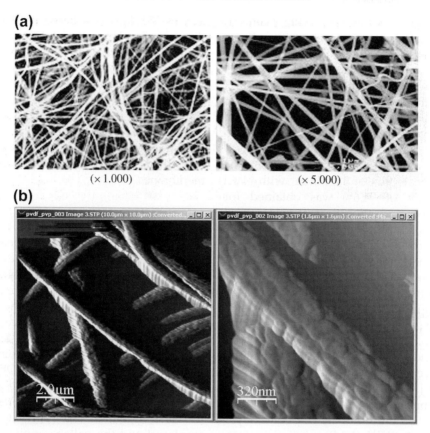

FIGURE 2.4 PVDF nanofibre membrane surface: (a) SEM image and (b) AFM image. *Source: Reprinted from [14]. Copyright 2008, with kind permission from Elsevier.*

permeate fluxes of the commercial MF membranes (5–28 kg/m$^2 \cdot$h) at transmembrane temperatures ranging from 25 to 83 °C and this approach may eventually enable the MD process to compete with conventional seawater desalination processes. These results may open up a new avenue of the research in MD membrane science and technology, both experimental and theoretical. Optimizing these new promising generation of MD membranes using other polymers, solvents and additives should be carried out to improve the MD performance of nanofibre membranes and the effects of the electrospinning parameters on the MD performance should be studied.

Track-Etched Membranes

Another type of membranes such as track-etched membranes (TEMs) can also be applied in MD. These membranes are prepared by irradiation of polymer films with heavy ions and subsequent etching of latent tracks. This membrane preparation technique can be applied to either hydrophilic or hydrophobic polymers as well as to copolymers. Uniform arrangements of pore sizes, regular and circular in shape, are achieved by this method as can be seen in Fig. 2.5, presented as an example [17,74]. The pore size distribution of this type of membranes is very narrow and sharp.

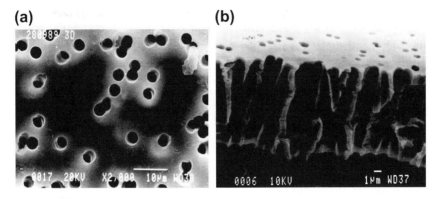

FIGURE 2.5 Surface (a) and cross-sectional (b) SEM images of polyethylene terephthalate (PET) track-etched membrane. Original figures were published in [17] (www.inderscience.com).

Despite all the above-cited researches in the field of MD membrane design engineering, detailed studies concerning the design of membranes for MD and systematic investigations of the effects of membrane parameters are still needed. More must be done before fabricating membranes that are suitable for different MD configurations and applications with outstanding performance.

MD MEMBRANE ENGINEERING AND MEMBRANE MATERIAL SELECTION FOR MD

The membranes should satisfy at least the following requirements for being used for MD:

i) The membrane may be comprised of a single layer or multi-layers. However, at least one of the layers should be made of a hydrophobic material and be porous.

ii) The pore size range may be from several nanometers to few micrometers. The pore size distribution should be as narrow as possible and the feed liquid should not penetrate into the pores. The *LEP* should be as high as possible. This is defined as the minimum transmembrane pressure that is required for distilled water or other feed solutions to enter into the pores by overcoming the hydrophobic forces. Otherwise, pore wetting will occur leading to the requirement of a large surface area for a given production rate and deterioration in the salt rejection. The *LEP* is characteristic to a given membrane. A high *LEP* can be achieved by material of low surface energy or high hydrophobicity (i.e. large contact angle to water or feed solutions) and small maximum pore size. On the other hand, a small maximum pore size parallels a small mean pore size and, consequently, low membrane permeability. Therefore, a compromise between the high *LEP* and the high productivity should be made by choosing an appropriate pore size and pore size distribution.

iii) The tortuosity factor (i.e. the measure of the deviation of the pore structure from straight cylindrical pores normal to the surface) should be small. This is inversely proportional to the MD membrane permeability. In MD studies, in order to predict the transmembrane flux, a value of 2 is frequently assumed for membrane tortuosity factor.

iv) The porosity (void volume fraction open to MD vapour flux) of the single-layer membrane or that of the hydrophobic layer in the case of the multi-layered membrane should be as high as possible. This is proportional to the MD membrane permeability. In fact, membranes with higher porosity can provide large spaces for evaporation. Therefore, it is generally agreed upon that the higher membrane porosity results in the higher permeate flux regardless of the MD configuration.

v) The thickness of the single-layer membrane should have an optimized value as the thickness is inversely proportional to the rate of mass and heat transport by conduction through the membrane. In case of a multi-layered membrane, the hydrophobic layer thickness should be as thin as possible. While a high mass transport is favoured for the MD process, a high heat transport is considered to be a heat loss. Therefore, compromise should be made, again, between the mass and the heat transfer, by properly adjusting the membrane thickness.

One advantage of the multi-layered membrane is that a high mass transport is enabled by making the hydrophobic layer as thin as possible, while a low heat transfer is enabled by making the overall membrane thickness (hydrophobic layer + hydrophilic layer) as thick as possible.

vi) The thermal conductivity of the membrane material should be as low as possible. It must be mentioned here that most of the hydrophobic polymers have similar thermal conductivity values within the same order of magnitude. The thermal conductivity of commercial membranes lies between 0.04 W/m·K and 0.06 W/m·K. It is possible to diminish the membrane heat transfer by conduction using membranes of high porosities, since the conductive heat transfer coefficients of the gases entrapped in the pores are an order of magnitude smaller than most of the used membrane materials. This possibility is parallel to the requirement (iv) in order to achieve high MD permeability. Another possibility is to use bi-layered or multi-layered composite porous membranes with hydrophobic and hydrophilic layers. As previously mentioned, the hydrophobic layer must be as thin as possible. Furthermore, supported membranes can be used as MD membrane. The purpose of using the hydrophilic layer is to enhance the resistance for the conductive heat transfer and to make the membrane strong enough to prevent its deflection and rupture. But the hydrophilic layer should not increase the mass transfer resistance considerably.

vii) The membrane surface contacting the feed solution should be made of a material of high fouling resistance, although fouling effect in MD is not as strong as it is in pressure-driven membrane separation processes. When hydrophobic layer is in contact with the feed solution, the membrane surface modification is needed. This can be achieved by coating the surface with a thin layer of a fouling-resistant material, depending on the feed solution to be treated.

viii) The membrane as a whole should exhibit good thermal stability. Long-term stability is required for MD membranes at temperatures as high as 100 °C.

ix) The membrane material should have excellent chemical resistance to various feed solutions. If the membrane has to be cleaned, resistance to acid and base is necessary. Generally, the membrane in MD acts only as a support of the vapour/

liquid interfaces and does not modify the VLE of the aqueous solutions in contact with it.

x) The membrane should have a long life with stable MD performance (permeability and selectivity) when used commercially.

xi) And finally, another important requirement is that the membrane should be cheap.

As can be seen, the MD membranes have to meet several conditions simultaneously. The above-cited 11 features considerably determine the morphology and microstructure of the MD membranes and whether they have high transmembrane fluxes and rejection factors together with high heat efficiencies (i.e. low heat transfer by conduction). To summarize, the main requirements for the MD membrane are that the membrane must exhibit low resistance to mass transfer, must not be wetted by the aqueous solutions in contact with and only vapour and non-condensable gases are present within its pores during the MD operation. As water is usually the major component in the feed solution, the membrane must be hydrophobic and therefore has to be made of polymers or inorganic materials with a low surface energy. It is generally agreed upon that the MD permeate flux will increase with an increase in the membrane porosity and pore size under some limitations, which means that the pore size should allow a sufficiently high LEP_w, and with a decrease of the membrane thickness and pore tortuosity. In other words, to obtain a high MD permeability, the hydrophobic layer that governs the MD transport should be as thin as possible and its porosity as well as pore size should be as large as possible.

Different membrane preparation techniques may be applied to fulfil the above-mentioned requirements depending on the properties of the materials to be used. The useful materials should be selected according to criteria including compatibility with the liquids involved, cost, ease of fabrication and assembly, useful operating temperatures and low thermal conductivity. As mentioned earlier, MD membranes can be made by sintering, stretching, phase inversion, thermally induced phase separation (TIPS), dry/wet spinning or wet spinning, electrospinning or membrane surface modification by physical or chemical techniques such as coating, grafting and plasma polymerization [6,75–77]. For example, PVDF membranes are made by the phase inversion method, PP membranes are made generally by stretching and TIPS, and PTFE membranes are made by sintering or stretching process. Among the above techniques, the phase inversion is the most popular one.

Briefly, MD polymeric membranes are prepared by:

i) Sintering technique: Powder of polymeric particles is pressed into a film or plate and sintered just below its melting point.

ii) Stretching a homogeneous polymer film made from a partially crystalline material: Films are made by extrusion of polymer at a temperature close to its melting point coupled with a rapid drawdown. Crystallites in the polymer are aligned in the direction of drawing. After annealing and cooling, a mechanical stress is applied perpendicularly to the direction of drawing. This manufacturing process gives a relatively uniform porous structure and porosity of about 90%.

iii) Phase inversion technique: The polymer is first dissolved in an appropriate solvent and the formed solution is then cast to a thickness of 20–200 μm on a proper support. The homogeneous solution is separated into two phases, a polymer-rich phase and a liquid-rich phase, when the cast film is immersed into a coagulant (non-solvent) bath via solvent–non-solvent exchange. A large variety of pore sizes can be obtained by varying the polymer concentration, the amount and type of additive(s), the coagulant type, the temperature, etc. Almost all soluble polymers can be precipitated in a non-solvent bath and a large part of MD membranes are prepared by the phase inversion technique.

iv) **Thermal phase separation technique:** Since some polymers such as PP are not readily dissolved into solvent at room temperature, a polymer solution is prepared at elevated temperatures in an adequate solvent and then cast into a film. Precipitation takes place not by the addition of non-solvent but by cooling the solution to a de-mixing point. Extraction of the solvent is typically done by low-molecular-weight alcohols.

References

[1] B. Li, K.K. Sirkar, Novel membrane and device for direct contact membrane distillation-based desalination process, Ind. & Eng. Chem. Res. 43 (2004) 5300–5309.

[2] B. Li, K.K. Sirkar, Novel membrane and device for vacuum membrane distillation-based desalination process, J. Membr. Sci. 257 (2005) 60–75.

[3] P. Peng, A.G. Fane, X. Li, Desalination by membrane distillation adopting a hydrophilic membrane, Desalination 173 (2005) 45–54.

[4] F.A. Rodgers, Stacked microporous vapor permeable membrane distillation system, United States Patent Serial No. 3,650,905 (1972).

[5] F.A. Rodgers, Compact multiple effect still having stacked impervious and previous membranes, United States Patent Serial No. Re.27,982 (1974); original No. 3,497,423.

[6] M. Mulder, Basic principles of membrane technology, Kluwer Academic Publishers, Dordrecht, 1996.

[7] C. Feng, B. Shi, G. Li, Y. Wu, Preliminary research on microporous membrane from F2.4 for membrane distillation, Separation and Purification Technology 39 (2004) 221–228.

[8] C. Feng, B. Shi, G. Li, Y. Wu, Preparation and properties of microporous membrane from poly(vinylidene fluoride-co-tetrafluoroethylene) (F2.4) for membrane distillation, J. Membr. Sci. 237 (2004) 15–24.

[9] M.C. García-Payo, M. Essalhi, M. Khayet, Preparation and characterization of PVDF-HFP copolymer hollow fiber membranes for membrane distillation, Desalination 245 (2009) 469–473.

[10] Z. Jin, D.L. Yang, S.H. Zhang, X.G. Jian, Hydrophobic modification of poly(phthalazinone ether sulfone ketone) hollow fiber membrane for vacuum membrane distillation, J. Membr. Sci. 310 (2008) 20–27.

[11] S.R. Krajewski, W. Kujawski, M. Bukowska, C. Picard, A. Larbot, Application of fluoroalkylsilanes (FAS) grafted ceramic membranes in membrane distillation process of NaCl solutions, J. Membr. Sci. 281 (2006) 253–259.

[12] M. Khayet, J.I. Mengual, T. Matsuura, Porous hydrophobic/hydrophilic composite membranes: Application in desalination using direct contact membrane distillation, J. Membr. Sci. 252 (2005) 101–113.

[13] D.E. Suk, T. Matsuura, H.B. Park, Y.M. Lee, Synthesis of a new type of surface modifying macromolecules (nSMM) and characterization and testing of nSMM blended membranes for membrane distillation, J. Membr. Sci. 277 (2006) 177–185.

[14] C. Feng, K.C. Khulbe, T. Matsuura, R. Gopal, S. Kaur, S. Ramakrishna, M. Khayet, Production of drinking water from saline water by air-gap membrane distillation using polyvinylidene fluoride nanofiber membrane, J. Membr. Sci. 311 (2008) 1–6.

[15] K.W. Lawson, D.R. Lloyd, Review: Membrane distillation, J. Membr. Sci. 124 (1997) 1–25.

[16] M.S. El-Bourawi, Z. Ding, R. Ma, M. Khayet, A framework for better understanding membrane distillation separation process: Review, J. Membr. Sci. 285 (2006) 4–29.

[17] M. Khayet, J.I. Mengual, G. Zakrzewska-Trznadel, Direct contact membrane distillation for nuclear desalination. Part I: Review of membranes used in membrane distillation and methods for their characterization, Int. J. Nuclear Desalination 1 (2005) 435–449.

[18] M. Gryta, Osmotic MD and other membrane distillation variants, J. Membr. Sci. 246 (2005) 145–156.

[19] A.M. Alklaibi, N. Lior, Membrane-distillation desalination: Status and potential, Desalination 171 (2004) 111–131.

[20] A. Burgoyne, M.M. Vahdati, Direct contact membrane distillation, Sep. Sci. & Tech. 35 (2000) 1257–1284.

[21] M. Khayet, A. Velázquez, J.I. Mengual, Modelling mass transport through a porous partition: Effect of pore size distribution, J. Non-Equilib. Thermodyn. 29 (2004) 279–299.

REFERENCES

[22] M.A. Izquierdo-Gil, M.C. García-Payo, C. Fernández-Pineda, Air gap membrane distillation of sucrose aqueous solutions, J. Membr. Sci. 155 (1999) 291–307.

[23] P.P. Zolotarev, V.V. Ugrosov, I.B. Volkina, V.N. Nikulin, Treatment of waste water for removing heavy metals by membrane distillation, J. Hazardous Mat. 37 (1994) 77–82.

[24] K.W. Lawson, D.R. Lloyd, Membrane distillation: II. Direct contact MD, J. Membr. Sci. 120 (1996) 123–133.

[25] M.I. Vázquez-Gónzález, L. Martínez, Nonisothermal water transport through hydrophobic membranes in a stirred cell, Sep. Sci. & Tech. 29 (1994) 1957–1966.

[26] C.H. Lee, W. Hong, Effect of operating variables on the flux and selectivity in sweep gas membrane distillation for dilute aqueous isopropanol, J. Membr. Sci. 188 (2001) 79–86.

[27] R.L. Calibo, M. Matsumura, J. Takahashi, H. Kataoka, Ethanol stripping by pervaporation using porous PTFE membrane, J. Ferment. Technol. 65 (1987) 665–674.

[28] B.R Bodell, Silicone rubber vapor diffusion in saline water distillation, United States Patent Serial No. 285,032 (1963).

[29] M.E. Findley, Vaporization through porous membranes, Ind. & Eng. Chem. Process Des. Dev. 6 (1967) 226–237.

[30] P.K. Weyl, Recovery of demineralized water from saline waters, United States Patent 3,340,186 (1967).

[31] D.W. Gore, Gore-Tex membrane distillation, Proceedings of the 10th Ann. Convention of the Water Supply Improvement Association, Honolulu, USA, 25–29 July, 1982.

[32] D.Y. Cheng, Method and apparatus for distillation, United States Patent Serial No. 4,265,713 (1981).

[33] D.Y. Cheng, S.J. Wiersma, Composite membrane for a membrane distillation system, United States Patent Serial No. 4,316,772 (1982).

[34] D.Y. Cheng, S.J. Wiersma, Composite membrane for a membrane distillation system, United States Patent Serial No. 4,419,242 (1983).

[35] D.Y. Cheng, S.J. Wiersma, Apparatus and method for thermal membrane distillation, United States Patent Serial No. 4,419,187 (1983).

[36] Catalogue of Enka AG presented at Europe-Japan Joint Congress on Membranes and Membrane Processes, Stresa, Italy, June 1984.

[37] J.M. Ortiz de Zárate, L. Peña, J.I. Mengual, Characterization of membrane distillation membranes prepared by phase inversion, Desalination 100 (1995) 139–148.

[38] M. Tomaszewska, Preparation and properties of flat-sheet membranes from polyvinylidene fluoride for membrane distillation, Desalination 104 (1996) 1–11.

[39] A. Bottino, G. Capannelli, S. Munari, A. Turturro, High performance ultrafiltration membranes cast from LiCl doped solutions, Desalination 68 (1988) 167–177.

[40] M. Khayet, T. Matsuura, Preparation and characterization of polyvinylidene fluoride membranes for membrane distillation, Ind. & Eng. Chem. Res. 40 (2001) 5710–5718.

[41] M. Khayet, K.C. Khulbe, T. Matsuura, Characterization of membranes for membrane distillation by atomic force microscopy and estimation of their water vapor transfer coefficients in vacuum membrane distillation process, J. Membr. Sci. 238 (2004) 199–211.

[42] M. Tomaszewska, M. Gryta, A.W. Morawski, Study on the concentration of acids by membrane distillation, J. Membr. Sci. 102 (1995) 113–122.

[43] M. Tomaszewska, Concentration of the extraction fluid from sulphuric acid treatment of phosphogypsum by membrane distillation, J. Membr. Sci. 78 (1993) 277–282.

[44] N. Chlubek, M. Tomaszewska, Some properties of hydrophobic membranes for membrane distillation, Environ. Prot. Eng. 15 (1989) 95–103.

[45] A.G. Chmielewski, G. Zakrewska-Trznadel, N.R. Miljevic, W.A. Van Hook, Membrane distillation employed for separation of water isotopic compounds, Sep. Sci. Technol. 30 (1995) 1653–1667.

[46] Y. Fujii, S. Kigoshi, H. Iwatani, M. Aoyama, Selectivity and characteristics of direct contact membrane distillation type experiment: I. Permeability and selectivity through dried hydrophobic fine porous membranes, J. Membr. Sci. 72 (1992) 53–72.

[47] Y. Fujii, S. Kigoshi, H. Iwatani, M. Aoyama, Y. Fusaoka, Selectivity and characteristics of direct contact membrane distillation type experiment: II. Membrane treatment and selectivity increase, J. Membr. Sci. 72 (1992) 73–89.

[48] B. Wu, K. Li, W.K. Teo, Preparation and characterization of poly(vinylidene fluoride) hollow fiber membranes for vacuum membrane distillation, J. Appl. Polymer Sci. 106 (2007) 1482–1495.

[49] B. Wu, X. Tan, K. Li, W.K. Teo, Removal of 1,1,1-trichloroethane from water using a poly(vinylidene fluoride) hollow fiber membrane module: Vacuum membrane distillation operation, Sep. & Purf. Tech. 52 (2006) 301–309.

[50] A. Bottino, G. Capannelli, A. Comite, Novel porous poly(vinylidene fluoride) membranes for membrane distillation, Desalination 183 (2005) 375–382.

[51] K.Y. Wang, T.S. Chung, M. Gryta, Hydrophobic PVDF hollow fiber membranes with narrow pore size distribution and ultra-skin for the fresh water

[52] S.P. Deshmuck, K. Li, Effect of ethanol composition in water coagulation bath on morphology of PVDF hollow fiber membranes, J. Membr. Sci. 150 (1998) 75–85.

[53] D. Wang, K. Li, W.K. Teo, Preparation and characterization of polyvinylidene fluoride (PVDF) hollow fiber membranes, J. Membr. Sci. 163 (1999) 211–220.

[54] T. Uragami, M. Fujimoto, M. Sugihara, Studies on syntheses and permeabilities of special polymer membranes. 28. Permeation characteristics and structure of interpolymer membranes from poly(vinylidene fluoride) and poly(styrene sulfonic acid), Desalination 34 (1980) 311–323.

[55] M. Khayet, C.Y. Feng, K.C. Khulbe, T. Matsuura, Preparation and characterization of polyvinylidene fluoride hollow fiber membranes for ultrafiltration, Polymer 43 (2002) 2879–2890.

[56] M. Khayet, C.Y. Feng, K.C. Khulbe, T. Matsuura, Study on the effect of a non-solvent additive on the morphology and performance of ultrafiltration hollow-fiber membranes, Desalination 148 (2002) 321–327.

[57] W.D. Benzinger, D.N. Robinson, Porous polyvinylidene fluoride membrane and process for its preparation. United States Patent Serial No. 4,384,047 (1982).

[58] H.C. Shih, Y.S. Yeh, H. Yasuda, Morphology of microporous poly(vinylidene fluoride) membranes studied by gas permeation and scanning electron microscopy, J. Membr. Sci. 50 (1990) 299–317.

[59] D. Wang, K. Li, W.K. Teo, Porous PVDF asymmetric hollow fiber membranes prepared with the use of small molecular additives, J. Membr. Sci. 178 (2000) 13–23.

[60] J. Li, Z. Xu, Z. Liu, W. Yuan, H. Xiang, S. Wang, Y. Xu, Microporous polypropylene and polyethylene hollow fiber membranes: Part 3. Experimental studies on membrane distillation for desalination, Desalination 155 (2003) 153–156.

[61] K. Ohta, I. Hayano, T. Okabe, T. Goto, S. Kimura, H. Ohya, Membrane distillation with fluoro-carbon membranes, Desalination 81 (1991) 107–115.

[62] Y. Kong, X. Lin, Y. Wu, J. Cheng, J. Xu, Plasma polymerization of octafluorocyclobutane and hydrophobic microporous composite membranes for membrane distillation, J. Appl. Polym. Sci. 46 (1992) 191–199.

[63] Y. Wu, Y. Kong, X. Lin, W. Liu, J. Xu, Surface-modified hydrophilic membranes in membrane distillation, J. Membr. Sci. 72 (1992) 189–196.

[64] M. Khayet, T. Matsuura, Application of surface modifying macromolecules for the preparation of membranes for membrane distillation, Desalination 158 (2003) 51–56.

[65] M. Khayet, J.I. Mengual, T. Matsuura, Porous hydrophobic/hydrophilic composite membranes: Application in desalination using direct contact membrane distillation, J. Membr. Sci. 252 (2005) 101–113.

[66] M. Khayet, T. Matsuura, J.I. Mengual, Porous hydrophobic/hydrophilic composite membranes: Estimation of the hydrophobic-layer thickness, J. Membr. Sci. 266 (2005) 68–79.

[67] M. Khayet, D.E. Suk, R.M. Narbaitz, J.P. Santerre, T. Matsuura, Study on surface modification by surface-modifying macromolecules and its applications in membrane-separation processes, J. Applied Polymer Sci. 89 (2003) 2902–2916.

[68] A. Larbot, L. Gazagnes, S. Krajewski, M. Bukowska, W. Kujawski, Water desalination using ceramic membrane distillation, Desalination 168 (2004) 367–372.

[69] M.P. Belleville, M. Dornier, J. Sánchez, F. Vaillant, M. Rynes, G. Rios, Procedé d'extraction para diffusion membranaire, French patent application No. 0205959, May 15th, 2002.

[70] F. Brodard, J. Romero, M.P. Belleville, J. Sánchez, C. Combe-James, M. Dornier, G.M. Rios, New hydrophobic membranes for osmotic evaporation process, Separation and Purification Technology 32 (2003) 3–7.

[71] J.B. Xu, S. Lange, J.P. Bartley, R.A. Johnson, Alginate-coated microporous PTFE membranes for use in the osmotic distillation of oily feeds, J. Membr. Sci. 240 (2004) 81–89.

[72] Z. Jin, D.L. Yang, S.H. Zhang, X.G. Jian, Hydrophobic modification of poly(phthalazinone ether sulfone ketone) hollow fiber membrane for vacuum membrane distillation, J. Membr. Sci. 310 (2008) 20–27.

[73] S. Bonyadi, T.S. Chung, Flux enhancement in membrane distillation by fabrication of dual layer hydrophilic-hydrophobic hollow fiber membranes, J. Membr. Sci. 306 (2007) 134–146.

[74] M. Buczkowki, B. Sartowska, D. Wawszczak, W. Starosta, Radiation resistance of track etched membranes, Radiation Measurements 34 (2001) 597–599.

[75] T. Matsuura, Synthetic Membranes and Membrane Separation Processes, CRC Press, Boca Raton, USA, 1994.

[76] I. Pinnau, B.D. Freeman, Membrane Formation, A.C.S. Modification, Symposium Series 744, American Chemical Society, Washington, DC, USA, 2000.

[77] D.R. Lloyd, K.E. Kinzer, H.S. Tseng, Microporous membrane formation via thermally-induced phase separation: I. Solid-liquid phase separation, J. Membr. Sci. 52 (1990) 239–261.

CHAPTER

3

Formation of Flat Sheet Phase Inversion MD Membranes

OUTLINE

Introduction	41	Effect of Polymer Concentration	50
Principles of Formation of Porous Hydrophobic Flat Sheet Membranes	43	Effect of Non-solvent Additive and its Concentration	51
Thermodynamics Considerations	44	Effect of Different Non-solvent Additives	54
Kinetics Considerations	47	Effect of Solvent or Solvent Mixture	54
Effects of Process Parameters on Membrane Structure	48	Effect of Coagulation Bath Temperature	54
		Effect of Solvent Evaporation Time	55
Effect of Polymer Type	48	Effect of Drying	56

INTRODUCTION

Before discussing about different membrane preparation techniques for membrane distillation (MD), their characteristics and the parameters affecting their surface and bulk morphologies, it is informative to respond to the following questions:

(i) What is a membrane?
(ii) What are the different types of membranes?
(iii) What are the different groups of membranes?

When two single and/or multi-components media (i.e. environments, phases, etc.) are separated by a barrier through which their components can be transferred at different rates, a set of membrane system is formed through which a number of transport phenomena may be applied. Briefly, a membrane is a device that selectively permits the transfer of one or more compounds from a liquid and/or gas media. Different transmembrane transport phenomena, heat, mass and/or electrical, may occur depending on the type of the membrane. One can find both biological (i.e. cell, intracellular, mucous, serous, mesothelia surrounding organs, etc.) and artificial or synthetic membranes. These last membranes can be made from different types of materials (i.e. polymers, ceramics, metals, supported-liquid, gel, etc.), different morphological structures (dense, porous, composite, symmetric, asymmetric, etc.) and acquire different shapes (flat sheet, hollow fibre, capillaries, etc.). Flat sheet membranes are utilized in the construction of plate-and-frame, disc and spiral wound modules, whereas cylindrical (hollow fibre, capillaries) membranes are utilized in the construction of tubular modules, e.g. shell-and-tube membrane modules and to a less extent plate-and-frame modules.

As is well known within membranologists, synthetic membranes are used in different isothermal and non-isothermal separation processes such as reverse osmosis (RO), microfiltration (MF), ultrafiltration (UF), nanofiltration (NF), pervaporation (PV), dialysis, gas permeation, etc. and are applied in different fields, such as pharmaceutical, food, biomedical, water treatment, desalination, etc. It is worth quoting that polymers are materials most frequently employed to prepare separation membranes and they are mostly organic compounds.

This chapter will be devoted to synthetic polymeric membranes. In particular, it will be focused on the preparation of polymeric synthetic membranes because of three reasons. First, most industrial membranes belong to this category. Second, polymeric materials are easy to process and are less expensive. Third, most of the used membranes in MD are polymeric.

As may be corroborated, the principal reason for the growth of membrane technology during the past decades is the success in the development of synthetic membranes from polymer materials. In fact, the first synthetic membranes were prepared in 1907 by Bechold [1] and the commercial application became feasible after the development of asymmetric skinned membranes by Loeb and Sourirajan in 1962 [2]. These latter membranes are particularly interesting because of the presence of a very thin (0.1 and 0.5 µm) and dense skin layer exhibiting selective properties. The mechanical stability of this type of membrane is acquired by a porous sub-layer supporting the skin layer and having a thickness between 0.1 and 0.2 mm. Ten years later, Rodgers [3,4] in his successive patents related to distillation presented a system and a method of desalination using a stack of flat sheet membranes separated by non-permeable corrugated heat transfer films and working under DCMD configuration. It was disclosed in the aforementioned patents that the suitable materials for the membranes are those which permit the formation of microporous membranes having high porosity, i.e. 70–80%, and uniform pore size distribution and must be either poorly wettable or non-wettable (i.e. hydrophobic) by the used liquids or treated to be non-wettable. The cited polymers were polycarbonates, polyesters, polyethylene, polypropylene and the halogenated polyethylenes, particularly the fluorocarbons. Particular mention has been made to polyvinylidene fluoride (PVDF) as a preferred membrane material and the solvent–non-solvent casting process (i.e. phase inversion technique) as the preferred method of membrane preparation. The use of cellulose nitrate, cellulose acetate and cellulose triacetate microscopic and hydrophilic porous filter media coated with a silicone water repellent was also mentioned to provide a non-wetting porous membrane.

The majority of today's membranes used in MD consist of a polymeric symmetric or asymmetric microporous structure carrying a less porous skin on the top surface. The membranes may differ considerably in their structure, their function and in the way they are formed. In fact, preparation procedures for the different membrane types are described in patents and publications in detailed recipes in some manuscripts.

Until now, basically all polymeric membranes and the majority of the proposed MD membranes, no matter how different their structures and their heat and mass transport modes may be, are made by the phase inversion process, in which a polymer is transformed from a liquid to a solid state. This involves the conversion of a homogeneous polymer solution of two or more components into a two-phase system with a polymer-rich phase, which turns into the rigid membrane structure, and a polymer-poor phase, which turns into the membrane pores.

A number of methods can be used to achieve phase inversion. The most commonly applied in membrane manufacturing are the dry−wet phase inversion technique and thermally induced phase separation (TIPS). The dry−wet phase inversion technique, also called the Loeb−Sourirajan technique, was used by Loeb and Sourirajan in their development of the first cellulose acetate membrane for seawater desalination as stated earlier [2].

In this chapter we will provide some information about the principle of formation of porous hydrophobic flat sheet MD membranes by phase inversion method and the effects of the parameters on the final structure of the MD membranes.

PRINCIPLES OF FORMATION OF POROUS HYDROPHOBIC FLAT SHEET MEMBRANES

As stated in the introduction section, phase inversion is a process in which a polymer is transformed in a controlled way from a liquid state (or solution) to a solid state. This is achieved by a number of methods (solvent evaporation, precipitation from vapor phase, thermal precipitation and immersion precipitation or dry−wet phase inversion). The major parts of the laboratory made phase inversion membranes for MD are prepared by immersion precipitation or dry−wet phase inversion. This process can be used to produce asymmetric membranes as well as symmetrical porous membranes. Therefore, this technique is described more in detail in the present chapter. According to the Loeb−Sourirajan method [2], a polymer solution is prepared by mixing a polymer, a solvent and sometimes non-solvent(s) or additive(s). The solvent and non-solvent additive must be miscible and the solvent should exhibit a strong dissolving power with high volatility. The solution is then cast by a doctor blade to a predetermined thickness on a suitable surface (i.e. glass or metal plate, support or a backing material such as a non-woven textile fabric). After partial evaporation of the solvent or even without solvent evaporation, the cast film is immersed in a bath containing a non-solvent coagulant, often called gelation medium. During the solvent evaporation named hereafter the first step, if any, and the solvent−non-solvent exchange in the gelation bath, named hereafter the second step, solidification of a polymer film takes place. Mass transfer of solvent and non-solvent should take place in such a way that the non-solvent concentration in the polymer film increases. During the first step a thin skin layer of solid polymer is formed at the top of the cast film due to the loss of solvent. In the second step of coagulation non-solvent diffuses into, while solvent diffuses out of, the polymer film through the thin solid layer. The porous thin layer that forms during the first step becomes the top skin layer governing the MD performance of the membrane, while the porous structure having larger pore size that forms during the second step becomes the porous sub-layer, providing the mechanical strength to the membrane. A porous top skin layer can be formed employing a low polymer concentration in the casting solution and a short solvent evaporation period.

It must be stated here that when the solvent content in the solution film becomes low the solvent no longer is able to hold the polymer in one phase and liquid−liquid-phase separation (i.e. l–l-phase separation) takes place forming droplets of one liquid phase dispersed in the other continuous liquid phase. As a result of the increase in non-solvent concentration, the polymer solution becomes thermodynamically unstable and phase separation occurs, which is followed by precipitation. The polymer solution composition and the type of solvent and non-solvent(s) additive(s) affect the size and the number of dispersed droplets as well as the initiation time of phase separation. Further, the number and the size of the droplets will also control the final structure of the MD phase inversion membrane. More explanations based on thermodynamic principles are given later on.

The l–l-phase separation, which is due to the occurrence of a cloud point, can be detected by turbidity measurements or light scattering methods. To localize the cloud point of a ternary system, a possible method is titration of polymer solution containing polymer and solvent with non-solvent or with a mixture of solvent and non-solvent at a given temperature until permanent turbidity is detected.

The membrane should be dried before being used for MD experiments. Solvent exchange procedure is frequently used for this purpose. The water in the pores of relatively hydrophilic membranes cannot be evaporated in air at ambient conditions since shrinkage of the membrane may occur. Hence, the water present in the membrane pores is replaced by a water-miscible liquid such as ethanol or methanol. Subsequently, the latter liquid is air-evaporated to obtain a dry membrane. The reason to replace water with ethanol or methanol is to reduce the capillary force inside the pore so that the pore will not collapse during the drying process by shrinkage of the membrane.

Thermodynamics Considerations

Remember that the thermodynamic parameter controlling miscibility of two or more components is the free energy of mixing (ΔG_m). This is given by Flory–Huggins lattice theory as follows [5]:

$$\Delta G_m = RT[n_1 \ln(\phi_1) + n_2 \ln(\phi_2) + n_3 \ln(\phi_3) \\ + \chi_{12} n_1 \phi_2 + \chi_{13} n_1 \phi_3 + \chi_{23} n_2 \phi_3] \tag{3.1}$$

where the subscripts refer to non-solvent (1), solvent (2) and polymer (3), n_i is the number of moles, ϕ_i is the volume fraction, χ_{ij} is the interaction parameter between component i and component j, R is the gas constant and T is the absolute temperature in Kelvin. The first three terms on the right-hand side of Eq. (3.1) are related to the ideal entropy of mixing of the polymer solution, whereas the last three terms are related to the enthalpy of mixing of the polymer solution when only binary interactions are considered. The interaction parameter (χ_{ij}) can be deduced from swelling experiments of pure liquids and binary mixtures at different concentrations and temperatures. When the temperature changes the interaction parameter (χ_{ij}) will also change. In most systems the interaction parameter will increase by decreasing the temperature. Other expressions of ΔG_m may be found in other books and published materials but to obtain numerical values of ΔG_m several parameters concerning each component is required, which are, however, hardly available.

As is well known, based on thermodynamic principles every system tends to minimize its free energy. The entropy terms in Eq. (3.1) are always negative but rather small, whereas the enthalpy terms are positive in the case of positive interaction parameters.

For a given system exhibiting a limited miscibility, the thermodynamic state can be described in terms of ΔG_m, at constant pressure and temperature, as follows:

(i) A stable state where the solution is in a single phase. This is thermodynamically described by

$$\Delta G_m > 0 \quad (3.2)$$

$$\left(\frac{\partial \mu_i}{\partial x_i}\right)_{T,P} > 0 \quad (3.3)$$

where μ_i is the chemical potential of component i and x_i is its mole fraction.

(ii) An unstable state where the solution separates spontaneously into two phases in equilibrium. This state is thermodynamically described by

$$\Delta G_m < 0 \quad (3.4)$$

$$\left(\frac{\partial \mu_i}{\partial x_i}\right)_{T,P} < 0 \quad (3.5)$$

(iii) An equilibrium state established by

$$\Delta G_m = 0 \quad (3.6)$$

$$\left(\frac{\partial \mu_i}{\partial x_i}\right)_{T,P} = 0 \quad (3.7)$$

Let us consider a binary system consisting of a polymer and a solvent. Figure 3.1 shows the variation of ΔG_m with the volume fraction of the polymer for two different temperatures.

At the temperature T_1 the system is completely miscible over the entire concentration range. For a certain temperature T_2, due to the increase of the enthalpy term ΔG_m presents an upward bend. All compositions between the points ϕ' and ϕ'' can lower their free energy by separating into two liquid phases with the composition ϕ' and ϕ'', respectively. These two phases are in equilibrium with each other because their chemical potentials are the same. When the temperature increases, the enthalpy term becomes smaller and miscibility is improved. In this case the two points ϕ' and ϕ'' will come closer to each other and finally coincide in the critical point, which is characterized by the following two expressions:

$$\frac{\partial^2 \Delta G_m}{\partial \phi_i^2} = 0 \quad (3.8)$$

$$\frac{\partial^3 \Delta G_m}{\partial \phi_i^3} = 0 \quad (3.9)$$

The behaviour of a binary mixture with temperature will be shown in Chapter 6. In a three-component system polymer (P), solvent (S) and non-solvent (NS) the phenomena are basically the same, but a change in temperature is not necessary to induce l–l-phase separation, a change in composition is sufficient.

Figure 3.2 shows an isothermal ternary system (P, S and NS), which can be considered as a cross-section of a three-dimensional ΔG_m plot. The pure components (P, S and NS) are

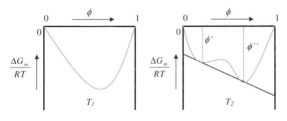

FIGURE 3.1 Dependence of ΔG_m on the volume fraction of the polymer (ϕ) for two temperatures.

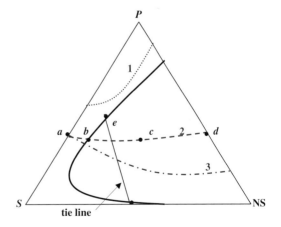

FIGURE 3.2 Schematic diagram showing the formation of a membrane from a three-component polymer mixture. 1, 2 and 3 represent possible composition paths.

localized at the corners of the triangle and any point within the triangle represents a mixture of all three components. Within a certain compositionally defined range of thermodynamic states, all three components are completely miscible, whereas in a different range the system decomposes into two distinct phases. The solid black line is the *l–l*-boundary, called the binodal, which separates the miscible region at the left from the immiscible region at the right. During membrane preparation, a thermodynamically stable polymer solution at the left side of the binodal is immersed in a non-solvent so that the final composition will be displaced to the right side in the phase diagram towards the P–NS axis. Three different composition paths (1, 2 and 3) can be distinguished as follows:

(i) Path 1 starts from a high initial polymer volume fraction on the P–S axis and does not cross the phase separation line. In this case, no phase separation occurs during the entire gelation process and the obtained membrane is dense and homogeneous through its entire cross-section.

(ii) Path 2 starts from a lower initial polymer volume fraction on the P–S axis and the ratio of the rate of solvent outflux to that of the non-solvent influx is relatively high. The obtained final membrane at the P–NS axis is asymmetric. As it is illustrated in Fig. 3.2 the film composition starts from the point 'a', passes through points 'b' and 'c' (polymer densification point), and finally reaches the point 'd' along the composition path. 'b' is the point where the phase separation starts to begin and 'c' is the point where the polymer solidification starts to occur. On the left side of point 'c', the polymer-rich phase, the composition of which is indicated by point 'e' on the phase separation line, is liquid. On the right side of point 'c', the polymer-rich phase is no longer liquid, but it is solid. If we consider the composition change near the film–gelation medium interface, the speed of the movement on the composition path is very fast and the points 'b' and 'c' are attained quickly. In this case there is not enough time for the solution to undergo the phase separation, and therefore, polymer and non-solvent are interdispersed, resulting in a dense skin layer with a composition indicated by the point 'c'. The solvent–non-solvent exchange proceeds further along the composition path, and the point 'd' is attained, while the dense structure of the surface layer is unperturbed. For the polymer solution far from the interface film–gelation medium, longer time is needed to achieve the solidification point 'c' and a polymer-rich phase and a polymer-poor phase are separated because there is enough time. If the amount of the polymer-rich phase is large relative to the amount of the polymer-poor phase, droplets of the latter phase are dispersed in the continuous polymer-rich phase and form pores in the continuous polymer matrix. As the distance from the interface film–gelation medium increases, the separation of the phases becomes clearer and the size of the pores increases. As a result an asymmetric structure of the membrane is formed (i.e. a dense surface layer at the film/gelation media interface followed by a porous sub-layer with increasingly larger pore sizes as the distance from the interface increases).

(iii) Path 3, in Fig. 3.2, starts from the same polymer volume fraction on the P–S axis as path 2. In this case the ratio of the rate of solvent outflux to that of non-solvent influx is smaller compared to path 2. The volume fraction of polymer at point 'c' is no longer high enough to form a continuous polymer-rich phase. Instead, a polymer-rich phase is dispersed in a non-solvent-rich phase. The obtained membranes following path 3 are too porous and mechanically too weak.

The membranes needed for MD applications should be hydrophobic and porous with high porosity and interconnectivity of the pores. The membrane should exhibit reasonably high mechanical properties, as well. The transmembrane hydrostatic pressure normally applied in MD is smaller than 2 10^3 kPa. These membranes may be formed following a path between paths 2 and 3 in Fig. 3.2. It should be pointed out that the solvent should be removed from the mixture at about the same rate as the non-sovlent enters. The pore size is normally fixed at the time of the polymer densification (point 'c'), and it remains the same while the solvent–non-solvent exchange proceeds along the composition path beyond the point 'c'. Once again, when the polymer system crosses the binodal two separate phases will begin to form. These are a polymer-rich phase represented by the upper end of the tie line and a polymer-poor phase represented by the lower end of the tie line. At a certain composition of the three-component mixtures, the polymer concentration in the polymer-rich phase will be high enough to be considered as solid. Further exchange of solvent and non-solvent will lead to the final formation of the membrane, the porosity of which is determined by the point on the P–NS line (e.g. point 'd'), representing the mixture of the solid polymer-rich phase and the liquid phase which is virtually free of polymer and solvent.

As can be seen, the membrane morphology can be controlled by manipulating the phase transition from liquid to solid. For instance, it is possible to prepare porous hydrophobic membranes as well as non-porous ones. The above discussions indicate that not only the composition path and its position relative to the binodal, but also the speed of the composition change along the path governs the structure of the membrane formed via gelation process. More detailed discussions of the thermodynamic aspects of membrane formation by the phase inversion technique may be found elsewhere [6].

As stated previously, the composition change on the ternary diagram is affected by the speed of solvent evaporation or by the speed of solvent–non-solvent exchange during the gelation step. Hence, kinetic consideration is also necessary to draw the composition path on the ternary diagram.

Kinetics Considerations

After the cast film of a polymeric solution is immersed in a non-solvent, the solvent and non-solvent are exchanged by diffusion processes. In some cases convection may play a role and can also be considered. The non-solvent will diffuse from the coagulation bath into the cast polymer solution (flux J_1), whereas the solvent will diffuse out of the polymer film into the coagulation bath (flux J_2). After a certain period of time a solid polymeric membrane is obtained. The fluxes J_1 and J_2 can be represented by phenomenological relations as follows:

$$J_1 = \sum_{j=1}^{2} L_{1j}(\phi_1, \phi_2) \frac{\partial \mu_j}{\partial x} \qquad (3.10)$$

$$J_2 = \sum_{j=1}^{2} L_{2j}(\phi_1, \phi_2) \frac{\partial \mu_j}{\partial x} \qquad (3.11)$$

where x is the distance from the interface polymer–gelation medium towards the bulk polymeric solution, L_{1j} and L_{2j} are the permeability coefficients of non-solvent and solvent, respectively; ϕ_1 and ϕ_2 are the volume fraction of non-solvent and solvent in the polymer solution, respectively; and the driving force for mass transfer of component i ($i=1, 2$) at any point of the polymeric solution is the gradient in chemical potential $\partial \mu_j/\partial x$. It must be pointed out here that the local composition at any point in the cast film is a function of time. These composition changes can be calculated as described in [7] and the results can be represented in a ternary phase diagram by the composition path as shown in Fig. 3.2. Kinetic parameters are rather difficult to determine for systems with more than two components (e.g. solvent and non-solvent). Therefore, for simplicity, the basic thermodynamic and kinetic relations of the phase separation process were discussed above for a binary system.

It is worth quoting that the kinetic effects depend on the system properties such as the diffusivities of the solvent and non-solvent components in the mixture, the viscosity of the solution and the chemical potential gradients, which act as driving forces for the diffusion of the components in the mixture. In fact, the chemical potential and diffusivities of the components in the system, and their dependencies on composition, temperature, viscosity, etc., are difficult to determine by independent experiments and therefore normally are not readily available. Therefore, a quantitative description of the membrane formation mechanism is nearly impossible and in general only a qualitative description is possible. Thus, rationalization of the membrane formation and correlation of the various preparation parameters with membrane structures and properties have been attempted by a large amount of experiments. Studies on the processes of membrane formation proved that the final structure of a membrane depends on many factors, i.e. composition of casting solution, temperature, support material, composition of gelation bath, etc. Therefore, the mechanism of membrane structure formation can be different in every specific case.

EFFECTS OF PROCESS PARAMETERS ON MEMBRANE STRUCTURE

As reported in the above section, there are many variables involved in the phase inversion technique that may affect the final membrane structure and MD performance. The polymer and the composition of the polymer solution, the non-solvent additive and its concentration, the solvent evaporation temperature and evaporation period, the nature of the gelation media and its temperature are the primary factors affecting the performance of the phase inversion membrane. Depending on the combination of variables, membranes of different polymeric materials with different structures, porosities and pore sizes can be prepared.

The experimental procedures described in this chapter are focused on the preparation of flat sheet membranes for MD from different polymer solutions using the dry/wet phase inversion process. The preparation of membranes by the phase inversion technique for other separation processes is described in detail in the literature and various patents. In what follows the effects of phase inversion process parameters on the MD membrane structure are reviewed.

Effect of Polymer Type

In the phase inversion technique not all polymers and copolymers can be employed for membrane preparation. Also, the final membrane structure, including bulk and surface morphologies, depends on the used polymer and copolymer although the same fabrication procedure is applied. As an example, in what follows, the characteristics of the membrane made from the polymer PVDF ($M_w = 1.02 \ 10^6$ g/mol) are compared to those of the membrane prepared from the copolymer tetrafluoroethylene and vinylidene fluoride named hereafter polyvinylidene fluoride-tetrafluoroethylene (PVDF-TFE)

($M_w = 1.57 \; 10^6 \, \text{g/mol}$) [8]. The same fabrication procedure is applied for both membranes. Figure 3.3 shows the scanning electron microscopy (SEM) images of PVDF and PVDF-TFE membranes. A significant difference can be observed between the top and bottom surfaces for both PVDF-TFE and PVDF membranes. Macrovoids appear through the cross-section of both polymeric membranes. Compared to the top membrane surfaces, larger pores are present at the bottom surfaces. However, in the case of PVDF-TFE, the bottom membrane surface exhibits smaller pore sizes than those of the PVDF. This may be attributed to the fact that before complete coagulation, PVDF-TFE separated from the glass plate during precipitation. The mean pore size of the PVDF-TFE membrane is 24 nm, while that of the PVDF membrane is 39 nm. Both are in the small pore size range for MD membranes. However, as will be shown later

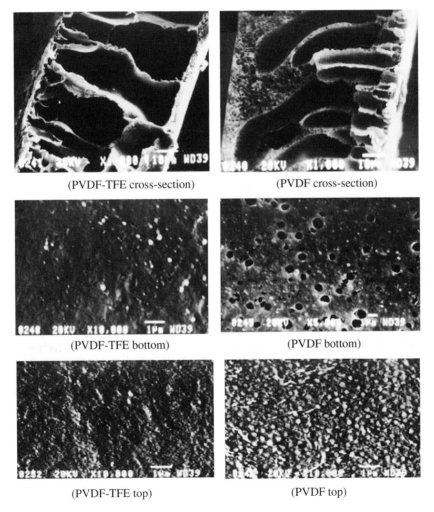

(PVDF-TFE cross-section)   (PVDF cross-section)

(PVDF-TFE bottom)   (PVDF bottom)

(PVDF-TFE top)   (PVDF top)

FIGURE 3.3 SEM images of PVDF-TFE and PVDF membranes prepared by the phase inversion technique using the same fabrication procedure. *Source: Reprinted from [8]. Copyright 2004, with kind permission from Elsevier.*

on, it is possible to fabricate membranes with larger pore sizes by adjusting other fabrication parameters. The porosity of the PVDF-TFE membrane was also found to be higher than that of the PVDF membrane. As a consequence, a lower permeate flux was observed for the PVDF-TFE membrane compared to that of the PVDF membrane under the same DCMD operating conditions. For example, for an aqueous feed solution containing sodium chloride (NaCl, 0.3 M), a feed temperature of 55 °C and a permeate temperature of 20 °C, the obtained DCMD permeate flux is 6.5 kg/m^2·h for the PVDF membrane and 3.5 kg/m^2·h for the PVDF-TFE membrane.

The PVDF-TFE membrane exhibits excellent mechanical properties and hydrophobicity than that of the PVDF membrane. The water contact angle of the PVDF-TFE membrane is 88.5° whereas that of the PVDF membrane is 80°. These data were obtained from dense films to avoid the effects of pores and surface roughness on the measured contact angle. The dense film was prepared from the solution containing 12 wt% of respective polymer in dimethyl acetamide (DMAC). Each polymer solution was cast on a glass plate and placed in a vacuum drying box maintained at 60 °C until the solvent was fully removed.

As far as mechanical properties are concerned, the membrane PVDF-TFE exhibits higher stretching strength and extension ratio at break (i.e. approximately 6—8 times higher) but a slightly inferior stretch elastic modulus than the membrane PVDF.

Effect of Polymer Concentration

The polymer concentration in the casting solution is one of the relevant factors affecting MD membrane structure. As the polymer concentration increases, both the pore size and porosity decrease. Thus the MD permeate flux is smaller for membranes prepared with higher polymer concentrations. Figure 3.4, shows an

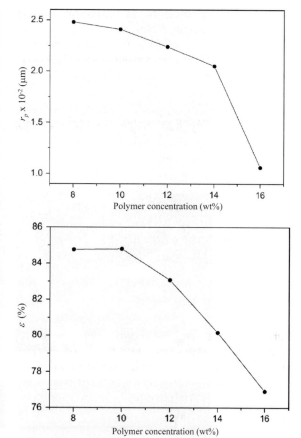

FIGURE 3.4 Effects of the copolymer PVDF-TFE concentration on the average pore size (r_p) and porosity (ε) of the phase inversion membranes. *Source: Reprinted from [9]. Copyright 2004, with kind permission from Elsevier.*

example for the copolymer PVDF-TFE. With increasing the copolymer concentration a denser surface layer forms at the top of the nascent membrane surface.

A gradual decrease of the pore size, porosity and permeability with increasing the polymer content in the casting solution was also observed for other polymers such as PVDF using different types of solvent (i.e. DMAC and dimethyl formamide, DMF) [10,11]. This result may be associated with the closer structures of the membranes cast from the solutions of higher

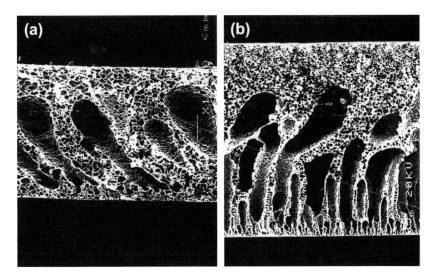

FIGURE 3.5 SEM images of PVDF membranes prepared from 10 wt% PVDF concentration in DMAC (a) and 25 wt% PVDF in DMAC (b). *Source: Reprinted from [11]. Copyright 1995, with kind permission from Elsevier.*

polymer concentration. Figure 3.5 shows the asymmetric membrane structure of PVDF membranes prepared from solutions of different polymer concentrations by a similar phase inversion procedure [11]. It can be seen that the thickness of the top layer having a sponge type structure increases with the increase of the polymer content in the casting solution, in such a way that the porosity becomes lower.

It must be pointed out that manufacturing MD membranes by the phase inversion technique from polymer solutions with a very low polymer concentration is practically impossible. For instance, when the polymer content is less than a threshold value, e.g. 10 wt% for PVDF, the resulting membranes become inconsistent and holes, which can be observed clearly against light, start to appear within the membrane [11].

It is worth quoting that the obtained membrane thickness is always smaller than the cast film thickness and for the same cast depth of the doctor blade it decreases as the polymer content in the solution becomes lower (Fig. 3.6). For example, when using DMAC as the solvent and for a casting depth of 400 μm, the formed membrane thickness varied from 85 to 54 μm as the PVDF concentration was decreased from 25 to 10 wt% [11].

Effect of Non-solvent Additive and its Concentration

In the phase inversion process, non-solvent additive in the casting solution acts as a pore-forming agent improving the MD permeability

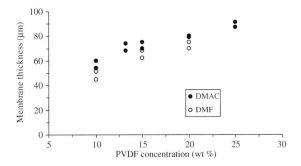

FIGURE 3.6 Effect of PVDF concentration on the obtained membrane thickness when using a casting depth of 400 μm and two solvents (DMAC, DMF). *Source: Data taken from [11].*

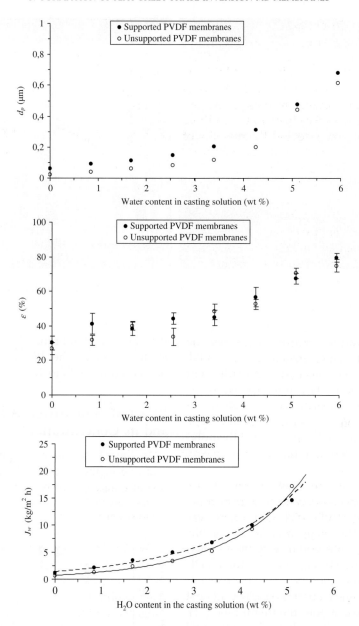

FIGURE 3.7 Pore size (d_p), porosity (ε) and *VMD* permeate flux (J_w) of supported and unsupported PVDF membranes versus non-solvent additive in the PVDF casting solution (VMD: feed temperature 25 °C; downstream pressure: 1666.5 Pa; distilled water used as feed). *Source: Reprinted from [12]. Copyright 2001, with kind permission from American Chemical Society.*

of the membranes. Various types of additives are used in different polymer solutions and with different concentrations. In general, when the concentration of the non-solvent additive is increased in the casting solution, the MD performance of the membrane increases. As an example, Fig. 3.7 shows the effects of the additive pure water on the pore size and the porosity of the prepared supported and unsupported PVDF membranes [12]. The polymer solutions were cast over a glass plate (i.e. unsupported membranes) or over a non-woven polyester backing material (i.e. supported membranes). These membranes were used for VOCs removal from water by vacuum membrane distillation (VMD). For both supported and unsupported PVDF membranes, the porosity and the pore size increased as the water content in the PVDF casting solution was increased and as a consequence the VMD permeate flux increased exponentially with the concentration of water in the casting solution as displayed in Fig. 3.7. It must be mentioned here that the pore sizes of the supported membranes are larger than those of the unsupported membranes for given water content in the PVDF solution. This is due to the shrinkage of pores that occurs in the unsupported membranes during membrane drying even after solvent exchange.

Figure 3.8 presents the change of the average pore size and porosity of both PVDF membrane and the copolymer PVDF-TFE membrane with the variation of the concentration of the salt additive lithium chloride (LiCl). Remember that MD permeability is better with high porosity and pore size.

The addition of LiCl to the casting solution increases the rate of precipitation during the immersion step of phase inversion process and a more open structure of the membrane is formed. The rapid precipitation of the polymer from the solutions containing LiCl is associated with the high miscibility of the additive with water and the interaction of LiCl-solvent and LiCl-PVDF [10,13].

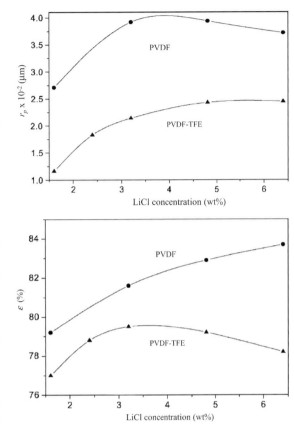

FIGURE 3.8 Effects of the additive LiCl on the average pore radius (r_p) and the porosity (ε) of the membranes prepared from PVDF and the copolymer PVDF-TFE by the phase inversion process (12 wt% polymer or copolymer in DMAC). *Source: Reprinted from [8]. Copyright 2004, with kind permission from Elsevier.*

It should be noted that the increase of the concentration of the non-solvent in the casting solution increases the size of the cavities inside the membrane and consequently reduces drastically the mechanical properties of the obtained MD membranes [10]. In fact, the high rate of polymer precipitation leads to the formation of cavities and macrovoids in the membrane. For example, as shown in Fig. 3.9, the addition of LiCl to the PVDF casting solution drastically decreases the strength at break of the PVDF membranes.

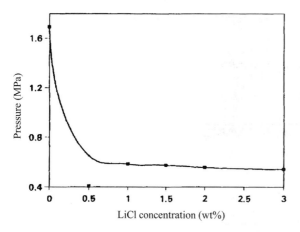

FIGURE 3.9 Effect of the additive LiCl on the strength at break of the PVDF membranes. *Source: Reprinted from [10]. Copyright 1996, with kind permission from Elsevier.*

Effect of Different Non-solvent Additives

As stated earlier, the addition of a non-solvent additive to a casting solution can drastically alter the membrane structure. Different morphologies of the resultant MD membranes are observed when using different additives although the same polymer, solvent, coagulant and phase inversion procedure are applied. For instance, as presented in Fig. 3.10, the SEM images of the PVDF-TFE membranes prepared using the LiCl additive show a cross-section containing a finger-like structure crossing the whole thickness of the membrane and no obvious difference can be detected between the top and bottom surface of the membrane. The addition of this additive to the casting solution seems to affect the rate of the copolymer PVDF-TFE precipitation equally at the top and bottom surfaces of the membrane. When lithium perchlorate trihydrate ($LiClO_4 \cdot 3H_2O$) was used as the non-solvent additive, the PVDF-TFE membrane demonstrated a remarkable difference between the top and the bottom surface with a cross-sectional structure possessing larger cavities and a clear skin layer. The PVDF-TFE membrane prepared with $LiClO_4 \cdot 3H_2O$ exhibits larger pore sizes and a higher porosity than the PVDF-TFE membrane prepared with LiCl under the same conditions. Consequently, the MD permeate flux of the membrane prepared from the additive $LiClO_4 \cdot 3H_2O$ was found to be higher than that of the membrane prepared from LiCl [9].

Effect of Solvent or Solvent Mixture

The solvent or the solvent mixture affects considerably the morphology of the resulted membrane and therefore the MD performance. The interactions between solvent–coagulant and between solvent–polymer play important roles affecting the rate of polymer precipitation. From the SEM images of the MD membranes prepared from PVDF using two different solvents, DMAC and DMF, the membrane cast from DMAC solution exhibits a sponge-like structure containing macrovoids without a clear top skin layer, whereas the membrane prepared from the DMF solution presents a macrovoid layer located underneath a dense skin of about 0.9 μm thickness [10]. The permeability of the latter PVDF membrane was found to be one third of the former PVDF membrane. The difference in the observed structures of the PVDF membranes can be justified by the solvent–water interaction. For instance, there is a greater tendency of water to mix with DMAC than with DMF [13].

Effect of Coagulation Bath Temperature

The coagulation speed of the cast polymer film affects the membrane morphology and the final structure of the MD membrane. This precipitation rate can be controlled by varying the temperature of the coagulant. As the coagulant temperature increases, the rate of solvent and non-solvent exchange becomes faster and a denser skin layer forms leading to a small MD performance. For example, Fig. 3.11 shows a continuous decrease of both the pore size and the porosity of PVDF-TFE membranes

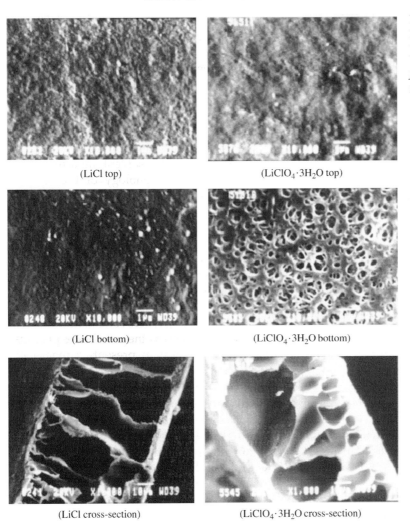

FIGURE 3.10 SEM images of PVDF-TFE membranes prepared using the non-solvents LiCl and $LiClO_4 \cdot 3H_2O$. *Source: Reprinted from [9]. Copyright 2004, with kind permission from Elsevier.*

prepared with two different additives (LiCl and $LiClO_4 \cdot 3H_2O$/trimethyl phosphate, TMP) [9].

When using PVDF membranes, the measured DCMD permeate flux was found to be lower (by nearly twofold) for the membranes precipitated from the coagulation bath temperature of 20 °C than for the membranes prepared at a lower temperature, 4 °C [10]. Figure 3.12 summarizes the decrease of the DCMD permeate flux of PVDF membranes with increasing coagulation bath temperature.

Effect of Solvent Evaporation Time

As stated earlier, there is a certain period between the casting of the polymer solution over a glass plate or over a baking material, and the immersion of the cast film into the coagulation bath. This period is tentatively called solvent evaporation time. The solvent evaporation time affects considerably the structure of the resulting membrane and therefore its MD performance. This effect may be attributed to

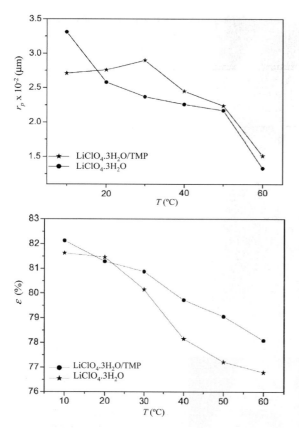

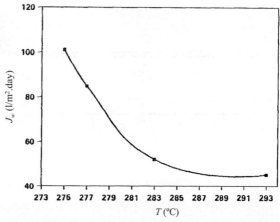

FIGURE 3.11 Effects of coagulation bath temperature (T) on pore radius (r_p) and porosity (ε) of PVDF-TFE membranes. *Source: Reprinted from [9]. Copyright 2004, with kind permission from Elsevier.*

FIGURE 3.12 Effects of coagulation bath temperature (T) on DCMD permeate flux (J_w) of PVDF membranes (feed temperature 60 °C, permeate temperature 20 °C, feed NaCl aqueous solution concentration 1–2%). *Source: Reprinted from [10]. Copyright 1996, with kind permission from Elsevier.*

the increase of polymer concentration in the top cast film layer and/or to the entanglement of macromolecules caused by polymer chains relaxation. As a consequence, a denser structure is formed depending on the exposure time.

As the solvent evaporation time is increased, the formed membrane will be free of defects and holes especially when low polymer concentrations are used to prepare the casting solution. Hence, denser membranes are produced inducing lower MD permeate fluxes. SEM images of both the bottom surface and the cross-section of two PVDF-TFE membranes with different solvent evaporation times are displayed as an example in Fig. 3.13 [9]. The pore sizes of both the bottom surface and the cross-section are smaller and a denser top skin layer is formed at the longer solvent evaporation time. This layer acts as a barrier for solvent/non-solvent exchange (i.e. diffusion of the coagulant inward and the solvent out from the casting solution).

Figure 3.14 presents the DCMD permeate flux of PVDF membranes prepared under different solvent evaporation times. As can be seen longer solvent evaporation time clearly decreases the permeate flux while the NaCl rejection factor was found to be practically 100%. The decrease of the DCMD permeate flux is attributed to the diminution of both pore size and porosity of the membrane with the solvent evaporation time.

Effect of Drying

After precipitation of the cast polymer film in a coagulation bath, the formed membrane must be dried before performing MD experiments. Solvent exchange procedure and/or post-treatment are used to dry the membranes. Air

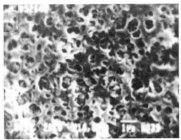 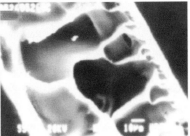

Solvent evaporation time 30 s

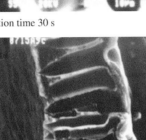

Solvent evaporation time 300 s

FIGURE 3.13 SEM images of bottom surface and cross-section of PVDF-TFE membranes prepared at different solvent evaporation times (30 s and 300 s), the solvent used was DMAC and the temperature was 30 °C. *Source: Reprinted from [9]. Copyright 2004, with kind permission from Elsevier.*

evaporation is frequently applied. This step may affect the MD membrane pore size and porosity. The applied drying procedure may reduce the maximum pore size close to twofold [10]. This is due to the possible shrinkage of the formed porous hydrophobic membrane. Figure 3.15 illustrates the effect of drying on the porosity of PVDF membranes. As can be seen, the applied drying procedure slightly decreases the membrane porosity.

In addition to the above mentioned parameters affecting the MD membrane morphology

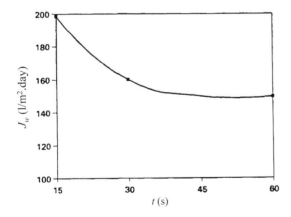

FIGURE 3.14 Effects of solvent evaporation time (t) on DCMD permeate flux (J_w) of PVDF membranes (feed temperature 60 °C, permeate temperature 20 °C, feed NaCl aqueous solution concentration 1–2%). *Source: Reprinted from [10]. Copyright 1996, with kind permission from Elsevier.*

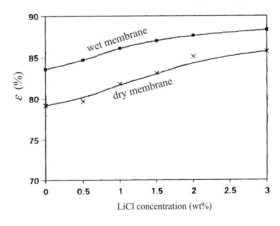

FIGURE 3.15 Effects of drying step and non-solvent additive concentration on the porosity (ϵ) of PVDF membranes. *Source: Reprinted from [10]. Copyright 1996, with kind permission from Elsevier.*

and performance, other procedures not mentioned here may also affect the membrane properties. These procedures are still not investigated in MD field. For example, membranes and membrane pores may be shrunk by applying an annealing step (heat treatment).

References

[1] H. Bechhold, Biochem. Z. (Biochemische Zeitschrift), in: Th. van den Boomgaard, A.J. Reuvers, C.A. Smolders (Eds.), Preparation, structure and properties of asymmetric membranes, in: A.M. Mika, T.Z. Winnicki, Advances in Membrane Phenomena and Processes, 6, Wroclaw Technical University Press, 1907, pp. 379–408. Wroclaw (1989).

[2] S. Loeb, S. Sourirajan, Seawater dimineralization by means of an osmotic membrane, Advances in Chemistry Series (ACS) 38 (1963) 117–132.

[3] F.A. Rodgers, Stacked microporous vapor permeable membrane distillation system, US Patent 3,650,905 (1972).

[4] F.A. Rodgers, Compact multiple effect still having stacked impervious and previous membranes, US Patent 3, 497,423 (Re. 27,982) (1974).

[5] P.J. Flory, Principles of Polymer Chemistry, Cornell University Press, Ithaca, 1953.

[6] T. Matsuura, Synthetic Membranes and Membrane Separation Processes, CRC Press, Boca Raton, FL, 1994.

[7] J.A. Reuvers, J.W.A. van den Berg, C.A. Smolders, Formation of membranes by means of immersion precipitation: Part I: A model to describe mass transfer during immersion precipitation, J. Memb. Sci. 34 (1987) 45–65.

[8] C. Feng, B. Shi, G. Li, Y. Wu, Preliminary research on microporous membrane from F2.4 for membrane distillation, Sep. Pur. Tech. 39 (2004) 221–228.

[9] C. Feng, B. Shi, G. Li, Y. Wu, Preparation and properties of microporous membrane from poly(vinylidene fluoride-co-tetrafluoroethylene) (F2.4) for membrane distillation, J. Memb. Sci. 237 (2004) 15–24.

[10] M. Tomaszewska, Preparation and properties of flat-sheet membranes from polyvinylidene fluoride for membrane distillation, Desalination 104 (1996) 1–11.

[11] J.M. Ortiz de Zárate, L. Peña, J.I. Mengual, Characterization of membrane distillation membranes prepared by phase inversion, Desalination 100 (1995) 139–148.

[12] M. Khayet, T. Matsuura, Preparation and characterization of polyvinylidene fluoride membranes for membrane distillation, Ind. Eng. Chem. Res. 40 (2001) 5710–5718.

[13] A. Bottino, G. Capannelli, S. Munari, A. Turturro, High performance ultrafiltration membranes cast from LiCl doped solutions, Desalination 68 (1988) 167–177.

CHAPTER 4

Formation of Hollow Fibre MD Membranes

OUTLINE

Introduction	59
Hollow Fibre Membranes with Single Porous Hydrophobic Layer	61
Hollow Fibre Membranes with Dual Porous Hydrophobic/Hydrophilic Layers	62
Copolymer Hollow Fibre Membranes	63
Principles of Formation of Porous Hydrophobic Hollow Fibre Membranes	64
Effects of Process Parameters on Hollow Fibre Membrane Structure	68
Influence of Polymer Concentration and Additive Content in Dope Spinning Solution	69
Effect of Different Coagulants on the Surface Morphology of the PVDF Membranes	71
Effects of Bore Fluid Flow Rate	73
Effects of the Air Gap Distance	74
Effects of the Take-up Speed	78
Effects of Polymer Solution Flow Rate	80
Effect of Post-treatment	81
Effects of Hollow Fibre Spinneret Design	81
Future Directions	85

INTRODUCTION

Since the first hollow fibre membranes were patented by Mahon in late 1960s [1,2], various types of hollow fibre membranes, hydrophobic or hydrophilic, dense or porous, single or double layered, have been proposed for different membrane separation processes leading to a fast growth of synthetic membrane technology. Nowadays, the hollow fibre membrane configuration is the most favoured membrane geometry in many membrane separation applications including membrane distillation (MD). Hollow fibre membrane modules normally exhibit large surface area per unit volume (e.g. the packing capacity of a hollow fibre membrane module may reach 500–9000 m^2/m^3) resulting in a high productivity per unit volume, are mechanically self-supporting, have good flexibility and are easy to assemble in module and to handle for different MD applications.

Similar to flat sheet membranes, hollow fibre membranes were prepared using different techniques and tested for MD. Most of them were prepared by the dry/wet spinning or the wet spinning technique using polyvinylidene fluoride (PVDF) polymer [3–9]. The process of fabricating hollow fibre membranes is more difficult than that of flat sheet membranes. The factors that control the morphology of hollow fibre membranes are quite different from those of flat sheet membranes. Different techniques are used to prepare hollow fibre membranes such as melt spinning plus stretching through thermally induced phase (TIP) inversion under high temperatures (i.e. melt-extruded/cold-stretching and melt spinning), dry spinning, wet spinning or dry/wet spinning [10].

It must be mentioned here that significant efforts have been made to develop new hollow fibre membranes with desirable structures and morphologies. However, until now, various membranologists admit that understanding the mechanisms of hollow fibre membrane formation is rather qualitative than quantitative. It is not easy to control the characteristics of hollow fibre membranes because various important spinning factors are involved simultaneously. Novel hollow fibre membranes that appeared in the membrane literature with different structures and morphologies are based mostly on trial-and-error experiments.

It is worth quoting that most commercially available hollow fibre membranes used in MD studies, such as polypropylene (PP), polyethylene (PE), polytetrafluoroethylene (PTFE) and PVDF membranes are fabricated by means of melt spinning. The fabrication procedure and basic understanding of these hollow fibre membranes are described in Chapter 5. Fabrication of hollow fibre membranes by the dry/wet phase inversion spinning technique at lower temperatures provides an alternative means to tailor membrane dimension and structure specifically for MD applications. This technique permits to prepare asymmetric hollow fibre membranes. Structures that contain large finger-like macrovoids and a sponge-like pore network allow holding a large volume of air, which can reduce the heat lost by conduction through the membrane section, providing less MD mass transfer resistance. All other design characteristics proposed in the previous chapters for MD flat sheet membranes are applicable to hollow fibre membranes. This chapter reviews the fundamental understanding of the hollow fibre membrane formation by dry/wet spinning and wet spinning techniques with desired characteristics and applications in MD.

As will be explained in the next section, the preparation of the hollow fibre membrane by the phase inversion technique often requires both internal and external coagulants and involves more controlling parameters than the flat sheet membrane. Shape and dimension of the spinneret, viscosity of the spinning dope (whether hollow fibres can be spun from the dope), temperature of the spinning dope, properties of the internal and external coagulants, flow rate of the bore fluid, dope extrusion rate, length and humidity of the air gap, wind-up speed, and fibre take-up speed are some of those examples. The effects of some of these parameters on hollow fibre membrane morphology and MD performance have been studied by many researchers as will be shown in the section 'Effects of process parameters on hollow fibre membrane structure' of the present chapter. It should also be pointed out that PVDF is the polymer most frequently chosen for the preparation of hollow fibre membranes for MD by the dry/wet or wet spinning technique. This is because it is a commercially available hydrophobic polymer that can be easily dissolved in common organic solvents, can form asymmetric membranes (i.e. MD mass transfer resistance is limited mainly in the skin layer of the asymmetric hollow fibre membrane), has good mechanical property and is chemically resistant to various organic solvents.

Hollow fibre membranes with a single porous hydrophobic layer, hydrophobic/hydrophilic porous layers, and even hydrophobic/hydrophilic/hydrophobic porous tri-layers, each layer having specifically designed characteristics can be used in MD and may be prepared by the dry/wet, wet or dry spinning technique. Polymers can be homopolymer or copolymer.

Hollow Fibre Membranes with Single Porous Hydrophobic Layer

Hollow fibre membranes with a single hydrophobic porous layer have been prepared from a single polymer by means of the dry/wet or wet spinning technique. Most of the studies have been reported as patents and papers published in refereed journals. These researches have been conducted to improve the membrane properties for membrane separation processes other than for MD.

Deshmuck and Li [11] and Wang et al. [12] introduced poly(vinylpyrrolidone) PVP in the PVDF casting solution as an additive in order to obtain highly porous PVDF hollow fibre membranes. However, trace quantities of PVP in the membrane affected the hydrophobicity of the PVDF membrane. These hollow fibre membranes could not be used in MD. Shih et al. [13] used ethanol to increase the gas permeability of dried PVDF hollow fibre membranes. Additives of small molecular weights such as water, ethanol and propanol were used by Wang et al. [14] for the preparation of PVDF asymmetric hollow fibre membranes for ultrafiltration (UF). Khayet et al. [15,16] proved that, for the formation of PVDF hollow fibres, the addition of a weak non-solvent such as ethanol to the internal and external coagulant delays the phase separation during gelation. The inner and the outer surfaces of the prepared hollow fibre membranes were characterized by atomic force microscopy (AFM) while their cross-section was observed by scanning electron microscopy (SEM) technique. A finger-like structure was formed when distilled water was used as a coagulant while a sponge-like structure appeared throughout the cross-section of the hollow fibre when 50% ethanol–water mixture (by volume) was used as the internal and the external coagulant. Generally, macro-voids and finger-like structures are formed when the coagulation process is fast, whereas the slow coagulation rate results in a sponge-like structure. It was observed that the pore size increased (13.0–35.6 nm) and the liquid entry pressure of water (LEP_w) decreased (>110.3 kPa) as the concentration of the non-solvent additive, ethylene glycol, was increased in the PVDF spinning solution and when ethanol was added either to the internal or to the external coagulant or both. Addition of ethanol either to the internal or external coagulant decreased the effective porosity. Furthermore, for the same prepared hollow fibre membrane, the pore size of the inner surface was larger than that of the outer surface. Khayet [17] investigated the influence of air gap, in a range 1–80 cm, on both the internal and the external morphology of PVDF hollow fibre membranes fabricated for UF. Based on AFM characterization technique, it was observed with the increase of the air gap that macromolecular nodules were aligned to the spinning direction, a simultaneous decrease of the pore size of both the inner and outer surfaces and an elongational stress because of gravity on both the internal and external surfaces of the PVDF hollow fibre membranes.

As can be seen from the above and other works reported in the membrane literature, a series of studies have been carried out in an effort to improve the properties of hollow fibre membranes, although these membranes were not necessarily prepared for MD purposes.

Despite a great number of works, the effects on morphological and structural characteristics and MD performance have not been fully elucidated due to the many interrelating spinning variables involved in the spinning technique. Hydrophobic homopolymers and copolymers other than PVDF and non-solvent additives other than those already studied should be used to design novel and competitive hollow fibre membranes for MD.

For MD process, Fujii et al. [3,4] were the first group who prepared hollow fibre membranes from different polymer solutions by the dry/wet spinning and PVDF was one of the tested polymers. Hollow fibre membranes with different pore sizes and porosities were prepared by varying the dope composition and spinning conditions. The fabricated fibres exhibited pore sizes of several nanometers but smaller than those of microfiltration (MF) membranes. For example, PVDF hollow fibre membranes were prepared with 4.0–24.8 nm mean pore size, 56–73% porosity, 0.675–0.844 mm internal diameter and 0.982–1.071 mm external diameter.

Recently, Wang et al. [8] prepared a PVDF hollow fibre membrane by the dry/wet spinning technique for DCMD using the solvent N-methyl-1-pyrrolidone (NMP) and ethylene glycol as a non-solvent additive. The PVDF concentration was 12 wt% and that of ethylene glycol was 8 wt%. The fabricated PVDF hollow fibre exhibited 0.16 μm mean pore size, a very narrow pore size distribution and an external ultra-skin layer over a porous support layer. When using an aqueous salt solution of 3.5 wt% as a feed, a feed temperature of 79.3 °C and a permeate temperature of 17.5 °C, the PVDF hollow fibre membrane produced 41.5 kg/m^2·h with a rejection as high as 99.99%.

Not only is the dry/wet spinning technique employed for preparation of hollow fibre membranes for MD, but also the wet spinning technique (i.e. without air gap distance). Various asymmetric microporous PVDF hollow fibre membranes with different pore sizes (i.e. 0.031–0.068 μm), effective porosities (71–1516 m^{-1}) and morphologies were prepared for vacuum membrane distillation (VMD) by the wet spinning technique using the solvent N,N-dimethylacetamide (DMAC) and the non-solvent additives LiCl and water [5–7]. Different pore sizes and porosities were achieved by varying the dope composition and the spinning conditions. It was found that the polymer dope composition was the most significant parameter controlling the morphology and permeation characteristics of the PVDF hollow fibre membranes. The increase of PVDF concentration yields reduction of the effective surface porosity and mean pore size of both the internal and the external surfaces of the hollow fibres. The PVDF hollow fibre membranes were employed to remove 1,1,1-trichloroethane (TCA) from aqueous solutions of different TCA concentrations as well as for toluene and benzene removal from water.

Hollow Fibre Membranes with Dual Porous Hydrophobic/Hydrophilic Layers

Similar to flat sheet porous composite hydrophobic/hydrophilic membranes reported previously in Chapter 3 [18,19], dual porous layers hydrophobic/hydrophilic hollow fibre membranes have advantages in DCMD over a single porous hydrophobic layer hollow fibre. These types of hollow fibre membranes can be prepared by a simultaneous co-extrusion method using dual layer spinneret design as will be described later on in the section 'Principles of formation of porous hydrophobic hollow fibre membranes'. One of the advantages of the co-extrusion method is the possibility to spin hollow fibre membranes with the inner and outer layers either hydrophobic or hydrophilic independently. Therefore, hollow fibre membranes with hydrophobic/hydrophilic or hydrophilic/hydrophobic layers can be fabricated by co-extrusion technique. Compared to other techniques to prepare composite hollow fibre membranes such as coating, the simultaneous co-extrusion approach is cost effective because it eliminates the second step of deposition of a selective layer upon hollow fibre membranes. One inconvenience of spinning dual hollow fibre membranes by the co-extrusion method is the risk of delamination and separation of the two layers due to polymer incompatibilities and/or different thermal expansion coefficients of the used polymers as well as to the repulsive force acting on the interface between the hydrophilic and the hydrophobic layers in an aqueous environment.

It must be mentioned here that dual layer hollow fibre membranes were first fabricated by Yanagimoto in late 1980s for MF and UF processes [20,21]. To improve water permeability, in 1989, Kuzumoto and Nitta simultaneously extruded dope solutions containing the same polymer but different solvents and additives [22]. Three years later, Ekiner et al. [23] fabricated dual layer hollow fibre membranes for gas separation. Since then, Li et al. [24] developed dual layer asymmetric hollow fibre membranes exhibiting similar selectivity as the single layer hollow fibre membranes but saved nearly 90% of material costs. Jiang et al. [25] and Li et al. [26], while studying the effects on the spinning conditions on dual layer hollow fibre membrane characteristics, produced various dual layer hollow fibre membranes with selective layer thicknesses smaller than 0.8 μm. Recently, Bonyadi and Chung [9] prepared dual layer hydrophilic/hydrophobic hollow fibre membranes for DCMD. The outer layer was a polymer solution made of 12.5 wt% PVDF in 87.5 wt% NMP and 30 wt% of hydrophobic cloisite 15A in the total polymer solution, whereas the inner layer was formed by 8.5 wt% of PVDF, 4 wt% of polyacrylonitrile (PAN) in 87.5 wt% NMP and 50 wt% (in total solid polymer solution) of the hydrophilic cloisite NA^+. The dual hydrophilic/hydrophobic hollow fibres were spun with 80 wt% methanol aqueous solution as the internal and external coagulants, at 3 cm air gap distance and by a free fall at which only gravitational force acted in the axial direction. The average pore size and porosity were 0.41 μm and 80%, respectively. A permeate flux as high as 55 kg/h·m^2 with a separation factor of 99.8% was achieved at 90 °C feed temperature and 16.7 °C permeate temperature for a 3.5 wt% sodium chloride (NaCl) aqueous solution.

Double layered hydrophobic/hydrophilic hollow fibre membranes or tri-layered hydrophobic/hydrophilic/hydrophobic hollow fibre membranes can also be prepared in a single step using one spinning dope by dry/wet spinning or dry spinning techniques using surface modifying macromolecules (SMMs) [18,19,27]. The hydrophobic SMMs are blended with a more hydrophilic polymer such as polyethersulfone (PES) to form the spinning dope. During spinning, the SMMs migrate towards the outer surface while travelling through the air gap, rendering the outer surface hydrophobic. SMMs can migrate to both the inner and outer surface, rendering both surfaces more hydrophobic than PES hollow fibre membranes fabricated without SMMs. This type of hollow fibre membranes has not yet been tested for MD and progress is expected in the future.

Copolymer Hollow Fibre Membranes

Copolymers can also be used to prepare MD hollow fibre membranes by dry/wet or wet spinning techniques. The copolymer poly(vinylidene fluoride-hexafluoropropylene) (PVDF-HFP) was employed by García-Payo et al. [28] for fabrication of porous hydrophobic hollow fibre membranes for DCMD by dry/wet spinning technique at a temperature of 40 °C. Different copolymer concentrations, ranging from 17 wt% to 24 wt%, were used. The solvent was DMAC and the non-solvent additive was polyethylene glycol (PEG). All the spinning parameters were kept constant except for the copolymer concentration. The effects of the copolymer concentration on the morphological properties of the PVDF-HFP hollow fibre membranes were studied by AFM, SEM and DCMD experiments. Different cross-sectional structures of the PVDF-HFP hollow fibre membranes and surface morphologies were observed with the increase of the copolymer concentration in the spinning solution. At high PVDF-HFP concentrations, the formed hollow fibre membranes exhibit a single sponge-like structure layer, whereas at low copolymer concentrations the cross-section of the prepared hollow fibre membranes has different layers of finger-like structure.

It must be pointed out that only few studies have been reported on the preparation of porous membranes and membrane modules specifically designed for MD applications. More efforts must be made to investigate the formation of hollow fibre membranes thoroughly and to understand the effects of spinning conditions on the membrane morphology in order to develop hollow fibre membranes of high MD performance that makes the MD process competitive with other membrane processes. Dual layer hydrophobic/hydrophilic or tri-layered hydrophobic/hydrophilic/hydrophobic hollow fibre membranes spun from a simultaneous extrusion process or in a single spinning process with blended SMMs should gain more attention in order to develop high MD performance and cost effective hollow fibre membranes.

PRINCIPLES OF FORMATION OF POROUS HYDROPHOBIC HOLLOW FIBRE MEMBRANES

After preparing the polymer spinning solution, which may contain a polymer, a solvent and a non-solvent as explained previously in Chapter 3, the solution should be filtered to remove impurities and insoluble contaminants. Subsequently, the resulted dope should be degassed before loading into the dope vessel. It must be pointed out that the spinning dope suitable for hollow fibre fabrication generally has a much greater viscosity and elasticity than that for the flat sheet membrane. The polymer solution can be finally loaded into the spinning system similar to the one shown schematically in Fig. 4.1. It consists of a gas cylinder (1), regulating pressure valve (2), pressure gauge (3), dope vessel (4), dope valve (5), bore liquid vessel (6), bore liquid pump (7), spinneret (8), air gap (9), coagulation bath (10), wind-up drum with different take-up speeds (11), fibre collecting reservoir (12) and wash water (13). In some spinning systems a circulation pump is employed instead of the gas cylinder (1) to drive the polymer solution through the spinneret. This has a tube-in-orifice structure as shown in Fig. 4.2. Typical dimensions of the spinneret are 0.5–1 mm inner diameter and 0.9–2 mm outer diameter. A polymer solution under gas pressure passes through the spinneret and is finally extruded from the annular space of the spinneret. The internal coagulant (bore fluid) comes out from the central tube of the spinneret and it is driven through the central tube either by the gravity force or by means of a circulation pump. The polymer solution, after being extruded from the spinneret, enters into a coagulation bath after traveling a certain distance of air gap (i.e. the distance between the spinneret and the coagulation bath), which may vary from 0 (wet spinning) to more than 1 m. Spinning parameters of hollow fibre MD membranes are summarized as an example in Table 4.1. In the dry/wet spinning process, coagulation of the internal surface of the nascent fibre starts immediately after its extrusion from the spinneret, whereas the external surface experiences coalescence and orientation of polymer aggregates through the air gap before gelation in the external coagulation bath takes place. After spinning, the nascent fibres are oriented by means of guiding wheels and finally pulled into a collecting

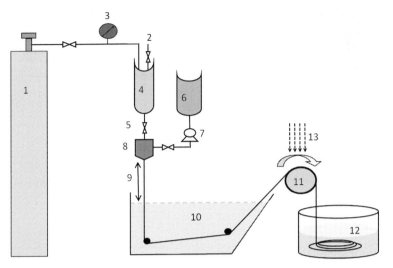

FIGURE 4.1 Schematic diagram of a typical hollow fibre spinning system: (1) spinning dope tank, (2) regulating pressure valve, (3) pressure gauge, (4) dope vessel, (5) dope valve, (6) bore liquid vessel, (7) bore liquid pump, (8) spinneret, (9) air gap, (10) coagulation bath, (11) wind-up drum (12) fibre collecting reservoir and (13) wash water. *Source: Reprinted from [43]. Copyright 2009, with kind permission from Elsevier.*

reservoir by a wind-up drum. During spinning, the take-up velocity is generally kept at the same speed as the free-falling velocity of the nascent fibre to prevent stretching of the membrane. It is worth noting that the hollow fibre membranes can be spun at room temperature (20–22 °C) or at a temperature higher than the ambient temperature. During spinning the take-up speed is normally maintained nearly the same as the dope extrusion speed so that no external elongational stresses, except gravity, can be applied to the nascent hollow fibre. It must be pointed out that a very large air gap distance may lead to flow instability of the fibre resulting in non-uniform fibre dimensions.

Figure 4.3 illustrates a dual layer spinneret design for preparation of double layered porous hydrophilic/hydrophobic or hydrophobic/hydrophilic hollow fibre membranes for MD. The outer layer dope, inner layer dope, and bore fluid are fed to the orifice by passing through three independent channels. In this case, it is necessary to use two pumps for the simultaneous extrusion of the spinning dopes through the dual layer spinneret. All the other parts of the spinning system are the same as the one presented in Fig. 4.1.

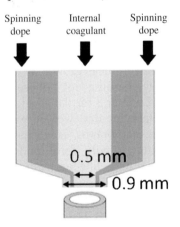

FIGURE 4.2 Spinneret used for preparation of single layer porous hydrophobic hollow fibre membranes. *Source: Reprinted from [43]. Copyright 2009, with kind permission from Elsevier.*

As can be seen in Fig. 4.1 two coagulation steps take place during hollow fibre spinning. These are the internal and external coagulation. Therefore, it is important to choose adequate coagulant liquids (i.e. type of the non-solvent liquid for phase inversion) and bore flow rate. The bore flow rate affects also the internal diameter of the spun hollow fibre membrane. Usually, water is the preferred external coagulant because

TABLE 4.1 Typical Spinning Parameters of MD Hollow Fibre Membranes

Dope solution composition (wt%) [a]	Bore flow rate (ml/min)	Length of air gap (cm)	Take-up speed (m/min)	Polymer flow rate (ml/min)	Polymer pressure (kPa)	Coagulant temperature (°C)	Reference
PVDF/DMAC/H$_2$O/LiCl (15/77.83/1.95/5.22)	1.1	0	4.5	–	138-207	25	[5,6]
PVDF/DMAC/H$_2$O/LiCl (15/77.83/1.95/5.2) PVDF/DMAC/H$_2$O/LiCl (18/75.9/1.67/4.43) PVDF/DMAC/H$_2$O/LiCl (20/75/1.37/3.63)	1.1 – 1.8	0	3.9 – 4.9	–	–	25	[7]
PVDF/NMP (12/88) PVDF/EG/NMP (12/8/80)	3	10	–	4	–	15.5	[8]
PVDF/PAN/NA$^+$/NMP (8.24/3.53/5.88/82.35) PVDF/15A/NMP (12.1/3.6/84.3)	0.5	3	Free fall	Inner: 2 Outer: 0.5	–	–	[9]

[a] PVDF, Polyvinylidene fluoride; PAN, Polyacrylonitrile; EG, Ethylene glycol; NA$^+$, Cloisite NA$^+$ clay hydrophilic particle; 15A, Cloisite hydrophobic clay particle; LiCl, Lithium chloride salt; DMAC, Dimethylacetamide; NMP, N-methyl pyrrolidone.

of its low cost and environmental friendliness. The proper choice of the coagulant is very important because the rate of demixing (i.e. phase separation) and the resultant inner and outer surface structures strongly depend on the chemistry and the composition of the coagulant as explained for flat sheet membranes in Chapter 3. The molecular sizes and solubility parameters of the used solvent and the internal and the external coagulant play important roles on membrane morphology. Solvent of a large size may have difficulties to leach out during the coagulation step. The difference in solubility parameters between the spinning dope and the internal or the external coagulant affects the coagulation rate and subsequently the porosity of the spun hollow fibre membrane. Depending on the initial dope composition and the ratio of solvent outflow to coagulant influx, the precipitation path may occur via spinodal decomposition or nuclei growth. The same qualitative understanding of phase inversion reported in Chapter 3 for flat sheet membranes can be applied for dry/wet or wet spinning hollow fibre membranes. However, in hollow fibre spinning, gravity force and elongation along the spinning line and applied stresses must be taken into consideration. Additional entropy terms have to be included in the Gibbs free energy equation (Eq. 3.1 in Chapter 3) when considering hollow fibre membrane spun isothermally. Thus, it is not appropriate to predict the phase inversion process by the traditional Flory–Huggins theory.

(a)

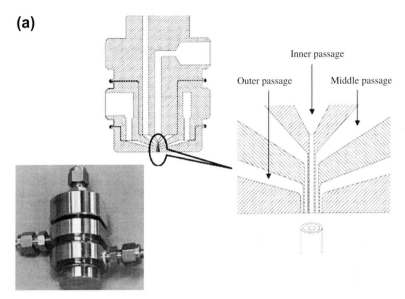

(b)

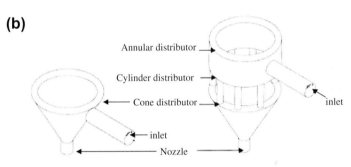

FIGURE 4.3 Dual spinneret designed for preparation of dual porous hydrophobic/hydrophilic or hydrophilic/hydrophobic hollow fibre membranes: (a) dual-layer spinneret and (b) outer passage of dual-layer spinneret. *Source: Reprinted from [24]. Copyright 2002, with kind permission from Elsevier.*

It is worth noting that when a polymer solution is extruded through a tube in orifice spinneret, shear stress is induced within the thin annular part of the spinneret. There are at least three forces (stresses) applied upon the dope spinning solution: (i) shear and elongation stresses within the spinneret, (ii) gravity induced by the weight of the spun hollow fibre and (iii) stress induced by the take-up system (i.e. wind-up drum shown in Fig. 4.1). It is known that the stress affects dramatically the polymer molecular orientation and relaxation at the outer surface of the nascent fibre membrane. If there is a certain distance of air gap before coagulation, the effects of elongation and relaxation will play important roles on final hollow fibre membrane structure and MD performance. Furthermore, the induced elongation stresses outside the spinneret from gravity through the air gap complicate the kinetics and dynamics of the phase inversion process. In addition, non-Newtonian polymer solutions may exhibit die swell and relaxation after exiting from spinneret as shown schematically in Fig. 4.4. Molecular chains tend to align themselves much better and this enhanced orientation causes the polymer molecules to pack closer to each other leading to a tighter skin fibre structure, which will affect hollow fibre membrane structure and MD performance. In fact,

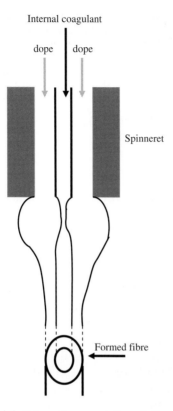

FIGURE 4.4 Schematic presentation of die swell occurring in the nascent MD hollow fibre membrane when extruded from a spinneret.

molecular orientation induced by shear stress within the spinneret might relax in the air gap region if the elongation stress along the spin line is small as spinning solution is a viscoelastic fluid or might be enhanced if the spin line stress is high. In other words, the elongation stress caused by gravity along the spinning line becomes more pronounced with increasing air gap.

The applied stresses may create extra phase instability during coagulation of internal and external surfaces of hollow fibre membranes and facilitate phase separation by shortening the time required for the transition from the binodal to the spinodal boundary. The stresses may induce orientation resulting in an oriented fibre structure. In general, the wet spun hollow fibre membranes exhibit a tight external surface morphology, while the dry/wet spun hollow fibre membranes in a long air gap may exhibit a more open cell structure. It must be informed that high-viscosity non-Newtonian liquids behave differently from low-viscosity Newtonian liquids inside the spinneret. The non-Newtonian liquid memorizes the shear and the elongation stresses, whereas the Newtonian liquid does not.

Usually the spun hollow fibre membranes have to go through solvent exchange and post-treatment before characterization tests. The spun hollow fibre membranes are normally subjected to solvent exchange treatment explained for the flat sheet membranes in Chapter 3. Once a hollow fibre membrane is formed, it can be stored in a non-solvent having low surface tension in order to remove residual solvent and to prevent possible deformation and shrinkage during the drying process, which may lead to collapse and closure of pores. Ethanol, methanol and hexane are frequently used. For example, PVDF hollow fibre membranes are immersed in pure ethanol or in 50% aqueous ethanol solution (by volume) for at least 4 h. Then, the hollow fibre membranes are dried in air at room temperature before characterization tests. Heat treatment of the spun porous hollow fibre membranes induces molecular relaxation and repacking of polymer chains, which will reduce pore sizes and remove membrane defects.

EFFECTS OF PROCESS PARAMETERS ON HOLLOW FIBRE MEMBRANE STRUCTURE

Many researchers have studied the effects of dope composition, properties of the internal and external coagulants and spinning conditions on the morphology and structure of hollow fibre membranes as well as on their performance in different membrane processes. Most of the studies have been summarized in membrane science and technology literature.

However, the fabrication of a porous hydrophobic hollow fibre membrane with a desirable MD performance is not a simple process and the effects on hollow fibre membrane morphology and permeation properties reported in the literature often provide conflicting observations. In the present section only those hollow fibre membranes, which were prepared for MD process are reviewed.

The primary process variables establishing the fibre geometry including the inner diameter and the fibre thickness are spinneret geometry, coagulant flow rate, dope extrusion rate, take-up velocity and air gap distance.

Influence of Polymer Concentration and Additive Content in Dope Spinning Solution

While studying membrane formation by the dry/wet or wet spinning technique, the effects of the polymer concentration and the additive content on the morphology and performance of hollow fibre membranes have been investigated by many researchers [7,11,15,16]. In general, the thickness of the hollow fibre membrane increases with increase in the polymer concentration in the dope. This was observed by Wu et al. [7] for PVDF hollow fibres. The thickness of the hollow fibre membranes is also affected by the fibre take-up speed and the flow rate of the internal coagulant.

It must be mentioned that a spinning dope with a low polymer concentration tends to result in membranes of high porosity in a non-solvent-induced phase separation. A significant reduction of the effective porosity was observed with increasing PVDF concentration in the spinning dope from 15 wt% to 20 wt% [7]. The effective porosity decreased from $1516\, m^{-1}$ to $71\, m^{-1}$. However, over this PVDF concentration range no apparent effect on the mean pore size of the prepared MD hollow fibre membranes was observed. A mean pore size was maintained at around 0.1 µm.

Since LEP_w is associated with the maximum pore size and the hydrophobicity of hollow fibre membranes, Wu et al. [7] observed an increase of LEP_w with an increase of the polymer concentration in the spinning solution. Values ranging from 0.6 to 0.8 MPa were obtained for hollow fibre membranes prepared for VMD using 15 wt% PVDF in the dope, whereas higher values up to 1.33 MPa were obtained for higher PVDF polymer concentrations, such as 20 wt%. It must be pointed out that these values are much higher than the applied transmembrane hydrostatic pressures in VMD experiments.

It must be mentioned that the decrease in the polymer concentration is related with the increase of the non-solvent additive in the dope, which may also increase the porosity and pore size of the spun hollow fibre membranes [11,15,16]. Figure 4.5 shows the SEM images of two types of hollow fibre membranes prepared with and without non-solvent additive.

Compared with the hollow fibre membrane spun without non-solvent additive, the hollow fibre membrane spun with the non-solvent additive exhibits a more uniform and porous inner skin containing smaller pores and a more porous outer surface skin with larger pores. Four types of structures may appear when non-solvent additives are added into the spinning dope: (i) asymmetric outer selective skin layer, (ii) finger-like macrovoids, (iii) sponge-like substructure and (iv) mesh-like structure of the inner skin layer. All these structures can be seen in Fig. 4.5(a). The formed sponge-like structure near the inner skin is due to the delayed phase separation, whereas thin selective layer formed at the outer surface is due to the external strong coagulant, which induces immediate phase inversion.

García-Payo et al. [28] studied the effects of the concentration of the copolymer PVDF-HFP on the cross-sectional structure and the structures at the inner and outer surfaces of DCMD hollow fibre membranes. Polymer dope, internal and external coagulants were all kept

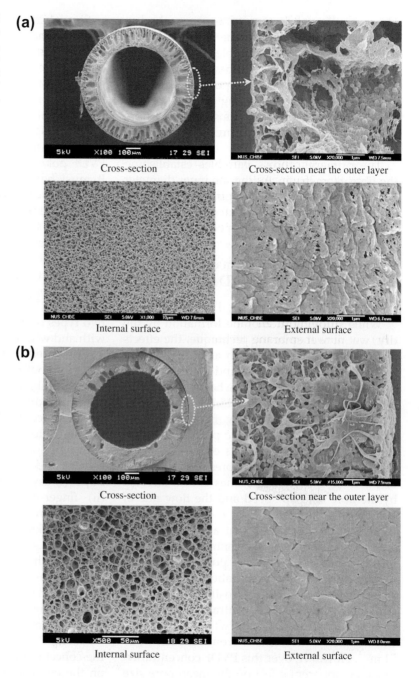

FIGURE 4.5 SEM morphology of PVDF hollow fibre membranes prepared from NMP solution with ethylene glycol used as additive (a) and without additive (b) (see Table 4.1 for more information). *Source: Reprinted from [8]. Copyright 2008, with kind permission from Elsevier.*

at 40 °C. About 30% increase of the internal and external diameter was observed with increasing copolymer concentration in the dope (i.e. the external diameter was 1635 ± 37 μm for the hollow fibre prepared with 17 wt% copolymer concentration, whereas it was 2099 ± 26 μm when prepared with 24 wt%). It was found that the inner diameters ranged from 1525 μm to 1989 μm. In contrast, the thickness did not change significantly, being maintained at around 80 ± 25 μm. As can be seen in Chapter 2 (Fig. 2.2) the cross-section of the hollow fibre membrane prepared with the lowest copolymer concentration, 17 wt%, exhibited a finger-like structure at both the external and internal layers. This finger-like structure disappeared from the internal layer of the hollow fibre membrane as the copolymer concentration was increased. Finally, a sponge-like structure was formed throughout all the cross-section of the hollow fibre membrane prepared with 24 wt%, the highest concentration. This result can be explained based on the decrease of the coagulation rate with the increase of the copolymer concentration in the dope. The DCMD performance was found to be better for the hollow fibres prepared with smaller copolymer concentrations due to their larger pore sizes as only a small increase of the porosity was detected with increasing the copolymer concentration. Figure 4.6 presents the AFM images of both the inner and outer surfaces of hollow fibre membranes prepared for DCMD with different PVDF-HFP copolymer concentrations. The nodules are seen as the high bright ridges, whereas the pores are seen as the deep dark depressions.

Effect of Different Coagulants on the Surface Morphology of the PVDF Membranes

When a strong coagulant such as water for PVDF polymer is used for the fabrication of the hollow fibre membrane, a dense and smooth surface with no obvious pores is formed.

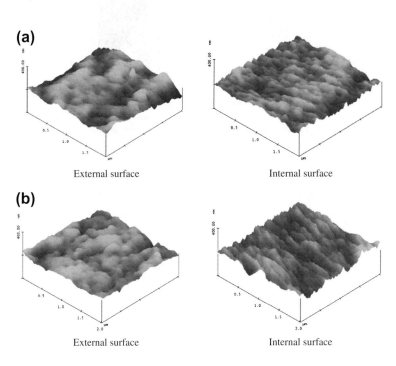

FIGURE 4.6 AFM images of PVDF-HFP hollow fibre membranes prepared with different copolymer concentrations 17 wt% (a) and 24 wt% (b). *Source: Reprinted from [28]. Copyright 2010, with kind permission from Elsevier.*

However, when a weaker coagulant is employed such as methanol, which may be mixed with water, the roughness of the formed membrane surface may increase considerably due partially to the formation of pores. This is the reason for using the coagulants: 80 wt% NMP aqueous solution by Wang et al. [8] for preparation of PVDF hollow fibre membranes for DCMD and 80 wt% methanol aqueous solution by Bonyadi and Chung [9] for fabrication of dual hydrophilic/hydrophobic hollow fibre membranes for DCMD.

Figure 4.7 shows the effects of four coagulation conditions on the cross-sectional structure of porous hydrophobic hollow fibre membranes [15]. Water and 50% (by volume) ethanol in water were used as internal and external coagulants. Different cross-sectional structures can be observed. In Fig. 4.7(a) prepared with water as internal and external coagulants, long finger-like voids are formed near the inner surface, while smaller cavities are formed near the outer surface. Between the inner and outer layers sponge-like structure appears. The inner layer is reduced when ethanol is added to the internal coagulant as can be seen in Fig. 4.7(c), whereas in Fig. 4.7(b) the outer layer is eliminated when ethanol is added to the external coagulant. A uniform sponge-like structure extends over the entire hollow fibre membrane cross-section, in Fig. 4.7(d), when ethanol is added to both the internal and external coagulants. The addition of ethanol delays the coagulation process and a long finger-like structure changes to a short finger-like structure and further to a sponge-like structure. It is worth quoting that the slow coagulation of the hollow fibre membranes can be explained on the basis of the mutual diffusivity of solvent/non-solvent exchange and solubility parameters of the materials

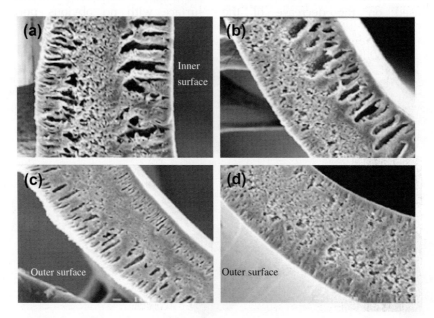

FIGURE 4.7 Cross-sectional SEM structure of PVDF hollow fibre membranes prepared with different coagulants: (a) water coagulant, (b) internal coagulant water and external coagulant 50% ethanol in water, (c) internal coagulant 50% ethanol in water and external coagulant water and (d) 50% ethanol in water coagulants. *Source: Reprinted from [15]. Copyright 2002, with kind permission from Elsevier.*

involved in hollow fibre spinning. The addition of ethanol in water reduces the diffusion of non-solvent into the nascent hollow fibre membrane and consequently decreases the rate of precipitation.

Furthermore, the presence of nodules and nodule aggregates was observed at both the inner and outer surfaces when hollow fibre membranes were prepared with water or ethanol/water mixtures as coagulants. Khayet et al. [15] found that the average nodule size of the inner surface was larger than that of the outer surface. This is because the coagulation of the internal surface of the nascent hollow fibres starts immediately after extrusion from the spinneret, whereas the external surface experiences coalescence and orientation of polymer aggregates while travelling through the air gap. However, the surface roughness increased when ethanol was added to the coagulant, which is due to the increase of the pore size. The pores at the inner surfaces of the hollow fibre membranes are larger than those at the outer surfaces and the pore size increased both at the inner and the outer surfaces as ethanol was added to the coagulants. Khayet et al. [15] observed a decrease of the effective porosity of PVDF hollow fibre membranes with the addition of ethanol in either the bore liquid or in the coagulation bath or both.

Effects of Bore Fluid Flow Rate

The flow rate of the internal coagulant affects the dimensions of the spun hollow fibre membranes considerably. Wu et al. [7] found that the thickness of the PVDF hollow fibre membranes prepared for VMD by wet spinning technique decreased with the increase of the internal coagulant flow rate. For example, when the internal coagulant flow rate was increased from 1.1 to 1.5 ml/min and further to 1.8 ml/min, the corresponding thickness of the PVDF hollow fibres membranes decreased from 188 to 176 μm and further to 155 μm. An enhancement of the inner diameter and a reduction of the outer diameter of the VMD hollow fibre membranes with the increase of the bore flow rate were detected.

Figure 4.8 shows a hydrophobic hollow fibre membrane with a wavy geometry, which might increase flow turbulence in MD applications reducing both temperature and concentration polarizations and therefore increasing the MD mass transfer. Very few studies have been conducted on this subject in not only MD field of research, but also in membrane science in general. This type of morphology may be regarded as irregularity and instability induced during spinning, leading to the deformation of the cross-section of hollow fibre membranes

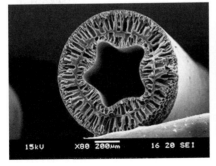 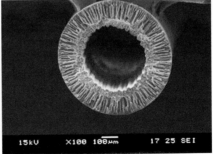

FIGURE 4.8 Irregular shape in the cross-section of PVDF hollow fibre membrane prepared by wet spinning technique using 20 wt% PVDF solution in NMP as the dope and 40 wt% NMP aqueous solution as the internal coagulant. *Source: Reprinted from [29]. Copyright 2007, with kind permission from Elsevier.*

fabricated by dry/wet or wet spinning techniques. One may believe that this instability is due to the bore fluid flow rate. Based on the experimental results using two types of polymers, i.e. PAN and PVDF, Bonyadi et al. [29] reported that the deformation happened in the fibre spinning line and could not be attributed to the drying or post-treatment steps of the spun hollow fibre membranes. By varying the bore fluid flow rate from 0 to 2 ml/min, Bonyadi et al. [29] confirmed that other factors might be responsible for the instability rather than the competitive force between the polymer solution and the bore fluid. As will be shown later, by increasing the air gap distance, it was observed that the number of corrugations in the inner contour of the hollow fibre membranes decreased and finally disappeared at a high air gap length of 20 cm when the dope PVDF concentration was 20 wt% in NMP and 40 wt% NMP aqueous solution was employed as the internal coagulant.

Effects of the Air Gap Distance

The air gap distance plays a very important role on the structure of the nascent hollow fibre and therefore may affect MD performance. An increase in air gap distance results in a significant decrease in permeance. This result may arise not only from the fact that different precipitation paths take place during the dry/wet and wet spinning processes, but also from the chain orientation and packing along the spinning line that is induced by elongation. However, if the air gap is too long, it may also create defects because of gravity and elongational stresses as reported in the section 'Principles of formation of porous hydrophobic hollow fibre membranes'. The hollow fibre membranes spun from a large air gap distance may have greater orientation and tighter molecular packing due to a high gravity-induced elongational stress than that of the wet spun hollow fibre membrane. For instance, the influence of the dry phase inversion process along the air gap distance is rather complicated and may vary depending on the spinning dope and the spinning conditions.

As reported earlier, macromolecules while exiting from the spinneret may experience die swell (Fig. 4.4) and relax, which will change the macromolecular orientation. The orientation may further change if there is an air gap between the spinneret and the external coagulation bath. This effect is probably stronger at the outer surface than at the inner surface. Moreover, the elongation stress outside the spinneret will increase with an increase of the air gap length due to the gravity and the spinning line stress will also be enhanced as the draw ratio is increased, resulting in an increase in the polymeric molecular orientation.

It was stated that the effect of the air gap length on the internal and external diameters is complex and experimental cases where the diameters decreased with the increase of the air gap length were observed as often as the increase of the diameters with the increase of the air gap [30].

It is worth quoting that no study has been reported yet on the effects of the air gap distance on the MD performance. Wang et al. [12] and Khayet [17] studied the effects of the air gap length on the PVDF hollow fibre membrane morphology, cross-sectional and surface structures, gas and UF permeation performance. Wang et al. [12] varied the air gap length from 5 cm to 15 cm and found that the water permeation flux tended to decrease with decreasing air gap distance while no significant effect on the separation factor was observed with the change of the air gap distance. Moreover, it was found that the effective porosity decreased when the air gap distance was increased while the change of the pore size with the air gap distance was not significant. On the other hand, Khayet [17] investigated more closely the influence of the air gap distance on both the internal and external morphology of PVDF hollow fibre membranes. The hollow fibre membranes were spun from polymer solution containing PVDF/ethylene glycol/DMAC with a weight ratio of 23/4/73, in a wide range of air gap lengths (i.e. 1–80 cm). Aqueous ethanol solution, 50% by volume, was used as the internal and external coagulant. The diameters and the wall thickness of the PVDF hollow fibre membrane spun with an air gap of 1 cm were greater than those of the other hollow fibre membranes prepared with higher air gap lengths. This result was attributed to the die swell of macromolecules when exiting from the spinneret due to the viscoelastic properties of the PVDF spinning solution, although Qin et al. [31] reported that the die swell might disappear and the shear-induced orientation might relax along an air gap distance of only 1 cm. Furthermore, the PVDF hollow fibre membranes prepared with air gap lengths between 5 and 80 cm showed a decrease of the outer diameter by about 4%, no significant variation in the inner fibre diameter was observed and hence the wall thickness decreased with the increase of the air gap length. This may be due to the elongation stress caused by the gravity. It must be mentioned that during hollow fibre membrane formation, gravity will influence the minimum jet-stretch ratio (i.e. ratio of take-up speed to dope extrusion speed) achievable in the spin line and introduces an elongational stress on the nascent hollow fibre membrane.

Khayet [17] found that the hollow fibre membranes spun with air gap lengths higher than 45 cm exhibit tighter cross-sectional structures (i.e. lower free volume) than the hollow fibre membranes spun with lower air gap lengths as can be seen in Fig. 4.9. At a low air gap distance, after exiting from the spinneret, the nascent hollow fibre membrane is immersed in the non-solvent coagulation bath more rapidly than at a larger air gap distance. This results in a greater amount of non-solvent and solvent trapped in the contracted polymer chains. Therefore, the hollow fibre may have a structure with a longer-range random and less oriented polymeric chain interaction, resulting in larger intermolecular voids or free volume.

It is well known that both surfaces of the porous hydrophobic hollow fibre membranes may play a role controlling the MD performance. The mean pore sizes and the pore size distributions of both the internal and external surfaces of the hollow fibre membranes can be determined by means of AFM analysis.

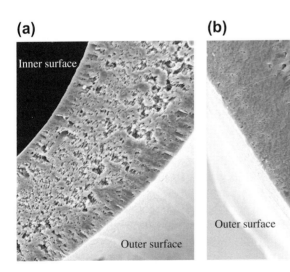

FIGURE 4.9 Cross-sectional SEM pictures of PVDF hollow fibre membrane prepared with air gap distance of 5 cm (a) and 80 cm (b). *Source: Reprinted from [17]. Copyright 2003, with kind permission from Elsevier.*

Figure 4.10 shows, as an example, the three-dimensional AFM images of the inner and outer surfaces of PVDF hollow fibre membranes prepared at different air gap lengths. These images indicate that the surface nodules appear to be randomly arranged when the air gap is small but form rows of nodule aggregates aligned in the spinning direction for the high air gap length. In general, the average nodule size at the inner and outer surfaces increases with an increase of the air gap distance. Molecular chains when subjected to a longer air gap tend to align themselves in much better way than those experiencing a shorter air gap length and this enhanced orientation will cause the polymer molecules to be packed closer to each other, resulting in a tighter structure.

The pore sizes and nodule sizes were determined by inspecting line profiles on the AFM images taken from different areas of the same hollow fibre membrane, as explained in Chapter 8. Figure 4.10 shows the pore sizes and the pore

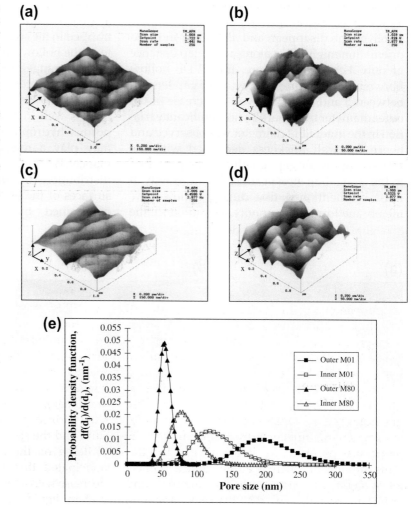

FIGURE 4.10 AFM images (a–d) and pore size distributions (e) of the inner and the outer surfaces of the PVDF hollow fibre membranes prepared with different air gap distances 1 and 80 cm. The number associated to M refers to the air gap distance. (a) inner surface of the membrane M01, (b) outer surface of the membrane M01, (c) inner surface of the membrane M80 and (d) outer surface of the membrane M80. *Source: Reprinted from [17]. Copyright 2003, with kind permission from Elsevier.*

size distributions at the inner and outer surfaces of the PVDF hollow fibre membranes prepared at different air gap distances.

It was observed that the mean nodule size of the outer surface increased with increasing the air gap length, while the mean pore size increased only slightly at the internal surface. This enhancement of the nodule size may be due to the increase of polymer interchain entanglement with the air gap distance. From phase separation stand point, the work induced by gravity may result in a change in the location of phase separation curves. It is believed that spinodal phase separation resulted in the nodular structure and took place under the condition of fast exchange of solvent and non-solvent. When the coagulation rate is slow, external stress due to gravity may either fasten the occurrence of spinodal decomposition, reduce the distance for a polymer solution between binodal and spinodal boundaries or may induce orientation as stated previously.

The pore size decreases simultaneously at the inner and outer surfaces of the PVDF hollow fibre membrane as the air gap distance is increased from 1 to 80 cm. There should be an elongation-induced pore deformation when the air gap is increased. A round pore should be elongated as an elliptical pore with a narrow fluid channel and the pore size distribution may shift from a broad distribution to a narrow one, as can be observed in Fig. 4.10. This result will eventually affect the mechanism of vapour transport through the hollow fibre membrane pores from Poiseuille type of flow to Knudsen type of flow.

Khayet [17] observed larger pore sizes at the outer surfaces of PVDF hollow fibre membranes than those at the inner surfaces for air gap lengths smaller than 25 cm. An opposite trend was observed for larger air gap lengths. This implies that there is an imbalance in effect of elongational stress on the internal and external surface. In the dry/wet spinning process, coagulation of the internal surface of the nascent hollow fibre starts immediately after extrusion from the spinneret and will bear more stress (gravity) at low air gap length than the external side. Thus, the inner pore size is gradually reduced and deformed with the increase of the air gap until the inner surface can no longer bear the stress. Then the outer side starts to bear more stress as the air gap increases and its pore size is reduced and deformed until it cannot bear more stress. After that a three-dimensional pore structure and an unstable spinning process may be created.

It was noticed that, in general, the roughness parameters decrease simultaneously at the internal and external surfaces of the porous hydrophobic hollow fibre membrane and the outer surfaces are smoother than the inner surfaces [17].

It is worth quoting that instabilities and irregularities may occur during the spinning of hollow fibres inducing corrugations at their external or internal contours. These corrugations can be eliminated increasing the air gap distance as can be seen in Fig. 4.11. It can be seen that the number of corrugations in the internal surface of the hollow fibre membranes is reduced approaching a final circular shape. Bonyadi et al. [29] stated that only a fraction of a second is needed for a nascent fibre to travel through the air gap and there is not enough time for magnification of the hydrodynamic instability effects. Therefore, it was concluded that the hydrodynamic instability, which lead to non-uniform cross-section and wavy geometry might be due to the pressure induced in the nascent fibre (as a result of diffusion/convection, precipitation, densification and shrinkage), which will buckle the rigid elastic shell formed at the interface between the bore fluid and the spinning dope.

Based on the above reviewed results, it is anticipated that the MD permeate flux will decrease with an increase in of the air gap length both for liquid and gas permeation [12,17]. Most

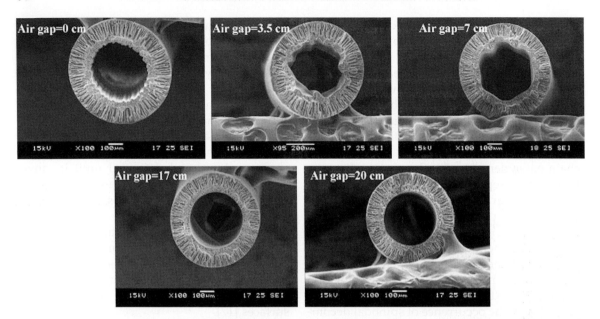

FIGURE 4.11 Cross-sectional structure of PVDF hollow fibre membranes prepared with wet spinning technique and dry/wet spinning technique using 20 wt% PVDF in the solvent NMP and 40 wt% NMP aqueous solution as internal coagulant. *Source: Reprinted from [29]. Copyright 2007, with kind permission from Elsevier.*

likely, it is caused by the greater orientation and tighter molecular packing that corresponds to the longer air gap (Fig. 4.9).

Although the effect of the air gap distance on the morphology and properties of porous hollow fibre membranes has been investigated intensively, further detailed and systematic studies need to be made for knowing if the trends observed in the foregoing works are also applicable for other hydrophobic spinning dopes.

Effects of the Take-up Speed

The effects of the take-up velocity on the characteristics of porous hydrophobic hollow fibre membranes prepared for VMD have been studied by Wu et al. [7]. Wet spinning technique was employed using different concentrations of PVDF, the solvent DMAC and the non-solvent additives, water and LiCl, in the spinning dope. Water was employed as the internal and external coagulant. During spinning, the take-up velocity was kept nearly the same as the free-falling velocity of the nascent fibre in the coagulation bath to prevent stretching and extension of the hollow fibre by mechanical drawing. It was found that the thickness of the hollow fibre membrane was affected by the fibre take-up velocity. This was increased with the increment of the take-up velocity. It must be pointed out that a higher take-up velocity reflects a higher dope extrusion rate, which means more polymer is extruded from the spinneret into the coagulation bath resulting in a thicker hollow fibre wall after coagulation. Furthermore, due to the enhancement of the hollow fibre membrane thickness and the reduction of the effective porosity with the increase of the take-up velocity, a decrease of permeability was observed. At low PVDF concentration in the dope solution, i.e. 15 wt%, the LEP_w of the prepared hollow fibre increased from 0.65 MPa to 0.8 MPa when the take-up velocity was increased from 3.9 m/min to 4.5 m/min. This means that the maximum pore size of the VMD hollow fibre membrane is reduced with the increase of the take-up velocity. In contrast, at high PVDF concentrations, i.e. 18 wt%, the LEP_w was maintained at the same level of around 1.3 MPa.

In various studies the elongational draw ratio, ϕ, is defined as:

$$\phi = \frac{(d_o^2 - d_i^2) \text{ spinneret}}{(d_o^2 - d_i^2) \text{ hollow fibre}} \quad (4.1)$$

where d_o and d_i are the outer and the inner diameters, respectively, of the spinneret and the spun hollow fibre membrane.

It is worth quoting that an increase in the elongational draw can diminish the number of macrovoids and macrovoid layers from the cross-section of hollow fibre membranes, change their dimensions and can also remove them [32,33]. This effect may be due to the rapid shrinkage of the hollow fibre diameter during the elongational stretch and to the simultaneous increase of the polymer extrusion rate, leading to a radial outflow of solvent towards the internal and external coagulants; and therefore hindering the diffusion of both coagulants into the nascent hollow fibre membrane, thus eliminating the formation of macrovoids.

In MD, presence of macrovoids in asymmetric hollow fibres may be desirable because the driving force is the transmembrane vapour pressure difference and the transmembrane pressure is very small, since both sides are near the atmospheric pressure. Moreover, the presence of macrovoids and macrovoid layers will decrease the thermal conductivity of the hollow fibre membrane matrix because gases will fill the void spaces. The only condition is that the wall of the macrovoids should be porous as can be seen in Fig. 4.12, which shows the SEM pictures of finger-like structures at the cross-sections near the internal and external surfaces with porous finger-like. It must be mentioned that the presence of macrovoids weakens the mechanical strength of the hollow fibre membranes and these fibres are not adequate for processes that require high hydrostatic pressures. Different studies have been carried out to reduce macrovoids or even eliminate them from the cross-section of hollow fibre membranes. The increase of polymer concentration and the use of high viscosity spinning solution, the spinning with high shear rates and the induction of delayed demixing or coagulation can be considered as the ways of reducing macrovoids.

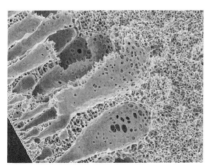
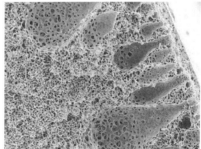

Internal cross-section   External cross-section

FIGURE 4.12 SEM structures of a hollow fibre membrane exhibiting finger-like structures with porous walls. *Source: Reprinted from [28]. Copyright 2010, with kind permission from Elsevier.*

Effects of Polymer Solution Flow Rate

The dope rheology plays a very important role in the process of formation of porous hydrophobic hollow fibre membranes by dry/wet or wet spinning techniques. When the polymer solution is pumped through a spinneret, shear stress will be produced within the thin annular space which reaches the highest value at the wall of the spinneret since the dopes are normally non-Newtonian fluids. The polymer flow rate is controlled in the spinning system by a circulation pump or by applying a controlled pressure over the dope and therefore different shear rates of the dope inside the annular region of the spinneret may be formed. With an increase in the dope flow rate, the shear rate of the dope within the annular region of the spinneret is also increased. The effect of shear rate or flow rate of the polymer solution on the MD performance of hollow fibre membranes is not studied yet. However, various studies have been carried out on this subject using different spinning dopes and the observed effects depend on the type of chosen polymer solution [34–36].

It is stated earlier that the high shear rate modifies the coagulation path and retards the formation of macrovoids or even eliminates the macrovoid layer. With an increase in the shear rate, the membrane structure becomes more compact affecting the whole cross-sectional structure as well as the surface morphology of the spun hollow fibre membranes.

Ren et al. [34] studied the effect of the shear rate on the structure of the PVDF hollow fibre membrane and water permeation using different polymer concentrations. It was observed that the water flux increased with the increase of the shear rate and then levelled off, whereas the rejection factor was almost the same. It was reported that when using PVDF hydrophobic polymer solutions, which are close to Newtonian fluids, the shear induced molecular chain orientation may occur but the degree of orientation is small. Larger pores may be formed in lower polymer concentration areas, leading to an increase in the permeability with the increase in the shear rate. Other authors [35,36] reported a decrease of the UF permeation flux and a decrease of the rejection factor by increasing the shear rate. When the shear rate is increased, the shear induced molecular chain orientation readily appeared and the molecular chains tended to pack more closely to each other leading to a denser skin and therefore lower permeability [35]. This opposite result was attributed to the type of the dopes used, which were non-Newtonian power-law fluids with a lower power law index, b in Eq. (4.2) (0.873–0.97). Ren et al. [34] found higher b values of 0.982 and 0.989.

$$\sigma = a\gamma^b \quad (4.2)$$

where σ is the shear stress (Pa), γ is the shear rate (s^{-1}), a is the rheological constant and b is the power-law index. The above equation was assumed to be applicable to high shear rates and to be able to describe the rheological behaviour of the spinning dopes within the spinneret.

Since the spinning dopes are normally non-Newtonian fluids, the stress may dramatically affect molecular orientation of polymer chains and chain packing. The increasing of shear rate may elongate and reduce pore size of hollow fibre membranes and their permeability [37].

Qin et al. [31] investigated the effects of shear induced orientation during UF hollow fibre fabrication by dry/wet and wet spinning techniques using different spinning polymer dopes. It was concluded that the hollow fibre membranes prepared with wet spinning technique exhibited greater shear induced molecular orientation than those prepared with the dry/wet spinning technique. In addition, Qin et al. [31] stated that the molecular orientation induced at the outer surface of the nascent hollow fibre, while the spinning dope is within the spinneret, can be frozen into the fibre when this goes into the coagulation bath immediately (wet-spinning); while it relaxes in the air gap before solidification if the elongation stress along the spin line is small. The wet spun hollow fibre membranes have smaller pore sizes and/or denser skins and lower water permeate fluxes than the dry/wet spun hollow fibre membranes. In contrast, the separation factor and mechanical properties were found to be greater for the dry/wet spun hollow fibre membranes. In another study, Qin et al. [35] observed a decrease of the pore size of the outer layer of UF hollow fibre membranes when the shear rate within the spinneret was increased and attributed this result to the enhanced molecular orientation.

Effect of Post-treatment

Prior to the drying step, the spun porous hydrophobic hollow fibre membrane usually go through the solvent exchange and other post-treatment procedures in order to remove residual solvents, prevent hollow fibre shrinkage, eliminate possible defects and reduce pore collapse. Conventional solvent exchange step before drying consists of immersion of the obtained hollow fibre membranes in one non-solvent or successive immersion in several non-solvent mixtures containing components with lower surface tensions than water. Mixtures of water with increased concentrations of lower surface tension components such as methanol and ethanol can be employed. Hexane can also be employed for further solvent exchange since its surface tension is even lower than alcohols and it is volatile.

Wu et al. [7] found that post-treatment of the spun PVDF hollow fibre membranes using ethanol caused a significant increment of the surface porosity and thus enhanced the VMD permeation flux. It was observed that the external surface of the hollow fibre membrane subjected to post-treatment was more porous than that of the hollow fibre membrane prepared without post-treatment.

It is recommended to dry the spun hollow fibre membranes at room temperature after the solvent exchange procedure. Heat treatment may induce molecule relaxation, which tends to minimize the surface defects and renders the hollow fibre membranes denser. Heat treatment is usually considered as an effective method to reduce pore sizes and remove defects of reverse osmosis (RO) and pervaporation (PV) membranes.

Effects of Hollow Fibre Spinneret Design

Hollow fibre spinneret (Figs. 4.2 and 4.3) is the most important part of the whole spinning setup. It is known that the primary process variables establishing hollow fibre diameters and thickness are spinneret geometry. Other spinning parameters such as the coagulant flow rate, the dope extrusion rate, the take-up velocity and the air gap distance can modify the geometrical parameters of the spun hollow fibre membranes.

Although the design and fabrication of hollow fibre spinneret are complicated tasks, various types of spinnerets (i.e. single or dual spinnerets, straight annular or conical spinnerets, micro-structured orifice in spinneret) with different configurations and dimensions are used in spinning hollow fibre membranes. Most information is maintained proprietary and one can hardly find a complete schema of a practical hollow fibre spinneret, which includes all needed dimensions and materials used. Usually the hollow fibre spinneret consists of a reservoir and an annular channel, which has a high annulus length per flow gap.

One of the most important requirements when designing hollow fibre spinnerets is the need to supply the bore fluid in the centre of the developing dope flow without disturbing the axis-symmetric flow upon leaving the spinneret. The most difficult parts of the spinneret design are precision and channel alignment. Designing spinnerets for preparation of dual

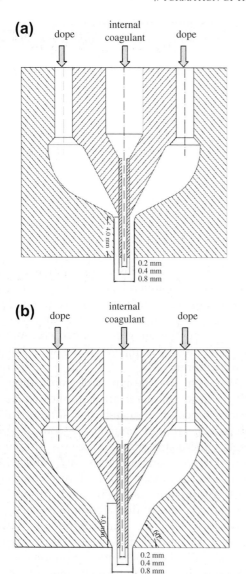

FIGURE 4.13 Design of two hollow fibre spinnerets with different dope flow angles, conical 60° (a) and straight 90° (b). *Source: Reprinted from [38]. Copyright 2004, with kind permission from Elsevier.*

or multi-layered hollow fibre membranes becomes even more complex.

The effect of hollow fibre spinneret design on the MD performance of porous hollow fibre membranes has not yet been studied. However, systematic studies on the effects of spinneret design and flow behaviour within the spinneret on the membrane performance in other membrane processes may be found in the literature.

Wang et al. [38] studied the effects of dope flow angle within a spinneret on the characteristics and UF performance of wet spun hollow fibre membranes. Two spinneret designs of different dope flow angles are shown in Fig. 4.13. It was observed that the hollow fibre membranes spun using a conical spinneret exhibited smaller pore sizes with larger geometric standard deviations and smaller water fluxes together with higher solute separations than the hollow fibre membranes prepared using a straight spinneret under the same spinning conditions. At low dope flow rates, similar cross-sectional morphology was observed for hollow fibre membranes prepared from both straight and conical spinnerets. However, at high dope flow rates macrovoids structure disappeared from the cross-section when the hollow fibre was spun through the straight spinneret, while macrovoids did not disappear when the hollow fibre was spun through the conical spinneret. Wang et al. [38] concluded that the flow angle in the spinneret is another variable affecting the structure and performance of the hollow fibre membranes.

The effects of both the elongation and shear rates induced by the spinneret geometry on the wet spun hollow membrane performance have been studied by Cao et al. [39]. Various spinnerets with different flow angles of 60°, 75° and 90° (i.e. straight spinneret) were used. A schema of a spinneret having a flow angle is shown in Fig. 4.14. Computational fluid dynamics (CFD) model was employed to simulate the flow profiles of the polymer solution within the spinneret together with the elongation and shear rates at the outermost point of the spinneret outlet. For a long straight spinneret (90° flow angle), the elongation rate can

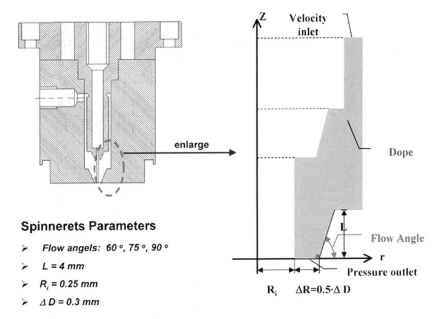

FIGURE 4.14 Schema of a spinneret with a flow angle. *Source: Reprinted from [39]. Copyright 2004, with kind permission from Elsevier.*

be disregarded (equal to zero) and the separation performance is affected only by the shear rate. It was stated that the elongation rate would primarily stretch the polymer chains and enhance chain packing in the outer skin of the hollow fibre membranes, while the shear rate will primarily align the polymer chains. The shear rate has a greater contribution to permeance than to selectivity of hollow fibre membranes spun by conical spinnerets (60° and 75°). Furthermore, the elongation rate has more impact on selectivity than on permeance.

Instead of a circular structure, attempts have been made by Nijdam et al. [40] and Çulfaz et al. [41] to prepare hollow fibre membranes with micro-metre scale corrugations at the inner or outer surfaces using micro-engineered spinnerets named also smart spinnerets (Figure 4.15). The modified spinneret can contain a micro-shaped orifice (insert in Fig. 4.15(a)) or a micro-structured needle (Fig. 4.15(b)). For example, to prepare a hollow fibre with a corrugated inner micro-structure, instead of a cylindrical needle the spinneret is fabricated with a structured needle. To spin a hollow fibre with a corrugated outer surface a micro-structured orifice is incorporated to the spinneret. Figure 4.16 shows SEM images of corrugated hollow fibre membranes.

It is worth to mention that spinning corrugated hollow fibre membranes is similar to spinning smooth cylindrical fibres the only difference is the use of a modified spinneret. The use of corrugated membrane surfaces is to enhance the heat and mass transfer by increasing the membrane surface area and the turbulence near the membrane surface. Corrugations act as turbulence promoters, but also may lead to an increase of pressure drop in the flow channel. It was observed that the size of the corrugations on the fibre was very sensitive to the air gap length and even disappeared from the external surface for high air gap lengths. When used in gas separation, the selectivity

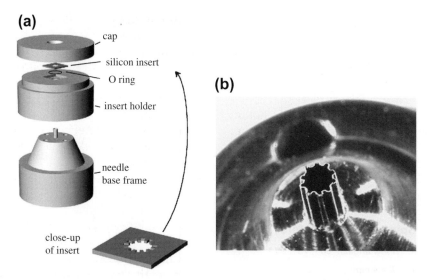

FIGURE 4.15 Details of spinnerets for preparation of corrugated hollow fibre membranes. *Source: Reprinted from [40]. Copyright 2005, with kind permission from Elsevier.*

(oxygen/nitrogen) and permeance of polyethersulfone/polyimide (PES/PI) blend corrugated hollow fibre membrane were found to be similar to those of smooth cylindrical fibres (i.e. circular shaped fibres) prepared under the same operating conditions. However, the gas permeate flow was increased 19%, which is the increase of membrane surface area due to corrugations [40]. It is worth quoting that an increase of membrane surface area of 89% can be obtained with a corrugated hollow fibre membrane [41]. Furthermore, PES/PVP blend hollow fibre membranes were used in UF and it was found that the water permeability, the molecular weight cut-off and pore size distribution of both the corrugated and circular shaped fibres were

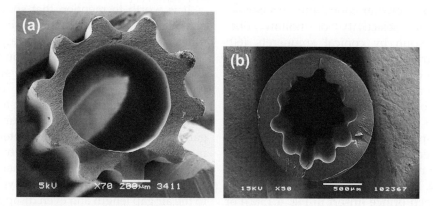

FIGURE 4.16 SEM images of polyethersulfone/polyimide (PES/PI) blend hollow fibre membranes with a corrugated outer (a) and inner (b) microstructure. *Source: Reprinted from [40]. Copyright 2005, with kind permission from Elsevier.*

similar [41]. This indicates a higher permeate flow of the corrugated hollow fibre compared to the circular shaped fibre and a similar UF separation factor. It is to be noted that corrugated membranes are not used yet in MD.

Dual hollow fibre spinneret design (Fig. 4.3) has been used for preparation of MD hollow fibre membranes. As commented earlier, Bonyadi and Chung [9] prepared hydrophilic/hydrophobic hollow fibre membranes for DCMD. In this case, co-extrusion of two dope solutions is necessary.

FUTURE DIRECTIONS

As reviewed, very few studies have been conducted on the preparation and characterization of dry/wet and wet spun hollow fibre membranes for MD process and very little attention has been paid to investigate the effects of the involved spinning parameters. Novel hollow fibre membranes that have either single porous hydrophobic layer or hydrophobic and hydrophilic porous multi-layers should be designed specifically for different MD configurations. Hollow fibre membranes with low thermal conductivity, high porosity and void volume fraction, large pore size, high LEP_w, low thickness and diameters (i.e. high fibre packing density per module) and reasonably high mechanical and thermal properties are desired.

There are yet some spinning parameters to which not much attention has been given. For example, systematic studies are required on the effects of the temperature of the internal and external coagulants as well as the temperature of the spinneret. The design of the spinneret and the shear rate working within the spinneret should further be investigated. Hydrophobic polymers and coagulants other than those already employed, a wider range of the air gap distance and the type of the gas along the spinning line may also be the topics of future investigation. It is worth quoting that the effects of the gas type on the morphology, structural parameters, surface characteristics and performance of PES hollow fibres were studied for UF by Khayet et al. [42]. The obtained hollow fibres spun with different gas types in the gas gap were classified into two groups based on the effects of the gas type on the external surface characteristics of the dry/wet spun hollow fibres and their UF performance. One group is the hollow fibres prepared with gases of high molecular mass and low thermal conductivity such as argon and carbon dioxide and the other group is the hollow fibres prepared with gases of lower molecular mass and higher thermal conductivity such as air, oxygen and nitrogen. A plausible explanation based on the thermal conductivity of the gases was reported.

Progress on the fabrication of dual or multi-layered hydrophobic/hydrophilic hollow fibre membranes and corrugated hollow fibre membranes should be done although in this case spinnerets with complex designs are necessary. The use of surface modifying macromolecules (SMMs) seems promising in MD as discussed previously.

References

[1] H.I. Mahon, Permeability separatory apparatus and membrane element, method of making the same and process utilizing the same, US Patent 3,228,876 (1966).
[2] H.I. Mahon, Permeability separatory apparatus and process using hollow fibers, US Patent 3,228,877 (1966).
[3] Y. Fujii, S. Kigoshi, H. Iwatani, M. Aoyama, Selectivity and characteristics of direct contact membrane distillation type experiment: Part I. Permeability and selectivity through dried hydrophobic fine porous membranes, J. Membr. Sci. 72 (1992) 53–72.
[4] Y. Fujii, S. Kigoshi, H. Iwatani, M. Aoyama, Y. Fusaoka, Selectivity and characteristics of direct contact membrane distillation type experiment: Part II. Membrane treatment and selectivity increase, J. Membr. Sci. 72 (1992) 73–89.

[5] B. Wu, X. Tan, W.K. Teo, K. Li, Removal of benzene/toluene from water by vacuum membrane distillation in a PVDF hollow fiber membrane module, Sep. Sci. & Tech. 40 (2005) 2679–2695.

[6] B. Wu, X. Tan, K. Li, W.K. Teo, Removal of 1,1,1-trichloroethane from water using a poly(vinylidene fluoride) hollow fiber membrane module: Vacuum membrane distillation operation, Sep. & Purf. Tech. 52 (2006) 301–309.

[7] B. Wu, K. Li, W.K. Teo, Preparation and characterization of poly(vinylidene fluoride) hollow fiber membranes for vacuum membrane distillation, J. Appl. Polymer Sci. 106 (2007) 1482–1495.

[8] K.Y. Wang, T.S. Chung, M. Gryta, Hydrophobic PVDF hollow fiber membranes with narrow pore size distribution and ultra-skin for the fresh water production through membrane distillation, Chem. Eng. Sci. 63 (2008) 2587–2594.

[9] S. Bonyadi, T.S. Chung, Flux enhancement in membrane distillation by fabrication of dual layer hydrophilic-hydrophobic hollow fiber membranes, J. Membr. Sci. 306 (2007) 134–146.

[10] M. Mulder, Basic principles of membrane technology, Kluwer Academic Publishers, Dordrecht, The Netherlands, 1992.

[11] S.P. Deshmuck, K. Li, Effect of ethanol composition in water coagulation bath on morphology of PVDF hollow fiber membranes, J. Memb. Sci. 150 (1998) 75–85.

[12] D. Wang, K. Li, W.K. Teo, Preparation and characterization of polyvinylidene fluoride (PVDF) hollow fiber membranes, J. Membr. Sci. 163 (1999) 211–220.

[13] H.C. Shih, Y.S. Yeh, H. Yasuda, Morphology of microporous poly(vinylidene fluoride) membranes prepared by gas permeation and scanning electron microscopy, J. Membr. Sci. 50 (1990) 299–317.

[14] D. Wang, K. Li, W.K. Teo, Porous PVDF asymmetric hollow fiber membranes prepared with the use of small molecular additives, J. Membr. Sci. 178 (2000) 13–23.

[15] M. Khayet, C.Y. Feng, K.C. Khulbe, T. Matsuura, Preparation and characterization of polyvinylidene fluoride hollow fiber membranes for ultrafiltration, Polymer 43 (2002) 3879–3890.

[16] M. Khayet, C.Y. Feng, K.C. Khulbe, T. Matsuura, Study on the effect of a non-solvent additive on the morphology and performance of ultrafiltration hollow-fiber membranes, Desalination 148 (2002) 31–37.

[17] M. Khayet, The effects of air gap length on the internal and external morphology of hollow fiber membranes, Chem. Eng. Sci. 58 (2003) 3091–3104.

[18] M. Khayet, T. Matsuura, Application of surface modifying macromolecules for the preparation of membranes for membrane distillation, Desalination 158 (2003) 51–56.

[19] M. Khayet, J.I. Mengual, T. Matsuura, Porous hydrophobic/hydrophilic composite membranes: Application in desalination using direct contact membrane distillation, J. Membr. Sci. 252 (2005) 101–113.

[20] T. Yanagimoto, Manufacture of ultrafiltration membranes, Japanese Patent 62,019,205 (1987).

[21] T. Yanagimoto, Method for manufacture of hollow-fiber porous membranes, Japanese Patent 63,092,712 (1988).

[22] E. Kuzumoto, K. Nitta, Manufacture of permselective hollow membranes, Japanese Patent 01,015,104 (1989).

[23] O.M. Ekiner, R.A. Hayes, P. Manos, Novel multicomponent fluid separation membranes, US Patent 5,085,676 (1992).

[24] D.F. Li, T.S. Chung, R. Wang, Y. Liu, Fabrication of fluoropolyimide/ polyethersulfone (PES) dual-layer asymmetric hollow fiber membranes for gas separation, J. Membr. Sci. 198 (2002) 211–223.

[25] L. Jiang, T.S. Chung, D.F. Li, C. Cao, S. Kulprathipanja, Fabrication of Matrimid/polyethersulfone dual-layer hollow fiber membranes for gas separation, J. Membr. Sci. 240 (2004) 91–103.

[26] D.F. Li, T.S. Chung, R. Wang, Morphological aspects and structure control of dual-layer asymmetric hollow fiber membranes formed by a simultaneous co-extrusion approach, J. Membr. Sci. 243 (2004) 155–175.

[27] K.C. Khulbe, C.Y. Feng, T. Matsuura, D.C. Mosqueda-Jimenez, M. Rafat, D. Kingston, R.M. Narbaitz, M. Khayet, Characterization of surface-modified hollow fiber polyethersulfone membranes prepared at different air gaps, J. Appl. Polymer Sci. 104 (2007) 710–721.

[28] M.C. García-Payo, M. Essalhi, M. Khayet, Effects of PVDF-HFP concentration on membrane distillation performance and structural morphology of hollow fiber membranes, J. Membr. Sci. 347 (2010) 209–219.

[29] S. Bonyadi, T.S. Chung, W.B. Krantz, Investigation of corrugation phenomenon in the inner contour of hollow fibers during the non-solvent induced phase-separation process, J. Membr. Sci. 299 (2007) 200–210.

[30] S.A. McKelvey, D.T. Clausi, W.J. Koros, A guide to establishing hollow fiber macroscopic properties for membrane applications,, J. Membr. Sci. 124 (1997) 223–232.

[31] J.J. Qin, J. Gu, T.S. Chung, Effect of wet and dry-jet wet spinning on the shear-induced orientation during the formation of ultrafiltration hollow fiber membranes, J. Membr. Sci. 182 (2001) 57–75.

[32] K.Y. Wang, D.F. Li, T.S. Chung, S.B. Chen, The observation of elongation dependent macrovoid evolution

in single-and dual-layer asymmetric hollow fiber membranes, Chem. Eng. Sci. 59 (2004) 4657—4660.

[33] Y. Xiao, K.Y. Wang, T.S. Chung, J. Tan, Evolution of nano-particle distribution during the fabrication of mixed matrix TiO_2-polyimide hollow fiber membranes, Chem. Eng. Sci. 61 (2006) 6228—6233.

[34] J. Ren, R. Wang, H.Y. Zhang, Z. Li, D.T. Liang, J.H. Tay, Effect of PVDF dope rheology on the structure of hollow fiber membranes used for CO_2 capture, J. Membr. Sci. 281 (2006) 334—344.

[35] J.J. Qin, R. Wang, T.S. Chung, Investigation of shear stress effect within a spinneret on flux, separation and thermomechanical properties of hollow fiber ultrafiltration membranes, J. Membr. Sci. 175 (2000) 197—213.

[36] J. Ren, Z. Li, F.S. Wong, D. Li, Development of asymmetric BTDA-TDI/MDI(P84) co-polyimide hollow fiber membranes for ultrafiltration: The influence of shear rate and approaching ratio on membrane morphology and performance, J. Membr. Sci. 248 (2005) 177—188.

[37] R. Wang, T.S. Chung, Determination of pore sizes and surface porosity and the effect of shear stress within a spinneret on asymmetric hollow fiber membranes, J. Membr. Sci. 188 (2001) 29—37.

[38] K.Y. Wang, T. Matsuura, T.S. Chung, W.F. Guo, The effects of flow angle and shear rate within the spinneret on the separation performance of poly-(ethersulfone) (PES) ultrafiltration hollow fiber membranes, J. Membr. Sci. 240 (2004) 67—79.

[39] C. Cao, T.S. Chung, S.B. Chen, Z.J. Dong, The study of elongation and shear rates in spinning process and its effect on gas separation performance of poly(ether sulfone) (PES) hollow fiber membranes, Chem. Eng. Sci. 59 (2004) 1053—1062.

[40] W. Nijdam, J. de Jong, C.J.M. van Rijn, T. Visser, L. Versteeg, G. Kapantaidakis, G.H. Koops, M. Wessling, High performance micro-engineered hollow fiber membranes by smart spinneret design, J. Membr. Sci. 256 (2005) 209—215.

[41] P.Z. Çulfaz, E. Rolevink, C. van Rijn, R.G.H. Lammertink, M. Wessling, Microstructured hollow fibers for ultrafiltration, J. Membr. Sci. 347 (2010) 32—41.

[42] M. Khayet, M.C. García-Payo, F.A. Qusay, K.C. Khulbe, C.Y. Feng, T. Matsuura, Effects of gas type on structural morphology and performance of hollow fibers, J. Membr. Sci. 311 (2008) 259—269.

[43] M. Khayet, M.C. García-Payo, F.A. Qusay, M.A. Zubaidy, Structural and performance studies of poly(vinyl chloride) hollow fiber membranes prepared at different air gap lengths, J. Membr. Sci. 330 (2009) 30—39.

CHAPTER 5

Thermally Induced Phase Separation for MD Membrane Formation

OUTLINE

Introduction	90
Principles of the Formation of Porous Hydrophobic TIPS Membranes	**92**
Thermodynamic Considerations	*93*
Liquid—Liquid Phase Separation	97
Solid—Liquid Phase Separation	98
Combined Liquid—Liquid and Solid—Liquid Phase Separation	100
Kinetic Considerations	*102*
Flat Sheet Membrane Preparation	*103*
Hollow Fibre Membrane Preparation	*104*
Effects of Process Parameters on the Membrane Structure	**105**
Flat Sheet Membranes	*105*
Effects of Polymer Type and Concentration in the Initial Polymer/Diluents System	105
Effects of Polymer Molecular Weight	108
Effects of the Type of Diluents, Its Concentration and Molecular Weight	108
Effects of Polymer/Diluent Melting Temperature and Time	109
Effects of Quenching Conditions	109
Effects of Nucleating Agents	110
Effects of Extractant	110
Effects of Drying	111
Effects of Stretching	111
Hollow Fibre Membranes	*112*
Effects of Polymer Type, Molecular Weight and Composition	112
Effects of Concentration and Type of Diluents	113
Effects of Spinning Temperature	115
Effects of Air Gap Distance	116
Effects of Water Bath Temperature and Take-up Speed	116
Effects of Spin Draw Ratio	116
Effects of Cold Stretching	117
Post-treatment	117
Special Considerations	117

INTRODUCTION

The concept of phase inversion membranes was introduced by Kesting and refers to a process in which a polymer solution is transferred in a controlled way from a liquid to a solid state [1]. The morphology of the membranes prepared by phase inversion can be controlled by manipulating the phase transition from liquid to solid. It is possible to prepare both porous and non-porous phase inversion membranes, and various different techniques can be distinguished: solvent evaporation, precipitation from vapour phase, precipitation by controlled evaporation, immersion precipitation or diffusion-induced phase separation (DIPS) and thermal precipitation, also known as thermally induced phase separation (TIPS). This chapter is devoted to TIPS technique.

The preparation of porous hydrophobic membrane structure via phase separation consists of two steps. In the first step the homogeneous polymer solution undergoes liquid–liquid demixing to obtain a polymer-rich continuous matrix and a dispersed polymer-lean phase. In the second step, the structure is fixed by crystallization, vitrification or gelation of the polymer-rich phase. Two major methods can be distinguished to induce phase separation:

(i) by adding a non-solvent (DIPS)
(ii) by lowering the temperature (TIPS).

Membrane distillation (MD) membranes can be categorized depending on their geometry, fabrication method and bulk structure. MD membranes may have either a symmetric (i.e. isotropic: uniform structure throughout the entire membrane thickness) or an asymmetric (i.e. anisotropic) structure. Chapters 3 and 4 covered phase inversion flat sheet and hollow fibre membranes exhibiting asymmetric porous structures. These membranes are prepared using immersion precipitation and a solution containing polymer, solvent(s) and non-solvent(s). In this case, the polymer precipitates as a result of solvent loss and non-solvent penetration forming a phase separated structure. In fact, asymmetric porous membranes can be grouped into:

(i) Integrally asymmetric membrane with a porous skin layer: Typically made by an immersion precipitation using a solution containing a polymer, a solvent and a non-solvent additive. Upon the immersion of the solution into a liquid (i.e. coagulant), which is a non-solvent for the polymer but miscible with the solvent, an asymmetric porous substructure with a skin layer containing pores of smaller size is formed. Details concerning the fabrication of this type of membranes can be found in Chapters 3 and 4.

(ii) Porous skinned asymmetric membranes: Made of at least two different materials and consist of a thin top porous layer over a porous support. Either the top or the support should be hydrophobic. The porous support provides mechanical strength, whereas the MD performance is dictated by the top layer. A multi-layered porous membrane with at least one hydrophobic layer can be included in this category. This type of membranes can be made by solution coating, plasma polymerization, use of surface modifying macromolecules, etc. [1–5].

Symmetric porous membranes can be prepared by a variety of techniques such as irradiation-etching, melt extrusion/stretching process and temperature-induced phase separation (TIPS). This last membrane preparation technique has been patented by several researchers and used by different companies such as 3M, Membrana/Akzona, Akzo-Nobel, Millipore and others to produce commercial membranes that are thermally stable and chemically resistant [6–14]. Some of these types of membranes are used in MD and are shown in Table 2.1. Particularly, TIPS technique has been used to form microporous hydrophobic polymeric membranes of controlled pore characteristics from a variety of crystalline and thermoplastic polymers, including polyolefins, copolymers and blends [9–12,15,16]. Commercially available TIPS polypropylene (PP) membranes have proved useful in MD [17–24]. For example, isotactic polypropylene (iPP) membranes are produced via TIPS on a commercial scale by Akzo and 3M Company and currently being used in a variety of applications other than MD, such as blood oxygenation, thermodialysis, microfiltration, ultrafiltration and breathable rainware [11,12,14,19,20].

In contrast to the solvent/non-solvent phase inversion membrane formation technique (i.e. immersion precipitation), in which the phase separation is driven by the change of the dope composition, the TIPS technique is caused by a temperature change and typically consists of five basic steps [25]:

(i) A homogeneous solution is first formed by melt blending a polymer and a liquid or solid having a low molecular weight and a high boiling point, referred to as the diluent (i.e. does not cause dissolution or swelling of the polymer at room temperature). The initial temperature must be less than the boiling point of the diluent and is typically 25–100°C greater than the melting temperature or glass transition temperature of the polymer. The polymer must be stable at this initial temperature. To avoid confusion, the term diluent will be used instead of solvent, hereafter. The diluent acts as a solvent only when it can dissolve all polymers. When the solvent power decreases sufficiently so that liquid–liquid demixing takes place, the term solvent is not appropriate anymore:

(ii) The hot solution is cast, extruded or spun in the desired shape (flat sheet, tube and hollow fibre).

(iii) The solution is cooled at controlled rate or quenched to induce phase separation and solidification.

(iv) The diluent(s) that is trapped in the polymer matrix during phase separation and solidification is removed. Solvent extraction procedure is frequently used in this step.

(v) The extractant is removed by evaporation to yield microporous structures.

(vi) Post-treatment processing might be applied to improve the desired separation characteristics of the TIPS membrane.

It must be pointed out that stretching might be included in step (vi). As can be deduced from the above cited steps, TIPS depends primarily on heat transfer. In contrast, solvent/non-solvent phase inversion membranes depend on multi-component mass transfer. The homogeneous solution from which TIPS membrane is formed is converted into a two-phase mixture via the removal of thermal energy rather than by the exchange of non-solvent and solvent for immersion precipitation-induced phase inversion membranes. TIPS can be applied to a wide range of polymers, including those that cannot be otherwise formed into membranes because of poor solubility in solvent. The TIPS process can be used to generate both dense and porous films, the latter with isotropic, anisotropic or asymmetric microstructures with an overall porosity as high as 90% as will be shown later on.

Following the six-step procedure outlined above, microporous hydrophobic polymeric membranes can be formed via liquid–liquid phase separation with subsequent solidification, solid–liquid phase separation or combined liquid–liquid and solid–liquid phase separation of a homogeneous melt blend [25–29]. The type of phase separation that is likely to occur for a given system (composition and temperature) is indicated by the phase diagram [30]. In fact, the phase diagram is a useful tool explaining the morphology and bulk structure of TIPS membranes. Detailed explanations are shown in the section 'Principles of the formation of porous hydrophobic TIPS membranes'. This chapter also includes both thermodynamic and kinetic considerations involved in the formation of the membrane structure via TIPS technique as well as the effects of different parameters. The structure variation of membranes formed via TIPS was investigated in terms of polymer/diluent thermodynamics and polymer crystallization kinetics focussing on both the polymer and diluent behaviours during TIPS process [27,31,32].

Briefly, liquid−liquid phase separation results from the thermodynamic instability of the polymer diluent system, which may be induced by lowering the temperature to increase unfavourable polymer/diluent interactions or by increasing the temperature to increase the free volume. In fact, the systems that exhibit an upper critical solution temperature undergo phase separation as the temperature is lowered, whereas the systems that undergo phase separation as the temperature is increased have a lower critical solution temperature. Solid−liquid phase separation usually results from the crystallization of the polymer from the homogeneous solution phase. The driving force for this phase separation is the difference in polymer chemical potential in the crystalline and solution phases [33]. The crystallization kinetics of the polymer, and in some cases the diluent, play important roles in determining the structure from the solid−liquid phase separation [31,32,34].

The advantages of the TIPS technique for membrane preparation are the following:

- The TIPS process is applicable to a wide range of polymers, including those that cannot be formed into membranes via DIPS because of their poor solubility.
- TIPS can be used to prepare membranes from semi-crystalline polymers.
- TIPS is capable of producing a variety of microstructures.
- TIPS is capable of producing relatively thick isotropic microporous structures. Formation of anisotropic microporous structures is also possible if a thermal gradient is induced in step (iii) of the membrane preparation procedure outlined earlier [15,16].
- Because the phase separation is thermally induced unlike solvent/non-solvent-induced phase inversion, there are fewer variables that need to be controlled.

Many studies have been carried out on the porous membrane formation by the TIPS technique using blend hydrophobic polymer/diluent systems. Within the MD literature, it was found that only one paper was dealing with the preparation and characterization of hollow fibre membranes for MD via TIPS technique [35−37]. However, various commercial flat sheet and hollow fibre TIPS membranes were used in different MD applications as reported earlier.

PRINCIPLES OF THE FORMATION OF POROUS HYDROPHOBIC TIPS MEMBRANES

As stated in the 'Introduction' section, the first step to prepare TIPS membranes is to choose an adequate polymer/diluent system to form a homogeneous melt blend, which will be cooled or quenched to induce phase separation (i.e. liquid/liquid phase separation, solid/liquid phase separation or combined liquid−liquid and solid−liquid phase separation). The steps to follow for preparation of porous hydrophobic membrane via TIPS technique were outlined in the 'Introduction' section.

The recommended diluent characteristics for the TIPS mechanism are the following:

(i) Miscible with the considered polymer at high temperatures

(ii) Relatively low molecular weight
(iii) Low volatility at high temperatures
(iv) Thermally stable at high temperatures

The polymer should exhibit good chemical and thermal resistance, physically strong and inexpensive. Table 5.1 shows some of the used polymer/diluents systems to prepare both flat sheet and hollow fibre TIPS porous hydrophobic membranes.

Thermodynamic Considerations

The thermodynamic description of the formation of porous hydrophobic membranes by TIPS technique is based on the assumption of thermodynamic equilibrium. It predicts under what conditions of temperature and composition a polymer/diluent system will separate into two phases and the ratio of the two phases in the heterogeneous mixture. As occurs for the solvent/non-solvent-induced phase inversion membranes, a quantitative description of the TIPS membrane formation mechanism is nearly impossible. However, a qualitative description of the membrane formation and correlation of the various involved parameters with membrane structures and properties are possible.

It was reported in Chapter 3 that the thermodynamic parameter describing the miscibility of two or more components (i.e. polymer and diluent in this case) is the Gibbs free energy of mixing (ΔG_m), for which an expression was given by the Flory–Huggins lattice theory [33] (Eq. 3.1). Under certain conditions of temperature and composition of a binary polymer/diluent system, ΔG_m is negative and the components of the system have a miscibility region over a defined concentration and temperature range. Following the thermodynamic considerations reported in Chapter 3 (section 'Principles of formation of porous hydrophobic flat sheet membranes'), a schematic phase diagram for a binary solution containing an amorphous polymer and a diluent (or solvent) displaying an upper critical solution temperature behaviour is shown in Fig. 5.1. Other phase diagrams for polymer/diluent systems are possible as will be presented below. The stable, meta-stable and unstable regions of the polymer/diluent solution are indicated in Fig. 5.1. The binodal line (continuous solid line in Fig. 5.1) is the boundary between the stable and the meta-stable regions of the polymer/diluent solution, whereas the spinodal (discontinuous line in Fig. 5.1) limits the meta-stable and the unstable regions. The spinodal curve divides the liquid–liquid region into meta-stable (between the binodal and the spinodal) and unstable (below the spinodal) regions. It is to be noted that the binodal and spinodal lines for binary polymer/diluent systems can be estimated using the Flory–Huggins theory [38,39]. ϕ^α and ϕ^β lie on the binodal curve and the two inflection points ϕ_1 and ϕ_2 are situated on the spinodal curve and are characterized by $\frac{\delta^2 \Delta G_m}{\delta \phi_i^2} = 0$.

Two different modes of the phase separation of a stable polymer/diluent solution can be distinguished: (i) spinodal decomposition and (ii) nucleation and growth.

At a temperature higher than the critical temperature, T_c (Point C in Fig. 5.1), the mixture forms a homogeneous solution for all compositions but at a lower temperature shows a miscibility region over a wide range of compositions. If the temperature of the polymer/diluent solution is decreased from T_1 (Point A) to T_2 (Point B), which is maintained constant, the solution A passes directly into the unstable region of the phase diagram and the resulting phase separation process is determined by spinodal decomposition. Between the two inflection points ϕ_1 and ϕ_2, the curvature of ΔG_m is such that $\frac{\delta^2 \Delta G_m}{\delta \phi_i^2} < 0$, which implies that the mixture is unstable and will separate spontaneously. Phase separation proceeds instantaneously and

TABLE 5.1 Examples of Polymer/Diluents Systems Used to Form Hydrophobic Porous Membranes via TIPS

Polymer	Diluent	Reference
Polyvinylidene fluoride (PVDF)	Dimethyl phthalate (DMP) Dioctyl phthalate (DOP) Dioctyl sebacate (DOS) Di(2-ehylhexyl) phthalate (DEHP) Dibutyl phthalate (DBP) γ-Butrolactone (γ-BA) Propylene carbonate (PC) Dibutyl sebacate (KD)	[67] [67] [67] [70] [25,70,76,77] [77] [77] [77]
Isotactic polypropylene (iPP)	Diphenyl ether (DPE) Mineral oil (MO) Tetradecane ($C_{14}H_{30}$) Dotriacontane ($C_{32}H_{66}$) Pentadecanoic acid ($C_{14}H_{29}COOH$) n,n-bis(2-hydroxylethyl)tallowamine Eicosane ($C_{20}H_{42}$) Eicosanoic acid ($C_{19}H_{39}COOH$) Hexamethylbenzene (HMB) Di-n-butyl phthalate (DBP) Dioctyl phthalate (DOP) Soybean oil	[30,43,45,47,61,73] [32] [31,32] [31,32,65] [31] [26,27] [27,31,32] [27,31] [34] [69,74] [69] [74]
Syndiotactic polypropylene (sPP)	Diphenyl ether (DPE)	[43,44]
Polyethylene (HDPE)	Diamyl phthalate (DAP) Diisodecyl phthalate (DIDP) Liquid paraffin (LP) Mineral oil (MO)	[59] [60] [60,78] [25]
Isotactic polystyrene (iPS)	Nitrobenzene	[29]
Poly(tetrafluoroethylene-co-perfluoro-propyl vinyl ether) (Teflon® PFA, Neoflon™ PFA)	Chlorotrifluoroethylene	[46]
Polychlorotrifluoroethylene (Kel-F)	Kel-F oil	[25]
Poly(2,6-dimethyl-1,4-phenylene ether) (PPE)	Cyclohexanol dioxane/isopropanol	[72]
Nylon 12	Poly(ethylene glycol)	[63]
Poly(4-methyl-1-pentene) (TPX)	Di-butyl phthalate (DBP)	[25]
Atactic polystyrene (aPS)[a]	Cyclohexane Cyclohexanol	[57] [40,50]

[a] Amorphous polymer.

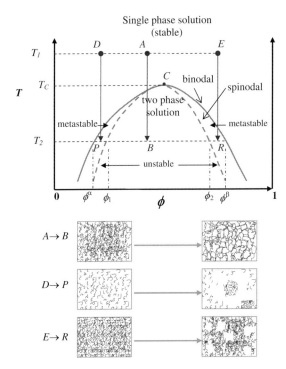

FIGURE 5.1 Schematic phase diagram of a binary polymer/diluent system (temperature vs. polymer volume fraction, ϕ) and structures of membranes resulting from: (A→B) spinodal decomposition; (D→P) nucleation and growth of polymer-poor phase; (E→R) nucleation and growth of polymer-rich phase.

results initially in a regular, highly interconnected structure, which tends to coarsen during the later stages of spinodal decomposition [40]. A typical morphology of the membrane made from the polymer/diluent solution A is schematically shown in Fig. 1 (A→B). The membrane produced following spinodal decomposition mechanism is well interconnected with highly uniform pore sizes and possess mechanical strength.

In the regions $\phi^\alpha < \phi < \phi_1$ and $\phi_2 < \phi < \phi^\beta$, $\frac{\delta^2 \Delta G_m}{\delta \phi_i^2} > 0$. This means that there is no driving force for spontaneous phase separation. A thermal quench of solution D and E in Fig. 5.1 will reside within the meta-stable region, and therefore, phase separation will occur by nucleation and growth. Solution D forms nuclei with composition ϕ^β, whereas solution E will form nuclei with composition ϕ^α. The resulting equilibrium phases are composed of ϕ^α and ϕ^β in both cases. The solution D consists of a small volume fraction of the polymer-rich phase, ϕ^β dispersed in a polymer-poor phase with composition ϕ^α, whereas the opposite applies for solution E. Demixing can only start if the concentration fluctuations have generated at least one stable nucleus. After nucleation, the nuclei will grow at the expense of the surrounding phase, which will eventually leads to phase separation. Related membrane structures are schematized in Fig. 5.1 (D→P, E→R). The membrane morphology associated with this phase separation is either a poorly interconnected, stringy and/or beady structure, which is mechanically fragile for solutions of low polymer concentration below the critical concentration; whereas for polymer concentrations above the critical concentration the produced membranes, with phase separation being initiated in the meta-stable region, exhibit interconnected structure with highly nonuniform pore sizes.

The above discussions deal with liquid–liquid phase separation mechanism and with its relation with membrane structures. For completely amorphous polymer-diluent systems, such as poly(methyl methacrylate) (PMMA)-1,4-butanediol and PMMA-sulfolane, phase separation can only occur via liquid–liquid phase separation as shown in Fig. 5.2(a) [15,16]. However, other possible phase diagrams may take place as shown in Fig. 5.2 depending on the polymer/diluent system.

Figure 5.2(b) shows a typical phase diagram of a semi-crystalline polymer/diluent system with favourable interactions between components. As can be seen, no liquid–liquid phase separation can take place in this kind of systems. In contrast, solid–liquid phase separation occurs throughout most of the composition range. These systems undergo solid–liquid phase

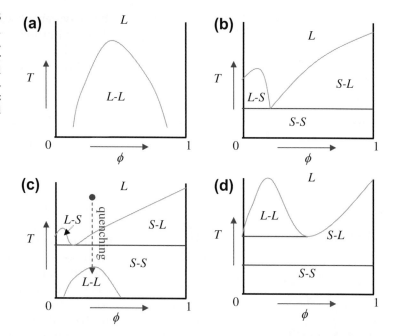

FIGURE 5.2 Schematic TIPS phase diagrams (temperature vs. volume fraction of polymer, ϕ). L, liquid; S, solid; L-L, liquid polymer plus liquid diluent; S-L, solid polymer plus liquid diluent; L-S, liquid polymer plus solid diluent; S-S, solid polymer plus solid diluent.

separation via polymer crystallization [25,27]. iPP/n-alkane is an example of these systems.

Figure 5.2(c) presents the phase diagram of a polymer/diluent system in which the interactions between components are less-favourable. In this case, there is a presence of an unstable region below the melting point depression curve, and both crystallization and liquid/liquid phase separation can take place [25,27,41]. As an example, iPP/n-fatty acid system is represented by Fig. 5.2(c).

In Fig. 5.2(d), the liquid−liquid phase separation curve and the melting point depression curve meet at the monotectic point (intersection between the liquid−liquid boundary and the dynamic crystallization curve). As can be seen, the initial composition of the polymer in the homogeneous melt blend together with the cooling conditions determines the sequence of the phase separation to be liquid−liquid or solid−liquid. The system iPP/n,n-bis(2-hydroxyethyl) tallowamine (TA) is represented by Fig. 5.2(d) [27].

Figure 5.3 shows as an example phase diagrams and interaction parameters for syndiotactic polypropylene (sPP) and diphenyl ether (DPE) system (sPP/DPE), and iPP and DPE system. To obtain phase diagrams, solid polymer/diluent samples with different concentrations were first prepared at an adequate temperature (i.e. over the critical temperature of the polymer/diluent system in order to get homogenous melt blend) and loaded in capillary tubes, which was purged with nitrogen and sealed to prevent oxidation. In the case of the phase diagrams presented in Fig. 5.3, the temperature is 453 K. After waiting for at least 5 min to ensure complete melting, the samples were cooled at a certain rate, in this case 10 K/min. The cloud point was determined visually by noting the first appearance of turbidity. Normally, for each melt blend concentration the experiment is performed three times [42−44].

The crystallization curve is determined calorimetrically by using dynamic crystallization temperatures. Differential scanning calorimeter

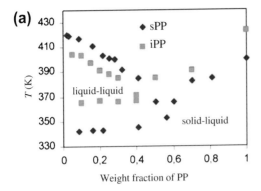

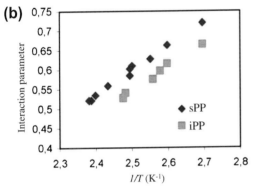

FIGURE 5.3 Phase diagrams for sPP/DPE and iPP/DPE systems (a) and interaction parameter versus inverse of absolute temperature (b). *Source: Reprinted from [43]. Copyright 2005, with kind permission from Elsevier.*

(DSC) is used. In this case the samples were prepared in the same way as above, but the cooling was done suddenly in liquid nitrogen to ensure a homogeneous solid solution. Then, a small portion of the solid solution was cut and used for the thermal analysis at a cooling rate of 10 K/min.

To understand the phase diagram, the interaction parameter (χ) was calculated based on Flory–Huggins theory [42,45]. By using the experimental data of the cloud point, a simple method for extrapolating the liquid–liquid phase boundary was reported based on the following two equations relating the polymer volume fraction of the phase ϕ_2^α and that of the polymer rich phase ϕ_2^β. These permit to describe the polymer's chemical potential in the two separate phases (i.e. binodal line):

$$[(\phi_2^\beta)^2 - (\phi_2^\alpha)^2]\chi$$
$$= \ln\left(\frac{1-\phi_2^\alpha}{1-\phi_2^\beta}\right) + \left(1 - \frac{1}{r}\right)(\phi_2^\alpha - \phi_2^\beta) \quad (5.1)$$

$$r[(1-\phi_2^\beta)^2 - (1-\phi_2^\alpha)^2]\chi$$
$$= \ln\left(\frac{\phi_2^\alpha}{\phi_2^\beta}\right) + (r-1)(\phi_2^\alpha - \phi_2^\beta) \quad (5.2)$$

where r is the ratio of the polymer's molar volume to the diluent molar volume (higher polymer molecular weight leads to higher r value). ϕ_2^α is the polymer's volume fraction in the polymer-poor phase, and ϕ_2^β the polymer volume fraction in the polymer-rich phase.

The interaction parameter can be calculated by converting first the experimental cloud point data from weight percent to volume fraction, which are used as ϕ_2^β and finally solving simultaneously (Eqs. (5.1) and (5.2)). Figure 5.3(b) shows the relationship of χ and the reciprocal of absolute temperature. Straight and parallel lines were obtained for both sPP/DPE and iPP/DPE systems with similar slopes and different intersections on the χ-axis. The χ parameters are shifted to higher values in the sPP/DPE confirming that the compatibility is lower in sPP/DPE system. It was concluded that the χ parameter is affected mainly by the entropy contribution due to the PP tacticity [43].

As can be seen, different phase separation mechanisms may occur as it is explained using Fig. 5.4.

Liquid–Liquid Phase Separation

The mechanism for this phase separation was described earlier. When a homogeneous melt blend is quenched to a temperature and a composition point below the binodal curve there is a driving force for the system to separate

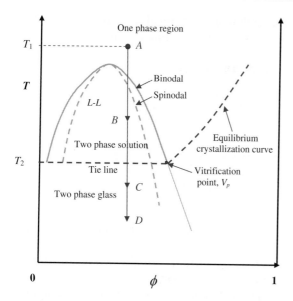

FIGURE 5.4 Example of a phase diagram of a binary polymer/diluent solution. The arrow AD represents a cooling trajectory to obtain a porous structure with the TIPS method. It goes from the homogeneous solution (A) at T_1 via the liquid–liquid demixing (L-L), B, to below the structure fixation temperature (T_2), C and D.

into polymer-rich and polymer-lean phases by one of two liquid–liquid phase separation mechanisms: spinodal decomposition or nucleation and growth. Both mechanisms result in the formation of droplets of the polymer-lean phase in a matrix of the polymer-rich phase when the concentration of the polymer in the melt blend is greater than the upper critical solution concentration [26,30,46,47]. As will be explained later, by decreasing the temperature below the horizontal crystallization line, discontinuous horizontal line at T_2 in Fig. 5.4, the polymer crystallizes, resulting in a membrane filled with diluent droplets. After extraction of the polymer-lean phase, microcells connected by pores are produced, as can be seen in Fig. 5.5, for iPP and DPE system that has undergone phase separation for different time periods.

For the TIPS membranes presented in Fig. 5.5, the temperature was dropped from 433 to 378 K at a rate of 130 K/min. The samples were held at 378 K for the indicated time periods and subsequently were quenched into liquid nitrogen in order to freeze the structure. Finally, the diluent DPE was extracted using methanol leaving behind empty spaces that formed the cells. As can be seen in Fig. 5.5, the average size of the cell increases with the increase of the holding time at 378 K, with a simultaneous decrease in the total number of cells per unit volume of the sample.

Solid–Liquid Phase Separation

When a homogeneous melt blend containing a semi-crystalline polymer is subjected to a temperature below the equilibrium melting temperature of the diluted polymer, the system undergoes solid–liquid separation via the crystallization of the polymer. The solid–liquid TIPS process can be performed following either an isothermal TIPS procedure in which the melt blend is driven to a constant temperature or a non-isothermal TIPS by decreasing the temperature at a controlled rate. When the concentration of the polymer in the melt blend is higher than the monotectic concentration and the temperature is decreased below the crystallization curve (Fig. 5.4) the polymer crystallizes depending on the magnitude of crystallization driving force, which is the difference in polymer chemical potential in the crystalline and solution phases. The magnitude of this driving force is directly proportional to the difference between the equilibrium melting temperature of the diluted polymer and the imposed crystallization temperature (i.e. degree of supercooling). In case of an isothermal TIPS, the driving force is set by the imposed degree of supercooling, which is typically high enough to induce rapid crystallization. In the case of the non-isothermal TIPS, the crystallization driving force is typically small and increases with time. Normally, the crystals form first as axialites that grow to form spherulites, which finally pack together as shown in Fig. 5.6 [25,32,48]. The cooling condition is a major

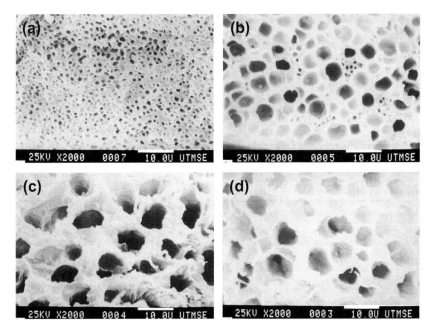

FIGURE 5.5 SEM of a 30 wt% iPP/DPE system, phase separated at 378 K for (a) 0 min (quenched directly in liquid nitrogen), (b) 1 min, (c) 4 min and (d) 8 min. *Source: Reprinted from [30]. Copyright 1994, with kind permission from Elsevier.*

kinetic parameter, which determines the spherulitic structure.

The PP membrane presented in Fig. 5.6(a) was prepared from 55 wt% PP in mineral oil solution, which was mixed and extruded at 245 °C and subsequently quenched in a 43 °C water bath. The resulted membrane exhibits spherulitic structure and has a structure characterized by polymer particles connected by tie fibrils. It was observed that slower cooling allows greater growth of the spherulites without qualitatively altering the structure. Similar experiments using different quench temperatures revealed that higher quench temperatures yield larger spherulites without qualitatively altering the spherulitic structure. The quenched samples have spherulites significantly smaller than the samples prepared via controlled cooling because of the greater nucleation rate associated with the quenching process [25].

The PVDF membrane was prepared from a 46 wt% solution in dibutyl phthalate (DBP) by compression moulding and quenching from 180 to 0 °C in water. The resulting fuzzy sphere structure is shown in Fig. 5.6(b). It was speculated that the fuzzy sphere structure is the product of nucleation and growth of the solid polymer from the melt. While the quenched PP-mineral oil samples (Fig. 5.6(a)) did not show irregular surface structure, the PVDF samples do because of the slower crystallization kinetics of the PVDF.

The different structures between the PP and PVDF membranes, shown here as an example, are attributed to the differences between the polymers in terms of molecular configuration, lamellar structure, spherulite formation and crystallization kinetics. The minor differences between the quenched and slowly cooled samples are the result of crystallization kinetics, which will be explained in detail later on.

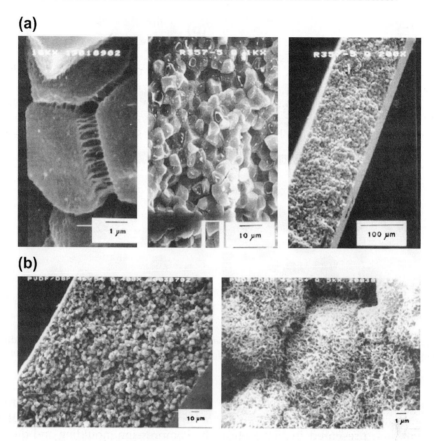

FIGURE 5.6 SEM images: (a) spherulitic structure resulting from solid—liquid phase separation of a 55 wt% PP in mineral oil solution quenched from 245 to 43 °C in water and (b) fuzzy sphere structure resulting from solid—liquid phase separation of a 46 wt% PVDF in dibutyl phthalate solution quenched from 180 to 0 °C in water. For both cases the diluents were extracted with 1,1,1-trichloroethane (TCE). *Source: Reprinted from [25]. Copyright 1990, with kind permission from Elsevier.*

Combined Liquid—Liquid and Solid—Liquid Phase Separation

In Fig. 5.4, if a polymer/diluent system is quenched to point B within the binodal curve, but above the equilibrium crystallization curve, liquid—liquid phase separation occurs via one of the two mechanisms reported previously but there is no driving force for crystallization. In this case the structure can be developed by droplet growth and coarsening (i.e. grow with time of droplet phase, which is diluent-rich, in matrix phase, which is polymer-rich) as will be explained in kinetics considerations [30,40,47]. The development of the structure or which is the same the droplet growth can be stopped by lowering the temperature to a point below the equilibrium crystallization curve for polymer crystallization. The points C and D in Fig. 5.4 are situated at different levels below the binodal and the equilibrium crystallization curves. Therefore, when a polymer/diluent melt blend is driven to these two points it will experience both liquid—liquid separation and polymer crystallization. The driving force and rate of crystallization is higher in point D than in point C. Thus, the kinetics of liquid—liquid phase separation and polymer crystallization differ at these two points and as consequence determine the final TIPS membrane structure [30]. These facts will be explained later on.

Figure 5.7 shows optical micrographs of a 30 wt% iPP/DPE system phase separating at 378 K. This holding temperature lies below the binodal and the equilibrium crystallization curve. Therefore, there is a driving force for both liquid—liquid phase separation and crystallization. In this case, the structure starts to develop via liquid—liquid phase separation prior to the formation of stable crystal nuclei. The micrograph taken at a holding time of 1 min (Fig. 5.7(a)) shows the presence of some small spherulites. As can be seen, although crystallization of the polymer has begun in some regions of the sample, the bulk of the sample has not yet crystallized and the liquid—liquid phase separated structure in these regions, continues to develop. At 4 min (Fig. 5.7(b)) the size of the spherulites has increased and most of the sample has crystallized. By 8 min (Fig. 5.7(c)) the entire sample has crystallized.

It is to note that related to the membrane formation procedure, thermodynamic considerations are not able to offer any explanation about structural variation within the membrane cross-section (i.e. symmetric, asymmetric or dense skin at the surface). These are determined by kinetic effects, which depend on system properties such as the diffusivities of the various components in the mixture, the viscosity of the solution and the chemical potential gradients, which act as driving forces for diffusion of the various components in the mixture and depend on both composition and temperature. These parameters change continuously during the phase separation (i.e. DIPS or TIPS), which constitutes the actual membrane formation process. In fact, the chemical potential and diffusivities of the various components in the system, and their dependencies on composition, temperature, viscosity, etc.,

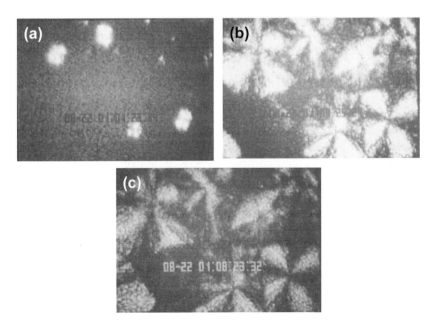

FIGURE 5.7 Optical micrographs of a 30 wt% iPP/DPE system phase separated at 378 K at holding times of (a) 1 min, (b) 4 min and (c) 8 min. *Source: Reprinted from [30]. Copyright 1994, with kind permission from Elsevier.*

are difficult to determine by independent experiments and therefore are not readily available. This makes a quantitative description of the membrane formation mechanism nearly impossible. A qualitative description, however, which allows rationalization of the membrane formation and correlation of the various preparation parameters with membrane structures and properties, is possible.

Kinetic Considerations

The final membrane structure depends also on the kinetics of the phase separation process and the local distribution of the polymer-rich phase at the point of vitrification. It was demonstrated that solidification of a binary polymer/diluent system occurs when the binodal line intersects the curve defining the glass transition, as shown in Fig. 5.4 [49,50]. In fact, the location of the vitrification point (V_p) is very important for the formation of membranes. If liquid–liquid phase separation occurs while cooling an initially stable blend solution, continuing phase separation and/or coarsening of the resulting phases will be arrested at the temperature where the tie-line intersects the vitrification point [50].

As stated above, when a polymer/diluent solution is cooled from the homogeneous one-phase region of the phase diagram to a temperature within the binodal curve, liquid–liquid TIPS occurs. In many systems, droplets of one phase form within a continuous matrix of a second phase. These two phases have compositions given by the binodal curve in the phase diagram. The droplet phase is diluent-rich phase and the matrix phase is polymer-rich. Once equilibrium composition is reached, these droplets grow with time. This process is known as coarsening. The thermodynamic driving for this droplet growth is the minimization of interfacial free energy via reduction of the interfacial area. It was observed that the droplets increase in size and decrease in number with time [30]. When the polymer is solidified by lowering the temperature and the diluent is extracted, the droplets become the cells in the membrane.

It was reported that three different mechanisms are responsible for the coarsening of the microstructure in the later stages of the phase separation: Ostwald ripening [51,52], coalescence [53] and the hydrodynamic flow mechanism [54,55]. The evolution of the coarsening mechanism from Ostwald ripening or coalescence to a hydrodynamic flow stage has been indicated experimentally in polymer solutions and membranes [57]. Furukawa [56] formulated the following scaling function:

$$r \propto t^\lambda \quad (5.3)$$

where λ refers to as the scaling or growth exponent that depends on the microscopic mechanism of particle growth. This scaling exponent was normally determined for several temperatures and polymer concentrations within the binodal curve. The domain size has been monitored by different characterization techniques such as electron microscopy, light scattering and optical microscopy [43,47,50,57,58]. A detailed review on the scaling experiments values in polymer/diluent systems may be found in [47]. As an example, Figs. 5.5 and 5.8 show cell growth with coarsening time.

In Fig. 5.8, the effect of the coarsening time was studied at 393 K for the system iPP/diamyl phthalate (DAP). For this system the temperature 393 K is above the crystallization temperature. Therefore, no solidification of polymer-rich phase occurs to restrain the growth of polymer-lean droplets and the pore size of the obtained TIPS membranes increases over the coarsening time when quenched at 393 K.

As in the DIPS technique (Chapters 3 and 4), both flat sheet and hollow fibre membranes were prepared by TIPS technique. Figures 5.9 and 5.10 show typical systems used for preparation of flat sheet and hollow fibre TIPS membranes, respectively.

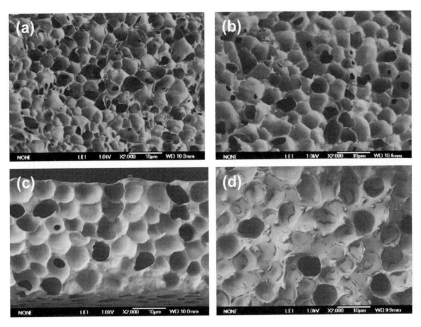

FIGURE 5.8 SEM cross-section images of iPP membranes as a function of coarsening time at 393 K (above crystallization temperature) in water with a polymer concentration of 20 wt%: (a) 10 s, (b) 30 s, (c) 60 s and (d) 120 s. *Source: Reprinted from [59]. Copyright 2009, with kind permission from Elsevier.*

Flat Sheet Membrane Preparation

Briefly, the polymer/diluent blend solution is filled in a test tube, which is normally purged with nitrogen and sealed to prevent oxidation. The test tube is then placed in an oven at a temperature higher than the melting temperature of the blend during a period of time so that a homogenous melt blend is formed. Subsequently, the solution is immersed in liquid nitrogen in order to induce solidification. A given amount of the solid solution is taken and placed in a thin film metallic mould with a given shape (Fig. 5.9). The mould is then covered with a Teflon or metallic thin plate, aluminium circular thin plate in Fig. 5.9. The use of different moulds allows the formation of different samples with different thicknesses. The whole assembly is then heated in the oven at a given temperature for a certain period until a melt blend is produced assuring the elimination of the possible influence of thermal history. To induce phase separation, the solution is immersed in liquid nitrogen or water at controlled temperature or simply the temperature of the oven is diminished at a given rate. The diluent present in the formed film is extracted with an adequate extractant, which is finally evaporated to produce microporous membranes.

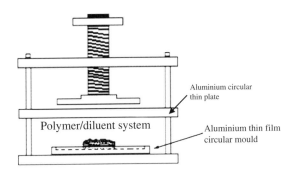

FIGURE 5.9 Schematic diagram of flat sheet membrane preparation system via TIPS technique. *Source: Reprinted from [44]. Copyright 2005, with kind permission from Elsevier.*

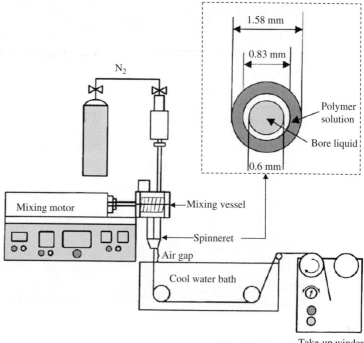

FIGURE 5.10 Schematic diagram of hollow fibre membrane preparation system via TIPS technique. *Source: Reprinted from [60]. Copyright 2003, with kind permission from Elsevier.*

The solvent exchange method of extracting the diluent prevents the collapse of the pores and film shrinkage. The membranes are normally dried in the fume hood with circulating air. The extraction method depends on the polymer/diluent system. The method of flat sheet membrane preparation via TIPS technique has been described by various researchers throughout the manuscripts reviewed in this chapter.

Hollow Fibre Membrane Preparation

Hollow fibre membranes were also prepared by TIPS technique. Figure 5.10 shows a typical example of the systems used. It is a batch-type extrusion apparatus (Imoto Co., BA-0) [60]. A given amount of a polymer/diluent system is mixed under a nitrogen atmosphere in the vessel heated at a temperature above the melting temperature of the blend. The homogeneous polymer/diluent melt blend is then fed to a spinneret by a gear pump under nitrogen pressure of about 0.2 MPa. The spinneret shows an outer and an inner tube. A diluent is also introduced into the inner tube of the spinneret to make a lumen of the hollow fibre. The fibre extruded from the spinneret enters in the water bath under a controlled temperature to induce phase separation and solidification by means of polymer crystallization. Simultaneously, the nascent fibre is wound by using a take-up winder at several speeds. The diluent present in the fibre is then extracted as indicated previously for flat sheet TIPS membranes.

The variables involved in the hollow fibre membrane preparation via the TIPS technique are the air gap distance, which is the distance from the spinneret to water bath, the water bath temperature, the take-up speed, the extrusion rate of the blend, the flow rate of the diluent in the inner tube of the spinneret, the temperature of the polymer/diluent blend and its composition etc.

EFFECTS OF PROCESS PARAMETERS ON THE MEMBRANE STRUCTURE

In MD process, the membrane pore size, porosity, thickness as well as the organization, connectivity and size of polymer affect the structural integrity and transport properties. A number of parameters are involved in TIPS technique and should be varied adequately to control flat sheet and hollow fibre membrane structure and morphology as well as MD performance. Phase diagram for the polymer/diluent system including cloud point curve and crystallization curve should be built first before starting preparation of TIPS membranes.

Flat Sheet Membranes

Effects of Polymer Type and Concentration in the Initial Polymer/Diluents System

In TIPS, by varying the composition and temperature, the polymer solutions can be brought into different regions in the phase diagram of the polymer/diluent system. Different mass transport mechanisms can occur during the membrane formation by liquid–liquid phase separation, solid–liquid phase separation, etc.

The polymer/diluent system and the type of polymer affect considerably the final morphology of TIPS membranes. For example, Fig. 5.11 shows two membranes prepared from sPP and iPP, both prepared under the same TIPS conditions. The pore size and the final structure of the sPP membrane may be influenced by the higher flexibility of the sPP polymer and the more amorphous character of the sPP than iPP [43]. During diluent extraction and subsequent drying step, the greater crystallinity of the iPP polymer may provide stability to the formed cellular structure. As shown in Fig. 5.3, for the sPP/DPE system, the cloud point curve was shifted towards higher temperatures compared to the iPP/DPE system and the equilibrium crystallization curve was shifted below that of the iPP/DPE system. Therefore, the droplet sizes were slightly larger in sPP/DPE system than in iPP/DPE system [43]. These results suggest that the pore size in the sPP membrane should be larger than in the iPP membrane and pore structure should be similar. The contrary is the case in Fig. 5.11, confirming that the physical characteristics of the polymer play important roles in the final structure and morphology of TIPS membranes.

In the case of liquid–liquid phase separation, polymer concentration is the most important factor affecting the membrane pore structure. For example, Fig. 5.12 shows the SEM cross-section morphology and tensile strength of iPP membranes prepared with different concentrations. Two types of structures (bicontinuous and cellular) can be observed without spherulites due to liquid–liquid phase separation prior to polymer crystallization. A change can be seen in Fig. 5.12 from a bicontinuous structure at low concentrations (10 wt%, 20 wt% and 30 wt%) due to spinodal decomposition with subsequent solidification of the polymer-rich phase to a cellular structure at higher concentrations due to nucleation and growth of polymer-lean droplets with further solidification of the polymer-rich phase. Moreover, the interconnectivity of the pores increases with the decrease of polymer concentration and the tensile strength increases with enhancing the polymer concentration in the iPP/DAP system.

In the case of solid–liquid TIPS membrane preparation, crystal nucleation and growth determine the morphology of the membrane. Upon the extraction of the diluents, the space between crystalline domains becomes the pores. To produce membranes with high porosity, it is better to use low polymer concentration in the polymer/diluent system. However, membranes with higher nucleation density are stronger because of increased integrity. In fact, self-nucleation depends on the pre-existence of polymer nuclei

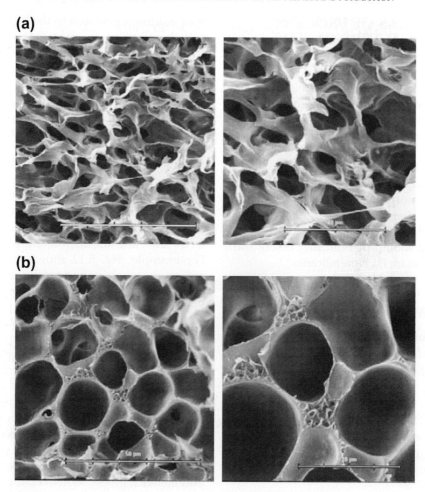

FIGURE 5.11 SEM structure of two TIPS membranes formed by different polymers (a) sPP, (b) iPP and similar preparation conditions. *Source: Reprinted from [43]. Copyright 2005, with kind permission from Elsevier.*

and nucleation density is directly proportional to the polymer content of the polymer/diluent system. The effects of polymer concentration on nucleation density and final TIPS membrane structure have been discussed in the literature [48,65]. As the polymer concentration increased, the temperature difference between polymer and diluent crystallization was increased and the spherulite had more time to change the size. It was observed that the spherulite size decreased with increasing the polymer concentration. Higher polymer concentration results in higher nucleation density. A proposed solution is to select appropriate heating time and rate of the polymer/diluent system to allow a lower polymer concentration to be used for preparing strong membranes [48]. Increased polymer concentration results in a greater number of nuclei, which in turn reduces the size of the polymer domains in the final membrane structure. Decreased polymer domain size results in smaller pores and membranes of greater integrity. It is reported that spherulites start by growing from a nucleus into lamella. The lamella is then spawn into

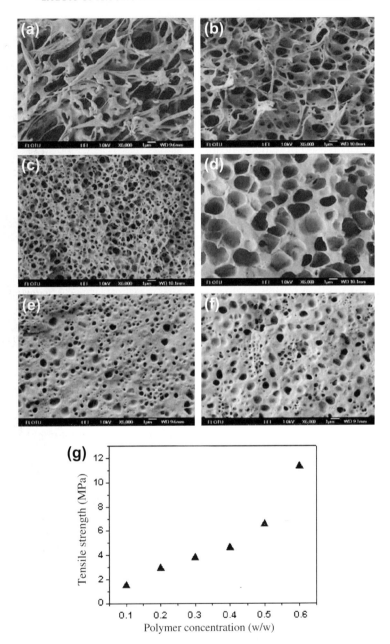

FIGURE 5.12 SEM cross-sectional structure (a–f) and tensile strength (g) of iPP membranes quenched in ice water using iPP/diamyl phthalate (DAP) blend at 473 K and different polymer concentrations: (a) 10 wt%, (b) 20 wt% (c) 30 wt%, (d) 40 wt%, (e) 50 wt% and (f) 60 wt%. *Source: Reprinted from [59]. Copyright 2009, with kind permission from Elsevier.*

sheaf-like structures called axialites, which then develop into spherulites. For example, poly(phenylene sulphide) (PPS) membranes formed using the diluent 4-benzoylbiphenyl (BBP) via solid–liquid TIPS show axialites to be the dominant structure for high polymer concentrations; whereas at lower polymer concentrations the dominant structure is spherulites [48]. Nucleation density increases with increased polymer concentration and final spherulite size decreases with increasing nucleation density.

When using other polymers such as PVDF, the crystallization temperature increased with the increase of polymer concentration and the size of spherical granules observed throughout the membrane cross-section increased whereas the number of granules diminished [77]. In this last case, solid–liquid phase separation induced granule-like structure in the cross-section of the formed membranes and no obvious liquid–liquid phase separation occurred before solid–liquid phase separation. It should be pointed out that the increase in degree of crystallization depends also on the concentration and type of diluent used. The effect of diluents will be shown later on.

Effects of Polymer Molecular Weight

Polymer molecular weight is one of the most important factors affecting thermodynamic and kinetic properties. With the decrease of the polymer molecular weight, the cloud point curve moves towards lower temperature region, whereas the dynamic crystallization temperature curve is generally not affected too much. Under the same cooling conditions, the membrane cell size decreases by enhancing the polymer molecular weight [61,62]. For example, Fig. 5.13 shows the difference in membrane structure of two porous membranes prepared from iPP of different molecular weights. The membrane prepared with lower polymer molecular weight exhibits large cellular pore structure, whereas a less interconnected structure

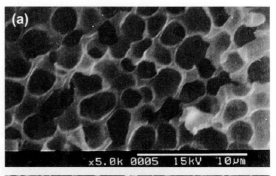

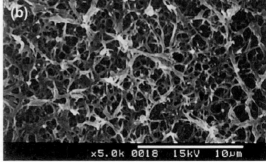

FIGURE 5.13 SEM cross-section of porous membranes prepared using iPP of different molecular weights (a) 12,000 and (b) 5,80,000. *Source: Reprinted from [61]. Copyright 2002, with kind permission from Elsevier.*

was observed for membranes prepared with higher polymer molecular weight. These results were attributed to the faster phase separation rate (i.e. latter stage of spinodal decomposition) in polymer/diluents systems with low polymer molecular weight. Both the pore size and shape are affected by the polymer molecular weight although the same polymer concentration and cooling conditions are adopted.

Effects of the Type of Diluents, Its Concentration and Molecular Weight

In TIPS technique, polymer dissolves in diluent at high temperature and a homogeneous melt blend is formed. Phase separation is induced by lowering the temperature. Upon diluent extraction and subsequent evaporation of the extractant, the diluent-rich phase becomes the membrane pores. It was observed that the domain size and therefore pore size of the formed TIPS membranes increased as the diluent molecular weight was increased [63]. It was also noted that the enhancement of the domain size was less pronounced for the case of lower content of diluents in the blend due to the interactions between polymer and diluent [63]. Because of the chain-like nature of macromolecules, they crystallize much more slowly than low molecular weight compounds. Kim et al. [31] concluded that diluent size affected crystallization temperature of both the polymer and the diluent, and consequently the crystallization rate. Alwattari and Lloyd [34] found that the pore size and morphology were controlled by the diluent crystal growth mechanism and crystallization kinetics relative to the crystallization kinetics of the polymer. McGuire et al. [65] reported on the effect of diluent concentration on spherulitic growth of the TIPS membranes (iPP) prepared by the solid–liquid phase separation and stated that the greater the diluent concentration, the more significant is the effect of diluent crystallization on the inhibition of spherulitic growth.

Effects of Polymer/Diluent Melting Temperature and Time

If the polymer/diluent system is held at a temperature higher than the critical temperature of the system for a sufficient period of time, the residual crystals (in case of semi-crystalline polymers) can be removed. In fact, the residual crystals may act as seeds for subsequent crystallization affecting nucleation density. As a consequence, high nucleation density results in smaller polymer domains and therefore the formed membranes will exhibit more distributed smaller pores and greater integrity. It was reported that increasing polymer/diluents temperature blend decreased nucleation density by destroying more nuclei. However, nucleation density typically levelled off after 5 min [48].

Effects of Quenching Conditions

Quenching conditions play an important role in determining the final membrane morphology. When the difference between the equilibrium melting temperature of the polymer/diluent system and the quenching temperature (i.e. supercooling degree) is high, the driving force of liquid–liquid and solid–liquid phase separation is also high. The magnitude of this driving force is directly proportional to the degree of supercooling. Liquid–liquid phase separation and solidification of polymer rich phase are two competitive factors influencing the resulting membrane morphology.

For quenching temperatures below the spinodal decomposition, it was found that the pore size increased with increasing quenching temperature [59]. By decreasing the supercooling degree, slower solidification rate of polymer-rich phase involves growth of polymer-lean phase leading to larger pore sizes. However, the rapid solidification of the polymer-rich phase restrains the growth of polymer-lean phase and few changes can be observed with the increase of quenching time. Little coarsening of the pore size was observed by Lin et al. [59] for iPP membranes when quenching was conducted at 333 K. The coarsening time effect on membrane morphology was explained in the section 'Principles of the formation of porous hydrophobic TIPS membranes'. When quenching temperature is above the crystallization temperature, no solidification of polymer-rich phase occurs to restrain the growth of polymer-lean droplets. As a consequence, the membrane pore sizes increase with enhancing coarsening time because the polymer-lean droplets have more time to grow.

For systems undergoing nucleation-controlled solid–liquid TIPS, the greater degree of supercooling, the faster the crystallization and the spherulitic growth rate, and greater the number of nuclei formed. For semi-crystalline polymer/diluent melt blends, slow cooling results in higher

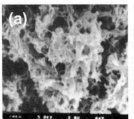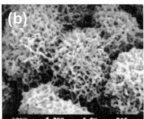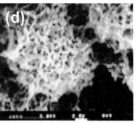

FIGURE 5.14 Cross-section SEM photographs of the PVDF/DMP membranes as a function of coarsening time: (a) 1 min, (b) 3 min, (c) 5 min and (d) 20 min. *Source: Reprinted from [67]. Copyright 2006, with kind permission from Elsevier.*

nucleation density [48]. However, fast cooling rate is normally preferred because it is representing more closely the actual TIPS process. The supercooling degree also affects nucleation density. Higher temperature difference results in higher driving force to form nuclei. At a fixed polymer concentration, nucleation density has been observed to increase with decreasing crystallization temperature and level off at sufficiently low crystallization temperature [48].

It is worth quoting that diluent crystallization does not play any role in determining the TIPS membrane structure if crystallization temperature is above the diluent crystallization temperature. However, diluent crystallization takes place together with the polymer crystallization if the TIPS technique is conducted at a temperature below the diluent crystallization temperature and therefore affects the membrane structure.

When quenching temperature was above the crystallization temperature, the polymer-rich phase droplets grew through the liquid–liquid phase region and congregated as presented in Fig. 5.14 for PVDF/dimethyl-phthalate (DMP), in which only 3 min is sufficient for formation of spherulites. It was observed that the size of spherulites increased with time, whereas the pore size did not change. In addition, the spherulite growth rate of PVDF was higher at higher quenching temperatures while the structure and size of pores were unvaried as coarsening or quenching depth was increased [67].

Effects of Nucleating Agents

Control of polymer crystallization kinetics by adding nucleating agent to the polymer/diluent system can facilitate the control of pore size, porosity and pores size distribution of TIPS membranes. By adding suitable nucleating agent, the membrane pore size may decrease as observed by Luo et al. [64] for dibenzyl sorbitol nucleated PP/soybean oil/DBP system. In this case, both the spherulite size of PP and the pore size of the membrane decreased, whereas the degree of crystallization increased. The addition of nucleating agent promotes the formation of narrow pore size distribution. Other components such as adipic acid and benzoic acid exerted practically no effect on nucleation and were not good nucleating agents for PP/soybean oil/DBP system. The effect of different nucleating agents on the structure of PP membranes is shown in Fig. 5.15. It was reported that spherulites were much smaller for TIPS membranes prepared by adding nucleating agent than for the non-nucleated membranes [65]. When nucleation increased, the space for the growth of spherulite decreased. By increasing the concentration of the nucleating agent in the polymer/diluent system, a slight decrease was observed for the TIPS membrane pore [64].

Effects of Extractant

Different extractants were tested for different polymer/diluent systems. For example, for

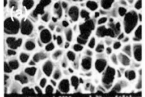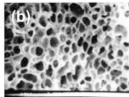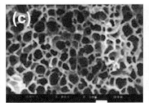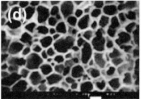

FIGURE 5.15 SEM photographs of the PP/soybean oil/DBP blended with different nucleating agents: (a) non-nucleating agent, pore size (1.95 μm), (b) dibenzyl sorbitol, pore size (1.72 μm), (c) adipic acid, pore size (1.80 μm) and (d) benzoic acid (1.97 μm). *Source: Reprinted from [64]. Copyright 2006, with kind permission from Elsevier.*

PP/soybean oil/DBP system, methanol, ethanol, formic acid, acetone, butane, n-hexane, dichloroethane and xylene were used as extractants [64]. Different extractant ratios (i.e. ratio of weight loss) were used and it was observed that adequate extractant for each polymer/diluent system must be selected. Ethanol, acetone, butane, n-hexane and dichloroethane were proper extractants for PP/soybean oil/DBP system and similar TIPS membrane porosities ($\approx$ 70%) were obtained using these extractants [64]. The solubility parameter can be used to select the suitable extractant for the polymer/diluent system. Close solubility parameter (i.e. high affinity) between extractant and diluent, and low affinity between extractant and polymer are desirable. Replacing the diluent with extractant can cause swelling or contraction (i.e. densification of the amorphous regions of the membrane and collapse of membrane pores) of the formed membrane. If strong affinity exists between the polymer and extractant than the polymer and diluent, the amorphous regions of the semi-crystalline film can swell. In addition, when choosing an extractant it is important also to consider its adequate properties for subsequent removal from the formed membrane (high volatility, high boiling point, low surface tension with the solid matrix of the membrane, etc.). Membrane pore size and porosity were observed to decrease with increasing surface tension and boiling point, while no discernable trend was observed between membrane shrinkage and solubility parameter of the extractant [66].

Effects of Drying

It is possible to manipulate membrane porosity and pore size prepared via TIPS technique by choosing the adequate extraction procedure and subsequent drying conditions for extractant removal. It was observed that the decrease of membrane pore size and porosity was less when the extractant was removed via freeze-drying compared to direct air-drying. Since the extractant is solid when freeze-drying is carried out, the pore does not collapse while the extractant is removed. In contrast, in air-drying the extractant is liquid and interacts with the amorphous phase of the polymer membrane matrix, softening the pore wall and leading to pores collapse after evaporation of the extractant [66].

Effects of Stretching

Stretching is a mechanical post-treatment that permits to alter physically the membrane skin and increase porosity. However, defects may be introduced by stretching and breaking the fragile cell walls. It was observed that stretching induced an increase of membrane pore size and porosity and a decrease of thickness of TIPS membranes prepared from poly(tetrafluoroethylene-*co*-perfluoro-(propyl vinyl ether)) (Teflon® PFA, NeoflonTM PFA)/chlorotrifluoroethylene [46]. Similar observations were presented for hollow fibre membranes prepared via TIPS technique and subsequent cold-stretching as will be shown below.

Hollow Fibre Membranes

During the formation of hollow fibre membranes by TIPS technique, there are many factors that influence the membrane morphology and structure. Different systems were used to prepare TIPS hollow fibres. A typical system to study the different possible parameters on hollow fibre structure, morphology and performance is shown in Fig. 5.10.

Effects of Polymer Type, Molecular Weight and Composition

Li et al. [35] studied the effect of the polymer type on the MD performance of TIPS hollow fibre membranes. Polyethylene (PE) and PP were considered and melt-extruded/cold-stretching method was used to spin hollow fibres for desalination by DCMD and VMD. Compared to PP hollow fibre membranes, higher water fluxes were obtained for the PE membranes in both DCMD and VMD. This was attributed to the larger pore size and porosity of the PE membranes. The average pore size of the PE membranes was 74 and 87 nm and the porosity was 53.3% and 66.3%; whereas the average pore size of the PP membranes was 44 and 56 nm and the porosity was 47.3% and 50%. The highest permeate flux reported was 0.8 $l/m^2 \cdot h$ in DCMD and about 4 $l/m^2 \cdot h$ in VMD.

To spin porous hydrophobic hollow fibre membranes via TIPS it is important to select proper materials. For example, PE with density higher than 0.96 g/cm^3 should be used to obtain sufficiently hard elasticity because this is the basis of microporosity of PE hollow fibre membranes [37]. Higher elastic recovery means better hard elasticity. Micropores were not formed when using polymers with low elastic recovery, whereas many interconnected micropores were formed when using polymers with higher elastic recovery [37].

Matsuyama et al. [60] also used different types of high density PE to prepare hollow fibre membranes via TIPS and found an asymmetric structure with smaller pores at the external surface of the hollow fibres (Fig. 5.16). This asymmetric structure was attributed to the fast cooling rate (i.e. from 200 to 50 °C), to the higher polymer concentration near the outer surface and to the diluent diisodecyl phthalate (DIDP) maintained at 200 °C, which was circulated through the lumen side of the nascent hollow fibre membrane. Kim et al. [68] also found an asymmetric structure of PP hollow fibre membranes. However, in this case, skin layer was formed at the inner surface due to the circulation of nitrogen gas through the lumen side of the nascent hollow fibre membranes and to the temperature gradient between the two surfaces, which caused different phase separation rates.

Polymer molecular weight affects the kinetics of pore and porosity formation. For the same type of polymer and concentration, the viscosity of the formed solution is low when the polymer molecular weight is low. Moreover, the faster evaporation of the diluent along the air gap leads to an increase of the polymer concentration at the outer surface of the nascent hollow fibre membrane. This was confirmed using high density PE of different molecular weights [60]. As the polymer molecular weight decreased, the cloud point also decreased and the phase separation structure obtained at the early stage of spinodal decomposition was small. Smaller polymer molecular weight led to the formation of hollow fibre membranes with smaller pore size and lower permeability.

Similar to TIPS flat sheet membranes, the initial composition of the polymer/diluent system also affects the pore size and porosity of the hollow fibre membranes prepared via TIPS. Since the diluent content in the polymer/diluent system also controls the porosity

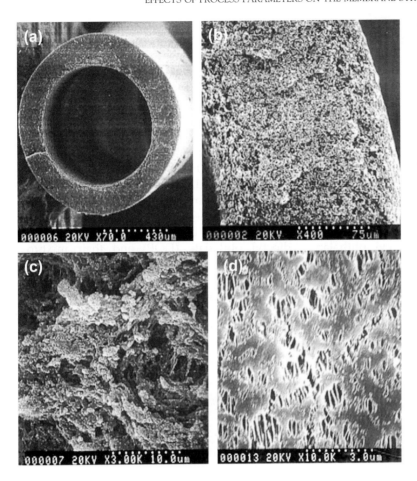

FIGURE 5.16 SEM images of the hollow fibre membrane prepared from high density PE polymer and the diluent diisodecyl phthalate (DIDP) with 0.238 m/s take-up speed, 50 °C water bath temperature and 5 mm air gap: (a) whole cross-section, (b) enlarged cross section, (c) inner surface and (d) outer surface. *Source: Reprinted from [60]. Copyright 2003, with kind permission from Elsevier.*

of the TIPS membrane, an increase of the polymer concentration decreases the porosity [68].

Effects of Concentration and Type of Diluents

Both thermodynamic and kinetic phenomena depend on the concentration and the type of diluents as explained in the section 'Principles of the formation of porous hydrophobic TIPS membranes'. When preparing flat sheet or hollow fibre membranes, the mode of phase separation (liquid−liquid phase separation, crystallization, etc.) depends on the polymer/diluent system and quenching conditions under consideration, resulting in different membrane structures, morphologies and performances. For example, in the case of liquid−liquid phase separation, the formed hollow fibre membranes have an interconnected sponge-like bulk microstructure and porous surfaces. On the contrary, the hollow fibre membranes prepared via solid−liquid phase separation show asymmetric spherulitic bulk structures and denser skin layers on both the outer and inner surfaces. As well, the structure of the inner surface of the hollow fibre membranes prepared via TIPS technique is affected by the type and temperature of the bore fluid because exchange between bore fluid and diluent or diluents mixture may occur. When the system proceeds via liquid—liquid phase separation and bore liquid has good compatibility with the diluent or diluents mixture used, the droplets of the polymer-lean phase formed near the inner surface may be replaced by the bore liquid, leading to the formation of bigger pores.

Sun et al. [78] prepared high-density polyethylene (HDPE) hollow fibre membranes via TIPS process of the HDPE/liquid paraffin (LP) system. Solid—liquid TIPS was carried out. In order to optimize the structure of HDPE hollow fibre membranes Matsuyama et al. [60] used two types of diluents (diisodecyl phthalate, DIDP and LP) to prepare the HDPE hollow fibre membrane. The HDPE hollow fibre membrane prepared via liquid—liquid phase separation with DIDP showed a porous structure at the outer surface and higher permeability than that prepared via only solid—liquid phase separation with LP. The water permeability of the high density PE membrane was found to be about three times higher using the diluent DIDP than using LP under the same operating conditions, except for the diluent type [60].

When using PE/DIDP system, liquid—liquid phase separation took place giving a porous structure at the outer surface of the hollow fibre membrane, whereas polymer crystallization occurred for PE/LP system and pores were hardly observed. The effect of diluent on TIPS membrane morphology is illustrated in Fig. 5.17 for the iPP hollow fibre membranes prepared using di-*n*-butyl phthalate (DBP) and dioctyl phthalate (DOP) and their mixtures (i.e. ternary system) as diluents. When the diluent was DBP, the iPP hollow fibre membranes exhibited cellular structure because of liquid—liquid phase separation (i.e. quenching temperature was higher than crystallization temperature). In contrast, when the diluent was

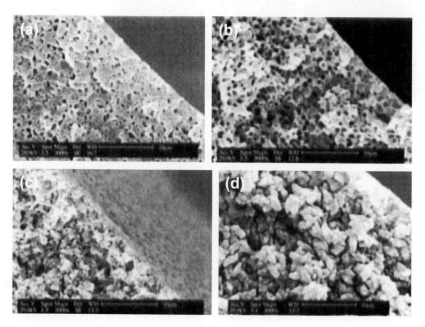

FIGURE 5.17 SEM images of iPP hollow fibre membranes prepared from different concentrations of diluents. (a) DBP, (b) 0.5 DOP mass fraction in diluent, (c) 0.8 DOP mass fraction in diluent and (d) DOP. *Source: Reprinted from [69]. Copyright 2006, with kind permission from Elsevier.*

DOP, the hollow fibre membrane exhibited spherulitic particulate structure originated from polymer crystallization. In this case, the mechanical strength of the resulting hollow fibre membrane was so weak that the pure water flux could not be measured. When the concentration of the diluent DOP was increased, the morphology of the hollow fibre membranes was changed from typical cellular structure to mixed structure (i.e. basically cellular but with particulate boundaries) and then particulate structure [69]. It was also observed for these hollow fibre membranes that the pure water flux increased, whereas the tensile strength and breaking elongation decreased with increasing concentration of DOP. When using the diluent DBP, the pure water flux of the hollow fibre membrane was very low because cellular pores were largely isolated and embedded in a continuous polymer matrix (i.e. absence of pore interconnectivity). In general, it was found that the mechanical properties of the hollow fibre membranes with typical cellular structure or mixed membrane structure were higher than those of the hollow fibre membranes with particulate structure, in which the liquid–liquid phase separation precedes polymer crystallization.

PVDF hollow fibre membranes were prepared via TIPS technique using different types and concentrations of diluents (i.e. diluent mixtures of DBP and di(2-ethylhexyl) phthalate, DEHP) [70]. Asymmetric structures of hollow fibres were observed for 40/60 wt% and 100/0 wt% DBP/DEHP systems. The entire cross-section consisted of spherulitic structure and the spherulites became bigger and more perfect as the DBP/DEHP ratio was increased. Spherulites near the outer surface were much smaller than near the inner surface. As stated earlier, in solid–liquid phase separation the crystal structure is affected by the cooling rate. The cooling rate at the outer surface was higher than that at the inner surface, which led to the formation of smaller spherulites near the outer surface. Moreover, as the DBP/DEHP ratio was increased, the skin layers near the outer and inner surface became thicker. In contrast, the PVDF hollow fibre membrane prepared using 30/70 wt% DPB/DEHP system exhibited smaller pore size, higher porosity and higher water permeability compared to the PVDF hollow fibre membranes prepared with 40/60 wt% and 100/0 wt% diluent systems. This was attributed to the interconnected sponge-like microstructure formed via liquid–liquid phase separation. PVDF hollow fibre membranes with spherulitic structures and thick skin layers near the outer and inner surfaces were found to have higher breaking stress as well as elastic modulus and lower elongation-at-break (strain). PVDF hollow fibre membranes with more uniform sponge-like microstructure formed via liquid–liquid phase separation had higher values of strain at break.

It must be pointed out that anisotropic flat sheet or hollow fibre membranes (i.e. the pore size varies throughout the membrane cross-section) were prepared by applying a controlled evaporation of the diluent prior to phase separation in the TIPS process, which is similar to the DIPS technique, or by imposing a thermal gradient across the cross-section of the membrane [15,16,62,72–76].

Effects of Spinning Temperature

The effect of spinning temperature is not as significant as the other parameters involved in spinning hollow fibre membrane via TIPS. Low spinning temperatures induce high viscosity of polymer melt solution and the increased stress of the polymer melt solution enhances the degree of orientation of the nascent hollow fibre membrane. It was reported that more oriented structure led to the formation of pore size larger than less oriented structure. PP hollow fibre membranes prepared with different spinning temperatures (liquid–liquid phase separation followed by solid–liquid phase separation) exhibited similar pore sizes [68].

The temperature of the spinneret also affects considerably the properties of the hollow fibre membranes. Care has to be taken not to decrease the spinneret temperature too much, since otherwise the spinneret will be blocked because of the solidification of the polymer/diluents solution. Berghmans et al. [72] stated that open skin of poly(2,6-dimethyl-1,4-phenelyene ether) (PPE) hollow fibre membranes can be produced by decreasing the spinneret temperature.

Effects of Air Gap Distance

Similar to DIPS, the air gap distance drastically affects the geometry, structure and performance of hollow fibre membranes prepared via TIPS technique. Small air gap length led to large pore size and high water permeability [60]. This was attributed to the evaporation of diluent along the air gap, which resulted in an increase of the polymer concentration at the external surface of the nascent hollow fibre membranes and a decrease of the corresponding pore size.

Effects of Water Bath Temperature and Take-up Speed

When the temperature of the bath was increased, the pore size and the water permeability also increased, whereas the solute rejection decreased [60,71]. This result was explained by the smaller cooling rate due to the high bath temperature [60].

When the bath temperature is high, the membrane is not likely to be solidified by the crystallization because the crystallization rate is slow. This makes the hollow fibre easy to be elongated by the take-up drum if it is used. Thus, the effect of the take-up speed on the hollow fibre membrane morphology and performance is stronger at high bath temperature (Fig. 5.18). The higher temperature hindered solidification process and a large amount of water penetrated into the polymer solution leading to larger pores near the outer surface. In addition, the droplet formed by liquid–liquid phase separation could sufficiently grow at higher temperature to form larger pores at the outer surface.

Effects of Spin Draw Ratio

It was reported that the hollow fibre membranes prepared by TIPS process and subsequent cold-stretching became thinner, more oriented in the spinning direction and more porous (especially at the outer surface)

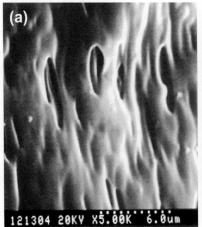 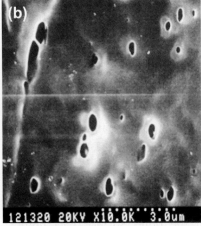

FIGURE 5.18 SEM images of the outer surface of poly(ethylene-*co*-vinyl alcohol) hollow fibre membranes prepared at different bath temperatures: (a) 55 °C and (b) 40 °C. *Source: Reprinted from [71]. Copyright 2003, with kind permission from Elsevier.*

by increasing the melt-draw ratio defined as the ratio of take-up speed to the extrusion rate of the polymer solution [37,68]. Higher spin-draw ratio can cause the formation of more perfect rows of nucleated crystals leading to higher elasticity recovery than the unstrained hollow fibre membrane. More perfectly oriented crystalline lamellae can contribute to the formation of more and larger micropores, which increases the membrane porosity and permeation rate [37]. Micropores of the hollow fibre membranes spun with higher spin—draw ratio are more and larger than those of the hollow fibre membranes spun with smaller spin—draw ratio.

Effects of Cold Stretching

The cold-stretching of the hollow fibre membranes prepared via TIPS technique remarkably increases the membrane porosity and reduces the internal and external diameters. During a typical straining process, the micropores of hollow fibre membranes are formed. The hollow fibre membranes are first strained to lower extensions under low temperatures to induce microvoids in the fibre wall. Then, the hollow fibres are strained to higher extensions under higher temperatures to form more and larger micropores without breaking of the hollow fibres. Shen et al. [37] followed this process for preparation of PE hollow fibre membranes. The porosity, pore size and gas permeation of the PE hollow fibre membranes increased with increasing extensions.

Kim et al. [68] prepared PP hollow fibre membrane from polypropylene/soybean oil mixture by the TIPS technique and subsequent cold-stretching. Tiny fibrils were developed throughout the cross-section of the hollow fibre membranes in the direction perpendicular to the spinning line. A decrease of the hollow fibre membrane density and an increase of porosity were observed with the increase of the cold-stretching ratio.

Post-treatment

It was reported that unstable defects were reduced and reordered to form more perfect structures by annealing of membranes under proper conditions. Annealing temperature and time are key factors in this case. Annealing hollow fibre membranes prepared via TIPS technique below the melting point of the polymer improved the elasticity of the hollow fibre membrane leading to the formation of micropores during straining step. Moreover, density of the hollow fibre membranes increased because of higher crystallinity [37]. As the annealing temperature was increased, better elasticity, greater porosity, larger pore size and higher gas permeation were achieved. In addition, annealing time also improved the elastic recovery, crystallinity, porosity and gas permeation during an initial period after which they levelled off (e.g. 2—4 h for PE hollow fibre membranes) [37].

Special Considerations

Many studies have been carried out on the formation of porous flat sheet and hollow fibre membranes by the TIPS process. Recently, studies have been carried out on the combined use of TIPS and DIPS (immersion precipitation). Isotropic and anisotropic membranes were formed with different pore sizes, porosities, thicknesses, etc. Various types of hydrophobic polymers, diluents, extractants as well as the effects of different fabrication parameters on the membrane properties were studied. However, very few studies are found in the literature using the obtained information for the preparation of membranes designed specifically for MD.

References

[1] R.E. Kesting, Synthetic polymer membranes, McGraw Hill, New York, 1972.
[2] T. Matsuura, Synthetic membranes and membrane separation processes, CRC Press, Boca Raton, FL, 1994.

[3] I. Pinnau, B.D. Freeman, Membrane formation and modification, acs symposium series 744, American Chemical Society, Washington DC, 2000.

[4] M. Khayet, T. Matsuura, J.I. Mengual, M. Qtaishat, Design of novel direct contact membrane distillation membranes, Desalination 192 (2006) 105–111.

[5] M. Khayet, J.I. Mengual, T. Matsuura, Porous hydrophobic/hydrophilic composite membranes: Application in desalination using direct contact membrane distillation, J. Membr. Sci. 252 (2005) 101–113.

[6] R. Mahoney, Microporous polyethylene hollow fibres and process of preparing them, United States Patent Serial No. 4,020,230 (1977).

[7] R. Mahoney, Process for preparing microporous polyethylene hollow fibres, United States Patent Serial No. 4,115,492 (1978).

[8] H. Tanzawa, Semi-permeable membranes, their preparation and their use, United States Patent Serial No. 3,896,061 (1975).

[9] A.J. Castro, Methods for making microporous products, United States Patent Serial No. 4,247,498 (1981).

[10] G.H. Vitzthum, Davis MA. 0.1 micron rated polypropylene membrane and method for its preparation, United States Patent Serial No. 4,490,431 (1984).

[11] G.H. Shipman, Microporous sheet material, Method of making and articles made therewith, United States Patent Serial No. 4,539,256 (1985).

[12] J.S. Mrozinski, Microporous materials incorporating a nucleating agent, United States Patent Serial No. 4,726,989 (1988).

[13] J.S. Mrozinski, Multi-layer laminates of microporous films, United States Patent Serial No. 4,863,792 (1988).

[14] K.E. Kinzer, Oriented microporous films, United States Patent Serial No. 4,867,881 (1989).

[15] G.T. Caneba, D.S. Soong, Polymer membrane formation through the thermal-inversion process, Part I, Experimental study of membrane structure formation, Macromolecules 18 (1985) 2538–2545.

[16] G.T. Caneba, D.S. Soong, Polymer membrane formation through the thermal-inversion process, Part II, Mathematical modeling of the membrane structure formation, Macromolecules 18 (1985) 2545–2555.

[17] K. Schneider, T.J. van Gassel, Membrandestillation, Chem. Ing. Tech. 56 (1984) 514–521 (in German).

[18] T.J. van Gassel, K. Schneider, An energy efficient membrane distillation process, in: E. Drioli, M. Nakagaki (Eds.), Membranes and membrane processes, Plenum Press, London, 1986.

[19] G.C. Sarti, C. Gostoli, Use of hydrophobic membranes in thermal separation of liquid mixtures: Theory and experiments, in: E. Drioli, M. Nakagaki (Eds.), Membranes and membrane processes, Plenum Press, London, 1986, pp. 349–360.

[20] E. Drioli, Y. Wu, V. Calabro, Membrane distillation in the treatment of aqueous solutions, J. Membr. Sci. 33 (1987) 277–284.

[21] R.W. Schofield, A.G. Fane, C.J.D. Fell, Heat and mass transfer in membrane distillation, J. Membr. Sci. 33 (1987) 299–313.

[22] M. Khayet, M.P. Godino, J.I. Mengual, Theoretical and experimental studies on desalination using the sweeping gas membrane distillation method, Desalination 157 (2003) 297–305.

[23] K.W. Lawson, D.R. Lloyd, Membrane distillation, Part I, Module design and performance evaluation using vacuum membrane distillation, J. Membr. Sci. 120 (1996) 111–121.

[24] M. Tomaszewska, M. Gryta, A.W. Morawski, Study on the concentration of acids by membrane distillation, J. Membr. Sci. 102 (1995) 113–122.

[25] D.R. Lloyd, Microporous membrane formation via thermally-induced phase separation, Part I, Solid-liquid phase separation, J. Membr. Sci. 52 (1990) 239–261.

[26] D.R. Lloyd, S.S. Kim, K.E. Kinzer, Microporous membrane formation via thermally-induced phase separation, Part. II, Liquid-liquid phase separation, J. Membr. Sci. 64 (1991) 1–11.

[27] D.R. Lloyd, K.E. Kinzer, H.S. Tseng, Microporous membrane formation via thermally-induced phase separation, Part III, Effect of thermodynamic interactions on the structure of isotactic polypropylene membranes, J. Membr. Sci. 64 (1991) 13–29.

[28] L. Broens, F.W. Altena, C.A. Smolders, D.M. Koenhen, Asymmetric membrane structure as a result of phase separation phenomena, Desalination 32 (1980) 33–45.

[29] J.H. Aubert, Isotactic polystyrene phase diagrams and physical gelation, Macromolecules 21 (1988) 3468–3473.

[30] A. Laxminarayan, K.S. McGuire, S.S. Kim, D.R. Lloyd, Effect of initial composition, phase separation temperature and polymer crystallization on the formation of microcellular structures via thermally induced phase separation, Polymer 35 (1994) 3060–3068.

[31] D.R. Lloyd, K.E. Kinzer, H.S. Tseng, Microporous membrane formation via thermally-induced phase separation, Part V, Effect of diluent mobility and crystallization on the structure of isotactic polypropylene membranes, J. Membr. Sci. 64 (1991) 41–53.

[32] G.B.A. Lim, S.S. Kim, Q. Ye, Y.F. Wang, D.R. Lloyd, Microporous membrane formation via thermally-induced phase separation, Part IV, Effect of isotactic polypropylene crystallization kinetics on membrane structure, J. Membr. Sci. 64 (1991) 31–40.

REFERENCES

[33] P.J. Flory, Principles of polymer chemistry, Cornell University Press, Ithaca, NY, 1965.

[34] A.A. Alwattari, D.R. Lloyd, Microporous membrane formation via thermally-induced phase separation, Part VI, Effect of diluent morphology and relative crystallization kinetics on polypropylene membrane structure, J. Membr. Sci. 64 (1991) 55–68.

[35] J.M. Li, Z.K. Xu, Z.M. Liu, W.F. Yuan, H. Xiang, S.Y. Wang, Y.Y. Xu, Microporous polypropylene and polyethylene hollow fiber membranes, Part III, Experimental studies on membrane distillation for desalination, Desalination 155 (2003) 153–156.

[36] Z. Xu, J. Wang, L. Shen, D. Men, Y. Xu, Microporous polypropylene hollow fiber membrane. Part I. Surface modification by the graft polymerization of acrylic acid, J. Membr. Sci. 196 (2002) 221–229.

[37] L.Q. Shen, Z.K. Xu, Y.Y. Xu, Preparation and characterization of microporous polyethylene hollow fiber membranes, J. Appl. Polym. Sci. 84 (2002) 203–210.

[38] H. Tompa, Polymer solutions, Betterworths, London, 1956.

[39] L. Yilmaz, A.J. McHugh, Analysis of nonsolvent-solvent-polymer phase diagrams and their relevance to membrane formation modelling, J. Appl. Polym. Sci. 31 (1986) 997–1018.

[40] S.W. Song, J.M. Torkelson, Coarsening effects on the formation of microporous membranes produced via thermally induced phase separation of polystyrene-cyclohexanol solutions, J. Membr. Sci. 98 (1995) 209–222.

[41] W.R. Burghardt, Phase diagrams for binary systems exhibiting both crystallization and limited liquid-liquid miscibility, Macromolecules 22 (1989) 2482–2486.

[42] W. Yave, R. Quijada, D. Serafini, D.R. Lloyd, Effect of the polypropylene type on polymer-diluent phase diagrams and membrane structure in membranes formed via the TIPS process, Part I, Metallocene and Ziegler-Natta polypropylenes, J. Membr. Sci. 263 (2005) 146–153.

[43] W. Yave, R. Quijada, D. Serafini, D.R. Lloyd, Effect of the polypropylene type on polymer-diluent phase diagrams and membrane structure in membranes formed via the TIPS process, Part II, Syndiotactic and isotactic polypropylenes produced using metallocene catalysts, J. Membr. Sci. 263 (2005) 154–159.

[44] W. Yave, R. Quijada, M. Ulbricht, R. Benavente, Syndiotactic polypropylene as potential material for the preparation of porous membranes via thermally induced phase separation (TIPS) process, Polymer 46 (2005) 11582–11590.

[45] K.S. McGuire, A. Laxminarayan, D.R. Lloyd, A simple method of extrapolating the coexistence curve and predicting the melting point depression curve from cloud point data for polymer-diluent systems, Polymer 35 (1994) 4404–4407.

[46] M.R. Caplan, C.Y. Chiang, D.R. Lloyd, L.Y. Yen, Formation of microporous Teflon® PFA membranes via thermally induced phase separation, J. Membr. Sci. 130 (1997) 219–237.

[47] K.S. McGuire, D.S. Martula, A. Laxminarayan, D.R. Lloyd, Kinetics of droplet growth in liquid-liquid phase separation of polymer-diluent systems: model development, J. Colloid Interface Sci. 182 (1996) 46–58.

[48] C.Y. Chiang, D.R. Lloyd, Effects of process conditions on the formation of microporous membranes via solid-liquid thermally induced phase separation, J. Porous Mat. 2 (1996) 273–285.

[49] J. Arnauts, H. Berghmans, Amorphous thermoreversible gels of atactic polystyrene, Polym. Com. 28 (1987) 66–68.

[50] R.M. Hikmet, S. Callister, A. Keller, Thermoreversible gelation of atactic polystyrene: phase transformation and morphology, Polymer 29 (1988) 1378–1388.

[51] I.M. Lifshitz, V.V. Slyozov, The kinetics of precipitation from supersaturated solid solutions, J. Phys. Chem. Solids 19 (1961) 35–50.

[52] C. Wagner, Theorie der Alterung von Niederschlägen durch Umlösen (Oswald-Reifung), Z. Elektrochem. 65 (1961) 581–591.

[53] K. Binder, D. Stauffer, Theory for the slowing down of the relaxation and spinodal decomposition of binary mixtures, Phys. Rev. Lett. 33 (1974) 1006–1009.

[54] E.D. Siggia, Latte stage of spinodal decomposition in binary mixtures, Phys. Rev. A. 20 (1979) 595–605.

[55] I.G. Voigt-Martin, K.H. Leister, R. Rosenau, R. Konningsveld, Kinetics of phase separation in polymer blends for deep quenches, J. Polym. Sci., Part B: Polym. Phys. 24 (1986) 723–751.

[56] H. Furukawa, A dynamic scaling assumption for phase separation, Adv. Phys. 34 (1985) 703–750.

[57] S.W. Song, J.M. Torkelson, Coarsening effects on microstructure formation in isopycnic polymer solutions and membranes produced via thermally induced phase separation, Macromolecules 27 (1994) 6389–6397.

[58] J. Lal, R. Bansil, Light-scattering study of kinetics of spinodal decomposition in a polymer solution, Macromolecules 24 (1991) 290–297.

[59] Y.K. Lin, G. Chen, J. Yang, X.L. Wang, Formation of isotactic polypropylene membranes with bicontinuous structure and good strength via thermally induced phase separation method, Desalination 236 (2009) 8–15.

[60] H. Matsuyama, H. Okafuji, T. Maki, M. Teramoto, N. Kubota, Preparation of polyethylene hollow fibre membrane via thermally induced phase separation, J. Membr. Sci. 223 (2003) 119–126.

[61] H. Matsuyama, T. Maki, M. Teramoto, K. Asano, Effect of polypropylene molecular weight on porous membrane formation by thermally induced phase separation, J. Membr. Sci. 204 (2002) 323–328.

[62] P.M. Atkinson, D.R. Lloyd, Anisotropic flat sheet membrane formation via TIPS: atmospheric convection and polymer molecular weight effects, J. Membr. Sci. 175 (2000) 225–238.

[63] B.J. Cha, K. Char, J.J. Kim, S.S. Kim, C.K. Kim, The effects of diluent molecular weight on the structure of thermally-induced phase separation membrane, J. Membr. Sci. 108 (1995) 219–229.

[64] B. Luo, J. Zhang, X. Wang, Y. Zhou, J. Wen, Effects of nucleating agents and extractants on the structure of polypropylene microporous membranes via thermally induced phase separation, Desalination 192 (2006) 142–150.

[65] K.S. McGuire, D.R. Lloyd, G.B.A. Lim, Microporous membrane formation via thermally-induced phase separation, Part VII, Effect of dilution, cooling rate, and nucleating agent addition on morphology, J. Membr. Sci. 79 (1993) 27–34.

[66] H. Matsuyama, M.M. Kim, D.R. Lloyd, Effect of extraction and drying on the structure of microporous polyethylene membranes prepared via thermally induced phase separation, J. Membr. Sci. 204 (2002) 413–419.

[67] M. Gu, J. Zhang, X. Wang, H. Tao, L. Ge, Formation of poly(vinylidene fluoride) (PVDF) membranes via thermally induced phase separation, Desalination 192 (2006) 160–167.

[68] J.J. Kim, J.R. Hwang, U.Y. Kim, S.S. Kim, Operation parameters of melt spinning of polypropylene hollow fibre membranes, J. Membr. Sci. 108 (1995) 25–36.

[69] Y. Zhensheng, L. Pingli, C. Heying, W. Shichang, Effect of diluent on the morphology and performance of IPP hollow fibre microporous membrane via thermally induced phase separation, Chin. J. Chem. Eng. 14 (2006) 394–397.

[70] G.L. Ji, L.P. Zhu, B.K. Zhu, C.F. Zhang, Y.Y. Xu, Structure formation and characterization of PVDF hollow fibre membrane prepared via TIPS with diluent mixture, J. Membr. Sci. 319 (2008) 264–270.

[71] M. Shang, H. Matsuyama, M. Teramoto, D.R. Lloyd, N. Kubota, Preparation and membrane performance of poly(ethylene-co-vinyl alcohol) hollow fiber membrane via thermally induced phase separation, Polymer 44 (2003) 7441–7447.

[72] S. Berghmans, H. Berghmans, H.E.H. Meijer, Spinning of hollow porous fibers via the TIPS mechanism, J. Membr. Sci. 116 (1996) 171–189.

[73] H. Matsuyama, M. Yuasa, Y. Kitamura, M. Teramoto, D.R. Lloyd, Structure control of anisotropic and asymmetric polypropylene membrane prepared by thermally induced phase separation, J. Membr. Sci. 179 (2000) 91–100.

[74] B. Luo, Z. Li, J. Zhang, X. Wang, Formation of anisotropic microporous isotactic polypropylene (iPP) membrane via thermally induced phase separation, Desalination 233 (2008) 19–31.

[75] H. Matsuyama, S. Berghmans, D.R. Lloyd, Formation of anisotropic membranes via thermally induced phase separation, Polymer 40 (1999) 2289–2301.

[76] X. Li, Y. Wang, X. Lu, C. Xiao, Morphology changes of polyvinylidene fluoride membrane under different phase separation mechanisms, J. Membr. Sci. 320 (2008) 477–482.

[77] Y. Su, C. Chen, Y. Li, J. Li, PVDF membrane formation via thermally induced phase separation, J. Macromol. Sci. Part A: Pure & Appl. Chem. 44 (2007) 99–104.

[78] H. Sun, K.B. Rhee, T. Kitano, S.I. Mah, HDPE hollow fibre membrane via thermally induced phase separation, Part II, Factors affecting the water permeability of the membrane, J. Appl. Polym. Sci. 75 (2000) 1235–1242.

CHAPTER 6

Membrane Modification for MD Membrane Formation

OUTLINE

Introduction	122
Membrane Modification Methods	126
Porous Hydrophobic/Hydrophilic Composite Membranes: Principles of the Membrane Formation	126
Radiation Graft Polymerization	126
Plasma Polymerization	127
Grafting Ceramic Membranes	129
Surface Coating	130
Hydrophobic Solution Coating	131
Hydrophilic Solution Coating	133
Casting Hydrophobic Polymer over Porous Supports	134
Surface Modification by SMMs	136
SMM Synthesis	138
SMM Characterization	138
SMM Membrane Preparation	140
Co-Extrusion Spinning	141
Effects of Process Parameters on the Membrane Structure and MD Performance	**141**
Radiation Graft Polymerization	141
Plasma Polymerization	142
Grafting Ceramic Membranes	143
Surface Coating	144
Hydrophobic Surface Coating	144
Hydrophilic Solution Coating	145
Casting Hydrophobic Polymer over Porous Supports	146
Surface Modification by SMMs	148
Effect of Hydrophilic Polymer Concentration	149
Effect of Hydrophilic Polymer Type	151
Effect of SMMs Type	152
Effect of SMM Concentration	153
Effect of Solvent	154
Effect of SMM Stoichiometric Ratio	154
Effect of Evaporation Time	155
Co-Extrusion Spinning	158

INTRODUCTION

In general, asymmetric membranes can be grouped into four basic structures [1–4]:

(i) an integrally asymmetric membrane with a porous skin layer;
(ii) an integrally asymmetric membrane with a dense skin layer;
(iii) a thin-film porous composite membrane; and
(iv) a thin-film dense composite membrane.

The membranes included in the groups (i) and (ii) are made of one material, whereas the membranes included in groups (iii) and (iv) are made of at least two different materials and consist of a thin top layer over a porous support or a backing material, which provides mechanical strength to the whole membrane while the membrane performance is controlled mainly by the top thin layer. Multi-layered composite membranes consisting of a porous support and several layers of different materials belong to the two last groups. The porosity, pore size and membrane material of thin-film composite membranes change through their cross-section. Because the selective layer (i.e. the top layer for flat sheet membranes and the inner and/or outer layer for hollow fibre and capillary membranes) is very thin, membrane permeate fluxes are high. This type of membrane was originally developed for reverse osmosis (RO) applications [4]. Nowadays, thin-film composite membranes are used in nanofiltration (NF), pervaporation (PV), gas separation and membrane distillation (MD) among others [5–9].

In general, the membrane top skin layer governs the performance of a separation process. The surface deposition of contaminants from feed solutions is also affected by the surface properties of the membrane leading to reduction of the membrane performance, especially in a long-term operation. In the membrane literature, to improve transport properties of polymeric membranes as well as their chemical resistance (i.e. solvent resistance, swelling, fouling resistance, etc.), various membrane modification techniques have been developed: physical, chemical or bulk modification (i.e. polymer blends). There are a number of methods to fabricate composite membranes with a dense or porous thin film, such as solution coating or dip coating, interfacial polymerization, graft polymerization, lamination, plasma polymerization or plasma deposition and use of surface-modifying macromolecules (SMMs) [3,7–10]. It is worth quoting that the first reported membrane modification method involved annealing of porous membranes by heat treatment [3].

One of the advantages of the asymmetric membranes grouped above in (iii) and (iv) is that the thin film is quite independent from the porous support. This allows to optimize and select an adequate material for the thin film as well as to control its thickness. Expensive materials can be used because only a small amount of material is necessary to form a layer whose thickness is as small as 1 μm. In fact, the goal of the preparation of composite membranes is to make the selective layer that governs the separation as thin as possible, while maintaining them in a defect-free form. The selective layer can be dense, as in RO or gas separation membranes, or porous with pore sizes in a range from 10 nm to 50 nm, as in ultrafiltration (UF) membranes, or with larger pore sizes as in MD. In order to achieve a high permeate flux, a thickness as low as 50–100 nm is required, but when the membrane thickness becomes so small, the problem of defect formation arises. To solve this problem, coating of a second thin layer of a highly permeable polymer such as silicone rubber was applied to seal the undesirable defects [6,11]. Heat treatment or solvent swelling techniques were applied for elimination of micro-defects in the thin separating layer [3].

It must be pointed out that when employing porous supports for preparation of either porous or dense thin-film composite membranes, the supports must be chemically resistant against the solvent or solvent mixture from which the thin layer is formed and should have a high surface porosity as well as optimum pore size. In fact, the support should not provide any significant resistance to mass transport during the separation process.

It is to be noted that the surface-modified membranes for MD applications have so far been developed in a laboratory scale by radiation graft polymerization [12], plasma polymerization [12–16], grafting ceramic membranes [17,18], hydrophobic or hydrophilic surface coating [19–23] or casting hydrophobic polymer over flat sheet or porous fibres as supports [24,25], use of SMMs [9,26–33] and co-extrusion spinning [34]. These methods are reviewed briefly in section 6.2.

The use of composite membranes in MD was described first by Cheng and Wiersma in a series of U.S. patents in late 1980s [35–38]. They claimed the use of composite membranes consisting of a thin porous hydrophobic layer and one or two thin hydrophilic layers. In the last case, a hydrophobic layer is sandwiched between two thin hydrophilic layers, which may be made of different materials such as cellulose acetate (CA), mixed esters of cellulose, polysulfone (PS) or polyallylamine. It was also stated that the composite membrane could be formed by coating the hydrophilic layer(s) on the hydrophobic layer. The composite membrane may be a microporous hydrophobic layer covered with a non-porous hydrophilic coating material. Plasma-polymerized allylamine, dehydrated polyvinyl alcohol (PAV) and polyacrylic acid were proposed for the hydrophilic layer. Fluoro-substituted vinyl polymers were considered as ideal materials for the microporous hydrophobic layer. More details are outlined in section 1.3. Since 1980s only few studies have been performed using their patented concept.

Rodgers [39,40] indicated in a couple of U.S. patents on direct contact membrane distillation (DCMD) that the suitable materials for the membranes were those which permitted the formation of microporous membranes having high porosity (i.e. 70–80%) and uniform pore size distribution. The cited polymers were polycarbonates, polyesters, polyethylene, polypropylene (PP), polyvinylidene fluoride and the halogenated polyethylenes, particularly the fluorocarbons. Moreover, it was mentioned that microscopic hydrophilic porous filters made of cellulose nitrate (CN), CA and cellulose triacetate could also be used for MD after coating with a silicone water repellent material to render them non-wettable porous membranes.

In 1991, Ohta et al. [41] reported on the application of a partially hydrophilic dense fluoro-carbon composite membrane for seawater desalination. DCMD configuration was used and the obtained permeate fluxes were of magnitude similar to those achieved with porous hydrophobic membranes ($<6 \, kg/m^2 \cdot h$). The effects of the DCMD operation parameters were similar to those observed for a single porous hydrophobic layer. The authors used the term MD for both porous and dense membranes. It was also found that the permeability and thermal efficiency of fluoro-carbon membrane were superior to those of silicon membrane (i.e. silicone polymer and PS composite membrane, hydrophilic, dense and semipermeable) [42]. However, no details were given on the transport mechanism through this type of membrane.

One year later, Kong et al. [13] employed a hydrophilic microporous CN membrane surface modified via plasma polymerization of octafluoro-cyclobutane (OFCB). A tri-layer membrane with a hydrophilic layer sandwiched between two hydrophobic layers was prepared and tested for desalination by DCMD. The effects of various polymerization conditions on the DCMD performance and the structure of the membrane were investigated. The same group, Wu et al. [12], proposed other composite membranes for MD prepared via radiation graft polymerization of styrene on CA to achieve the required hydrophobicity and CN membrane modified by plasma polymerization using two types of monomer systems, vinyltrimethylsilicon (VTMS)/carbon tetrafluoride (CF_4) and OFCB.

In the same year, polymer-coated membranes and heat-treated membranes were proposed by Fujii et al. [19,20]. Silicone rubber (Si-LTV, Dow Corning Toray Silicone Co. Ltd.), poly(1-trimethylsilyl-1-propyne) (PMSP) and polyketone (Honshu Kagaku Co. Ltd.) were used as the coating materials of PVDF hollow fibre membranes. The coating was carried out from the bore side of the hollow fibre. The membranes were employed for the treatment of organic (ethanol, acetone, acetonitrile, n-butanol) aqueous solutions by DCMD.

Recently, Khayet et al. [9,26] proposed a new type of porous composite hydrophobic/hydrophilic flat sheet membranes for DCMD application. The feed solution was brought into contact with the hydrophobic side of the membrane, whereas the permeate cold liquid was circulated in contact with the hydrophilic layer. The membranes were prepared by the simple phase inversion method using fluorinated SMMs. These SMMs are oligomeric fluoropolymers synthesized by polyurethane chemistry and tailored with fluorinated end groups. The membranes were prepared in a single casting step from a blend dope containing a hydrophilic polymer such as polyetherimide (PEI) and an SMM. During the membrane formation, the SMM migrates towards the top air/polymer interface, rendering the membrane surface hydrophobic. It was observed that these membranes were promising for desalination by DCMD. During the past few years, polyethersulfone (PES) and PS membranes were further modified by the collaboration of Khayet and Matsuura groups (Department of Applied Physics I, Faculty of Physics, University Complutense of Madrid, Spain & Industrial Membrane Research Laboratory, University of Ottawa, Canada) using different types of SMMs, different solvents, different additives and different membrane preparation conditions in order to optimize the MD performance of this type of composite hydrophobic/hydrophilic membranes [28–33].

Composite hydrophobic/hydrophilic hollow fibre membranes were also proposed by Bonyadi and Chung [34] for desalination by DCMD configuration and Jin et al. [21] for desalination by vacuum membrane distillation (VMD) configuration. The membranes fabricated by Bonyadi and Chung [34] were prepared by co-extrusion dry/jet wet spinning method. The outer hydrophobic layer was prepared from a 12.5/87.5 wt/wt PVDF/N-methyl pyrrolidione (NMP) solution containing 30 wt% hydrophobic cloisite 15A, whereas the inner layer was formed by a solution of lower polymer concentration, 8.5/4/87.5 wt/wt PVDF/polyacrylonitrile (PAN)/NMP containing 50 wt% hydrophilic cloisite NA^+. The clay particles were used to modify the hydrophobicity of the PVDF hollow fibre membrane matrix. Methanol/water (i.e. 80/20 wt/wt) was used as internal and external coagulants because it is a weak coagulant and can induce a high porous structure on PVDF membranes with high surface and bulk porosities, large pore size and narrow pore size distribution. However, the membranes had weak mechanical properties. A permeate flux as high as 55.2 kg/m^2·h was reached with 99.8% NaCl separation using a 3.5 wt% NaCl aqueous solution, a feed temperature of 90.3 °C and a permeate temperature of 16.5 °C. The hydrophobic/hydrophilic membranes prepared by Jin et al. [21] were poly(phthalazinone ether sulfone ketone) (PPESK) hollow fibre composite membranes coated by silicone rubber and sol-gel polytrifluoropropylsiloxane. First, the PPESK hollow fibre UF membranes were prepared by the dry/jet wet spinning technique, and then their internal surfaces were coated. The effects of coating conditions (i.e. coating time, coating temperature and the concentration of silicone rubber) on the VMD performance were studied and the membrane stability was evaluated based on long-term experiments (up to 14 days). Lower permeate fluxes than those reported by Bonyadi and Chung [34] were obtained, but their VMD performance was stable in a long-term experiment.

In 2004, Li and Sirkar [14] reported on novel hollow fibre membranes and modules for use in DCMD and a year later for VMD configuration [15]. Their novel membranes were commercial porous PP hollow fibres (Accurel MEMBRANA, Wuppertal, Germany) exhibiting different dimensions and thicknesses. The external surfaces of these hollow fibre membranes were coated via plasma polymerization using various ultrathin microporous silicone-fluoropolymers (fluorosilicone). The reason for applying the coating layer was to provide an additional porous layer having higher hydrophobicity than PP, which itself is one of the polymeric materials with very low surface energy.

Larbot et al. [17] proposed ceramic tubular fibre membranes grafted by fluoroalkylsilanes (FASs) for water desalination by DCMD. The ceramic membranes were alumina (Al_2O_3) and zirconia (ZrO_2) of pore sizes 200 nm and 50 nm. The water contact angles of the grafted membranes were higher than the original ceramic membranes, indicating that the grafted membranes were more hydrophobic. Lower DCMD fluxes (~95 $l/m^2 \cdot day$) and about 100% salt rejection factors were observed for the grafted ZrO_2 membrane having smaller pore size (i.e. 50 nm). By contrast, higher fluxes (~202 $l/m^2 \cdot day$) with lower rejection factors (99.5%) were obtained for the grafted ZrO_2 membrane having higher pore size (i.e. 200 nm). It was concluded that graft ceramic hollow fibre membranes would be of great interest in MD. A similar type of membranes was patented previously in 2002 by Belleville et al. [43] and proposed for osmotic evaporation process. A year later, the same group reported on their study [44]. The membranes consisted of tubular porous alumina, provided by Pall Corporation (Exekia Division) with mean pore sizes of 0.2 μm and 0.8 μm, which were grafted by siloxane compounds. It was reported that the obtained permeate flux was independent of the membrane pore size, concluding that a limiting water vapour diffusion occurred through the membrane pores.

Composite hydrophilic/hydrophobic membrane was proposed recently for MD to prevent pore wetting of the hydrophobic layer by the presence of a hydrophilic layer. In this case, the feed solution was brought into contact with the hydrophilic side of the composite membrane. In fact, one of the limitations of the MD process is the risk of hydrophobic membrane pore wetting, which reduces both the permeate flux and the separation performance of the membrane. In 2004, Xu et al. [22] proposed the coating of hydrophobic porous membranes with hydrophilic sodium alginate hydrogel. A microporous polytetrafluoroethylene (PTFE) membrane (Poreflon 020-40, Sumitomo Electric Fine Polymer, Osaka, Japan) having 0.2 μm pore size was alginate coated and then cross-linked by a water-soluble carbodiimide (WSC). The observed reduction of the overall mass transfer coefficient due to coating was less than 5% and the coated membranes were resistant to wetting for at least 300 min when exposed to orange-—oil—water mixture. A year later, Peng et al. [23] cast a dense hydrophilic polymer solution on porous PVDF membrane (GVHP, Millipore see Table 2.1). The polymer solution was a blend of PVA and polyethylene glycol (PEG) cross-linked by aldehydes and sodium acetate. The composite hydrophilic/hydrophobic membranes were tested for desalination by DCMD and it was observed that the behaviour of the coated membranes was similar to that of the uncoated membrane. Separation factors of more than 99% were achieved with permeate fluxes only 9% lower for the coated membranes compared to the corresponding uncoated one, which was 23.7 $kg/m^2 \cdot h$ at 70 °C feed temperature and 22 °C permeate temperature. For these membranes, it was speculated that the feed liquid first dissolved in the hydrophilic layer, followed by diffusion through a continuous pathway along the polymer chain and finally evaporated at the interface between the hydrophobic and hydrophilic layers. It was concluded that the hydrophilic layer could prevent wetting of the hydrophobic membrane pores.

It is interesting to note that compared with the modified membranes used in other membrane separation processes, the studies on membrane surface modification for MD applications have not been extensive and mature yet. More theoretical (heat and mass transport through modified membranes in MD) and experimental work in the field of membrane bulk and surface modification for MD is required. This would certainly expand the material resource for MD membranes and bring about a great advance in the development of MD process.

MEMBRANE MODIFICATION METHODS

Porous Hydrophobic/Hydrophilic Composite Membranes: Principles of the Membrane Formation

In general, there are various surface and bulk membrane modification methods applied to improve the performance of both dense and porous membranes as well as flat sheet and hollow fibre membranes used in different membrane separation processes such as PV, RO, NF, UF and gas separation. For example, interfacial polymerization consists of deposition of a thin selective layer on top of a porous substrate membrane by interfacial *in situ* polycondensation. A large number of modification procedures were carried out depending on the monomer type [5]. The first interfacially polymerized thin-film composite membranes were developed by Cadotte and co-workers of Film Tech in 1970s, which made a breakthrough in membrane performance for RO applications [5]. This technique is used mostly for preparation of high-performance RO and NF membranes [5,45]. In what follows in this section, only those membrane modification techniques used to develop membranes for MD are included. Special attention will be paid to porous hydrophobic/hydrophilic composite membranes.

Radiation Graft Polymerization

Radiation-induced grafting is an interesting method that has been used for the past 50 years to prepare a variety of separation membranes most noticeably for dialysis, electrodialysis, RO and electrolysis processes [47–52]. A variety of grafting techniques that include chemical grafting [46], photografting [47,48], plasma grafting [49], thermal grafting [50] and radiation grafting with γ-rays have been proposed [51,52].

An example of grafting procedure, using styrene monomer, is as follows. The membrane support is cut into pieces of known weight and size, washed with acetone and then dried in a vacuum oven at 60 °C for 1 h. The clean film is then placed into a glass ampoule and styrene monomer of known concentration diluted with dichloromethane is added. The styrene concentration can be varied from 20 vol.% to 100 vol.%. The grafting mixture is subsequently flushed with purified nitrogen for 8 min to remove the air and then the ampoule is tightly sealed. The ampoule is subjected to γ-rays from a Co^{60} source for 15 h at dose rate of 1.32 kGy/h. After the reaction is completed, the grafted film is removed from the ampoule, washed with toluene and soaked therein overnight to remove the residual monomer and the homopolymer occluded in the film surfaces. The grafted membrane support is finally dried under vacuum at 80 °C and weighed. The process of soaking and drying can be repeated few times until a constant weight is obtained to ensure the complete removal of the homopolymer.

In general, the surface of a porous substrate membrane is irradiated with γ-rays, which causes the generation of radicals on the membrane surface. Then, the membrane is immersed in a monomer solution. The graft polymerization of the monomers is initiated at the membrane surface. By choosing a very hydrophobic monomer, the hydrophobicity of the surface will increase considerably.

Wu et al. [12] applied radiation graft polymerization of styrene on microporous flat sheet CA membrane to prepare polystyrene-grafted modified membrane for MD. Polystyrene is a hydrophobic material having a water contact angle around 91°. The CA membrane was prepared from a casting solution by the phase inversion technique (see Chapter 3) with a thickness of 150 µm, a mean pore size of 0.15 µm and a porosity of 78%. First, the grafting solution was prepared using the monomer styrene (St), the swelling agent pyridine (Pyd) and the chain transfer agent carbon tetrachloride (CCl_4). Subsequently, the CA membrane was immersed in the grafting solution and the sample was then irradiated by Co^{60} with a dose rate of 0.91×10^4 rad/h for a certain time under nitrogen atmosphere. Polymerization of styrene occurred at the membrane, forming polystyrene. Finally, the membrane was washed with alcohol and dried at room temperature.

The degree of grafting can be gravimetrically determined as the percentage of the weight increase using Eq. (1):

$$\text{Degree of grafting (\%)} = (W_g - W_o)/W_o \times 100 \quad (6.1)$$

where W_g and W_o are the weights of grafted and un-grafted membranes, respectively.

The amount of polystyrene grafted onto the membrane surface can be calculated as a grafting yield (GY):

$$GY = (W_a - W_b)/A \quad (6.2)$$

where W_a and W_b represent the weight of the membrane before and after grafting, respectively, and A is the area of the membrane.

Plasma Polymerization

Plasma polymerization is a process in which organic monomers are split and decomposed into various active particles under glow discharge. The active particles are then recombined to form polymers on the surface of the substrate. Various studies have been carried out on the preparation of selective membranes with high performance for different applications using this method, although it was initially used to form electrical insulation and protective coatings [53–55]. In plasma polymerization, all volatile organic molecules, even those without functional groups, can be reacted to form a polymer [56].

Figure 6.1 shows a schema of a prototype system used to perform plasma polymerization. When vacuum is maintained inside the tubular reactor and a high-frequency electric field is applied outside, a glow discharge is generated inside the reactor. Plasma may consist of various ions, radicals, electrons or molecules and is formed in the glow discharge. A typical plasma may be initiated using helium, argon, or another inert gas under an adequate pressure. When a porous substrate membrane is placed in the plasma field, its surface will be changed depending on the plasma properties. The membrane surface can be etched and/or chemically active sites can be formed. Upon contact with organic compounds, an irregular polymerization can occur therein [2]. Monomer polymerization proceeds by a complex mechanism involving ionized molecules and radicals and is completely different from conventional polymerization reactions. It is worth quoting that plasma polymerization is a complex chemical reaction process of various active particles. Both deposition on the surface of the membrane caused by polymerization and ablation from the

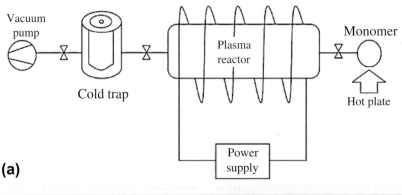

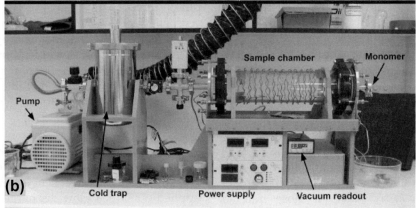

FIGURE 6.1 (a) Schematic diagram of a system used to perform plasma polymerization and (b) photo of plasma reactor. *Source: Reprinted from [10]. Copyright 2011, with kind permission from Elsevier.*

surface of the membrane caused by the bombardment with particles may take place. The vapour pressure of the monomers, the power and voltage used in the discharge reaction, and the type and temperature of the substrate all affect the polymerization reaction.

Wu et al. [12] applied plasma polymerization to a commercial porous hydrophilic CN substrate in order to prepare membranes for MD. The thickness of the flat sheet CN membrane was 100 µm, its nominal pore size was 0.45 µm and its porosity was 80%. The plasma polymerization was carried out in a capacitance coupling reaction system with external electrodes and the monomer was placed between those electrodes. A 7 cm diameter of the CN membrane was placed under the monomer inlet. The applied discharge frequency to the plasma chamber was 13.6 MHz. The monomer vapour was introduced in the reaction system after the reactor was evacuated to a pressure of 0.39 Pa. Two types of monomer systems were used. Those were VTMS/CF_4 and OFCB. In each case, the pressure was adjusted, i.e. 12 Pa for OFCB monomer, 3.9 for VTMS and 15.6 Pa for CF_4. Finally, glow discharge was initiated. The effects of discharge power and discharge time on the properties of the modified membranes as well as on their MD performance were studied as will be shown in section 6.3. It was found that the plasma-modified membranes presented good MD permeability. The OFCB plasma-modified CN membrane exhibited DCMD permeate fluxes ranging between 12.0 and 32 kg/m^2·h and the NaCl rejection factor was found to be between 89.6% and 92.1% when the feed solution was 0.5 M NaCl solution, the feed temperature was 70 °C and the permeate temperature was 25 °C. The VTMS/CF$_4$ plasma-modified CN membrane exhibited higher DCMD permeate flux (37.2–84.5 kg/m^2·h with a rejection factor of 81.0–96.4% when the feed solution was 0.5 M NaCl solution, the feed temperature was 70 °C and the permeate temperature was 25 °C). In general, it was found that the DCMD performance of the plasma-modified CN membrane made by the monomer system VTMS/CF$_4$ was better than that of the monomer system OFCB and the DCMD permeate flux of both membranes decreased, while the rejection factor increased when the discharge time was getting longer. This was attributed to the gradual decrease of the pore size with increasing discharge time until a dense layer was formed on the membrane surfaces.

The same authors, Kong et al. [13], employed plasma polymerization using the monomer system OFCB to prepare a tri-layer flat sheet membrane for MD consisting of a hydrophilic microporous CN membrane sandwiched between two hydrophobic layers. The membranes were tested in DCMD employing 0.3–0.5 M NaCl feed aqueous solutions. The effects of various polymerization conditions on the DCMD performance and structures of the membranes were investigated. It was observed that the tri-layer membrane exhibited similar DCMD behaviour to the typical single-layer porous hydrophobic membranes commonly used in MD.

It should be pointed out that both flat sheet and hollow fibre membranes can be modified by plasma polymerization. Hollow fibre membranes were adopted by Li and Sirkar [14–16]. The external surface of the commercial porous PP hollow fibres (Accurel MEMBRANA, Wuppertal, Germany; Table 2.2) of different dimensions and thicknesses were coated with a variety of microporous plasma-polymerized silicone-fluoropolymer (fluorosilicone). PP is a hydrophobic material. Therefore, the reason for using plasma polymerization is to provide an additional porous layer having higher hydrophobicity than PP. Details on plasma polymerization of PP hollow fibre membranes were not reported in [14–16]. It was reported that the DCMD permeate flux of the plasma-polymerized hollow fibre PP membranes was remarkably high (41–79 kg/m^2·h) with a concentration of NaCl in the permeate less than 8 ppm when 1 wt% NaCl aqueous solution was used as feed at a temperature ranging from 85 °C to 90 °C and a permeate temperature of 15–17 °C. No membrane pore wetting was detected over 400 h DCMD experiments. When these modified hollow fibre membranes were used in VMD desalination process, permeate fluxes as high as 71 kg/m^2·h were obtained for a feed temperature of 85 °C and a vacuum pressure of 60–66 mmHg.

The results observed in the above-cited studies indicated that plasma-polymerized membranes with good performance in MD (i.e. high permeate flux) could be obtained. In fact, plasma polymerization, in which many different kinds of monomers can be polymerized onto the surface of porous materials, has become an efficient method to prepare hydrophobic porous membranes of high performance from hydrophilic membranes. This presents a bright future for MD in practical applications.

Grafting Ceramic Membranes

Ceramic membranes are commonly made from metal oxides such as alumina, zirconia, titania (TiO$_2$), silica (SiO$_2$) or a combination of these materials and exhibit high chemical and thermal stabilities [57]. Therefore, this type of membranes are more appropriate for modification as compared to polymeric membranes. However, due to the presence of hydroxyl groups (−OH) at their surface, ceramic membranes show a hydrophilic behaviour. It is worth quoting that in the field of ceramic membranes, surface modification is generally used to reduce the pore size and to render the membrane hydrophobic [17]. Several studies have been carried out on the surface modification of different ceramic membranes and FASs are often used for modification [58−63]. In fact, FASs are the group of compounds that efficiently enhance the hydrophobic character of different hydrophilic surfaces. As shown in Fig. 6.2, grafting process can be performed by reaction between hydroxyl groups (−OH) at the membrane support and ethoxy groups (O−Et) presented in organosilane compounds [63]. In this case, the process leads to the formation of a monomolecular layer of organosilane compound at the surface of the membrane support. Thus, the hydrophobic character of the membrane support can be enhanced by using organosilane compounds containing hydrophobic fluorocarbon chains (FAS). This grafting process can be applied to both flat sheet and hollow fibre, capillary or tubular membrane supports.

Larbot et al. [17] and Krajewski et al. [18] used ceramic tubular membranes grafted by FASs for water desalination by DCMD and air gap membrane distillation (AGMD), respectively. The ceramic membranes were alumina and zirconia of pore sizes 200 nm and 50 nm, purchased from Pall Exekia. The grafting procedure consists of immersing the sample into the FAS solution for different periods of time. The concentration of silane solution in chloroform can be varied. About 10^{-2} mol/l of 1H,1H,2H,2H-perfluorodecyltriethoxysilane, $C_8F_{17}C_2H_4Si(OEt)_3$ (C8) in chloroform, was used as grafting solution [17,18]. This solution must be prepared under argon atmosphere in order to avoid polycondensation of C8 in presence of water vapour. Other compounds with shorter fluoro-alkyl chains such as $C_6F_{13}C_2H_4Si(OMe)_3$ (C6) and $CF_3C_2H_4Si(OEt)_3$ (C1) were also used. The grafting solution was kept at room temperature. Successive soakings of the sample in the grafting solution followed by drying at 100 °C for 12 h between each soaking were carried out. The grafting rate can be controlled by choosing the appropriate grafting time, number of soaking and drying time. The effects of these parameters on the characteristics of MD ceramic membranes were studied and some of the obtained results are presented in section 6.3. In general, it was observed that the measured water contact angles of the membranes were higher after grafting, indicating that the grafted membranes were more hydrophobic.

Surface Coating

This method is also named solution coating or dip coating. A polymer solution, which in most cases dilute, is deposited directly onto the surface of a porous support (i.e. substrate

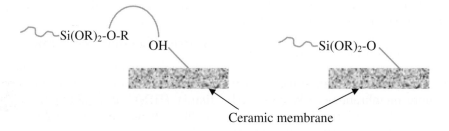

FIGURE 6.2 Schematic representation of grafting ceramic membranes.

or backing material). This can be a flat sheet or a hollow fibre support. Before coating, the porous substrate is cleaned and dried. Then, the surface is dipped into a bath containing a dilute polymer solution. After the membrane is taken out of the bath, the solvent is removed by evaporation, leaving a thin layer of the latter polymer on top of the substrate membrane. Because the coating layer is very thin reaching values of less than 2 µm, it is difficult to prepare membranes without defects [3]. Moreover, penetration of the coating polymer solution into the pores of the substrate membrane may occur. Ultrahigh-molecular-weight polymers were considered for the formation of the coating layer [64]. Surface coating may be carried out by a simple casting of the polymer solution over a backing material, which was made by the phase inversion technique detailed in Chapter 3.

Different types of surface coating solutions have been considered for the fabrication of MD membranes.

Hydrophobic Solution Coating

Coated PVDF hollow fibre porous membranes were proposed for DCMD process by Fujii et al. [20]. Silicone rubber (low-temperature vulcanization, Si-LTV, Toray-Dow-Corning Silicone Co. Ltd.), PMSP and polyketone (Honshu Kagaku Co. Ltd.) were used as coating materials. Coating was carried out at the internal surface of the PVDF hollow fibre membranes. The coated hollow fibre membranes were applied for the treatment by DCMD of organic/aqueous solutions (i.e. ethanol, acetone, acetonitrile and n-butanol). It is worth quoting that PVDF is a hydrophobic polymer; however, the PVDF hollow fibre membranes were coated with silicone rubber to render their hydrophobicity higher to be used for organic/aqueous solutions preventing wetting of the membrane pores.

To prepare this type of membranes, PVDF hollow fibre membranes were first spun applying the spinning procedure presented and detailed in Chapter 4. After coagulation, the spun PVDF hollow fibre membranes were washed in water, soaked in 70 wt% glycerol aqueous solution, before being stored. Dried hollow fibre membranes were prepared as follows. They were washed in water to remove glycerol and then soaked in methanol. Methanol was further replaced by n-hexane using the solvent exchange method. Finally, the hollow fibre membranes were dried in air at room temperature.

Silicone-coated PVDF hollow fibre membranes were prepared as follows [20]. The coating solution was prepared by dissolving 2 wt% Si-LTV in Freon 113. The silicone prepolymer mixture contains poly(dimethylsiloxane), triacetoxymethylsilane and diacetoxydibutyltin. The hollow fibres were inserted into a glass tube, which was sealed at both ends with epoxy resin. The hollow fibres were left in contact with methanol for 1 h. Then methanol was removed by blowing nitrogen gas for few seconds. The silicone prepolymer solution was introduced using a syringe through the bore side and drained after several seconds. Subsequently, the coated hollow fibre membranes were dried at room temperature by permeating nitrogen gas through the membrane from the inner to the outer side of the hollow fibre membranes for 1 h. Finally the hollow fibres were cured at 70 °C for 4 h, dried under vacuum at 35 °C for 12 h and removed by cutting open the glass tube.

The PMSP- and polyketone-coated hollow fibre membranes were prepared in a similar way as for silicone coating. To carry out the PMSP coating, a dope solution containing 1 wt% polymer in cyclohexane was used. In this case just before coating, methanol soaking was not applied. For polyketone coating, the solutions were prepared by dissolving polyketone A (acetophenone-type polymer, Halon 80) in toluene (1 wt% coating solution) and polyketone B (hydrogenated polymer of polyketone A, Halon 110) was dissolved in ethanol (1 wt% coating solution).

It was observed that ethanol permeability through both silicone-coated and uncoated PVDF hollow fibre membranes was similar but that of water decreased remarkably after coating treatment [20]. Thus the ethanol selectivity for the coated membranes was higher and in many cases exceeded the vapor/liquid equilibrium. The PMSP-coated PVDF membranes yielded lower ethanol and water permeability; however, the decrease in the water permeability was much greater and thus the selectivity was increased. Similar results were obtained with the polyketone-coated PVDF hollow fibre membranes. Moreover, when using acetone, acetonitrile and n-butanol aqueous solutions as feed, the selectivity of the same coated PVDF hollow fibre membranes was found to be higher than their relative volatility. This was attributed to the decrease of the water permeate flux. On the other hand, the heat treatment of the PVDF membrane had the effect of increasing the hydrophobicity of the membrane. The contact angle increased from 94.3° to 102° after heat treatment at 80 °C for 89 h. Therefore, water permeate flux was observed to decrease, while the ethanol permeate flux was not varied and therefore the selectivity was increased and became higher than the relative volatility. These results were explained by the fact that ethanol molecule has a hydrophobic ethyl group and a hydrophilic OH group, and a change of hydrophobicity has little effect on the diffusion of this molecule as the effects on both groups may be compensated. By contrast, water molecule has only hydrophilic OH groups and a change of hydrophobicity has stronger effects on the sorption parameter, while the structural changes have less effect because of its small molecular size.

Hydrophobic coating was also applied at the internal surface of a hydrophilic hollow fibre surface, PPESK, for preparation of hydrophobic/hydrophilic membranes for desalination by VMD [21]. The coated material was silicone rubber (i.e. middle-temperature vulcanizing addition-type silicone rubber, Kehua new Material Co., Beijing, China) and sol-gel polytrifluoropropylsiloxane (i.e. hydroxyl-terminated polytrifluoropropylsilxane oil, Shanghai 3F New Materials Co., Ltd.). First, the PPESK hollow fibre UF membranes were prepared by the dry/jet wet spinning technique (see Chapter 4). PPESK was purchased from Dalian Polymer New Material Co. Ltd. (Liaoning, China), the solvent used was dimethyl acetamide (DMAC) and the additive was diethylene glycol (DEG).

To prepare the coated PPESK hollow fibre membrane with silicone rubber, 5–20 g/l of silicone rubber was dissolved in petroleum ether and the obtained solution was circulated through the lumen side of the PPESK hollow fibres membranes. This last procedure was repeated several times. The coated membranes were then heated at different temperatures (40, 60 and 100 °C) for different heating periods of time.

PPESK hollow fibre membrane coated with sol-gel polytrifluoropropylsiloxane was prepared as follows. A solution containing trifluoroacetic acid (TFA), water and methyltrimethoxysilane (MTMS) was prepared so that the concentration of water in TFA is 5% and 400 µl of TFA/water was mixed with 600 µl of MTMS. After polymerization of MTMS was taken place, 600 µl of polytrifluoropropylsiloxane oil was added and vortexed quickly and the whole mixture was centrifuged at 3000 rpm for 5 min. The polymerization time was up to 30 min. A top clear solution was formed and this was used for hollow fibre coating. It was circulated through the inner side of the PPESK hollow fibre membrane several times. Then the coated membrane was dried at 25 °C. It was observed that the MTMS prepolymerization before adding polytrifluoropropylsiloxane oil could significantly affect gelation time and therefore the morphology of the coating layer. Long prepolymerization time resulted in a rapid gelation on membrane support, reducing the risk of membrane pores blockage. However, excessive prepolymerization time induced an uneven coated layer.

It must be pointed out that preparation of composite membranes by sol-gel method as the one shown above has two advantages: (i) sol-gel reaction occurs at low temperature, 25 °C, avoiding heating and possible membrane shrinkage and (ii) no solvent is needed, avoiding membrane support damage.

The effects of coating conditions (i.e. coating time, coating temperature, concentration of silicone rubber, prepolymerization time of MTMS) on the MD performance are shown in section 6.3.

It was found that the higher VMD permeate flux was obtained at the lower coating temperature and lower concentration of silicone rubber in the coating solution. A permeate VMD flux as high as $3.5 \, l/m^2 \cdot h$ with a salt rejection factor of 99% was reached using a feed NaCl aqueous solution having a concentration of $5 \, g/l$, a feed temperature of 40 °C, a downstream pressure of 0.078 MPa and employing the coated hollow fibre with 60 °C coating temperature, 9 h coating time and $5 \, g/l$ silicone rubber. Moreover, the studied composite hollow fibre membrane showed stable VMD performance in a long-term experiment. In the case of polytrifluoropropylsiloxane-coated hollow fibre membranes, the highest permeate flux of $3.7 \, l/m^2 \cdot h$ with a NaCl rejection factor of 94.6% was observed for 30 min prepolymerization time. It was proved that the prepolymerization time was an important factor, affecting VMD performance of this type of membranes. Higher prepolymerization time increased the permeability of the membrane [21].

Hydrophilic Solution Coating

One of the limitations of the MD process is the risk of hydrophobic membrane pore wetting, which reduces both the permeate flux and the solute rejection. Composite hydrophilic/hydrophobic membranes were proposed to prevent pore wetting of the hydrophobic layer by coating a hydrophilic layer over the hydrophobic layer [22,23]. In this case, the feed solution to be treated by MD is brought into contact with the hydrophilic layer of the composite membrane.

Xu et al. [22] used sodium alginate hydrogel (Manugel GMB) as a hydrophilic coating solution. A microporous PTFE membrane (Poreflon 20–40, Sumitomo Electric Fine Polymer, Osaka, Japan) of pore size 0.2 μm was alginate coated and then cross-linked by a WSC (Sigma Aldrich) or the cross-linking agent 1-ethyl-3-(3-dimethylaminopropyl) carbodiimide.

The system used for membrane coating is presented in Fig. 6.3. It is a similar set-up used for osmotic distillation (OD). Coating was carried out at about 23 °C. First, the lower compartment was filled with a calcium chloride ($CaCl_2$) aqueous/ethanol solution (25 wt% $CaCl_2$, 10 wt% ethanol and the balance water). The upper compartment was filled with $5 \, cm^3$ or $10 \, cm^3$ of a solution containing 0.1 wt% sodium alginate and 0 wt%, 10 wt% or 20 wt% ethanol. Water transport started to occur through the PTFE membrane from the upper to the lower compartment and the sodium alginate precipitated as soon as water was removed. Ethanol was used in the $CaCl_2$ aqueous solution in order to reduce the ethanol loss from the alginate coating solution. In all cases, coating time was 24 h. The solutions were drained from the cell and the coated membrane was then cross-linked with carbodiimide solution for 24 h. The optimum cross-linking solution was found to be 100 mM of WSC and ethanol concentration of 60 vol.% at pH 4, which was achieved by using hydrochloric acid. The cross-linking solution was finally drained and the membrane was rinsed with water and ethanol to remove residual WSC and acid. It was observed that membranes with incomplete coatings exhibited transparent spots when ethanol was applied.

Figure 6.4 shows the SEM images of the surfaces of the uncoated PTFE membrane, a partially coated membrane without cross-linking step and a fully coated membrane

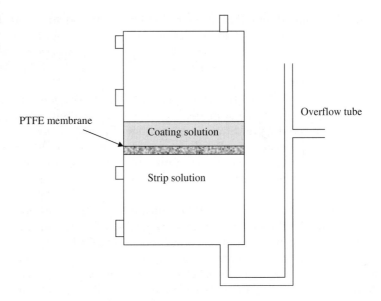

FIGURE 6.3 OD cell used for membrane coating. *Source: Reprinted from [22]. Copyright 2004, with kind permission from Elsevier.*

with cross-linking step using a coating solution containing 10 wt% ethanol. It can be seen that the uncoated membrane consists of a structure of parallel fibres interconnected with fine fibrils. When ethanol was added to the coating solution, ethanol penetrated into the PTFE structure but uncoated places are still visible.

Peng et al. [23] cast a dense hydrophilic polymer solution on porous PVDF (GVHP) membrane (see Table 2.1). Before casting, GVHP membrane was soaked in a potassium dichromic-sulphuric acid solution. Then, it was washed with water and dried at room temperature. The casting polymer solution was a blend of PVA and PEG cross-linked by aldehydes and sodium acetate. Three per cent PVA was blended with 20% PEG (i.e. mixture of two PEGs of different molecular weights: 2.000 and 10.000). After incubation at 40 °C for 2 min, the prepared polymer gel was cast on the surface of the GVHP pre-treated membrane using a gardener knife of 20 μm. Then the cast layer together with the GVHP support was heated at a temperature ranging from 30 °C to 60 °C for 24 h to form the composite hydrophilic/hydrophobic MD membrane. A sodium salt such as sodium acetate was introduced into the polymer network for promoting microphase separation. The obtained membranes were tested for desalination by DCMD configuration using 3.5 wt% NaCl aqueous solution. The effects of the feed temperature and salt concentration were investigated. The behaviour of the coated membranes was found to be similar to that of the uncoated membrane and the DCMD permeate flux of the coated membrane was only 9% lower than that of the uncoated membrane with a separation factor of 99% (i.e. 6–10 μS/cm electrical conductivity). It was supposed that the feed liquid was first dissolved in the hydrophilic layer of the membrane due to swelling effect. Water molecules were then diffused and finally evaporated at the interface between the hydrophobic and hydrophilic layers. It was concluded that the hydrophilic layer could prevent wetting of the hydrophobic membrane pores.

Casting Hydrophobic Polymer over Porous Supports

Porous hydrophobic/hydrophilic composite membranes can be prepared by simple casting

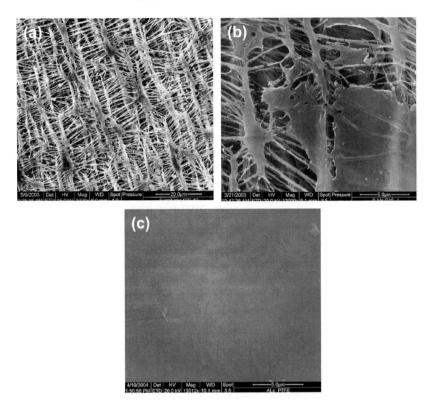

FIGURE 6.4 SEM images of (a) uncoated PTFE porous membrane support, (b) a partially coated PTFE membrane without cross-linking step and (c) fully coated membrane with cross-linking step. *Source: Reprinted from [22]. Copyright 2004, with kind permission from Elsevier.*

of a hydrophobic polymer solution over a hydrophilic porous support (i.e. backing material) or a hydrophobic support having large pores permitting penetration of liquid water. Casting of a polymer solution over a membrane support and subsequent coagulation is a simple method for the preparation of porous and supported flat sheet, tubular, capillary or hollow fibre membranes. An additional advantage of this type of membranes lies in their improved mechanical properties due, obviously, to the support. Furthermore, the porous support acts as a pore-forming agent by simple stretching of the nascent membrane during its coagulation in the water bath. The free shrinkage of the cast solution is hindered by the support and as a consequence the nascent membrane is stretched causing the formation of pores. The sizes of the pores formed by this stretching action may reach the range of a few hundred nanometers. Non-solvent additives can also be used in the polymer solution to form the pores. It is however necessary to choose an adequate support to prevent the support from destruction by the coating solution as well as from the pore blocking.

Khayet and Matsuura [24] cast a PVDF/DMAC/water solution over a non-woven polyester backing material (Osmonics, Inc. Minnesota, USA) and prepared supported membranes following the phase inversion method as reported in Chapter 3. The PVDF concentration was 15 wt%, whereas the water content was varied from 0 wt% to 6.8 wt%.

The membranes were tested for the removal of volatile organic compounds (e.g. chloroform) from water by VMD configuration. Various membrane characterization techniques were applied to determine the porosity, the pore size, the pore size distribution and the liquid entry pressure of water (LEP_w).

Bottino et al. [25] cast binary PVDF/NMP solution on the outer surface of porous tubular supports in order to prepare porous PVDF hollow fibres and capillaries for desalination by sweeping gas membrane distillation (SGMD) (see Fig. 6.5). The PVDF concentration in the dope was varied from 10 wt% to 20 wt%. Various PVDF membranes with different structures, porosities and permeation properties were fabricated [25]. It was concluded that among the different casting parameters studied, the PVDF solution concentration was the most relevant parameter affecting PVDF membrane properties. Diluted and less viscous solutions yielded a more homogeneous distribution of the PVDF coating on the braid support and permitted the formation of membranes with higher permeability. The dilute casting solutions penetrated into the support and formed thinner and more fragile coatings that were more affected by the stretching action of the support and consequently more porous and highly permeable PVDF membranes were formed.

Surface Modification by SMMs

Many of the surface modification methods as those described in the preceding sections are complicated and require at least one additional step for membrane fabrication. Also an answer should be given to the following: *How can we maintain the pore structure after surface-coating process?*

One of the answers to this question, and probably the simplest one, is to introduce active additives that can migrate to the air/film interface and change its chemistry while leaving the bulk properties intact. This method was followed to prepare both porous and dense composite membranes using fluorinated SMMs [4,7–9,27–33]. These are oligomeric fluoropolymers synthesized by polyurethane chemistry and tailored with fluorinated end groups as will be shown later on. According to this method, membranes can be prepared by the phase inversion technique in only one casting step employing a polymer solution containing the host hydrophilic polymer and an SMM with/without another additive (see Chapter 3). Only a small quantity of SMM is required. When the solution of a polymer blend is equilibrated with air, with which the solution is in contact, the polymer having the lowest surface energy (hydrophobic polymer) will concentrate at the air/solution interface and reduce the system's interfacial tension as a consequence. In fact, the segregation of a polymer of lower surface tension at the surface was

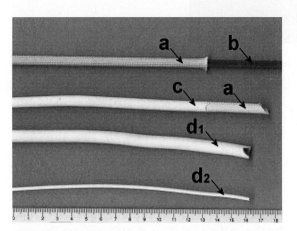

FIGURE 6.5 Porous supports and membranes: (a) tubular glass fibre braid (wall thickness 0.5 mm, inner diameter 4.5 mm); (b) stainless steel bar (this was inserted into the lumen of the braid in order to impart it a given stiffness); (c) supported membrane (PVDF coating on outer surface of the braid); (d_1) Accurel PP membrane (nominal pore size 0.5 μm; wall thickness 1 mm, inner diameter 5 mm); (d_2) Accurel PP membrane (nominal pore size 0.2 μm; wall thickness 450 mm, inner diameter 1.8 mm). *Source: Reprinted from [25]. Copyright 2005, with kind permission from Elsevier.*

confirmed by a number of researchers for the miscible blend of two different polymers [65].

It must be pointed out that various parameters affect the host polymer or polymer additive migration towards the air/film interface such as the surface-free energy of the involved polymers (i.e. host polymer and polymer additive), the casting temperature, the interactions between the base polymer and the polymer additive and those between each individual polymer and the solvent, the molecular weights of the polymers, the solvent evaporation rate, the film thickness and the difference in chemical potential created at the interface between the polymer blend and the phase in contact with the polymer (i.e. air, water). Other than the composition of the casting polymer solution, which involves polymer, solvent and nonsolvent additive(s), the parameters responsible for the ultimate membrane physical properties are evaporation time, evaporation temperature, humidity of the casting atmosphere, thickness (casting bar thickness), type of coagulation medium and its temperature in case of wet phase inversion, post-treatment (i.e. heat treatment, solvent exchange method), etc.

During the past few years, hydrophobic/hydrophilic porous composite membranes were developed by the collaboration of Khayet, Matsuura and co-workers using SMMs and they were tested in desalination by DCMD [9,26–28]. These membranes were found to be promising for desalination by DCMD as they combine the low resistance to mass flux, achieved by the diminution of the water vapour transport path through the hydrophobic thin top layer, and a low conductive heat loss through the membrane, obtained by using a thicker hydrophilic sub-layer. The membranes were prepared by the phase inversion technique in a single casting step from PEI polymer solutions containing SMMs, the solvent DMAC and the nonsolvent γ-butyrolactone (GBL). Only 2 wt% SMM was added to the casting solution. SMM preparation and characterization will be shown later on.

During membrane formation, the SMM migrates towards the top air/polymer interface, rendering the membrane hydrophobic. This was confirmed by contact angle measurements and X-ray photoelectron spectroscopy (XPS) analysis, which indicated the gradient in fluorine concentration across the membrane cross-section as a result of the migration of fluorinated end groups to the airside surface during membrane formation. The effect of the PEI concentration in the casting solution on the permeate flux and on the membrane characteristics was investigated. For each PEI concentration, it was found that the LEP_w of the SMM-modified membrane was higher than that of the unmodified membrane, while the pore sizes were smaller and decreased as the concentration of PEI concentration was increased. The pore sizes of the SMM-modified PEI membranes determined by the gas permeation test were 12–23 nm, which are smaller than the pore sizes of the commercial membranes used in MD. However, their DCMD permeabilities were found to be of the same order of magnitude to those of the commercial PTFE membranes (TF200 and TF450), even though the latter membranes have higher porosities and an order of magnitude greater pore sizes. Furthermore, the DCMD permeate flux of the SMM/PEI membranes prepared with 12 wt% PEI was found to be greater than those of the commercial PTFE membranes. Very high separation factor (>99.7%) was obtained for the SMM-modified PEI membranes when the feed was aqueous NaCl solution. A theoretical model was developed to estimate the thickness of the two layers of these composite membranes using the DCMD experiments together with other membrane parameters and the heat and mass transfer equations. The thickness of the hydrophobic layer was calculated to be lower than 8 µm, which is an order of magnitude lower than that of the PTFE layer of the commercial membranes [27].

Recently, PS and PES flat sheet membranes were further modified by the Khayet and Matsuura group using different types of

SMMs, different solvents, different additives and different membrane preparation conditions in order to optimize the MD performance of the composite hydrophobic/hydrophilic membranes [29–33]. When SMMs were blended in PES, it was found that the contact angles of the top surface of flat sheet membranes increased from 76° of PES to 116°. This value is nearly equal to that of PTFE (Teflon). It is worth quoting that the use of SMMs can also be applied for the preparation of composite hollow fibre membranes for MD. Double-layered hydrophobic/hydrophilic hollow fibre membranes or tri-layered hydrophobic/hydrophilic/hydrophobic hollow fibre membranes can be prepared in a single step using one spinning dope by dry/wet spinning or dry spinning techniques using SMMs. The hydrophobic SMMs are blended with a hydrophilic polymer to form the spinning dope. During spinning, the SMMs will migrate towards the outer surface while fibre is travelling through the air gap, rendering the outer surface hydrophobic. In fact, SMMs can migrate to both the inner and outer surfaces. This type of hollow fibre membranes has not yet been tested for MD and progress is expected in the future.

SMM Synthesis

The system used for SMM synthesis is shown schematically in Fig. 6.6. Firstly, air should be purged from the reactor vessel by introducing nitrogen (N_2) gas. The SMMs were synthesized by a two-step solution polymerization method following the reactions shown schematically in Fig. 6.7, which is given as an example for the synthesis of the SMM (methylene bis(p-phenyldiisocyanate) (MDI)/DEG/oligomeric fluoroalcohols BA-L) for a 3:2:2 stoichiometric ratio. The initial step involved the reaction of a diisocyanate with a polyol in a common solvent (DMAC). The solvent was distilled before use. MDI and DEG reacted to form a urethane prepolymer. The reaction was then terminated by the addition of an oligomeric fluoroalcohol to end-cap the prepolymer resulting in SMM with hydrophobic end groups. The SMM was precipitated from the solution with distilled water, washed with 30/70 v/v acetone/water mixture 3 times to leach out unreacted monomer and finally dried in an oven at 50 °C. It is necessary that the two polymerization steps be performed in a controlled atmosphere of a pre-purified nitrogen. Temperature, solvent volume, reactant mole ratio, reactant concentration and stir rate are important parameters in determining the molecular weight and molecular-weight distribution of the SMMs. Various SMM formulations were synthesized with different combinations of monomers and stoichiometries. A list of the SMMs together with the reactants and the diisocyanate/polyol/fluoroalcohol molar ratio is given in [7]. SMMs with different molecular weights can be synthesized by properly adjusting the reaction conditions.

The structures of some SMMs are given in Fig. 6.8. To synthesize these SMMs, MDI was commonly used for the hard segment of the prepolymer (polyurethane or polyurea). Oligomeric fluoroalcohol BA-L was also commonly used for end-capping of the prepolymer. The only difference is the soft segment of the prepolymer, which was diphenylsulfone (DPS), DEG, polypropylene glycol (PPG) and α,ω-aminopropyl poly(dimethyl siloxane) (PDMS), respectively.

SMM Characterization

The elemental analysis of fluorine content in SMMs was carried out using the standard method in ASTM D3761. An accurate weight (10–50 mg) of sample was placed into oxygen flask bomb combustion (Oxygen Bomb Calorimeter, Gallenkamp). After pyro-hydrolysis, the fluorine (ion) was measured by an ion chromatography (Ion Chromatograph, Dionex DX-1000) [7,31–33]. The glass transition

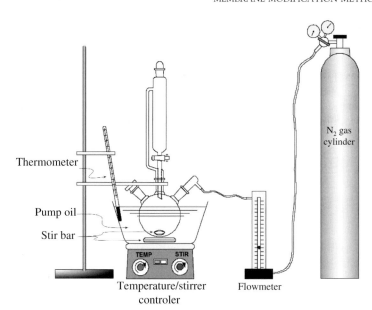

FIGURE 6.6 Schematic presentation of apparatus for SMM synthesis.

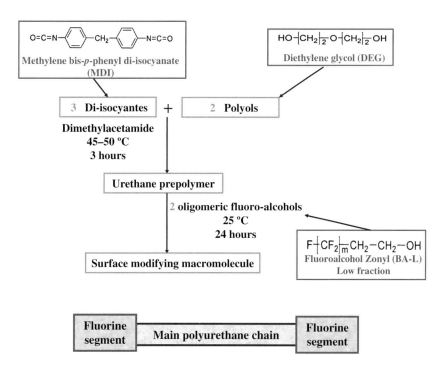

FIGURE 6.7 Reaction scheme for the synthesis of SMM (MDI/DEG/BA-L) with a 3:2:2 stoichiometric ratio.

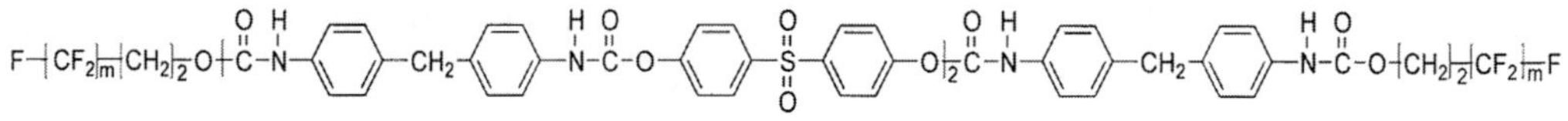

MDI/DPS/BA-L

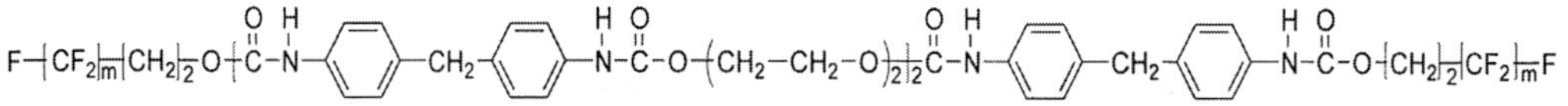

MDI/DEG/BA-L

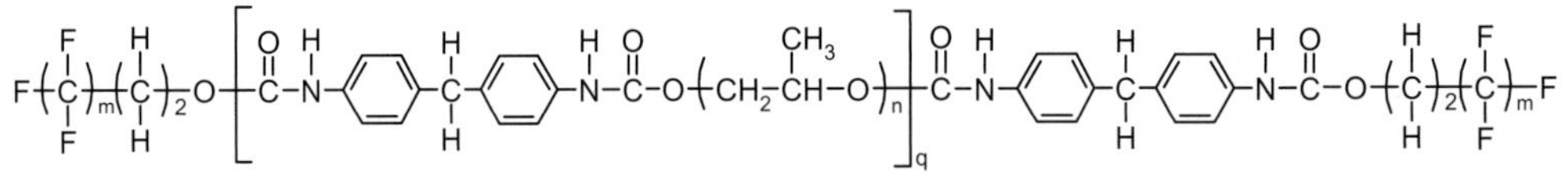

MDI/PPG/BA-L

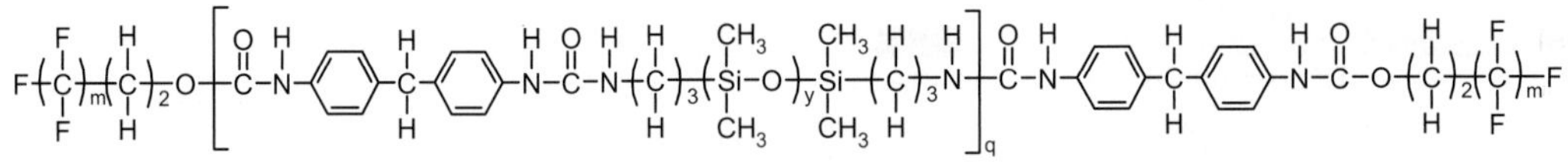

MDI/PDMS/BA-L

FIGURE 6.8 Chemical formula of some SMMs.

temperature (T_g) was examined by differential scanning calorimeter (DSC) equipped with universal analysis 2000 program (DSC Q1000, TA Instruments, New Castle, DE). About 10 mg of polymer was crimped into aluminum pan. The SMMs were annealed at 280 °C for 10 min, then quenched to −50 °C and scanned at a heating rate of 10 °C/min. The T_g value was recorded at the mid-point of the corresponding heat capacity transition [7,31–33]. The weight average molecular weight (M_w) and the number average molecular weight (M_n) of the synthesized SMMs were measured by gel permeation chromatography (GPC) using Waters Associates GPC chromatograph equipped with Waters 410 refractive index detector. Three Waters UltraStyragel™ packed columns were installed in series. Tetrahydrofuran (THF) was filtered and used as the carrier solvent at 40 °C and a flow rate of 0.3 ml/min. First, the calibration of the system was performed using polystyrene (Shodex, Tokyo, Japan) standards with different molecular weights between 1.3×10^3 g/mol and 3.15×10^6 g/mol. The standards and SMM samples were prepared in a THF aqueous solution (0.2%, w/v) and filtered prior to injection through 0.45 μm filter to remove high-molecular-weight components. Millenium 32 software (Waters) was used for data acquisition [7,31–33].

SMM Membrane Preparation

A small amount of an SMM was blended in a host hydrophilic polymer solution such as PES, PEI or PS. The solvents used were NMP or DMAC. After casting a thin film of the polymer solution, the cast film was kept in an oven for a predetermined period to evaporate the solvent. The solvent evaporation temperature was varied between room temperature and 100 °C. The film was then immersed in ice-cold water for coagulation or in water kept at room temperature. The variables involved in the membrane preparation are the hydrophilic polymer concentration, SMM concentration in the casting solution, evaporation temperature, evaporation period, coagulant temperature, casting thickness, etc. The effects of some of these parameters are presented in section 6.3.

Co-Extrusion Spinning

Recently, some attempts were made to spin composite hollow fibre membranes for MD. Porous composite bi-layered hydrophilic/hydrophobic hollow fibre membranes or tri-layered hollow fibre membranes can be prepared by co-extrusion spinning using adequate spinneret as reported in Chapter 4. The hydrophobic layer can be formed on the inside, the outside or both sides of the hollow fibre membranes. Bonyadi and Chung [34] prepared dual-layer hydrophilic/hydrophobic hollow fibre membranes for desalination by DCMD. The outer layer was a polymer solution made of 12.5 wt% PVDF in NMP. Hydrophobic cloisite 15A was added into the PVDF solution so that cloisite 15A concentration in the solution became 30 wt%. The inner layer was formed by 8.5 wt% PVDF, 4 wt% PAN, 87.5 wt% NMP and hydrophilic cloisite NA$^+$ (50 wt% in the polymer solution). The dual hydrophilic/hydrophobic hollow fibres were spun with 80 wt% methanol aqueous solution as the internal and external coagulants, at 3 cm air gap distance and by free fall. Therefore, only gravitational force acted in the axial direction, while the hollow fibre was travelling in the air gap. The average pore size and porosity were 0.41 μm and 80%, respectively. A permeate flux as high as 55 kg/m^2·h with a separation factor of 99.8% was achieved at 90 °C feed temperature and 16.7 °C permeate temperature for a 3.5 wt% sodium chloride (NaCl) aqueous solution.

EFFECTS OF PROCESS PARAMETERS ON THE MEMBRANE STRUCTURE AND MD PERFORMANCE

As shown in the previous section, various techniques have been adopted to modify membranes for the application in MD process. This section is devoted to the effects of some of the parameters on the membrane structure as well as the MD performance.

Radiation Graft Polymerization

This method was applied by Wu et al. [12] to prepare polystyrene-grafted modified membrane for MD. Microporous flat sheet CA was used as the membrane support. Initially, the effect of styrene/pyridine composition on the degree of grafting and water contact angle of the grafted CA membrane was studied. The radiation time was 7 h. It was observed that both the grafting percentage and the contact angle showed a maximum at 10% pyridine (90% styrene) in the grafting solution as shown in Fig. 6.9. The maximum contact angle measured was 69.5° corresponding to the high

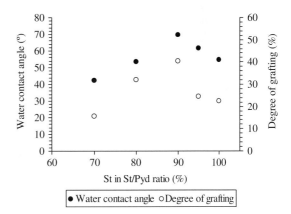

FIGURE 6.9 Effect of St in St/Pyd ratio on water contact angle and degree of grafting of CA porous membrane. *Source: Data taken from [12]. Copyright 1992, with kind permission from Elsevier.*

degree of grafting 40.4 and St/Pyd ratio of 90:10. Therefore, Pyd content was fixed at 10% in the grafting solution and the effect of St/CCl$_4$ was studied maintaining the radiation time at 15 h. The maximum water contact angle observed was 88.6° when using a concentration of 5% CCl$_4$ in the grafting solution.

Three types of polystyrene grafting modified CA membranes were applied for desalination by DCMD. The membranes were prepared maintaining the concentration of pyridine at 10% and the amount of CCl$_4$ was 0% and 5%. The highest water contact angle, 96°, was observed for the membrane prepared with 85% St, 10% Pyd, 5% CCl$_4$ and 23 h radiation time. For the same grafting solution, the water contact angle was greater when the radiation time was longer. However, the DCMD permeate flux was lower, although the NaCl rejection factor was higher. Moreover, the membrane prepared without CCl$_4$ in the grafting solution exhibited the lowest permeate flux. This result could probably be attributed to the short grafting time of polystyrene onto the CA membrane surface induced by the addition of the chain transfer agent CCl$_4$, resulting in higher hydrophobicity of the surface of the modified membrane. In any case, the benefit of the chain transfer agent is not clear and more studies should be carried out. Compared to the asymmetric PVDF membrane having a permeability of 0.68×10^{-3} kg/m^2·h·Pa and a rejection factor of 99.9%, it was found that the permeability of the polystyrene-grafted CA membrane ranged from 0.14×10^{-3} to 0.96×10^{-3} kg/m^2·h·Pa with rejection factor decreasing from 99.1% to 66.7%. These results indicated that the radiation-grafted membranes with good performance in MD could be obtained if the grafting conditions were properly controlled.

Plasma Polymerization

Wu et al. [12] and Kong et al. [13] applied plasma polymerization to a commercial porous hydrophilic CN substrate in order to prepare bi-layered flat sheet (hydrophobic/hydrophilic) and tri-layered (hydrophobic/hydrophilic/hydrophobic) membranes for MD, respectively. The effects of discharge power and discharge time on the properties of the modified membranes as well as on their MD performance were studied. In this method, both deposition onto the membrane surface due to polymerization and ablation from the membrane surface due to bombardment with particles occur simultaneously, depending on the discharge power. Moreover, particles can also vapourize from the membrane surface and contribute to polymerization. Therefore, the surface characteristics of the modified membrane will depend on the dominant phenomena, ablation and deposition. Based on SEM studies, the pore sizes of the OFCB plasma-modified CN membranes prepared with different discharge powers were maintained similar; however, surface properties were different. The water contact angle decreased with an increase in discharge power down to a minimum value of 100° for 150 W and then increased up to 120° for 200 W. No clear trends were observed between the DCMD permeate flux and the discharge power. The NaCl rejection factor also decreased with increasing discharge power to the lowest value 89.6% and then increased up to 92%. It was explained that the layer of the deposited polymer was not thick enough to prevent moistening of the membrane. Similar behaviours were observed for the VTMS/CF$_4$ plasma-polymerized CN membranes prepared with different discharge power values.

It was observed that both the thickness and the pore size of the plasma-modified membranes could be controlled by discharge time. The pore size gradually decreased with increasing discharge time until a dense surface is formed. This is the reason of the observed decrease of the DCMD permeate flux and the increase of the NaCl rejection factor approaching 100% with increasing discharge time as shown in Fig. 6.10 for two types of plasma-modified membranes.

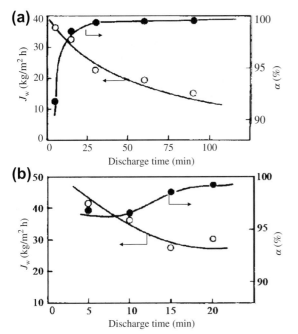

FIGURE 6.10 Effects of discharge time on DCMD permeate flux (J_w) and NaCl rejection factor (α) of (a) OFCB plasma modified CN membrane (70 °C feed temperature, 25 °C permeate temperature, 0.5 M NaCl feed solution) and (b) VTMS/CF$_4$ plasma modified CN membrane (60 °C feed temperature, 25 °C permeate temperature, 0.5 M NaCl feed solution). *Source: Reprinted from [12]. Copyright 1992, with kind permission from Elsevier.*

Grafting Ceramic Membranes

As stated in the previous section, grafting of the ceramic alumina and zirconia of different pore sizes (200 nm and 50 nm) was carried out by means of successive supports soaking in the grafting solution (i.e. 1H,1H,2H,2H-perfluoro-decyltriethoxysilane, $C_8F_{17}C_2H_4Si(OEt)_3$ (C8) in chloroform) followed by drying after each soaking step [17,18]. The concentration of FAS in the grafting solution, type of FASs, number of soaking, drying temperature and grafting time may affect the characteristics of the grafted ceramic membranes to different extents as well as to their MD performances. Larbot et al. [17] studied the effects of grafting conditions on the characteristics of the resulted MD membranes. For example, the effect of grafting time was studied maintaining all other grafting parameters the same (constant concentration of FAS, C8, in chloroform: 10^{-2} mol/l; grafting at room temperature; drying temperature: 100 °C and 12 h between successive soakings). Based on thermal gravimetric analysis (TGA) carried out to compare the grafted and ungrafted ZrO_2, it was observed that the sample weight loss increased with grafting time and 118 h was enough for ZrO_2 grafting. Moreover, it was found that the water contact angle was higher for the grafted ceramic membranes using silanes with larger fluoro-alkyl chains, but it was independent of the pore size of the ceramic membrane support as can be observed in Fig. 6.11. Therefore, it seems better to use FAS with longer fluoro-alkyl chains such as C8 for MD membrane design.

Compared to the ungrafted ceramic membranes, the permeability of the grafted membranes was drastically decreased. For example, for ZrO_2 of 200 nm pore size, the permeability was decreased from 2700 l/m²·h·bar corresponding to the ungrafted

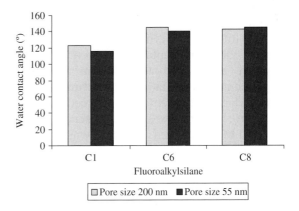

FIGURE 6.11 Water contact angle of ceramic membranes (ZrO_2) of different pore sizes grafted with different FASs. *Source: Data taken from [17]. Copyright 2004, with kind permission from Elsevier.*

support to $2.10 l/m^2 \cdot h \cdot bar$ of C8-grafted support. The trends observed for the grafted ceramic membranes between the permeate flux and the temperature difference as well as salt concentration were the same as other types of MD membranes [17,18].

It was also found that when using the grafted Al_2O_3 tubular membrane having pore size of 200 nm, the NaCl rejection factor ranged from 90% to 96% at low NaCl concentrations (10^{-3}–10^{-2} M), whereas for higher NaCl concentrations higher rejection factors up to 99% were achieved. DCMD permeate flux of $163.2 l/m^2 \cdot day$ was reached when the feed temperature was 95 °C. It must be pointed out that no explanations were given for the cause of the lower rejection factors observed when using lower NaCl concentrations. When the grafted ZrO_2 membrane having smaller pore size (i.e. 50 nm) was used, lower DCMD permeate fluxes ($\approx 95 l/m^2 \cdot day$) and near 100% rejection factor were obtained; whereas higher DCMD permeate fluxes ($\approx 202 l/m^2 \cdot day$) with lower rejection factors (99.5%) were obtained for the grafted ZrO_2 membrane having higher pore size (i.e. 200 nm).

When the grafted ceramic membranes were employed in AGMD process for desalination, Krajewski et al. [18] observed that the rejection of NaCl was close to 100%. A permeate flux as high as $7.0 l/m^2 \cdot h$ was obtained for grafted zirconia membranes with an alumina support at a feed temperature of 95 °C, a condensation surface temperature of 5 °C and 1 M NaCl concentration of feed aqueous solution. Lower fluxes were observed when using 1 M NaCl feed solution was used. It was concluded that the use of grafted ceramic membranes would be of great interest in MD [17,18]. The pore size and porosity of ceramic membranes do not affect MD performance of the grafted membranes significantly. Based on the LEP_w, the ceramic membrane that has a smaller pore size exhibited better grafting stability.

Surface Coating

Hydrophobic Surface Coating

Various factors, such as coating time, coating temperature, concentration of coating material in the coating solution, can directly affect the membrane characteristics as well as the performance of hydrophobic coated membranes. Figure 6.12 shows, as an example, the morphology of the uncoated and silicone rubber-coated PPESK hollow fibre membrane. The morphology of the inner surface of PPESK hollow fibre membrane was modified and the coating layer remained unchanged after VMD application, indicating that this hydrophobic coated membrane exhibited good thermal stability [21].

Jin et al. [21] studied the effects of coating time, coating temperature and concentration of silicone rubber solution on VMD desalination performance. As can be seen in Fig. 6.13, no clear trends could be observed between the permeate flux and the coating time, whereas a slight increase was detected for the NaCl rejection with increasing coating time. After about 3 h coating time, the salt rejection factor reached 99%. These results were attributed to the viscosity of the coating solution and the shrinkage of the membrane support. In fact, the viscosity of the silicone rubber increased with coating time and the risk of pore blocking was much higher, leading to diminution of VMD permeate flux and enhancement of the salt rejection factor. When the silicone rubber became gelatinous, PPESK membrane pores were hardly blocked and VMD permeate flux increased. It is worth quoting that shrinkage of the membrane reduces the pore size of the coating layer.

The coating temperature and the concentration of the coating solution affect considerably the VMD performance of the coated PPESK hollow fibre membranes. It was observed that higher coating temperatures increased silicone rubber vulcanization velocity, changing its state and accelerating membrane shrinkage, which

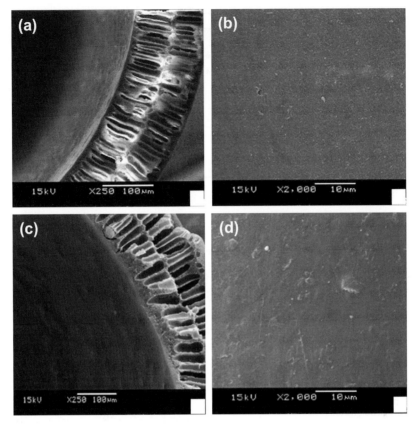

FIGURE 6.12 SEM images of uncoated (a, b) and silicone rubber-coated (c, d) PPESK hollow fibre membrane (coating time 7 h, coating temperature 100 °C, b and d are the inner surfaces of the hollow fibre membranes). *Source: Reprinted from [21]. Copyright 2008, with kind permission from Elsevier.*

resulted in the reduction of the pore sizes. Furthermore, when the concentration of the coating solution was lowered, thinner coating layers were formed. As a result, less pores were blocked and higher VMD permeate fluxes were obtained with greater salt rejection factors. It should be noted that the risk of blockage of membrane pores increases, when a coating solution of low concentration is used because solidification of the coating layer is not fast at low temperatures. Jin et al. [21] could achieve a high VMD performance of PPESK hollow fibre membrane with high stability using 5 g/l silicone rubber solution, 60 °C coating temperature and 9 h coating time.

Figure 6.14 shows the effect of prepolymerization time of MTMS on MD performance of hydrophobic coated membranes by means of sol-gel polytrifluoropropylsiloxane method [21]. Higher VMD permeate fluxes could be achieved with larger prepolymerization time. However, a decline of the salt rejection factors was observed for an increase in prepolymerization time. In this case, the highest permeate flux was observed for 30 min prepolymerization time.

Hydrophilic Solution Coating

Coating hydrophobic porous membranes with a hydrophilic coating solution was carried out in MD applications in order to diminish the risk of

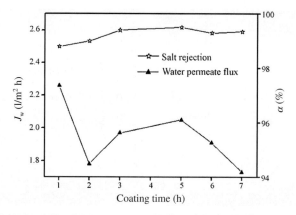

FIGURE 6.13 Effects of coating time on VMD permeate flux (J_w) and NaCl rejection factor (α) of silicone rubber-coated PPESK hollow fibre membranes (coating temperature 100 °C, 10 g/l silicone rubber solution, feed temperature 40 °C, vacuum pressure 0.078 MPa). *Source: Reprinted from [21]. Copyright 2008, with kind permission from Elsevier.*

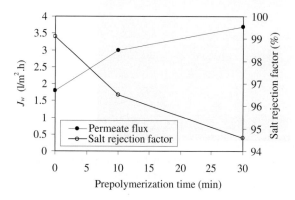

FIGURE 6.14 Effects of MTMS prepolymerization time on VMD permeate flux and NaCl rejection factor of polytrifluoropropylsiloxane-coated PPESK hollow fibre membranes (feed temperature 40 °C, vacuum pressure 0.078 MPa). *Source: Data taken from [21]. Copyright 2008, with kind permission from Elsevier*

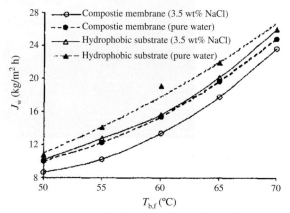

FIGURE 6.15 Effect of feed temperature ($T_{b,f}$) on DCMD permeate flux (J_w) of hydrophilic coated and uncoated hydrophobic porous membrane support (GVHP). *Source: Reprinted from [23]. Copyright 2005, with kind permission from Elsevier.*

membrane pore wetting. The trends of the curves presenting the MD permeate flux of the coated membranes as a function of each operating parameter of the MD process (feed temperature, permeate temperature, feed concentration, feed flow rate, permeate flow rate, etc.) were similar to those of the uncoated MD membranes but with less than 10% decline. Figure 6.15 shows, e.g., the DCMD permeate flux of both hydrophilic coated and uncoated membranes as function of the feed temperature for distilled water and 3.5 wt% NaCl feed solutions [23].

Casting Hydrophobic Polymer over Porous Supports

The support can not only enhance mechanical properties of the formed membranes, but

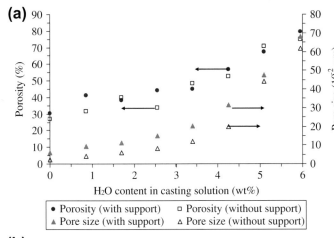

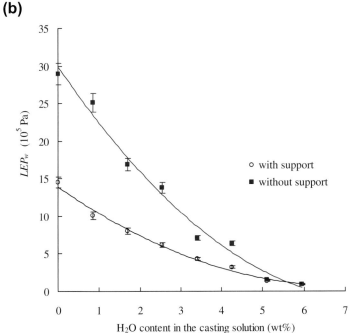

FIGURE 6.16 Effects of the membrane support on the pore size, porosity and LEP_w of PVDF flat sheet membranes prepared by solvent/non-solvent induced phase inversion technique. *Source: Data published in [24]. Copyright 2001, with kind permission from American Chemical Society.*

also diminish the shrinkage of the cast membrane that reduces the size of the membrane pores. Khayet and Matsuura [24] prepared flat sheet hydrophobic porous membranes with and without support for MD. It was observed, as presented in Fig. 6.16, that the supported membranes exhibited greater pore size, although the composition of the casting solution and the membrane preparation conditions were exactly the same. This result was attributed to the shrinkage of pores that occurred in the unsupported membranes during the solvent exchange procedure for membrane drying.

The LEP_w values of the supported membranes were found to be smaller than those of the unsupported membranes. This is because of the larger pore size of the supported membranes. In fact, the pore sizes of the supported membranes were larger than those of the unsupported membranes when the water concentration in the PVDF casting solution was below 5.1 wt%. By contrast, when the water concentration in the casting solution was either 5.1 wt% or 5.95 wt%, the LEP_w values and the pore sizes were almost the same for both the supported and unsupported membranes. Moreover, it seems that the support induced higher membranes porosities. The VMD permeate flux and the overall mass transfer coefficient were higher and the separation factor was lower, in the removal of chloroform from water by MD, for the supported membrane than for the unsupported membrane when the PVDF concentration in the casting solution was equal to or lower than 4.3 wt% [24].

Bottino et al. [25] studied the effects of casting parameters (PVDF polymer concentration in the casting solution, immersion time of the braid, extraction speed of the braid from the casting solution) on the characteristics and SGMD performance of porous tubular supports. It was observed that the PVDF weight deposited on the braid increased when higher polymer concentrations were used, increased linearly with the braid length, from the top (i.e. the part first extracted from the polymer solution) to the bottom, increased with increasing immersion time and by lowering the extraction speed. It was also found that the supported membranes prepared with lower PVDF polymer concentrations in the casting solution were less permeable than those prepared with higher PVDF contents. However, the variation of the immersion time had only a limited effect on the permeability of the supported membrane. When using 3 w/v% NaCl aqueous solution, high salt rejection factors were observed (i.e. 99.5–99.9%) and the permeate flux was slightly higher than those of some commercial membranes but lower than 2.6 l/m²·h.

Surface Modification by SMMs

As it is shown in the next chapter, various techniques have been applied for the characterization of surface-modified membranes by SMMs such as scanning electron microscopy (SEM), atomic force microscopy (AFM), contact angle measurements, XPS, LEP_w measurements, solute transport and gas permeation among others. The pore size, pore size distribution, porosity, surface roughness, MD performance, etc. were determined. The effects of various membrane preparation parameters have been investigated, although the use of these membranes in MD was begun only recently. As will be shown later on, the effects of the hydrophilic host polymer type and its concentration in the casting solution, SMM concentration in the casting solution, evaporation temperature, evaporation period, coagulant temperature, casting thickness, etc. have been studied. The results are summarized briefly as follows:

- The surface of the SMM-modified membranes was enriched with fluorine groups associated with SMM and, therefore, were more hydrophobic than the unmodified membranes.
- XPS analysis with different take-off angles showed that the surfaces of the unmodified membranes, which contained no SMM, had no fluorine. By contrast, fluorine was present at the surfaces of all the SMM-modified membranes. Furthermore, a significant depletion in carbon content and a slight increase in nitrogen content on the SMM-modified membrane surfaces were observed.
- The SMM-blended membranes exhibited lower liquid and gas permeation fluxes compared to the unmodified membranes

prepared under the same conditions without SMM. This was attributed to their higher hydrophobicity and smaller pore size.
- The mean pore sizes of the SMM-modified membranes were smaller than those of the unmodified membranes prepared using the same host polymer and similar conditions.
- The pore densities and surface porosities of the SMM-modified membranes were higher than those of the unmodified membranes prepared using the same host polymer and similar conditions.
- The SMM-modified membranes exhibited top surfaces smoother than the corresponding unmodified membranes.
- The LEP_w values of the SMM-modified membranes were higher than those of the unmodified ones.

In what follows, the effects of some parameters on the SMM-modified membranes are outlined.

Effect of Hydrophilic Polymer Concentration

Khayet et al. [9,26,27] and Suk et al. [29] fabricated PEI and PES membranes whose surfaces were modified by SMMs. They have studied the effects of the concentration of the hydrophilic polymers PEI and PES in the casting solutions on various membrane characteristics as well as on the MD performance. It was observed that the pore sizes of the SMM-modified membranes decreased as the concentration of the hydrophilic polymer in the casting solution was increased. This conclusion was supported by various characterization methods based on the solute transport, the gas permeation and the AFM. Moreover, it was found that both the pore density and the surface porosity, determined from the solute transport and the AFM method, increased with the increase of the hydrophilic polymer concentration. Based on the AFM analysis, the top surface of the SMM-modified membranes became smoother with an increase of the hydrophilic polymer concentration. Figure 6.17 shows the AFM images of the top and bottom surfaces of the SMM-modified and unmodified membranes. The AFM images of the top and bottom membrane surfaces are different and the nodules and nodule aggregates are observed at the surfaces of both SMM-modified and unmodified membranes. The nodules are seen as bright high peaks in the AFM pictures, whereas the pores are seen as dark depressions. Larger pores were obtained from the AFM images at the top and bottom surfaces of the SMM-modified and the unmodified PEI membranes prepared with smaller PEI concentrations in the casting solutions.

In addition, as presented in Fig. 6.18, the water contact angles of the SMM-modified membranes decreased with the increase of the hydrophilic base polymer (PEI) concentration. The SMM migration depends on the type of the SMM as will be explained later on. The decrease of water contact angle with the increase of the base polymer concentration was attributed to the increase of the viscosity of the polymer solution, which slowed down SMM migration to the top membrane surface. The hydrophobic layer thickness of the SMM-modified membranes was evaluated by Khayet et al. [27]. It was found that the hydrophobic layer thickness became smaller when the membrane was prepared with lower PEI concentration in the casting solution, as can be seen in Fig. 6.19. This indicates that the SMM migration towards the top membrane surface is faster when the host hydrophilic polymer concentration is smaller in the casting solution, leading to a higher SMM concentration in the top membrane surface and a thinner hydrophobic layer. This result is in accordance with the water contact angle shown in Fig. 6.18. As a consequence, the SMM-modified membranes exhibited better MD performance (i.e. higher DCMD permeate flux and salt rejection factors) when the membrane

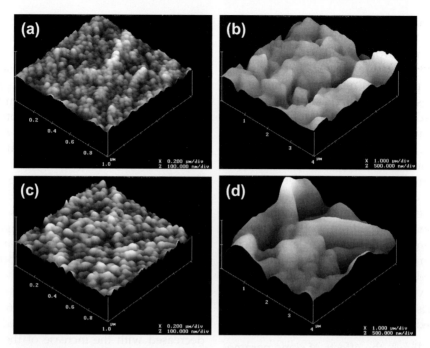

FIGURE 6.17 Three-dimensional AFM pictures of the top (a, c) and bottom (b, d) surfaces of unmodified (a, b) and SMM-modified (c, d) PEI membranes prepared with 12 wt% PEI concentration in the casting solution, 10 wt% of GBL and 2 wt% of SMM (MDI/DPS/BA-L, shown in Fig. 6.8) for the SMM modified membrane. *Source: Reprinted from [66]. Copyright 2003, with kind permission from Elsevier.*

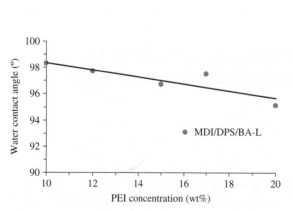

FIGURE 6.18 Advancing water contact angle versus PEI concentration of the SMM (MDI/DPS/BA-L)-modified membranes (2 wt% SMM concentration in the casting solution and 0 min DMAC solvent evaporation time).

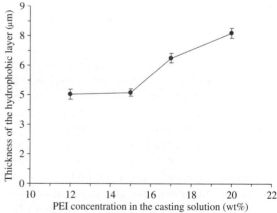

FIGURE 6.19 Effect of the hydrophilic PEI concentration in the casting solution on the hydrophobic layer thickness of the SMM modified membranes. *Source: Reprinted from [27]. Copyright 2005, with kind permission from Elsevier.*

was prepared with lower concentration of the hydrophilic host polymer [26].

Effect of Hydrophilic Polymer Type

SMM-modified membranes were prepared using different types of hydrophilic host polymers such as PES, PEI, PS and PVDF. It was observed that the SMM migration towards the top surface during membrane formation depended on the type of the host polymer used. For example, membranes involving two types of SMMs (MDI/PPG/BA-L and MDI/PDMS/BA-L in Fig. 6.8) were prepared under similar conditions but with two different hydrophilic polymers, i.e. PES (15 kDa molecular weight) and PEI (30.8 kDa molecular weight). The SMM concentration in the casting solution was maintained at 1.5 wt% and the membranes were prepared without NMP solvent evaporation time. In general, it was found that the membranes made of PES host polymer were better than those prepared with PEI host polymer. The water contact angles of both types of membranes were similar but cross-sectional SEM images were different (Fig. 6.20). The finger-like structure of the SMM/PES membrane reached the bottom side where small macro-voids were formed in vertical direction, whereas the finger-like structure of the SMM/PEI membrane became more irregular in the middle of the cross-section and large macro-voids were formed in horizontal direction.

In addition, it was observed that the SMM/PES membranes exhibited smaller pore size, narrower pore size distribution, smoother top surface (from AFM analysis), lower LEP_w value but higher porosity compared to SMM/PEI membranes. The DCMD permeate fluxes of the two membranes together with a commercial membrane (FGLP, Millipore see Table 2.1) were measured at different average temperatures (i.e. average temperature of feed and permeate solutions). Both the commercial membrane and the SMM-blended PES and PEI membranes exhibited an exponential increase of the DCMD permeate flux with an increase in average temperature and the permeate fluxes were higher for the SMM-modified membrane than the commercial membrane. Among the SMM-modified membranes, the PES membrane showed higher permeate flux than the PEI membrane. These results were attributed to the higher effective porosity of the SMM-modified PES membrane compared to that of SMM-modified PEI membrane. In fact, an increase of the effective porosity means an increase in either the porosity and/or pore radius or a decrease in effective pore length. For the three tested

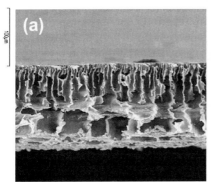 

FIGURE 6.20 Cross-sectional SEM images of SMM (MDI/PDMS/BA-L): (a) modified PES membrane and (b) modified PEI membrane. *Source: Reprinted from [67]. With kind permission from DWT Editorial Office.*

membranes, the solute separation factor was higher than 99.9% (i.e. the permeate conductivity was smaller than 25 µS/cm). Smaller DCMD permeate fluxes were observed in the presence of NaCl feed solution due to the decrease in water vapour pressure with an increase of the salt concentration leading to a diminution of the driving force.

It is worth quoting that, in general, SMM migration towards the top membrane surface depends on the type of the host hydrophilic polymer and its molecular weight.

Effect of SMMs Type

The segregation of SMMs at the top membrane surface depends considerably on the formulation, molecular weight and fluorine content of SMMs. In fact, the most significant contribution to the molecular weight of SMMs comes from the size of the prepolymer chain generated in the first step of the polymerization reaction and not the size of the fluorine tail. This is because the addition of the fluoroalcohol is a chain-terminating step and the fluorine-containing reactant is mono-functional and theoretically cannot allow for significant increase in the molecular weight. It is worth quoting that generally fluorine content of SMMs decreases with the increase in the molecular weight of SMM [7]. For example, by comparing the two types of SMM, i.e. MDI/DPS/BA-L and MDI/DEG/BA-L (presented in Fig. 6.8), the fluorine content of both SMMs were similar, 20.0 wt% and 19.8 wt%, respectively; however, the molecular weight of MDI/DEG/BA-L was more than 2 times higher than that of the MDI/DPS/BA-L. Generally, the membranes containing MDI/DPS/BA-L showed higher contact angles (i.e. advancing and receding values) than those containing MDI/DEG/BA-L for both PEI and PVDF membranes. This result was attributable in part to the low molecular weights of MDI/DPS/BA-L, which facilitate SMM migration to the surface as fluorine content of both SMMs was found to be similar. From XPS analysis, it was found that fluorine concentration was higher for the MDI/DPS/BA-L-modified surfaces than for MDI/DEG/BA-L, especially at low hydrophilic host polymer concentrations. In addition, Khayet et al. [66] observed that the pore size of the MDI/DPS/BA-L-blended membrane was smaller than that of the MDI/DEG/BA-L-blended membrane at low PEI concentrations in the casting solution (i.e. less than 17 wt%); whereas at higher PEI concentrations an opposite trend was observed.

By SEM study, it was also found difference in SMM-modified membranes prepared under similar conditions but with different SMMs types. For example, Fig. 6.21 shows

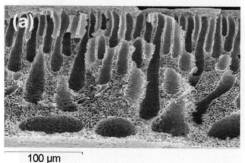 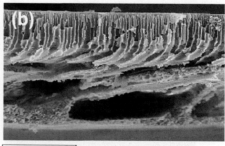

FIGURE 6.21 Cross-sectional SEM images of SMM modified PEI membranes using two different SMMs: (a) MDI/PPG/BA-L = 3:2:2 (PUP); (b) MDI/PDMS/BA-L = 3:2:2 (PUDU). PEI concentration, 12 wt%; GBL concentration, 10 wt%; NMP solvent, 78 wt%; gelation bath temperature, 20 °C. *Source: Reprinted from [31]. Copyright 2009, with kind permission from Elsevier.*

cross-sectional SEM images of two SMM-modified membranes. As can be seen, the membranes are asymmetric with finger-like structure at the top surface, whereas the structure of the bottom surface varies depending on the type of SMMs [31]. This affects the MD performance of the SMM-modified membranes. It was observed that the DCMD permeate fluxes of the SMM-blended PEI membranes using the SMMs (MDI/PPG/BA-L = 3:2:2) were higher than those of PEI membranes using the SMMs (MDI/PDMS/BA-L = 3:2:2) [31]. In fact, the fluorine content of the two SMMs are 11.45 wt% and 11.75 wt% for MDI/PPG/BA-L and MDI/PDMS/BA-L, respectively; whereas their weight average molecular weights (M_w) are 3.61×10^4 g/mol and 2.71×10^4 g/mol, respectively. The water contact angle of the modified PEI membranes using these two SMMs were found to be higher for the membrane modified with MDI/PDMS/BA-L compared to that of the membrane modified with MDI/PPG/BA-L. Both membranes were prepared under the same conditions except the type of SMMs. Furthermore, for the same SMM concentration, MDI/PDMS/BA-L-blended PEI membrane exhibited more fluorine than MDI/PPG/BA-L-blended PEI membrane. This was related to the fluorine concentration of the SMMs, which was found to be higher for MDI/PDMS/BA-L compared to MDI/PPG/BA-L. In addition, the LEP_w value of the MDI/PPG/BA-L-modified PEI membrane was smaller than that of the MDI/PDMS/BA-L-modified PEI membrane, indicating that this last membrane was more hydrophobic and/or had smaller maximum pore size than the membrane prepared with MDI/PPG/BA-L.

Effect of SMM Concentration

Although the amount of SMMs in the blend polymer solution is frequently maintained below 3 wt%, its effect on the membrane characteristics as well as on the MD performance cannot be ignored. In general, the water contact angle of the top membrane surface was found to be higher than that of the bottom surface and increased with the SMM concentration in the polymer solution. This indicates that the hydrophobicity of the SMM-modified membrane increases with increasing SMM concentration [7,31]. The effects of the SMM concentration on the water contact angle for some SMM-modified membranes are shown in Fig. 6.22. Similar results were observed by means of XPS analysis. Fluorine was not detected in the unmodified membrane since fluorine is associated only with SMMs. For all the SMM-blended membranes, fluorine contents at the top side were found to be higher than those at the bottom side, indicating SMM migration towards the top layer of the membranes. The fluorine content of the SMM-modified membranes increased by increasing the SMM concentration, as shown in Fig. 6.22.

On the other hand, no clear trends were observed between the DCMD permeate flux and the SMM concentration. This may be attributed to the simultaneous change of the pore size, porosity, pore tortuosity, etc. Figure 6.23 shows the DCMD permeate fluxes of three SMM-modified membranes prepared under the same conditions except for the SMM concentration. The DCMD permeate flux of all SMM-modified membranes increased exponentially

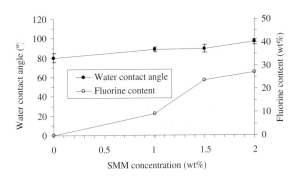

FIGURE 6.22 Effects of SMM concentration on water contact angle and fluorine content of SMM (MDI/PPG/BA-L = 3:2:2) modified PEI membranes.

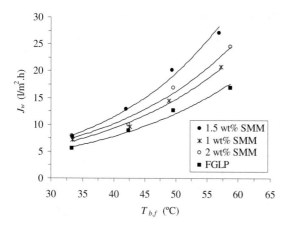

FIGURE 6.23 DCMD permeate flux (J_w) versus bulk feed temperature ($T_{b,f}$) of SMM (MDI/PPG/BA-L = 3:2:2) modified PEI membranes with different SMM concentrations (1 wt%, 1.5 wt%, 2 wt%) under the same preparation conditions and the commercial FGLP membrane (temperature difference 10 °C).

with the feed temperature and the SMM-modified membrane prepared with 1.5 wt% SMM showed the best performance among the tested membranes, despite the fact that the other membranes prepared with 1 wt% and 2 wt% SMM showed permeate flux enhancements by 20% and 40%, respectively, compared to the commercial membrane (FGLP). For all these membranes, the salt rejection factor was found to be more than 99%. The order in DCMD permeate fluxes was in good agreement with the measured gas (i.e. air) permeances, i.e in both cases the order was 1.5 wt% SMM > 2 wt% SMM > 1 wt% SMM [31]. These results validated the conclusion that the MD membranes with higher pore size and effective porosity will exhibit higher permeate fluxes. Another important observation is that the SMM-modified membranes exhibited higher permeate fluxes than the commercial membrane (FGLP), regardless of the SMM concentration used. This is probably due to the sponge-like structure of the hydrophilic layer, which seems a favourable structure for having higher MD permeate flux, and to the smaller thickness of the top hydrophobic layer.

Effect of Solvent

The selection of the adequate solvent for preparation of polymer blend containing SMMs, host polymer and additive(s) is an important initial task to carry out in order to prepare a homogeneous polymer solution. Until now the DMAC and NMP were the solvents used for the preparation of the SMM-modified membranes for MD. The effects of these two solvents on membrane characteristics and DCMD performance were investigated. It was observed that the SMM (MDI/PDMS/BA-L = 3:2:2)-modified PEI membranes using NMP exhibited higher DCMD permeate fluxes (i.e. around 11% higher) than those prepared using DMAC, with lower LEP_w value, smaller water contact angle and higher pore size and effective porosity [31]. These results are attributed to the speed of the SMM migration, which is faster in DMAC than in NMP, leading to higher water contact angle, greater fluorine content and smaller pore size. Based on cross-sectional SEM study, it was observed that the SMM-modified PEI membranes consisted of a bottom layer with fully developed macro-pores of finger-like structure, an intermediate layer of finger-like structure, and a skin layer on the top, regardless of the solvent. However, the macro-pores of the membrane prepared with DMAC were smaller. This can also explain why the permeate flux of SMM-modified PEI membrane became lower for DMAC solvent than for NMP solvent.

Effect of SMM Stoichiometric Ratio

As stated earlier, the type of SMMs affects considerably the characteristics of the SMM-modified membranes and their MD performance. Improvement of the DCMD membrane performance can be attempted by changing the stoichiometric ratio of the monomers involved in the SMM synthesis and therefore SMM structures. The two-step polymerization reaction presented in Fig. 6.8 can be followed to prepare SMMs with the same chemicals but different stoichiometries. For example,

(i) SMM1: 2(MDI)/1(PDMS)/2(BA-L)
(ii) SMM2: 3(MDI)/2(PDMS)/2(BA-L)
(iii) SMM3: 4(MDI)/3(PDMS)/2(BA-L).

The fluorine content as well as the molecular weight of the SMMs changed with the stoichiometric ratio. The measured fluorine contents of these SMMs were 16.21 wt%, 11.75 wt% and 10.06 wt%, respectively; and the molecular weights were 2.95×10^4 g/mol, 2.71×10^4 g/mol and 3.30×10^4 g/mol. The results showed that fluorine content decreased with increasing the ratio of PDMS to MDI.

PEI membranes were modified following the same procedure but using the three different SMMs (SMM1, SMM2 and SMM3). It was observed that the water contact angle of the modified PEI membranes with SMM1 (100.2°) and SMM3 (93.6°) were greater than that of the modified membrane with SMM2 (91.9°). In fact, the surface hydrophobicity of the SMM-blended membrane is determined by the interplay of the chemical formula and migration of SMMs. Hence, it is difficult to find a simple relationship between the water contact angle and the chemical formula of SMMs.

Fluorine content was found to be higher at the surface of the modified membrane with SMM1, followed by that prepared with SMM2 and then SMM3. This result could be related to the order in the fluorine contents of the SMMs (i.e. SMM1 > SMM2 > SMM3). It was also observed that the fluorine concentration at the top side of SMM1- and SMM2-blended PEI membranes was significantly higher than at their bottom sides. On the other hand, SMM3 blended PEI membrane exhibited only a small difference between the top and bottom sides. This indicated that the migration of SMM1 and SMM2 to the top membrane surface was much faster than that of SMM3.

The LEP_w values were found to be 4.0 bar for the SMM2-blended PEI membrane, 4.5 bar for the SMM1-blended PEI membrane and 4.7 bar for the SMM3-blended PEI membrane. Moreover, the order in the pore size as well as in the effective porosity were found to be SMM2 > SMM1 > SMM3. It was corroborated that the SMM-modified PEI membranes with SMM2 (i.e. 3:22) yielded the best DCMD performance among the tested membranes.

Effect of Evaporation Time

The effect of solvent evaporation time (i.e. period between casting blend polymer solution and coagulation of the cast film) on the SMM-modified membrane characteristics and DCMD performance was investigated using different solvents, host hydrophilic polymers and SMM types [28,29,31]. Figure 6.24 presents schematically SMM migration towards the top membrane surface, conformational rearrangement of SMMs at the surface and finally surface coverage with SMMs. Both ends of the SMM containing fluorine are oriented perpendicularly to the polymer solution/air interface, leading to exposure of these parts to air. Migration and conformational rearrangement of SMMs are kinetic processes

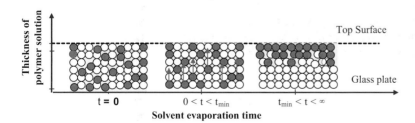

FIGURE 6.24 Schematic diagram illustrating SMM migration (filled circle: SMM macromolecule; open circle: polymer; t_{min}: minimum evaporation time to reach the equilibrium concentration at the membrane surface).

that require certain amount of times for completion. It seems reasonable to assume that the surface migration is slower than the conformational rearrangement.

It was observed that an increase in fluorine concentration and a significant depletion in carbon content took place at the top surface of the SMM-modified membranes with increased solvent evaporation time. In fact, high surface hydrophobicity could be achieved under mild membrane preparation conditions such as lower evaporation temperature (room temperature) and shorter evaporation period. SMM-modified membranes of high hydrophobicity were prepared even without solvent evaporation time, indicating SMM migration during coagulation step [26,31,66].

Figure 6.25 shows the effects of solvent evaporation time on the surface characteristics of the SMM membranes. As can be observed, the surfaces of all membranes are not smooth and possess nodule-like structure and nodule aggregates. The mean pore size of the unmodified membrane was larger than those of the SMM-modified membranes and decreased with the increase in the solvent evaporation time. It was also found that both the pore density and the surface porosity increased with the increase in the solvent evaporation time. The mean roughness of the unmodified membrane was higher than that of the SMM-modified membranes and decreased with the solvent evaporation time. The reduction in surface roughness may be attributed in part to the reduction of the

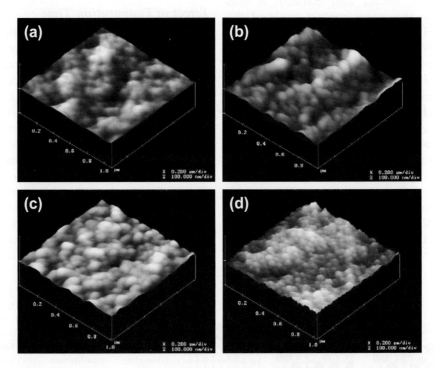

FIGURE 6.25 AFM images of the unmodified PEI membrane (a) and SMM (MDI/DPS/BA-L) modified PEI membranes prepared with different solvent evaporation times: (b) 0 min, (c) 3 min, (d) 5 min. *Source: Reprinted from [28]. Copyright 2004, with kind permission from Elsevier.*

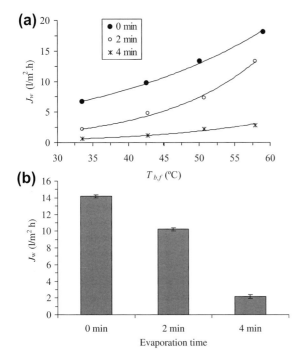

FIGURE 6.26 Effect of solvent evaporation time on DCMD permeate flux (J_w) of SMM (MDI/PDMS/BA-L = 3:2:2) modified PEI membranes: (a) feed temperature effect ($T_{b,f}$) on DCMD flux of distilled water; (b) 0.5 M NaCl in feed solution, feed temperature 65 °C and permeate temperature 15 °C.

pore size due to the migration of SMM towards the membrane surface.

The effect of the evaporation time on the DCMD permeate flux of the SMM-modified membranes was studied. (For example, Fig. 6.26 shows a decrease in DCMD flux with an increase in solvent evaporation period, when the other membrane preparation parameters are kept constant.). For all the tested membranes, the salt rejection factor was found to be higher than 99%. Moreover, it was observed that the LEP_w increased with solvent evaporation time, indicating that the SMM-modified membrane prepared with longer solvent evaporation time exhibited smaller maximum pore size. Both pore size and effective porosity were found to be higher without solvent evaporation, corroborating the highest DCMD permeate flux of the membrane prepared with direct coagulation after casting. Therefore, with respect to PEI as the host polymer and this particular SMM (MDI/PDMS/BA-L = 3:2:2), practically no evaporation time was necessary to make the membrane surface sufficiently hydrophobic to be used in MD.

Figure 6.27 shows the cross-sectional SEM images of the SMM (MDI/PDMS/BA-L) modified membranes prepared without solvent evaporation and with 4 min evaporation time. Both images consist of a bottom layer of fully developed macro-voids with a finger-like structure, an intermediate layer with a finger-like structure and a skin layer on the top. The size and number of macro-pores decreased with an increase in the evaporation time and the

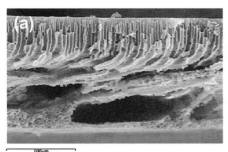
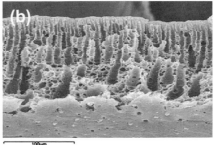

FIGURE 6.27 Cross-sectional SEM images of SMMs (MDI/PDMS/BA-L = 3:2:2) modified PEI membranes: (a) 0 min, (b) 4 min. *Source: Reprinted from [31]. Copyright 2009, with kind permission from Elsevier.*

sponge-like structure between macro-voids became more evident. The macro-voids were almost totally eliminated when the evaporation time was 4 min. This may also explain the dramatic decrease in DCMD permeate flux with the increase in the evaporation time, although the thin top hydrophobic layer governs the DCMD permeate flux of this type of membranes.

The effects of other membrane preparation parameters (coagulation temperature, solvent evaporation temperature, etc.) on MD performance of SMM-modified membranes should be further studied.

Co-Extrusion Spinning

Co-extrusion dry/jet wet spinning method was employed by Bonyadi and Chung [34] to prepare composite hydrophilic/hydrophobic hollow fibre membranes for DCMD. The method and the system for membrane preparation were outlined earlier in section 6.2 and Chapter 4. The outer hydrophobic layer was formed from a PVDF/NMP solution (12.5/87.5 wt/wt) containing 30 wt% hydrophobic cloisite 15A, had a thickness of about 50 μm and showed a contact angle of 136°. The hydrophilic inner layer was formed from a PVDF/PAN/NMP solution (8.5/4/87.5 wt/wt) containing 50 wt% hydrophilic cloisite NA$^+$ with a contact angle as low as 50°. Methanol/water (i.e. 80/20 wt/wt) solution was chosen as the coagulant after studying the effects of different coagulant compositions on the morphology of PVDF membranes. It was found that a weak coagulant such as water/methanol (20/80 wt/wt) can induce a three-dimensional porous structure on PVDF membranes with high surface and bulk porosities, big pore sizes, sharp pore size distributions, high surface contact angles and high permeabilities but rather weak mechanical properties. To carry out characterization experiments, including the contact angle measurements, flat sheet membranes were prepared by the phase inversion technique. Figure 6.28 shows the SEM images of the top surface of PVDF flat sheet membranes prepared using different coagulant compositions. When a strong coagulant, i.e. water, was employed, a dense and smooth surface was formed. The roughness of the membrane surface as well as the number and size of the pores increased when weak coagulant, i.e. 40% methanol in water, was used. When the coagulant was 80 wt% methanol in water, a three-dimensional porous structure was observed with larger pore sizes (i.e. mean pore size 0.41 μm). This change in membrane surface structure is due to the delay of demixing, spinodal decomposition and coarsening as explained in Chapters 3 and 4. The apparent spherulitical structure upon the fibre network may be due to coarsening effects. Briefly, for PVDF/NMP/water system, the rapid phase separation of

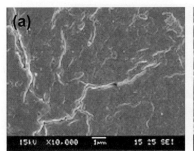  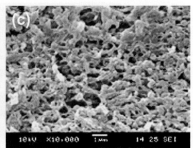

FIGURE 6.28 Effect of coagulant on PVDF membrane surface morphology: (a) water; (b) 40 wt% methanol in water, 80 wt% methanol in water. *Source: Reprinted from [34]. Copyright 2007, with kind permission from Elsevier.*

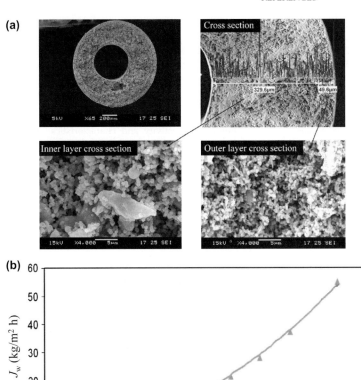

FIGURE 6.29 SEM images of (a) composite hydrophobic/hydrophilic hollow fibre membranes and ($T_{b,f}$) effect of bulk feed inlet temperature on DCMD permeate flux (J_w). *Source: Reprinted from [34]. Copyright 2007, with kind permission from Elsevier.*

polymer solution may occur via nucleation growth producing a rough and dense membrane surface. By contrast, for the system PVDF/NMP/(80 wt% methanol in water), demixing is delayed and phase separation is slow. When increasing methanol concentration in coagulant, the coagulation path may go through spinodal decomposition forming a three-dimensional structure with larger pore sizes [34].

The MD performance of the composite hydrophobic/hydrophilic hollow fibre membrane was similar to single-layer hollow fibre membranes (Fig. 6.29), i.e. the water vapour flux increased exponentially with an increase in feed inlet temperature. However, the composite hollow fibre membrane exhibited much higher DCMD permeation flux than most of MD membranes. In fact, the DCMD permeate flux was as high as 55.2 kg/m²·h with a salt separation factor of 99.8% using a 3.5 wt% NaCl aqueous solution, a feed temperature of 90.3 °C and a permeate temperature of 16.5 °C.

References

[1] R.E. Kesting, Synthetic polymer membranes, McGraw Hill, New York, 1972.
[2] T. Matsuura, Synthetic membranes and membrane separation processes, CRC Press, Boca Raton, USA, 1994.

[3] I. Pinnau, B.D. Freeman, Membrane formation and modification, ACS symposium series 744, American Chemical Society, Washington, DC, USA, 2000.

[4] M. Khayet, T. Matsuura, J.I. Mengual, M. Qtaishat, Design of novel direct contact membrane distillation membranes, Desalination 192 (2006) 105−111.

[5] R.J. Peterson, Composite reverse osmosis and nanofiltration membranes, J. Membr. Sci. 83 (1993) 81−150.

[6] J.M.S. Henis, M.K. Tripodi, A novel approach to gas separations using composite hollow fiber membranes, Sep. Sci. Tech. 15 (1980) 1059−1068.

[7] M. Khayet, D.E. Suk, R.M. Narbaitz, J.P. Santerre, T. Matsuura, Study on surface modification by surface-modifying macromolecules and its applications in membrane-separation processes, J. Appl. Poly. Sci. 89 (2003) 2902−2916.

[8] M. Khayet, G. Chowdhury, T. Matsuura, Surface modification of polyvinylidene fluoride pervaporation membranes, AIChE J. 48 (2002) 2833−2843.

[9] M. Khayet, T. Matsuura, Application of surface modifying macromolecules for the preparation of membranes for membrane distillation, Desalination 158 (2003) 51−56.

[10] L. Zou, I. Vidalis, D. Steele, A. Michelmore, S.P. Low, J.Q.J.C. Verberk, Surface hydrophilic modification of RO membranes by plasma polymerization for low organic fouling, J. Membr. Sci. 369 (2011) 420−428.

[11] W.R. Browall, Method for sealing breaches in multilayer ultrathin membrane composites, U.S. Patent 3,980,456 (1976).

[12] Y. Wu, Y. Kong, X. Lin, W. Liu, J. Xu, Surface-modified hydrophilic membranes in membrane distillation, J. Membr. Sci. 72 (1992) 189−196.

[13] Y. Kong, X. Lin, Y. Wu, J. Cheng, J. Xu, Plasma polymerization of octafluorocyclobutane and hydrophobic microporous composite membranes for membrane distillation, J. Appl. Polym. Sci. 46 (1992) 191−199.

[14] B. Li, K.K. Sirkar, Novel membrane and device for direct contact membrane distillation-based desalination process, Ind. & Eng. Chem. Res. 43 (2004) 5300−5309.

[15] B. Li, K.K. Sirkar, Novel membrane and device for vacuum membrane distillation-based desalination process, J. Membr. Sci. 257 (2005) 60−75.

[16] L. Song, B. Li, K.K. Sirkar, J.L. Gilron, Direct contact membrane distillation-based desalination: Novel membranes, devices, larger-scale studies, and a model, Ind. & Eng. Chem. Res. 46 (2007) 2307−2323.

[17] A. Larbot, L. Gazagnes, S. Krajewski, M. Bukowska, W. Kujawski, Water desalination using ceramic membrane distillation, Desalination 168 (2004) 367−372.

[18] S.R. Krajewski, W. Kujawski, M. Bukowska, C. Picard, A. Larbot, Application of fluoroalkylsilanes (FAS) grafted ceramic membranes in membrane distillation process of NaCl solutions, J. Membr. Sci. 281 (2006) 253−259.

[19] Y. Fujii, S. Kigoshi, H. Iwatani, M. Aoyama, Selectivity and characteristics of direct contact membrane distillation type experiment: I. Permeability and selectivity through dried hydrophobic fine porous membranes, J. Membr. Sci. 72 (1992) 53−72.

[20] Y. Fujii, S. Kigoshi, H. Iwatani, M. Aoyama, Y. Fusaoka, Selectivity and characteristics of direct contact membrane distillation type experiment: II. Membrane treatment and selectivity increase, J. Membr. Sci. 72 (1992) 73−89.

[21] Z. Jin, D.L. Yang, S.H. Zhang, X.G. Jian, Hydrophobic modification of poly(phthalazinone ether sulfone ketone) hollow fiber membrane for vacuum membrane distillation, J. Membr. Sci. 310 (2008) 20−27.

[22] J.B. Xu, S. Lange, J.P. Bartley, R.A. Johnson, Alginate-coated microporous PTFE membranes for use in the osmotic distillation of oily feeds, J. Membr. Sci. 240 (2004) 81−89.

[23] P. Peng, A.G. Fane, X. Li, Desalination by membrane distillation adopting a hydrophilic membrane, Desalination 173 (2005) 45−54.

[24] M. Khayet, T. Matsuura, Preparation and characterization of polyvinylidene fluoride membranes for membrane distillation, Ind. & Eng. Chem. Res. 40 (2001) 5710−5718.

[25] A. Bottino, G. Capannelli, A. Comite, Novel porous poly(vinylidene fluoride) membranes for membrane distillation, Desalination 183 (2005) 375−382.

[26] M. Khayet, J.I. Mengual, T. Matsuura, Porous hydrophobic/hydrophilic composite membranes: Application in desalination using direct contact membrane distillation, J. Membr. Sci. 252 (2005) 101−113.

[27] M. Khayet, T. Matsuura, J.I. Mengual, Porous hydrophobic/hydrophilic composite membranes: Estimation of the hydrophobic-layer thickness, J. Membr. Sci. 266 (2005) 68−79.

[28] M. Khayet, Membrane surface modification and characterization by X-ray photoelectron spectroscopy, atomic force microscopy and contact angle measurements, Appl. Surf. Sci. 238 (2004) 269−272.

[29] D.E. Suk, T. Matsuura, H.B. Park, Y.M. Lee, Synthesis of a new type of surface modifying macromolecules (nSMM) and characterization and testing of nSMM blended membranes for membrane distillation, J. Membr. Sci. 277 (2006) 177−185.

[30] M. Qtaishat, M. Khayet, T. Matsuura, Guidelines for preparation of higher flux hydrophobic/hydrophilic composite membranes for membrane distillation, J. Membr. Sci. 329 (2009) 193−200.

[31] M. Qtaishat, D. Rana, M. Khayet, T. Matsuura, Preparation and characterization of novel hydrophobic/

hydrophilic polyetherimide composite membranes for desalination by direct contact membrane distillation, J. Membr. Sci. 327 (2009) 264—273.

[32] M. Qtaishat, D. Rana, T. Matsuura, M. Khayet, Effect of surface modifying macromolecules stoichiometric ratio on composite hydrophobic/hydrophilic membranes characteristics and performance in direct contact membrane distillation, AIChE J. 55 (2009) 3145—3151.

[33] M. Qtaishat, M. Khayet, T. Matsuura, Novel porous composite hydrophobic/hydrophilic polysulfone composite membranes for desalination by direct contact membrane distillation, J. Membr. Sci. 341 (2009) 139—148.

[34] S. Bonyadi, T.S. Chung, Flux enhancement in membrane distillation by fabrication of dual layer hydrophilic-hydrophobic hollow fiber membranes, J. Membr. Sci. 306 (2007) 134—146.

[35] DY. Cheng, Method and apparatus for distillation, United States Patent Serial No. 4,265,713, (1981).

[36] DY Cheng, SJ. Wiersma, Composite membrane for a membrane distillation system, United States Patent Serial No. 4,316,772, (1982).

[37] DY Cheng, SJ. Wiersma, Composite membrane for a membrane distillation system, United States Patent Serial No. 4,419,242, (1983).

[38] DY Cheng, SJ. Wiersma, Apparatus and method for thermal membrane distillation, United States Patent Serial No. 4,419,187, (1983).

[39] FA. Rodgers, Stacked microporous vapor permeable membrane distillation system, United States Patent Serial No. 3, 650,905, (1972).

[40] F.A. Rodgers, Compact multiple effect still having stacked impervious and previous membranes, United States Patent Serial No. Re. 27,982 (1974). original No. 3,497,423.

[41] K. Ohta, I. Hayano, T. Okabe, T. Goto, S. Kimura, H. Ohya, Membrane distillation with fluoro-carbon membranes, Desalination 81 (1991) 107—115.

[42] K. Ohta, K. Kikuchi, I. Hayano, T. Okabe, T. Goto, S. Kimura, H. Ohya, Experiments on sea water desalination by membrane distillation, Desalination 78 (1990) 177—185.

[43] Belleville MP, Dornier M, Sánchez J, Vaillant F, Rynes M, Rios G., Procedé d'extraction para diffusion membranaire, French patent application No. 0205959, 2002.

[44] F. Brodard, J. Romero, M.P. Belleville, J. Sánchez, C. Combe-James, M. Dornier, G.M. Rios, New hydrophobic membranes for osmotic evaporation process, Sep. Pur. Tech. 32 (2003) 3—7.

[45] L.T. Rozelle, J.E. Cadotte, K.E. Cobian, C.V. Kopp, Nonpolysaccharide membrane for reverse osmosis: NS-100 membranes, in: S. Sourirajan (Ed.), Reverse Osmosis and Synthetic Membranes: Theory, Technology, Engineering, National Research Council of Canada, Ottawa, 1977.

[46] T. Yamaguchi, F. Miyata, S. Nakao, Pore-filling type polymer electrolyte membranes for a direct methanol fuel cell, J. Membr. Sci. 214 (2003) 283—292.

[47] A.M. Mika, R.F. Childs, J.M. Dickson, B.E. McCarry, D.R. Gagnon, A new class of polyelectrolyte-filled microfiltration membranes with environmentally controlled porosity, J. Membr. Sci. 108 (1995) 37—56.

[48] A. Wenzel, H. Yanagishita, D. Kitamoto, A. Endo, K. Haraya, T. Nakane, N. Hanai, H. Matsuda, N. Koura, H. Kamusewitz, D. Paul, Effects of preparation condition of photoinduced graft filling-polymerized membranes on pervaporation performance, J. Membr. Sci. 179 (2000) 69—77.

[49] T. Yamaguchi, S. Nakao, S. Kimura, Plasma-graft filling polymerization: preparation of a new type of pervaporation membrane for organic liquid mixtures, Macromolecules 24 (1991) 5522—5527.

[50] W. Jiang, R.F. Childs, A.M. Mika, J.M. Dickson, Pore-filled cation-exchange membranes containing poly(styrenesulfonic acid) gels, Desalination 159 (2003) 253—266.

[51] R. Simons, J. Zuccon, M.R. Dickson, M. Shaw, Pervaporation and evaporation characteristics of a new type of ion exchange membrane, J. Membr. Sci. 78 (1993) 63—67.

[52] M. Khayet, M.M. Nasef, J.I. Mengual, Radiation grafted poly(ethylene terephthalate)-graft-polystyrene pervaporation membranes for organic/organic separation, J. Membr. Sci. 263 (2005) 77—95.

[53] H. Yasuda, Plasma polymerization for protective coatings and composite membranes, J. Membr. Sci. 18 (1984) 273—284.

[54] H. Yasuda, Composite reverse osmosis membranes prepared by plasma polymerization, in: S. Sourirajan (Ed.), Reverse osmosis and synthetic membranes: Theory, technology, engineering, National Research Council Canada, Ottawa, 1977.

[55] A.F. Stancell, A.T. Spencer, Composite permselective membrane by deposition of an ultrathin coating from a plasma, J. Appl. Polym. Sci. 16 (1972) 1505—1514.

[56] N. Inagaki, J. Ohkubo, Plasma polymerization of hexafluoropropene/methane mixture and composite membranes for gas separations, J. Membr. Sci. 27 (1986) 63—75.

[57] K. Li, Ceramic membranes for separation and reaction, John Wiley & Sons, Ltd, Chichester, England, 2007.

[58] J. Font, R.P. Castro, Y. Cohen, On the loss of hydraulic permeability in ceramic membranes, J. Colloid Interface Sci. 181 (1996) 347—350.

[59] M. Rovira-Bru, F. Giralt, Y. Cohen, Protein adsorption onto zirconia modified with terminally grafted polyvinylpyrrolidone, J. Colloid Interface Sci. 235 (2001) 70–79.

[60] R.P. Castro, Y. Cohen, H.G. Monbouquette, The permeability behavior of polyvinylpyrrolidone-modified porous silica membranes, J. Membr. Sci. 84 (1993) 151–160.

[61] J.D. Jou, W. Yoshida, Y. Cohen, A novel ceramic-supported polymer membrane for pervaporation of dilute volatile organic compounds, J. Membr. Sci. 162 (1999) 269–284.

[62] W. Yoshida, Y. Cohen, Removal of methyl tert-butyl ether from water by pervaporation using ceramic-supported polymer membranes, J. Membr. Sci. 229 (2004) 27–32.

[63] S.R. Krajewski, W. Kuhawski, F. Dijoux, C. Picard, A. Larbot, Grafting of ZrO2 powder and ZrO2 membrane by fluroalkylsilanes, Colloids & Surf. A: Phys. Eng. Aspects 243 (2004) 43–47.

[64] M.E. Rezac, W.J. Koros, Preparation of polymer-ceramic composite membranes with thin defect-free separating layers, J. Appl. Poly. Sci. 46 (1992) 1927–1938.

[65] M. Khayet, M.V. Álvarez, K.C. Khulbe, T. Matsuura, Preferential surface segregation on homopolymer and copolymer blend films, Surf. Sci. 602 (2007) 885–895.

[66] M. Khayet, C.Y. Feng, T. Matsuura, Morphological study of fluorinated asymmetric polyetherimide ultrafiltration membranes by surface modifying macromolecules, J. Membr. Sci. 213 (2003) 159–180.

[67] M. Qtaishat, T. Matsuura, M. Khayet, K.C. Khulbe, Comparing the desalination performance of SMM blended polyethersulfone to SMM blended polyetherimide membranes by direct contact membrane distillation, Desalination & Water Treatment 5 (2009) 91–98.

CHAPTER 7

Formation of Nano-Fibre MD Membranes

OUTLINE

Introduction	163	Electrical Conductivity of the Solution	176
Principles of the Formation of Nano-Fibre Membranes	166	Effect of Process Parameters	178
		Applied Voltage	178
Effects of Electro-Spinning System Parameters and Process Parameters on the Membrane Structure	170	Polymer Flow Rate	179
		Temperature	180
		Needle Tip	180
Effect of System Parameters	172	Gap Distance Between Needle Tip and Collector	180
Polymer Concentration	172		
Molecular Weight	174	Post-Treatment	182
Solvent	175	Ambient Parameters	184

INTRODUCTION

Nano-scale materials can be rationally designed to exhibit novel and significantly improved physical and chemical properties. Polymer nano-fibres, an important class of nano-materials, have attracted increasing attentions in the last ten years because of their high surface-to-mass (or volume) ratio and special characteristics attractive for advanced applications [1–16]. Within the connotation of nanotechnology and nano-structured materials, a nano-fibre generally refers to a fibre having a diameter less than 100 nm. The high ratio of surface area to mass is a primary characteristic of nano-fibres. These can be fabricated by several techniques such as electro-spinning or electro-static spinning [1–16], melt-blown [17,18], phase separation [19], molecular self assembly [20–22] and template synthesis [23,24].

Electro-spinning, also called electrostatic spinning, is the most popular and preferred technique to use for preparation of polymeric nano-fibres. It is simple, cost-effective and able to produce continuous nano-fibres of various materials. It is also important to organize nano-fibres of various types (e.g. porous, hollow and core/sheath) into well-defined arrays or hierarchical architectures in three-dimensional network. In addition, electro-spinning seems to be the only method that can be further developed for large-scale production of continuous nano-fibres for industrial applications. It must be pointed out that for fibres spun from polymer solutions, the presence of residual solvent in the electro-spun fibres facilitates bonding of intersecting fibres, creating a strong cohesive porous structure with high void volume.

Unlike conventional fibre spinning techniques (wet spinning, dry spinning, melt spinning and gel spinning), which are capable of producing polymer fibres with diameters down to the micrometer range, electro-spinning is a process capable of producing polymer fibres in the nanometer diameter range through the application of an external electric field, and thus high surface-to-volume ratio and a high density of pores can be formed. The surface area of nano-fibres could be further increased through the formation of pores in the matrix of each individual fibre by controlling the solutions and parameters for electro-spinning [1,2]. Due to the inherent properties of the electro-spinning process, which can control the deposition of polymer fibres onto a target substrate, nano-fibres with complex and seamless three-dimensional shapes could be formed. When no support (i.e. backing material) is used to collect the nano-fibres, the obtained membrane is named non-woven mat or fibrous mat.

It must be pointed out that polymeric nano-fibres produced by electro-spinning have become a topic of great interest for the past few years, although the process of electro-spinning has been known for almost 70 years (the first patent was issued to Formhals in 1934 [25]), wherein an experimental setup was outlined for the production of polymer filaments using electro-static force. Until 1993, this technique had been known as electrostatic spinning, and there were only few publications dealing with its use in the fabrication of thin fibres [26].

Typically, electro-spinning is applicable to a wide range of polymers like those used in conventional spinning (i.e. polyolefin, polyamides, polyester as well as biopolymers). A rich variety of electro-spun homopolymer and copolymer fibres (i.e. cellulose acetate, polyvinylidene fluoride, polyethylene oxide, polyurethane, polyethylene terephthalate, polystyrene, poly(2-hydroxyethyl methacrylate), styrene-butadiene-styrene triblock copolymer, etc.) with diameters ranging from several tens of nanometers to several microns are being fabricated and studied for applications in such diverse fields as affinity membrane, tissue engineering scaffold, drug deliver carrier, biosensor, electronic and semi-conductive materials and reinforced nanocomposite, etc., as can be seen in Fig. 7.1. For example, polyvinylidene fluoride (PVDF) nano-fibre web was used as a polymer electrolyte binder or a separator for a battery [9]. The obtained diameters of the electro-spun PVDF nano-fibres ranged from 100 to 800 nm. PVDF with carbon nano-tubes were also electro-spun successfully through the control of the solution viscosity and surface tension [8].

It is worth quoting that the nano-fibres assembled into a membrane-like structure exhibit among others good tensile strength, excellent moisture vapour transport, good resistance to the penetration of chemical and biological agents, large surface-area per unit mass, highly ordered polymer chains, more controllable membrane parameters (pore size, porosity, thickness), etc. For instance, nano-fibres with a diameter of 100 nm have a ratio of geometrical surface area to mass of approximately 100 m^2/g (the very large surface area to volume ratio for a nano-fibre can be as large as 10^3 times of that of a micro-fibre). Moreover, electro-spun nano-fibre membranes may be produced over a wide range of porosity and pore size values, from nearly non-porous polymer coatings, to

INTRODUCTION

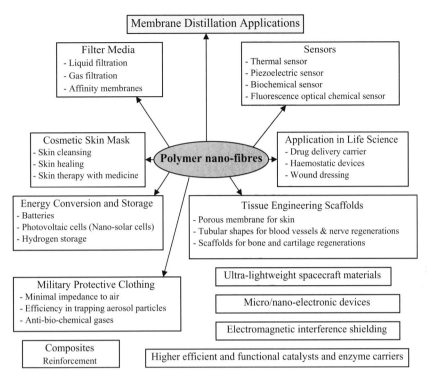

FIGURE 7.1 Application fields of polymer nano-fibres.

very porous and delicate fibrous structures. All these characteristics make nano-fibres interesting candidates for a wide variety of applications and ideal substrate for separation processes including desalination by membrane distillation (MD).

Recently, nano-fibre membranes were proposed for MD [27]. A PVDF membrane was prepared by the electro-spinning method. Scanning electron microscopy (SEM) and atomic force microscopy (AFM) images were presented in Chapter 2 (Fig. 2.4). The PVDF nano-fibre membrane was applied for desalination of saline water by air gap membrane distillation (AGMD) at different NaCl feed concentrations, feed temperatures and permeate temperatures. Similar trends to those of the other types of MD membranes were observed in AGMD experiments. A permeate flux as high as 11.5 kg/m²h with a NaCl rejection factor higher than 98.5% was obtained. It was also found that the nano-fibre membrane was intact and unplugged after many days of operation. These results opened up a new avenue of the research in MD membrane science and technology, both experimental and theoretical. Studies of this new promising generation of MD membranes using other polymers, solvents and additives should be carried out to improve the MD performance of nano-fibre membranes. The effects of the electro-spinning parameters on the MD performance should also be studied. Membranes with a high and controlled void volume will be achieved by designing nano-structured membranes based on nano-fibres and micro-fibres. The structural properties of the electro-spun non-woven fibrous membranes include high surface area to volume ratio, micro scaled interstitial space, high void volume and inter-connectivity, good morphology, stability and controllable mesh parameters. This option seems to be a relatively simple solution that fulfils all the conditions needed for a MD membrane outlined in Chapter 2 and for achieving high permeability and low thermal conductivity.

It is worth quoting that nano-structured MD membranes with different parameters and materials are needed to help understanding the physical nature of mass transport in different MD configurations and to perform systematic studies on the effects of membrane parameters on MD performance. There are many parameters that will influence the morphology of the resultant electro-spun fibres as well as their arrangements in membranes.

These parameters may be broadly classified into polymer solution parameters and processing conditions, as shown in Chapter 6, Section 'Effects of process parameters on the membrane structure and MD performance'.

PRINCIPLES OF THE FORMATION OF NANO-FIBRE MEMBRANES

The schematic for a typical electro-spinning system is depicted in Fig. 7.2. The electro-spinning process, in its simplest form, consists of a syringe to hold the polymer solution connected to a circulation pump, two electrodes (a spinneret or a metallic needle and a grounded conductor collector) and a DC voltage supply in the kV range. Direct current (DC) power

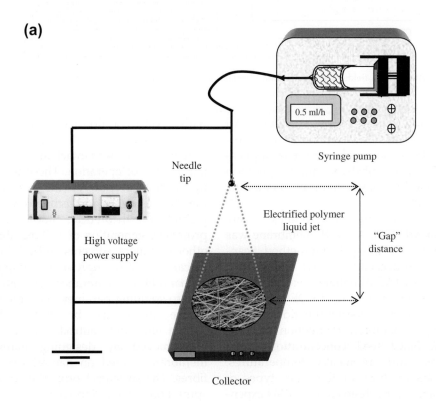

FIGURE 7.2 Schematic diagram of electro-spinning systems (a) plate collector, (b) rotating grounded collection drum.

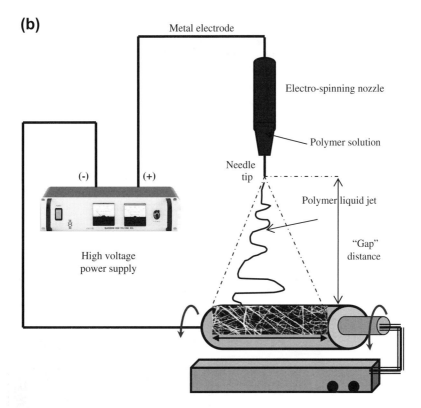

FIGURE 7.2 (*continued*).

supplies are usually used for electro-spinning although the use of alternating current (AC) potentials is also feasible. The polymer drop from the tip of the needle connected to the syringe by a Teflon tube is drawn into a fibre due to the high voltage. The jet is electrically charged and the charge causes the fibres to bend. Every time the polymer fibre loops in the air gap, its diameter is reduced. Finally, the fibre is collected as a web of fibres on the surface of a grounded metallic target or collector.

As can be seen in Fig. 7.2, electro-spinning is realized by applying a high voltage to a capillary filled with the polymer fluid to be spun with help of an electrode. The arrows in Fig. 7.3(a) indicate the direction of the electrostatic field lines and the length of the lines are a qualitative indication of the field strength. When an external electric field is applied to a polymer solution, ions will aggregate around the electrode of opposite polarity. When a positive voltage is applied to the metal syringe, the ions of the polymer solution of like-polarity will be forced to aggregate at the surface of the drop suspended at the tip of the metal syringe. The electric field generated by the surface charge will cause the pendant drop of the polymer solution at the capillary tip to deform into a conical shape. The drop will experience two major types of electrostatic forces, the electrostatic repulsion between the surface charges and the Coulombic force exerted by the external electric field. When a critical

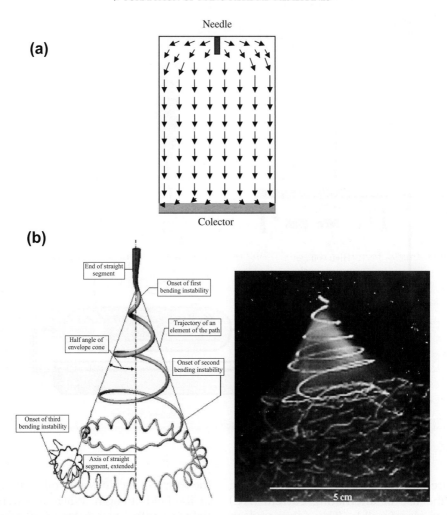

FIGURE 7.3 (a) Profile of the electrostatic field and (b) a picture together with a schematic representation of an electrospinning jet. *Source: Reprinted from [16]. Copyright 2008, with kind permission from Elsevier.*

voltage (typically 5 kV depending on the polymer solution used) is exceeded, the electrostatic forces act in opposition to, and overcome the solution surface tension of the fluid and a charged thin jet of solution is erupted from the surface of the cone. In fact, the surface tension of the fluid favours sphere-like shapes with smaller surface area per unit mass. The jet moves towards a ground plate acting as a counter electrode or collector (plate, rotating metal cylinder, or conducting coagulation bath). Due to the viscosity of the polymer solution and the presence of entanglements, the jet does not transform into spherical droplets as expected for a liquid cylindrical thread. As the electrified jet travels through the air, the solvent evaporates while the polymer fibre is stretched, elongated, whipped and finally deposited on the form of a non-woven mat or over a support (i.e. backing material). Figure 7.3(b) shows the random motion of the electro-spinning jet indicating the chaotic (i.e. instability) nature of the electro-spinning jet motion. This jet instability is the result of different interaction variables including electrostatic forces, gravity, air friction, polymer viscosity and surface tension. Reneker et al. [28] developed a model of jet motion taking into consideration these variables. However, detailed understanding of the jet motion remains elusive. The typical path of the jet, as schematized in Fig. 7.3(b), is composed of a straight segment followed by a coil of increasing diameter driven by the resultant lateral repulsive electric force in the radial direction with respect to the straight jet. After some turns are formed, a new electrical bending instability forms a smaller coil on a turn of the larger coil. The perimeter of each turn of the coils increases progressively and the turns become smaller and thinner and so on until the elongation of the jet is stopped by its solidification. The irregular and broken lines that appear in Fig. 7.3(b) near the bottom of the image are glints from the multitude of smaller coils produced by high bending instabilities [16].

Fibre packing is affected by the conductivity of the collectors. An electrically conductive collector dissipates electric charges and reduces the repulsion among fibres, thus favouring a tightly packed and thick membrane structure. On a non-conductive collector, the presence of electrostatic charges causes fibres to repel each other, giving a more loosely packed fibrous network [29].

It must be mentioned here that the setup for electro-spinning can be modified to directly generate nano-fibres with core-shell or hollow structures, and as uniaxially aligned arrays or layer-by-layer stacked films [1,2]. Core/sheath nano-fibres or hollow nano-fibres (i.e. nano-tubes) could be fabricated by co-electro-spinning two different polymer solutions through a spinneret comprising two coaxial capillaries. With the use of conventional setup for electro-spinning, it is possible to fabricate core/sheath nano-fibres or hollow structure with a polymer solution containing two polymers that will phase separate as the solvent is evaporated. For example, Megelski et al. [6] observed core/sheath structures when a solution containing poly(carbonate) and poly(butadiene) was electro-spun into thin fibres. By controlling the spinning conditions (e.g. electric field strength, concentration of sheath liquid, and the feeding rates for both liquids), the sheath thickness and inner diameter of the nano-fibres could be varied in the range from tens of nanometers to several hundred nanometers. The surface-area of a nano-fibre can be greatly increased when its structure is switched from a solid to a porous one. It is to be noted that an increase of surface-areas is beneficial to many applications that include catalysis, filtration, absorption, fuel cells, solar cells, batteries, tissue engineering and MD.

Phase separation processes discussed in previous chapters also take place during electro-spinning and nano-fibres of spinodal or binodal phase morphologies were obtained [5]. Nucleation and crystallization occurred inside nano-fibres. Two slightly different approaches have been reported for introducing a porous structure into the bulk of an electro-spun nano-fibre. One of them was based on the selective removal of a component from nano-fibres made of a composite or blend material. The other one involved the use of phase separation of different polymers during electro-spinning under the application of proper spinning parameters [4,5].

In general, control of the electro-spinning process produces fibres with nanometer scale diameters, along with various cross-sectional shapes, beads, branches and buckling coils or zigzag. The addition to the polymer solution to be electro-spun of different chemical reagents, other polymers, dispersed particles, proteins as well as the application of different post-treatments of nano-fibres and nano-structured membranes such as chemical or thermal treatments, broaden the usefulness of the nano-structured membranes.

The following are some important features of electro-spinning:

(i) Suitable solvent should be available for dissolving the polymer.
(ii) The vapour pressure of the solvent should be suitable so that it evaporates quickly enough for the fibre to maintain its integrity when it reaches the target but not too quickly to allow the fibre to harden before it reaches the nanometer range.
(iii) The viscosity and surface tension of the solvent must neither be too large to prevent the jet from forming nor be too small to allow the polymer solution to drain freely from the needle.
(iv) The power supply should be adequate to overcome the viscosity and surface tension of the polymer solution to form and sustain the jet from the needle.
(v) The gap between the needle and grounded surface should not be too small to create sparks between the electrodes and should be large enough for the solvent to evaporate in time for the fibres to form.

EFFECTS OF ELECTRO-SPINNING SYSTEM PARAMETERS AND PROCESS PARAMETERS ON THE MEMBRANE STRUCTURE

As stated earlier, up to now the only published study using electro-spun nano-fibre membranes in MD was carried out by Feng et al. [27]. Only one nano-fibrous mat was prepared and tested in MD. The process parameters were as follows. The polymer dope was prepared by dissolving 18 wt% PVDF (Kynar®761, Elf-Chem USA) in dimethylformamide (DMF). The voltage applied was 18 kV. The flow rate of the dope solution was 2 ml/h. The air gap distance (i.e. length from the needle tip to the metal plate collector) was 18 cm. Under such process parameters, 6 ml of polymer solution was consumed to prepare 0.15 mm thick nano-fibrous mat. The nano-fibre membrane was further dried in a fume hood for 24 h at room temperature.

The determined diameter of the nano-fibre membrane was about 500 nm and the contact angle of the PVDF nano-fibre membrane was found to be much higher (130°) than that of the dense PVDF membrane prepared by the phase inversion technique. The contact angle of nano-fibre membrane was greater than the dense membrane also for nano-fibrous membranes prepared from other materials. The enhanced hydrophobic property of the MD membrane surface is particularly important, since it prevents the penetration of the feed and/or permeate liquid into the membrane. This characteristic of nano-fibrous membranes makes them more attractive for MD applications than the phase inversion membranes. Figure 2.4 shows the SEM and AFM images of the nano-fibres. The rough surface observed for individual nano-fibres is due to the presence of nodules oriented in the direction of the jet axis. This may also explain together with the air contained in the nano-fibre mat the higher water contact angles observed for the electro-spun nano-fibre membrane.

The PVDF nano-fibre membrane was applied for desalination of saline water by AGMD at different NaCl feed concentrations (1–6 wt%) and temperature differences between the feed and permeate varying from 15 °C to 60 °C. Similar AGMD trends as those of the other types of MD membranes were observed. A permeate flux as high as 11.5 kg/m^2 h with a NaCl rejection factor higher than 98.5% was obtained. It was also found that the nano-fibre membrane was intact and unplugged after many days of operation. As presented in Fig. 7.4, the NaCl rejection did not change considerably during the 25 days operation and no decrease in the permeate flux with time was observed, showing

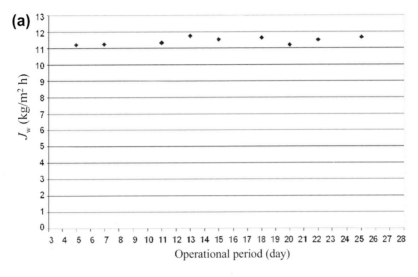

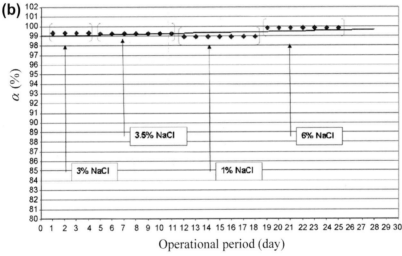

FIGURE 7.4 MD performance of PVDF nano-fibre membrane: (a) AGMD permeate flux (J_w) *versus* operational time at the NaCl feed aqueous solution of 3.5 wt% and a transmembrane temperature difference of 60 °C, (b) NaCl rejection factor (α) *versus* operational time for different concentrations of NaCl feed aqueous solutions and a transmembrane temperature difference of 60°C. *Source: Reprinted from [27]. Copyright 2008, with kind permission from Elsevier.*

that the membrane was still applicable after a long-term testing. It was claimed that the MD permeate flux was comparable to the permeate fluxes of the commercial microfiltration (MF) membranes (5–28 kg/m² h) at transmembrane temperatures ranging from 25 to 83 °C and this approach would eventually enable the MD process to compete with conventional seawater desalination processes. These results opened up a new avenue of the research in MD membrane science and technology, both experimental and theoretical.

It was observed that the following parameters and processing variables affect the morphological structure of electro-spun fibres [1,2]:

(i) System parameters such as type, molecular weight, molecular-weight distribution and architecture (branched, linear, etc.) of the polymer, polymer concentration, solvent type and polymer solution properties (viscosity, conductivity and surface tension).

(ii) Process parameters such as electric potential, flow rate of polymer solution, distance between the capillary and collection screen, ambient parameters (temperature, humidity and air velocity in the chamber) and finally motion of collector.

In what follows, only brief descriptions of the effects of some of these parameters on the physical characteristics of the nano-fibres are given. With the understanding of both the system and process electro-spinning parameters, it will be possible to prepare novel nano-fibrous MD membranes with various forms and arrangements and create nano-fibres with different morphologies by varying the electro-spinning parameters.

The properties of the polymer dope have the most significant influence on the resultant nano-fibre morphology and nano-structured supported and unsupported membranes (nano-fibrous mat). For example, the viscosity of the polymer solution together with its electrical properties affects the extent of elongation of the formed jet and as a consequence affects the diameter of the resultant nano-fibre. The surface tension plays an important role in the formation of beads along the fibre length.

Effect of System Parameters

Polymer Concentration

One of the most important parameters related to electro-spinning process is the fibre diameter. Since nano-fibres are resulted from evaporation or solidification of polymer fluid jets, the fibre diameters depend primarily on the jet sizes as well as on the polymer contents in the jets. In general, higher polymer concentration dissolved in a solvent results in a higher viscosity of the solution, and a higher viscosity results in a larger fibre diameter. In fact, formation of nano-fibre depends on both the polymer concentration and the viscosity of the dope solution. Both the viscosity of the polymer solution and the electro-spun nano-fibre diameter increase with increasing the polymer concentration within a certain range. This may be attributed to the greater resistance of the solution to be stretched by the charges on the jet. Deitzel et al. [30] pointed out that the fibre diameter increased with increasing polymer concentration according to a power law relationship. Demir et al. [31] found that the fibre diameter was proportional to the cube of the polymer concentration.

Polymer solutions with low viscosity lead to more instability of the jet and the risk of breaking into droplets is high. Drops and beads may be formed and those are reduced with increasing polymer concentration. As an example of hydrophobic polymers used in electro-spinning, Fig. 7.5 shows the effects of the PVDF polymer concentration and voltage on electro-spun fibre morphology [13]. For both low and high molecular weight PVDF polymers, beads and inhomogeneities are observed at low polymer concentrations. As the polymer concentration was increased, the number of beads decreased and the amount of fibre formation was increased. In this case, the optimum conditions for electro-spun nano-fibre formation were found to be 30 wt% and 15 kV for low molecular weight PVDF, with an average fibre diameter of 213 nm ± 70 nm; whereas for high molecular weight PVDF, the optimum conditions were 25 wt% and 20 kV and the obtained electro-spun fibres exhibited an average fibre diameter of 340 nm ± 150 nm.

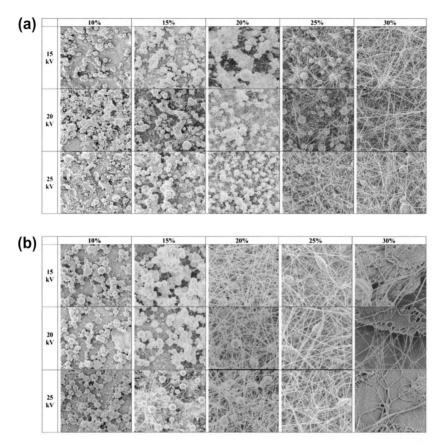

FIGURE 7.5 SEM images of electro-spun PVDF solutions prepared in the solvent mixture DMF/acetone (8/2 volume ratio) at different polymer concentrations and voltages for low molecular weight PVDF polymer (Kynar®710, Arkema) (a) and high molecular weight PVDF polymer (Kynar®740, Arkema) (b) (1.5 ml/min polymer flow rate, 15 cm air gap). *Source: Reprinted from [13]. Copyright 2010, with kind permission from Elsevier.*

In fact, beads formation is affected by several factors such as polymer concentration, solution viscosity, net charge density and surface tension [32]. Beads, which are the product of instabilities of the jet under an electric field, are formed mainly because the viscosity of the solution at a low concentration is lower than the surface tension. This has the effect of decreasing the surface area per unit mass of a fluid. In fact, for a solution with a large amount of solvent molecules, these will congregate adopting a spherical shape due to surface tension. In general, beads disappear as the polymer concentration in the dope increases. Doshi and Reneker [33] reported that by decreasing surface tension of a polymer solution, fibres could be obtained without beads. However, the surface tension is more likely a function of solvent compositions and is less dependent on polymer concentration. Not necessarily, a lower surface tension of a solvent will always be more suitable for electro-spinning [1].

When using polystyrene (PS) polymer dissolved in tetrahydrofuran (THF), an increase of fibre diameter from 0.8 μm to 20 (±10) μm with increasing PS concentration from 18 wt%

to 35 wt% was also observed [6]. The PS fibres were twisted, exhibiting a ribbon-shaped cross-section and surfaces with a nano-porous structure as presented in Fig. 7.6. The shape and the size of the nanopores were observed to change with the PS concentration.

It must be pointed out that higher polymer concentration induced smaller nano-fibrous membrane deposition area. An increase of the polymer concentration means that the viscosity of the solution is strong, discouraging the bending instability to set in for a longer distance. As a result, the jet path is reduced and the bending instability spreads over a smaller area [34].

Molecular Weight

One of the factors affecting the viscosity of the solution is the molecular weight of the

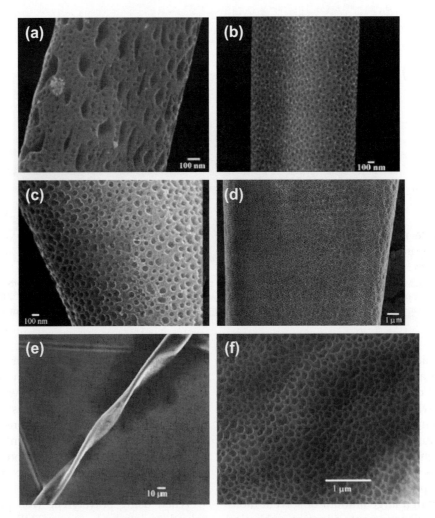

FIGURE 7.6 Field emission scanning electron microscopy (FESEM) of fibres electro-spun from different PS concentrations: (a) 18 wt%, (b) 25 wt%, (c) 28 wt%, (d) 30 wt%, (e) 35 wt%, (f) 35 wt% higher magnification. *Source: Reprinted from [6]. Copyright 2002, with kind permission from American Chemical Society.*

polymer. In general, when using the same solvent, the viscosity of a polymer solution is higher when the molecular weight of the same polymer is higher. Increasing both the polymer concentration and the molecular weight results in greater polymer chain entanglements within the solution, which is necessary to maintain the continuity of the jet during electro-spinning and avoid jet breaking into small droplets or formation of beads. When low viscosity solutions are used, the high content of solvent molecules and few chain entanglements make the surface tension effect predominant along the electro-spinning jet, causing beads formation along the fibre. When the viscosity is increased, there will be a higher amount of polymer chains entanglement and the charges on the electro-spinning jet will be able to fully stretch the solution. When the viscosity of the solution is too high, the solution may solidify at the tip of the needle before initiation of electro-spinning [35].

Solvent

The physical properties of the solvents play an important role in electro-spinning process. In this case, the electrolytic character of the solvent is an important parameter. The solvent used for preparation of the dope plays an important role to decide both the dope properties such as the surface tension and the viscosity, as well as the obtained nano-fibre characteristics. In fact, the surface tension of the dope depends on both the polymer and solvent. For example, it was observed that the poly(vinyl chloride) (PVC) solutions prepared using mixtures of two solvents, tetrahydrofuran (THF) and N,N-dimethylformamide (DMF), with the volume ratios 60/40, 50/50 and 40/60 did facilitate electro-spinning process because of the adequate surface tension, boiling point and viscosity of the solutions [36]. When THF was used as solvent, the capillary tip was often closed because of the low boiling point of this solvent (65 °C, rapid evaporation). THF has a low dielectric constant (7.6 at 25 °C), a dipole moment of 1.7 D and does not show polyelectrolyte behaviour although it is an excellent solvent for PVC and its compact structure leads to van der Waals attraction between molecules. In contrast DMF is a poor solvent for PVC compared to THF. However, DMF is a dipolar aprotic solvent having a high dielectric constant (36.7 at 25 °C), a dipole moment of 3.8 D and randomly dissociates in positive and negative charges in solution exhibiting a polyelectrolyte behaviour. The diameter of the obtained fibre using only THF as solvent was found to be greater than 3 μm and presented a broad distribution ranging from 500 nm to 6 μm [36]. In the case of DMF solvent, the average diameter of the obtained electro-spun fibres was smaller, 200 nm. For the mixed solvents (THF/DMF) at different volume ratios, the average diameters of the prepared electro-spun fibres were found to be quite similar and smaller than 1 μm.

Figure 7.7 shows the SEM images of electro-spun PVDF fibres prepared using 20 wt% PVDF in DMF/acetone mixtures of different weight ratios [37]. As can be seen, the average fibre diameter was decreased while the size and number of beads were increased by increasing the amount of DMF in the solvent mixture DMF/acetone. This was attributed to the higher polarity and boiling point of DMF over acetone.

The solvent volatility plays a critical role in the pore formation process of electro-spun fibres. The electro-spun fibres showed an increase in surface roughness and pores by increasing the boiling point or decreasing the vapour pressure of the solvents or solvent mixtures [6]. A high density of pores was observed on PS nano-fibres prepared using THF solvent and the resulting microtexture at the fibre surface enhanced the surface area by as much as 20–40% depending on the fibre diameter [6]. Figure 7.8 presents the effect of the decreasing vapour pressure of the solvent mixture THF/DMF on the microstructure of PS electro-spun fibres. The electro-spun fibres

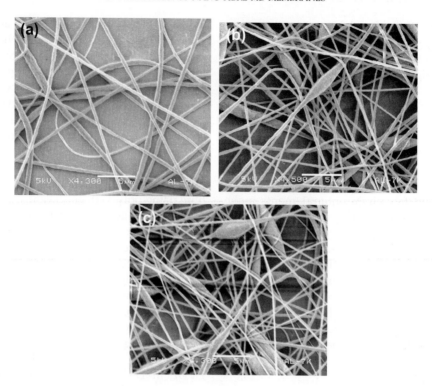

FIGURE 7.7 SEM images of PVDF fibres electro-spun in DMF/acetone mixtures with different weights ratios, 60/40 (a), 70/30 (b) and 80/20 (c) using 15 kV voltage and 20 wt% PVDF polymer concentration in the dope. *Source: Reprinted from [37]. Copyright 2007, with kind permission from Elsevier.*

prepared with THF only exhibit a ribbon-like shape and have a diameter of about 10 μm. When DMF was added (75/25%), the pore sizes became larger and shallower. At an equal volume ratio of solvent mixture (50/50%), the surface roughness and microstructure of the fibres were observed and disappeared from the surface of the fibres electro-spun using the solvent DMF only. Therefore, it was stated that the microstructure of nano-fibres diminished and finally disappeared leaving a smooth surface with decreasing solvent volatility. When the solvent was carbon disulfide (CS_2) having very high vapour pressure (48.2 kPa at 25 °C) and low boiling point (46 °C), the PS polymer droplet on the needle formed a skin on its surface immediately and hardly any PS electro-spun fibres could be obtained requiring change of the needle.

The polymer solution with higher dielectric property reduces the beads formation and the diameter of the resultant electro-spun fibres due to the increase in jet path [38]. Solvents such as DMF can be added to a solution to increase its dielectric constant [39]. In fact, the use of polymer solutions with higher dielectric constant results in an increase of the bending instability of the electro-spinning jet due to the enhancement of the deposition area of the nano-fibres [40].

Electrical Conductivity of the Solution

This is a key factor determining the electro-spinning current. If the polymer solution is

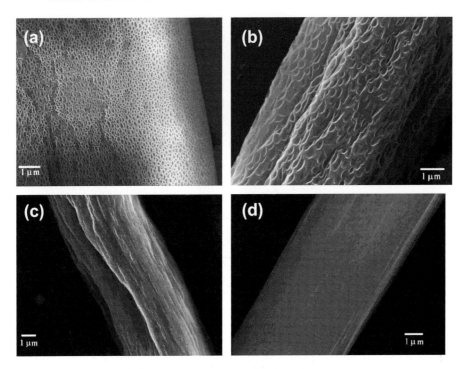

FIGURE 7.8 FESEM images of PS fibres electro-spun using different solvent mixtures THF/DMF: (a) 100 % THF, (b) 75/25 %, (c) 50/50 %, (d) 100 % DMF. *Source: Reprinted from [6]. Copyright 2002, with kind permission from American Chemical Society.*

absolutely insulating, or the applied voltage is not high enough so the electrostatic force cannot overcome the surface tension, fibre cannot be produced by electro-spinning. However, if a small amount of salt or polyelectrolyte is added in the polymer solution, electrical conductivity of the solution will increase. This will result in a higher charge density on the surface of the ejected jet during electro-spinning and more electric charges are carried by the jet, leading to higher elongation forces imposed to the jet under electrical field. The stretching of the solution will increase resulting in smooth fibres with smaller diameters and the risk of formation of beaded fibres become high [32]. Addition of a salt to the uncharged solution preserves the electrical neutrality because salt may dissociate into positive and negative ions.

Although all ionic solutions contain charged molecules or ions, the solution is electrically neutral because the number of positive and negative ions is exactly equal. The essential excess ions are usually created near the interface between a metallic conductor and the molecules in the solution. It must be mentioned that Demir et al. [31] observed a high increase of the mass flow with a small increase of the salt concentration (triethylbenzylammonium chloride) in the copolymer polyurethaneurea dissolved in DMF.

It was also observed that the presence of ions increased the electrical conductivity of the polymer solution, reduced the critical voltage for electro-spinning, yielded a greater bending instability and resulted in larger deposition area of the fibres [41,42]. It is to be noted that most of the organic solvents are non-conductive.

It must be mentioned that the size of the ions affects the electro-spun morphology. Ions with smaller atomic radius have a higher charge density and therefore a faster mobility under an external electric field. When using NaCl salt having smaller atomic radii of sodium and chloride ions than those of potassium and phosphate ions, the obtained electro-spun fibres exhibited the smallest diameters, whereas the electro-spun fibres prepared using KH_2PO_4 had the largest diameters and those prepared using the salt NaH_2PO_4 exhibited intermediate diameters [43].

In electro-spinning process, it is interesting to increase the electrical conductivity of the solution and at the same time reduce its surface tension. In this case, ionic surfactants such as triethyl benzyl ammonium chloride were used [44]. This was found to cause also a decrease of electro-spun fibre diameter. Care should be taken because the addition of ionic salt may cause an increase in the viscosity of the solution. The viscoelastic force will be stronger than the Coulombic force, resulting in an increase of the fibre diameter [34].

Effect of Process Parameters

Applied Voltage

An important parameter in electro-spinning technique is the application of a high voltage to the dope. The high voltage will produce the necessary charges on the polymer solution initiating electro-spinning process when the electrostatic force in the solution overcomes the surface tension of the solution. When the applied voltage is higher, the greater amount of the induced charges will cause faster acceleration of the electro-spinning jet and then more quantity of polymer solution will be drawn from the needle tip. These will result in a larger fibre diameter.

Depending on the feed flow rate of the dope and the polymer concentration, a high voltage may be required so that the Taylor cone is stable. The Coulombic repulsive force in the jet will then stretch the viscoelastic solution. In various cases, a higher electric voltage causes greater stretching of the polymer solution, reducing in this way the diameter of electro-spun fibres. Lee et al. [36] observed a gradual decrease of the diameter of the fibre electro-spun from 15 wt% PVC dissolved in 50% volume ratio of THF/DMF, with the electric voltage varying from 8 to 15 kV. No remarkable change in the morphology was observed. Similarly, the PS fibre diameter was reduced with increasing the electro-spinning voltage from 5 to 12 kV [6]. The surface of these PS fibres was rough with nanopores densely packed on the surface as shown above in Figs. 7.6 and 7.8. The pore size distribution was not affected by varying the voltage.

When using 15 wt% polyvinylacetate (PVAC) in ethanol solution, the diameter of the electro-spun fibre decreased, as can be observed in Fig. 7.9, with the applied voltage and then increased at a voltage higher than 17.5 kV. At this high voltage, there was not enough time for the electro-spinning solution to be developed and this is why the fibre diameter was increased [45]. When a higher PVAC concentration (20 wt%) was used, a gradual increase of the electro-spun fibre diameter was observed by increasing the electric voltage from 10 to 15 kV. These results may be related with the viscosity of the polymer solution and the distance between the needle tip and the collector. When the polymer concentration is small, the viscosity is low and a high voltage may induce the formation of secondary jets, which results in reduction of fibre diameter [31]. Moreover, a large distance between the needle tip and the collector will allow more time for the electro-spun fibre to stretch and elongate along the gap distance. Therefore, when the applied voltage is decreased, the jet acceleration will also decrease and its flight time through the gap distance will increase favouring the formation of fibres with smaller diameters.

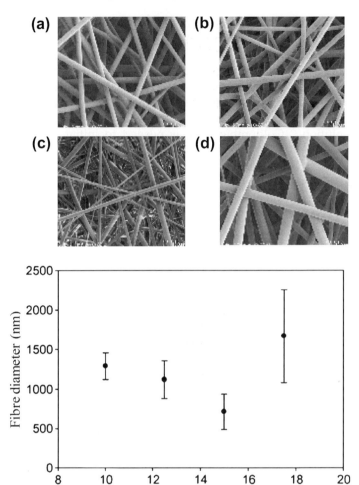

FIGURE 7.9 Effect of the applied voltage on the diameter of electro-spun fibre from 15 wt% PVAC in ethanol. SEM: (a) 10 kV, (b) 12.5 kV, (c) 15 kV, (d) 17.5 kV. *Source: Reprinted from [45]. Copyright 2008, with kind permission from Elsevier.*

Other than the variation of the diameter of the electro-spun fibres, it must be pointed out that the crystallinity of the fibres also was affected by the change of the electric voltage. The polymer molecules tend to be more ordered during electro-spinning, resulting in a greater crystallinity with the increase of the voltage. However, above a certain value of the voltage, the crystallinity of the fibre diminished [35]. This may be also related to the flight time of the electro-spinning jet in the gap distance between the needle tip and the collector, since the orientation of the polymer molecules requires some time. Short flight time means that the polymer molecules will reach the collector before alignment.

Polymer Flow Rate

The flow rate of the polymer solution, which is generally controlled by a circulation pump as shown in Fig. 7.2, determines the quantity of the solution available for electro-spinning. This will affect the volume charge density and the electrical current of the polymer solution, which increase or decrease depending on the polymer solution [46]. In order to maintain a stable Taylor cone for a given applied electric voltage, a corresponding polymer solution flow rate should be adjusted. When the feed flow rate is increased, the diameter of the electro-spun fibre is also increased. This is expected as there is a greater amount of polymer solution that is drawn from the needle tip (Fig. 7.10). However, there is a certain limit. If the polymer flow rate is too high, greater volume of polymer solution will be drawn from the needle tip and the electro-spinning jet will take more time to dry. As a result, the solvent(s) in the deposited fibres over the collector may not have enough time to evaporate. Therefore, the residual solvent may cause the fibres to fuse together forming denser fibrous membrane. It must be indicated also that when the polymer flow rate is increased, there is a corresponding increase in charges enhancing the stretching of the polymer solution, as stated earlier, which counters the expected increase of the fibre diameter due to increase of the volume of the polymer solution. Nasir et al. [46] observed that the PVDF fibre diameter decreased with increasing polymer flow rate up to 5 µl/min and then remained constant for higher flow rates. When a high polymer flow rate is applied, beads may be formed as claimed by Megelski et al. [6], for the electro-spinning of PS fibres. In addition, it was also reported that bead formation is a function of the PS concentration, the spinning voltage and the gap distance between the needle tip and the collector.

Temperature

Enhancing the temperature of the polymer solution will increase the evaporation rate along the flight distance of the electro-spinning jet and will reduce the viscosity of the polymer solution [35]. More uniform fibre diameters were prepared when the copolymer polyurethaneurea was electro-spun at a higher temperature [31].

Needle Tip

The internal diameter of the needle or spinneret affects also electro-spinning process. A needle with small inner diameter was found to reduce both the risk of bead formation on the electro-spun fibres and their diameter [35]. In fact, the size of the polymer droplet formed at the needle tip is smaller when a needle with a smaller internal diameter is used and the surface tension is higher. Therefore, greater Coulombic force is needed to initiate electro-spinning jet. As well, for smaller internal diameter of the tip, the flight time for the solution becomes larger (i.e. acceleration of the jet is decreased) leading to more stretching and elongation of the solution.

Gap Distance Between Needle Tip and Collector

As mentioned above, the flight time of the electro-spinning jet along the gap distance between the highly charged needle tip and the grounded collector may affect considerably the fibre's characteristics. Decreasing the gap distance has the same effect as increasing the electrical voltage inducing higher electric field strength. When the gap distance is too short, the instability of the jet increases and the spinning solution cannot be fully stretched, resulting in greater fibre diameter. Moreover, the instability of the jet due to the strong electric field may result in the formation of beads and sometimes fully developed fibre cannot be made. In addition, when the gap distance is small, the electro-spinning jet may not have enough time for the solvent(s) to evaporate before it reaches the collector and the fibres may merge together as seen in Fig. 7.11 [47].

When the gap distance is too large, the strength of the electric field becomes weak resulting in an increase of fibre diameter and sometimes electro-spinning is hard to be accomplished. Longer gap distance indicates that there is longer flight time for the polymer solution to be stretched and elongates before reaching the collector.

It was observed that depending on the polymer solution parameters, the effect of varying

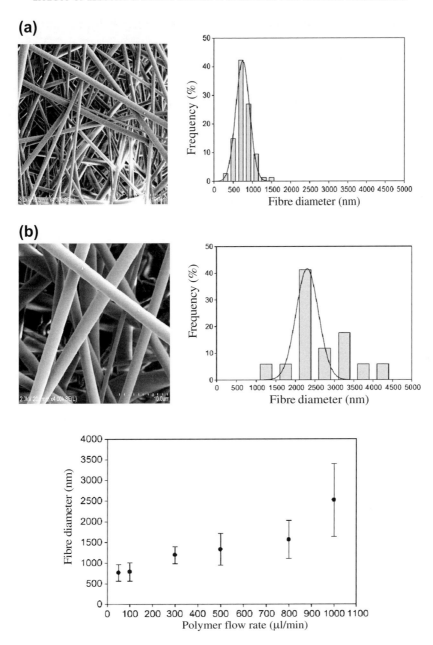

FIGURE 7.10 Effect of polymer flow rate on the diameter of electro-spun fibre from 15 wt% PVAC in ethanol. SEM: (a) 50 μl/min, (b) 1000 μl/min. *Source: Reprinted from [45]. Copyright 2008, with kind permission from Elsevier.*

FIGURE 7.11 SEM images of electro-spun fibres prepared under different gap distances: (a) 2 cm, (b) 0.5 cm. *Source: Reprinted from [47]. Copyright 1999, with kind permission from Elsevier.*

the distance may or may not have a significant effect on the fibre morphology. Nasir et al. [46] reported that the gap distance had no significant effect on the PVDF fibre diameter and explained that the increase of the gap distance induced a decrease of the electrical field strength when a constant electrical voltage was applied, whereas the solvent evaporation time of the polymer jet increased. Megelski et al. [6] also observed no significant change of the electro-spun PS fibre size with the change of the gap distance. However, inhomogeneous distribution of elongated beads took place when the gap distance was reduced. When PVC fibres were electro-spun, Lee et al. [36] detected a slight decrease of fibre diameter with increasing gap distance. Park et al. [45] observed a decrease of the diameter of electro-spun PVAC fibre with increasing the gap distance down to a minimum value followed by a gradual increase of the fibre diameter. This is due to the decrease in the electrostatic field strength resulting in less stretching of the fibres and indicates that there is an optimal electrostatic field strength below which the stretching of the solution will decrease, resulting in increased fibre diameter.

Post-Treatment

Similar to the MD membranes prepared by other techniques, post-treatment is an important step affecting electro-spun nano-fibrous membrane parameters when they are formed. This was discussed in the previous chapters. For example, Fig. 7.12 shows the effects of heat treatment (145 °C for 18 h) on the PVDF nano-fibre membrane characteristics [12]. During the electro-spinning process, the fibres overlap each other in a completely random manner and an open membrane structure is obtained. When heat treatment is applied to the nano-fibre membranes at temperatures below the material's melting point, overlapped nano-fibres tend to fuse together as can be seen in Fig. 7.12(a), improving the structural integrity of the electro-spun nano-fibrous membranes and enhancing their mechanical strength. Therefore, thermal treatment of nano-fibrous membranes above

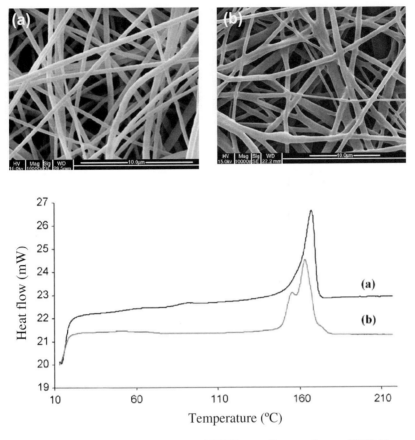

FIGURE 7.12 Effects of heat treatment on the structure of PVDF nano-fibre membranes: FESEM images and differential scanning calorimetry (DSC): (a) before and (b) after heat treatment. *Source: Reprinted from [12]. Copyright 2006, with kind permission from Elsevier.*

the glass transition temperature and below the melting temperature permits to join the electro-spun fibres at the different nodes (i.e. interfibre bonding) and improves their connectivity. This makes the nano-fibrous membrane rigid and mechanically strong, enhancing its tensile strength. Conditions of the heat treatment depend on the type of the polymer and its molecular weight. For example, the heat treatment conditions to promote interfibre bonding in electro-spun nano-fibres were reported to be 240 °C for 1 h for poly(etherimide) nano-fibre web and 188 °C for 6 h for polysulfone [48,49]. Most electro-spun nano-fibre webs do not have interfibre bonding, although stronger physical properties were observed for the electro-spun fibres having interfibre bonding.

In addition, it is worth quoting that heat treatment allows the nano-fibres to attain a crystalline structure. In the case of PVDF nano-fibre membranes, the heat treated membrane exhibited two melting peaks, as shown in Fig. 7.12(b). However, only one melting peak was observed for the untreated membrane indicating the more ordered molecular structure of the heated nano-fibre membrane [12].

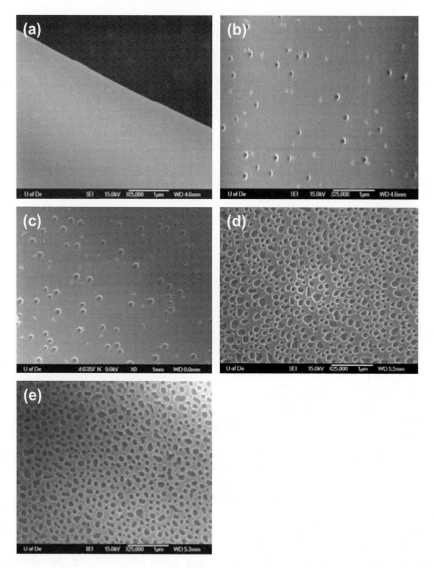

FIGURE 7.13 Effect of humidity on the surface structure of electro-spun PS(M_w = 190 kg/mol) nano-fibre using the solvent tetrahydrofuran (THF); humidity level: (a) <25%, (b) 31–38%, (c) 40–45%, (d) 50–59%, (e) 60–72%. *Source: Reprinted from [50]. Copyright 2004, with kind permission from American Chemical Society.*

Ambient Parameters

Electro-spinning environment affects the morphology and structure of the nano-fibre. This may be due to the change of the external electric field. Parameters such as the humidity, temperature, type of atmosphere (composition of the air) and vapour pressure can interact with the polymer solution along the gap distance between the needle tip and the collector and also after deposition of the nano-fibres over the collector. For example, the solvent vapour pressure and the humidity in the atmosphere surrounding the electro-spun nano-fibres were found to affect strongly the formation of pores of electro-spun nano-fibre membranes [35,50]. It was suggested that the cooling effect due to rapid evaporation of a highly volatile solvent might induce the polymers to separate into different phases in the liquid jet. Because of evaporative cooling and condensation, water droplets could also be formed within the fibres to promote the formation of porous nano-fibres [6,50]. A higher humidity level was found to cause an increase in the pore density, pore size, pore shape and distribution of the pores as can be seen in Fig. 7.13.

Electro-spun fibres from polystyrene (PS), polycarbonate (PC) and poly(methyl methacrylate) (PMMA) were found to exhibit submicron surface features when the polymers were dissolved in a volatile solvent and electro-spun in a humid environment [6].

It must be pointed out that different gases under high electrostatic field have different behaviour. Baumgarten [26] found that no spinning could be done in helium atmosphere and the fibres electro-spun in Freon®-12 gas exhibited diameters 1.4–2.6 times greater than those of the fibres electro-spun in air at identical conditions.

The effect of ambient parameters is still poorly investigated. The effects of humidity on the electrostatic charges, electrical discharge during electro-spinning and accumulation of residual charges on the nano-fibres need to be studied in details [35]. The effect of the pressure and temperature on the electro-spinning jet may be of great interest. Both parameters will determine the rate of solvent evaporation along the electro-spinning jet and at the tip needle affecting the fibre structure.

Before concluding this chapter, it should also be mentioned that nano-fibre membranes fabricated by melt-electro-spinning have to be utilized as MD membranes. It would also be interesting to study nano-fibres of different structures such as porous nano-fibres, flattened or ribbon-like nano-fibres, branched nano-fibres, hollow nano-fibres and core-shell nano-fibres as MD membranes [5,51–54]. The presence of pores on the surface of electro-spun fibres can serve to increase the surface area of membranes for MD applications and the use of hollow nano-fibres will serve to decrease the thermal conductivity of the nano-structured MD membranes.

References

[1] Z.M. Huang, Y.Z. Zhang, M. Kotaki, S. Ramakrishna, A review on polymer nanofibers by electrospinning and their applications in nanocomposites, Comp. Sci. Tech. 63 (2003) 2223–2253.

[2] D. Li, Y. Xia, Electrospinning of nanofibers: Reinventing the wheel? Adv. Mater. 16 (2004) 1151–1170.

[3] I.S. Chronakis, Novel nanocomposites and nanoceramics based on polymer nanofibers using electrospinning process: A review, Mat. Proc. Tech. 167 (2005) 283–293.

[4] M. Bognitzki, T. Frese, M. Steinhart, A. Greiner, J.H. Wendorff, A. Schaper, M. Hellwig, Preparation of fibers with nanoscaled morphologies: Electrospinning of polymer blends, Polym. Eng. Sci. 41 (2001) 982–989.

[5] M. Bognitzki, W. Czado, T. Frese, A. Schaper, M. Hellwig, M. Steinhart, A. Greiner, J.H. Wendorff, J.H. Wendorf, Nanostructured fibers via electrospinning, Adv. Mater. 13 (2001) 70–72.

[6] S. Megelski, J.S. Stephens, D.B. Chase, J.F. Rabolt, Micro- and nanostructured surface morphology on electrospun polymer fibers, Macromolecules 35 (2002) 8456–8466.

[7] X. Wang, C. Drew, S.H. Lee, K.J. Senecal, J. Kumar, L.A. Samuelson, Electrospun nanofibrous membranes for highly sensitive optical sensors, Nano Lett. 2 (2002) 1273–1275.

[8] C. Seoul, Y.T. Kim, C.K. Baek, Electrospinning of poly(vinylidene fluoride)/dimethylformamide solutions with carbon nanotubes, J. Polym. Sci. Part B: Polym. Phys. 41 (2003) 1572–1577.

[9] S.S. Choi, Y.S. Lee, C.W. Joo, S.G. Lee, J.K. Park, K.S. Han, Electrospun PVDF nanofiber web as polymer electrolyte or separator, Electrochim. Acta 50 (2004) 339–343.

[10] Z. Ma, M. Kotaki, S. Ramakrishna, Electrospun cellulose nanofiber as affinity membrane, J. Membr. Sci. 265 (2005) 115–123.

[11] M. Ma, R.M. Hill, J.L. Lowery, S.V. Fridrickh, G.C. Rutledge, Electrospun poly(styrene-block-dimethylsiloxane) block copolymer fibers exhibiting superhydrophobicity, Langmuir 21 (2005) 5549–5554.

[12] R. Gopal, S. Kaur, Z. Ma, C. Chan, S. Ramakrishna, T. Matsuura, Electrospun nanofibrous filtration membrane, J. Membr. Sci. 281 (2006) 581–586.

[13] K. Magniez, C.D. Lavigne, B.L. Fox, The effects of molecular weight and polymorphism on the fracture and thermo-mechanical properties of a carbon-fibre composite modified by electrospun poly(vinylidene fluoride) membranes, Polymer 51 (2010) 2585–2596.

[14] R.S. Barhate, S. Ramakrishna, Nanofibrous filtering media: Filtration problems and solutions from tiny materials: Review, J. Membr. Sci. 296 (2007) 1–8.

[15] M.E. Helgeson, N.J. Wagner, A correlation for the diameter of electrospun polymer nanofibers, AIChE J. 53 (2007) 51–55.

[16] D.H. Reneker, A.L. Yarin, Electrospinning jets and polymer nanofibers, Polymer 49 (2008) 2387–2425.

[17] X. Yan, G. Liu, F. Liu, B.Z. Tang, H. Peng, A.B. Pakhomov, C.Y. Wong, in: Angew. Chem. Int. (Ed.), Superparamagnetic tribloc copolymer/Fe_2O_3 hybrid nanofibers, 40, 2001, pp. 3593–3596.

[18] S. Borkar, B. Gu, M. Dirmyer, R. Delicado, A. Sen, B.R. Jackson, J.V. Badding, Polytetrafluoroethylene nano-microfibers by jet blowing, Polymer 47 (2006) 8337–8343.

[19] R.G. Flemming, C.J. Murphy, G.A. Abrams, S.L. Goodman, P.F. Nealey, Effects of synthetic micro- and nano-structured surfaces on cell behavior, Biomaterials 20 (1999) 573–588.

[20] T.A. Desai, Micro-and nanoscale structures for tissue engineering constructs, Med. Eng. Phys. 22 (2000) 595–606.

[21] A. Curtis, C. Wilkinson, Nanotechniques and approaches in biotechnology, Trends Biotechnol. 19 (2001) 97–101.

[22] H.G. Graighead, C.D. James, A.M.P. Turner, Chemical and topographical patterning for directed cell attachment, Curr. Opin. Solid State Mater. Sci. 5 (2001) 177–184.

[23] C.T. Laurencin, A.M. Ambrosio, M.D. Borden, J.A. Cooper Jr., Tissue engineering: Orthopedic applications, Ann. Rev. Biomed. Eng. 1 (1999) 19–46.

[24] H. Li, Y. Ke, Y. Hu, Polymer nanofibers prepared by template melt extrusion, J. Appl. Polym. Sci. 99 (2006) 1018–1023.

[25] A. Formhals, Process and apparatus for preparing artificial threads, U.S. Patent No 1,975,504 (1934).

[26] P.K. Baumgarten, Electrostatic spinning of acrylic microfibers, J. Colloid Interface Sci. 36 (1971) 71–79.

[27] C. Feng, K.C. Khulbe, T. Matsuura, R. Gopal, S. Kaur, S. Ramakrishna, M. Khayet, Production of drinking water from saline water by air-gap membrane distillation using polyvinylidene fluoride nanofiber membrane, J. Membr. Sci. 311 (2008) 1–6.

[28] D.H. Reneker, A.L. Yarin, H. Fong, S. Koombhongse, Bending instability of electrically charged liquid jets of polymer solutions in electrospinning, J. Appl. Phys. 87 (2000) 4531–4547.

[29] H. Liu, Y.L. Hsieh, Ultrafine fibrous cellulose membranes from electrospinning of cellulose acetate, J. Polym. Sci. Part B: Polym. Phys. 40 (2002) 2119–2129.

[30] J.M. Deitzel, J. Kleinmeyer, J.K. Hirvonen, T.N.C. Beck, Controlled deposition of electrospun poly(ethylene oxide) fibers, Polymer 42 (2001) 8163–8170.

[31] M.M. Demir, I. Yilgor, E. Yilgor, B. Erman, Electrospinning of polyurethane fibers, Polymer 43 (2002) 3303–3309.

[32] H. Fong, I. Chun, D.H. Reneker, Beaded nanofibers formed during electrospinning, Polymer 40 (1999) 4585–4592.

[33] J. Doshi, D.H. Reneker, Electrospinning process and applications of electrospun fibers, J. Electrostatics 35 (2–3) (1995) 151–160.

[34] C. Mit-uppatham, M. Nithitanakul, P. Supaphol, Ultrafine electrospun polyamide-6 fibers: Effect of solution conditions on morphology and average fiber diameter, Macromol. Chem. Phys. 205 (2004) 2327–2338.

[35] S. Ramakrishna, K. Fujihara, W.E. Teo, T.C. Lim, Z. Ma, An introduction to electrospinning and nanofibers, World Scientific Pub. Co. Ltd., Singapore, 2005.

[36] K.H. Lee, H.Y. Kim, Y.M. La, D.R. Lee, N.H. Sung, Influence of a mixing solvent with tetrahydrofuran and N, N-dimethylformamide on electrospun poly(vinyl chloride) nonwoven mats, J. Polym. Sci. Part B: Polym. Phys. 40 (2002) 2259–2268.

[37] W.A. Yee, M. Kotaki, Y. Liu, X. Lu, Morphology, polymorphism behavior and molecular orientation of electrospun poly(vinylidene fluoride) fibers, Polymer 48 (2007) 512–521.

[38] W.K. Son, J.H. Youk, T.S. Lee, W.H. Park, The effects of solution properties and polyelectrolyte on elctrospinning of ultrafine poly(ethylene oxide) fibers, Polymer 45 (2004) 2959–2966.

[39] K.H. Lee, H.Y. Kim, Y.M. Ra, D.R. Lee, Characterization of nanostructured poly(e-caprolactone) nonwoven mats via electrospinning, Polymer 44 (2003) 1287–1294.

[40] C.M. Hsu, S. Shivakumar, N, N-dimethylformamide additions to the solution for the electrospinning of poly(e-caprolactone) nanofibers, Macromol. Mater. Eng. 289 (2004) 334–340.

REFERENCES

[41] W.K. Son, J.H. Youk, T.S. Lee, W.H. Park, Electrospinning of ultrafine cellulose acetate fibers: studies of a new solvent system and deacetylation of ultrafine cellulose acetate fibers, J. Polym. Sci. Part B: Polym. Phys. 42 (2004) 5–11.

[42] J.S. Choi, S.W. Lee, L. Jeong, S.H. Bae, B.C. Min, J.H. Youk, W.H. Park, Effect of organosoluble salts on the nanofibrous structure of electrospun poly(3-hydroxybutyrate-co-3-hydroxyvalerate), Int. J. Biol. Macromol. 34 (2004) 249–256.

[43] X. Zong, K. Kim, D. Fang, S. Ran, B.S. Hsiao, B. Chu, Structure and process relationship of electrospun bioadsorbable nanofiber membranes, Polymer 43 (2002) 4403–4412.

[44] J. Zeng, X. Xu, X. Chen, Q. Liang, X. Bian, L. Yang, X. Jing, Biodegradable electrospun fibers for drug delivery, J. Control. Release 92 (2003) 227–231.

[45] J.Y. Park, I.H. Lee, G.N. Bea, Optimization of the electrospinning conditions for preparation of nanofibers from polyvinylacetate (PVAc) in ethanol solvent, J. Ind. Eng. Chem. 14 (2008) 707–713.

[46] M. Nasir, H. Matsumoto, T. Danno, M. Minagawa, T. Irisawa, M. Shioya, A. Tanioka, Control of diameter, morphology, and structure of PVDF nanofiber fabricated by electrospray deposition, J. Polym. Sci. Part B: Polym. Phys. 44 (2006) 779–786.

[47] C.J. Buchko, L.C. Chen, Y. Shen, D.C. Martin, Processing and microstructural characterization of porous biocompatible protein polymer thin films, Polymer 40 (1999) 7397–7407.

[48] S.S. Choi, S.G. Lee, C.W. Joo, S.S. Im, S.H. Kim, Formation of interfiber bonding in electrospun poly(etherimide) nanofiber web, J. Mater. Sci. 39 (2004) 1511–1513.

[49] M. Zuwei, M. Kotaki, S. Ramakrishna, Surface modified nonwoven polysulphone (PSU) fiber mesh by electrospinning: a novel affinity membrane, J. Membr. Sci. 272 (2006) 179–187.

[50] C.L. Casper, J.S. Stephens, N.G. Tassi, D.B. Chase, J.F. Rabolt, Controlling surface morphology of electrospun polystyrene fibers: Effect of humidity and molecular weight in the electrospinning process, Macromolecules 37 (2004) 573–578.

[51] S. Koombhongse, W. Liu, D.H. Reneker, Flat polymer ribbons and other shapes by electrospinning, J. Polym. Sci. Part B: Polym. Phys. 39 (2001) 2598–2606.

[52] A.L. Yarin, W. Kataphinan, D.H. Reneker, Branching in electrospinning of nanofibers, J. Appl. Phys. 98 (2005) 064501.

[53] D. Li, Y. Xia, Direct fabrication of composite and ceramic hollow nanofibers by electrospinning, Nano Lett. 4 (2004) 933–938.

[54] Z. Sun, E. Zussman, A.L. Yarin, J.H. Wendorff, A. Greiner, Compound core-shell polymer nanofibers by co-electrospinning, Adv. Mater. 15 (2003) 1929–1932.

CHAPTER

8

MD Membrane Characterization

OUTLINE

Introduction: Characteristics Needed for MD Membranes 189

Determination of Pore Size (Mean Pore Size, Pore Size Distribution) by Different Physical Methods 191
    Gas Permeation Test 191
    Wet/Dry Flow Method 193
    Mercury Porosimetry 199
    Electron Microscopy 200
    Atomic Force Microscopy 203

Porosity and Effective Porosity: Techniques Used 209

Evaluation of Membrane Pore Tortuosity 211

Penetration Pressure Determination 211

Thermal Stability Tests 216

Mechanical Stability 218

Chemical Stability 220

Other Characterization Techniques 221
    Thickness of Bilayer Hydrophobic/hydrophilic Composite MD Membranes 221
    Mean Pore Size and Pore Size Distribution 221
    Optical Techniques 221

INTRODUCTION: CHARACTERISTICS NEEDED FOR MD MEMBRANES

The MD separation performance is directly associated with the overall membrane morphology. Therefore, the successful application of MD may be aided by a good knowledge of the different membrane parameters. Before conducting MD experiments, various techniques must be used for characterization of the MD membranes in order to avoid pore wetting and to ensure high MD efficiency.

As stated in Chapter 2, the main requirements for the MD process are that the membrane must not be wetted by the aqueous solutions and only vapor and non-condensable gases are present within its pores. As water is the major component used in MD applications, the membrane must be hydrophobic and must be made by polymers or inorganic materials with low surface energies. Capillary, hollow fibre or flat sheet porous hydrophobic membranes have been used in MD experiments. However, the choice of a membrane for a given MD application is a compromise between various membrane characteristics such as liquid entry pressure, mean pore size, pore size distribution, thickness of the whole membrane and thickness of the thin-skin layer controlling water vapor flux in the case of asymmetric or composite membranes, porosity, pore tortuosity and material or membrane thermal conductivity. However, the membrane manufacturers do not provide all these characteristics. Moreover, for the same commercial membranes, the values for most of their parameters differ depending on the characterization method followed and the observed variations of the membrane characteristics lead to inaccuracies in the predicted MD permeate flux and separation factors.

Hydrophobic porous membranes can be prepared by different techniques depending on the properties of the materials to be used. The useful materials should be selected according to the criteria that include compatibility with the liquids involved, cost, ease of fabrication and assembly, useful operating temperatures and thermal conductivity. In general, MD membranes have to meet several requirements simultaneously:

- Good thermal stability up to temperatures as high as 100 °C.
- Excellent chemical resistance to various feed solutions — acids and bases — especially if the membrane has to be cleaned.
- High entry pressure of liquid solutions in contact with the membrane (LEP).
- High permeability: In general, the MD permeate flux is proportional to the membrane pore size and porosity, but it is inversely proportional to the membrane thickness and pore tortuosity. In other words, to select a membrane with high MD permeability, the surface layer that governs the membrane transport must be as thin as possible so that the vapor transport distance across the membrane is as short as possible and its surface porosity as well as pore size must be as large as possible.
- Narrow pore size distribution: To avoid wetting of the maximum pore sizes or a fraction of the number of pores, which leads to a decrease of both the membrane surface area for MD and the membrane performance.
- Low thermal conductivity: Heat transfer by conduction through the membrane (material & pores) from the feed to the permeate side is heat loss in MD. This conductive heat loss is greater for thinner membranes. One interesting idea is to select a membrane with high porosity since the conductive heat transfer coefficients of the gases entrapped within the membrane pores are an order of magnitude smaller than that of the membrane matrix. Simultaneously, a membrane with a high porosity also exhibits high MD permeability. In contrast, if a thicker membrane is selected, both the heat flux through the membrane and the MD permeability will be low.

The committee formed during the 'Workshop on Membrane Distillation' held in Rome on 5 May 1986 discussed about the characteristics of the membranes used in MD. It was reported that the MD membranes should be characterized by the following five membrane 'performance' parameters: 'polymer' or other membrane material, thickness, porosity, nominal pore size and liquid-entry-pressure of water (LEP_w) [1,2]. Information about the thermal conductivity of the membrane, its pore size, distribution, shape, pore tortuosity, porosity or void volume fraction and membrane surface properties, such as roughness, are therefore of importance to both membrane manufacturers and users, so as to allow a meaningful prediction of MD performances of the membranes. Therefore, proper characterization of MD membranes is crucial for the development of commercial membranes for different MD applications. As will be shown, specialized characterization techniques are required. In this chapter, various characterization techniques will be discussed.

In general, the characterization techniques of MD membranes are physical methods, which can be divided in two main groups: (i) the techniques related to membrane permeation such as liquid and gas flow tests permitting to obtain various permeation related parameters and (ii) the techniques permitting to obtain directly the morphological and structural properties of the membranes including scanning electron microscopy (SEM), field emission scanning electron microscopy (FESEM) and atomic force microscopy (AFM) among others. These techniques must be considered as mutually complementary rather than competitive in order to better understand morphology and performance-related parameters. For example, surface membrane pore size is different from bulk pore size and for various membrane types pores are not cylindrical as assumed in various theoretical analyses. Moreover, the characterization technique in relation to membrane permeation should be chosen in such a way that the characterization method and the final application of the membrane are similar.

DETERMINATION OF PORE SIZE (MEAN PORE SIZE, PORE SIZE DISTRIBUTION) BY DIFFERENT PHYSICAL METHODS

An important factor governing the permeability and separation factor of MD membranes is pore size and its distribution. In the membrane literature, there are several well-established techniques to determine the pore size of flat sheet, capillary and hollow fibre membranes. Most of the techniques are explained in detail in review articles [3–5].

Gas Permeation Test

In MD process only vapor is transported through the membrane pores. Therefore, this technique is suitable to determine the pore size of the membranes used in MD. In fact, this technique was used to determine not only the mean pore size (d_p) but also the effective porosity (ε_e) defined as the ratio of the porosity and the effective pore length that takes into account the tortuosity of the membrane pores [6–9]. The gas permeation method was originally given by Yasuda and Tsai [10]. This method is as simple as to measure only the gas flow rate through a dry porous membrane at different transmembrane pressures. Different types of gases such as air and nitrogen can be employed. The permeate pressure can be either atmospheric pressure or lower than the atmospheric pressure. Figure 8.1 shows some experimental systems used for gas permeation tests of flat sheet, capillaries and hollow fibre membranes.

The total gas permeation flow rate (J_g) through the MD membrane can be expressed as:

$$J_g = B_g A_m \Delta P \qquad (8.1)$$

where B_g is the gas permeance, A_m is the total membrane area and ΔP is the pressure difference through the membrane.

For a porous medium, J_g can also be written in terms of both diffusive or Knudsen flow and viscous or Poiseuille flow as follows:

$$J_g = (P_k + P_v) A_p \frac{\Delta P}{L_p} \qquad (8.2)$$

where A_p is the porous area of the membrane surface, L_p is the effective pore length that takes into consideration the pore tortuosity (τ), P_k and P_v are the Knudsen and Poiseuille gas permeability coefficients, respectively.

Rearranging Eqs. (8.1 and 8.2), the gas permeance can be written as:

$$B_g = (P_k + P_v) \frac{\varepsilon}{L_p} \qquad (8.3)$$

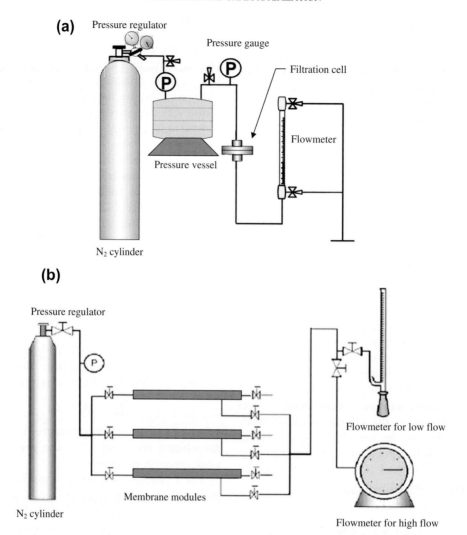

FIGURE 8.1 Schematics of the experimental systems used for gas permeation tests of (a) flat sheet membranes. *(Source: Reprinted from [32]. Copyright 2009, with kind permission from Elsevier)* and (b) capillaries and hollow fibre membranes *(Source: Reprinted from [9]. Copyright 2002, with kind permission from Elsevier).*

where ε is the surface porosity of the membrane defined as:

$$\varepsilon = \frac{A_p}{A_m} \quad (8.4)$$

Knudsen flow (J_k) through a capillary tube with a diameter d_p is expressed as [11]:

$$J_k = \frac{2}{3}\left(\frac{2}{\pi MRT}\right)^{0.5}\frac{d_p}{L_p}\Delta P \quad (8.5)$$

where R is the gas constant, T is the absolute temperature and M is the molecular weight of the gas.

Poiseuille flow (J_v) through a capillary tube is expressed as [11]:

$$J_v = \frac{1}{16\mu RT} \frac{d_p^2}{L_p} P_m \Delta P \qquad (8.6)$$

where μ is the gas viscosity and P_m is the mean pressure within the capillary tube.

From Eqs. (8.5 and 8.6), the gas permeability coefficients P_k and P_v can be written as:

$$P_k = \frac{2}{3}\left(\frac{2}{\pi MRT}\right)^{0.5} d_p \qquad (8.7)$$

and

$$P_v = \frac{1}{16\mu RT} d_p^2 P_m \qquad (8.8)$$

Inserting Eqs. (8.7 and 8.8) into Eq. (8.3), the following expression of the gas permeance for a porous membrane is obtained [12]:

$$B_g = \frac{2}{3}\left(\frac{2}{\pi MRT}\right)^{0.5} d_p \varepsilon_e + \frac{P_m}{16\mu RT} d_p^2 \varepsilon_e$$
$$= I_0 + S_0 P_m \qquad (8.9)$$

where ε_e is the effective porosity defined as:

$$\varepsilon_e = \frac{\varepsilon}{L_p} \qquad (8.10)$$

By plotting a linear dependence between B_g and P_m, the intercept (I_0) and the slope (S_0) can be determined and consequently d_p and ε_e can be calculated using the following equations:

$$d_p = \frac{32}{3} \frac{S_0}{I_0} \left(\frac{8RT}{\pi M}\right)^{0.5} \mu \qquad (8.11)$$

$$\varepsilon_e = \frac{32\mu RT}{d_p^2} S_0 \qquad (8.12)$$

This method was originally developed for symmetric membranes. However, in the case of an asymmetric membrane with a skin layer, the measured pore size is characteristic of the skin layer [13–16].

Khayet et al. [5,17] used this technique to determine the mean pore size and the effective porosity of various commercial flat sheet membranes and compared the results to those supplied by the manufacturers (see Table 2.1). The measured d_p and ε_e of the most used commercial membranes in MD by the gas permeation test are summarized in Table 8.1.

The mean pore size of the membrane TF200 (198.96 nm) was found to be almost the same compared to the value given by the manufacturer (Table 8.1), while those obtained for the other commercial membranes TF450 and TF1000 were found to be smaller than the values given by the manufacturer Gelman. On the other hand, for Millipore membranes (GVHP and HVHP), the mean pore size obtained from the gas permeation test is larger than that given by the manufacturer.

It must be mentioned that one of the limitations of the gas permeation test is that it cannot determine the pore size distribution, which is one of the important MD membrane characteristics to be known. Other techniques and procedures can be used to determine the pore size distribution as will be shown later on. For instance, it is to be noted that Kong and Li [16] reported a procedure to determine the pore size distribution, mean pore size and standard deviation using the gas permeation test. Both standard normal distribution (Gaussian) and log-normal distribution functions were used to represent the pore size distribution. However, it was found that log-normal distribution function was better to produce more realistic predictions.

Wet/Dry Flow Method

The combined bubble point and gas permeation test is known as the wet/dry flow method or liquid displacement method and it can be used to determine the maximum pore size, the mean pore size and the pore size distribution of MD membranes. Various other names have also been used to describe this method that involves displacement of a liquid from wetted membranes using a purified gas. Figure 8.2

TABLE 8.1 Flat Sheet Commercial Membranes Commonly used in MD

Membrane trade name	Manufacturer	Material	Characteristics indicated by the manufacturer				Gas permeation test		Wet/dry flow method				Coulter porosimetry [19,20]	
			δ (μm)	d_P (μm)	ε (%)	LEP_w (kPa)	d_P (nm)	ε_e (m^{-1})	d_P (nm)	σ_P	ε_s (%)	τ	d_P (μm)	ε_e (m^{-1})
TF200	Gelman	PTFE/PP[a]	178	0.20	80	282	198.96	7878.1	233.38	1.07	43.18	1.59	0.31	7900
TF450				0.45		138	418.82	7439.0	491.67	1.10	44.65	1.44	0.47	8900
TF1000				1.00		48	844.32	9744.5	736.84	1.09	56.90	1.18	0.65	11100
GVHP	Millipore	PVDF[b]	110	0.22	75	—	283.15	2781.9	265.53	1.12	32.74	2.14	0.32	3350
HVHP			140	0.45			463.86	2904.7	451.23	1.19	33.64	2.12	0.66	3790

membrane thickness, δ; mean pore size, d_P; porosity, ε; liquid entry pressure of water, LEP_w; effective porosity, ε_e; surface porosity, ε_s; geometric standard deviation, σ_P; pore tortuosity, τ
[a] Flat sheet polytetrafluoroethylene, PTFE, membranes supported by polypropylene, PP.
[b] Flat sheet polyvinylidene fluoride, PVDF, membranes.
Source: Khayet et al. [5]; www.inderscience.com

DETERMINATION OF PORE SIZE (MEAN PORE SIZE, PORE SIZE DISTRIBUTION) BY DIFFERENT PHYSICAL METHODS

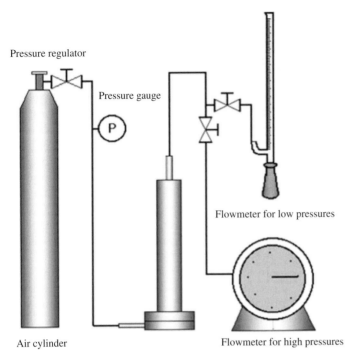

FIGURE 8.2 Schema of the experimental system used for bubble point and wet/dry flow method. *Source: Reprinted from [8]. Copyright 2001, with kind permission from American Chemical Society.*

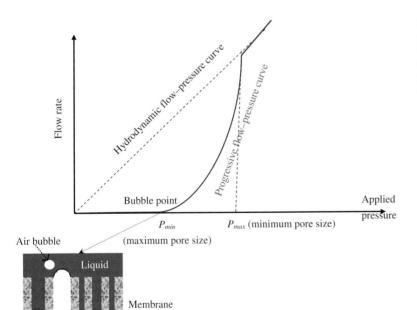

FIGURE 8.3 Typical gas flow–pressure curve obtained by the wet/dry flow method and schematic presentation of the bubble point corresponding to the maximum pore size.

shows an example of the systems used for this method. First, gas permeation rate is measured through a dry membrane. Generally a straight line is observed between the gas permeation rate and the transmembrane pressure difference as illustrated in Fig. 8.3 (broken line). Subsequently, the membrane is wetted by a liquid with a low surface tension such as isopropyl alcohol (IPA), which is assumed to fill all the pores and again the gas permeation rate is measured at increasing transmembrane pressures [8,17]. In this case the dependence of the gas flux on the applied transmembrane pressure is not linear as the membrane pores are plugged with a wetting liquid.

The bubble point test is actually to measure the diameter of the largest pore of the membrane. However, when the pore is small an extremely high pressure is required for this method, which may collapse the membrane pore structure. The pressure required to force the gas bubble through the wet pore is described by the Laplace equation:

$$d_P = \frac{4\sigma}{\Delta P} \cos\theta \qquad (8.13)$$

where ΔP is the pressure difference applied across the membrane, σ is the surface tension at the liquid–gas interface and θ is the contact angle of the liquid and membrane material.

As the pressure increases, it will reach a point where it can overcome the surface tension of the wetting liquid in the largest pores and will drive the liquid out of the pores. This point is called the bubble point. At transmembrane pressures lower than the bubble point, the pores remain filled with the wetting liquid and the gas flux is practically zero. Above the bubble point, the gas flow rate keeps increasing with an increase in the pressure because smaller pores are opened progressively with the increase of the pressure until all pores become empty at the pressure that corresponds to the minimum pore size. The range of pressures applied on the upstream side, therefore, depends on the range of the pore size of the membrane. Both the wet and dry test must be carried out under the same temperature, in most cases at room temperature, maintaining the downstream side of the system at atmospheric pressure. As shown in Fig. 8.3, the generated curve is usually S shaped. The methods suggested by Kesting [18] and Khayet et al. [17] are both applicable to obtain the pore size distribution of commercial membranes and MD membranes fabricated by different preparation techniques. For example, Fig. 8.4 shows the ratio of the gas permeation rate through wet membrane (J_w) to that of dry membrane (J_d) versus pore size for various MD membranes and the pore size distribution obtained by the Kesting method.

In Fig. 8.4 the ratio between the wet and dry flow rates are plotted versus the pore radius of polyvinylidene fluoride (PVDF) membranes. For each wetted membrane, the pores are filled with IPA at low values of ΔP, and the gas permeation rate, J_w, is practically zero. At a certain value of ΔP, corresponding to the bubble point of the membrane, the largest pores are opened, and the gas permeation rate, J_w, starts to increase. Smaller pores are opened progressively as ΔP increases according to the Laplace equation (Eq. 8.13). Finally, at the pressure corresponding to the minimum pore size, all the pores are empty and the ratio J_w/J_d becomes unity. This ratio remains unity thereafter. From the curves presented in Fig. 8.4(a), one can determine the maximum pore size from the bubble point and the mean pore size corresponds to J_w/J_d equal to 0.5. The pore size distribution can also be obtained as suggested by Kesting [18].

It is worth quoting that the mean pore sizes calculated by the two different methods, namely, the gas permeation test and the wet/dry flow method are slightly different. The ratio between the pore radius determined by the gas permeation test and that determined by the wet/dry flow method varies between 0.8 and 1.3 [8].

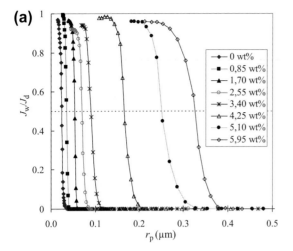

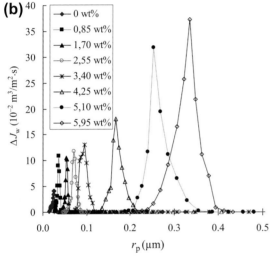

FIGURE 8.4 Ratio of wet membrane gas permeation rate, J_w, to dry membrane gas permeation rate, J_d, (a) and pore size distribution, (b) of PVDF supported membranes prepared for MD by the phase inversion method using different concentrations of the additive, water (0, 0.85, 1.70, 2.55, 3.40, 4.25, 5.10 and 5.95 wt%) in 15 wt% PVDF/dimethylacetamide (DMAC) casting solution (ΔJ_w is the variation of the gas permeation rate with the applied pressure and r_p is the pore radius). *Source: Reprinted from [8]. Copyright 2001, with kind permission from American Chemical Society.*

An improved procedure has been developed by Khayet et al. [17] to determine the mean pore size, the geometric standard deviation and the pore size distribution using the two experimental dry and wet curves as described in the following paragraphs:

The cumulative flow rate ratio for the pores with sizes below the size of the jth pores $d_p(j)$, $(j = 1, n)$, is:

$$g_a(j) = 1 - g'_a(j) = 1 - \frac{J_w(j)}{J_d(j)} \quad (8.14)$$

The incremental flow rate ratio occurring through the jth pores is:

$$g_d(j) = \frac{g'_a(j+1) - g'_a(j-1)}{2} \quad (8.15)$$

Taking into account that the flow rate is proportional to the pore area, the number of the jth pores (with size $d_p(j)$) is:

$$n_d(j) = K \frac{g_d(j)}{d_p(j)^2} \quad (8.16)$$

where K is a normalization factor that can be evaluated as:

$$K = \frac{g_a(n)}{\sum_{j=1}^{n} g_d(j)/d_p(j)^2} \quad (8.17)$$

Finally, the cumulative distribution of number of pores is:

$$n_a(j) = \sum_{k=1}^{j} n_d(k) \quad (8.18)$$

The pore size distribution can be expressed by the probability density function; e.g. log-normal distribution described by the following equation:

$$\frac{df(d_p)}{d(d_p)} = \frac{1}{d_p \ln \sigma_p (2\pi)^{0.5}} \exp\left[-\frac{(\ln d_p - \ln \mu_p)^2}{2(\ln \sigma_p)^2}\right] \quad (8.19)$$

where d_p is the pore size, μ_p is the mean pore size and σ_p is the geometric standard deviation.

Other distributions can also be used. The obtained cumulative distribution of relative number of pores $n_a(j)$ can be fitted to the function $f(d_p)$ and the mean pore size μ_p, together with the geometric standard deviation σ_p, can be evaluated for each MD membrane. The mean pore size will correspond to 50% of the cumulative number of pores and the geometric standard deviation can be calculated from the ratio of 84.13% of the cumulative number of pores to that of 50%.

In addition, the surface porosity (ε_s) defined as the ratio between the area of the pores to the total membrane surface area can be calculated from the following equation.

$$\varepsilon_s = \frac{N\pi}{4}\sum_{j=1}^{n} f_j d_j^2 \quad (8.20)$$

where N is the number of pores per unit area, known as pore density, and f_j is the fraction of the number of pores with size d_j.

It should be noted here that the surface porosity (ε_s) is different from the void volume or porosity (ε) which may be determined as:

$$\varepsilon = \varepsilon_s \tau \quad (8.21)$$

where τ is the pore tortuosity. Care should be taken using this equation for asymmetric membranes because a mean value is obtained for the tortuosity of the pores.

If the effective membrane porosity, which takes into account the tortuosity of the membrane pores (ε/τ) is known, the number of pores per unit area can be calculated from the following equation.

$$N = \frac{\varepsilon/\tau}{\sum_{j=1}^{n} f_j \pi r_j^2} \quad (8.22)$$

Both the cumulative pore size and the probability density curves for various commercial membranes (TF200, TF450, TF1000, GVHP and HVHP indicated in Table 8.1) used in MD were determined following the above procedure. Figure 8.5 shows the results.

Table 8.1 also summarizes the characteristics of some commercial membranes determined by the wet/dry flow method. It can be seen that the pore size given by the manufacturer lies within the range of the pore size distribution for all the membranes. When compared to the membrane TF200 (Gelman), the pore size distribution curves of the other membranes shift to the right and are lower and spread broader around their corresponding mean pore sizes. Except for the membrane TF450, the mean pore size of all other commercial membranes is smaller, when measured by the wet/dry flow method, as compared to the mean pore size measured by the gas permeation test.

It must be pointed out that a Coulter Porometer II Manufactured by Coulter Electronics Ltd. is based on wet/dry flow method. In this system the membrane sample is wetted with a Coulter Porofilm of low surface tension, low vapor pressure and low reactivity, which is assumed to fill all the membrane pores. Coulter porosimetry has been employed to characterize the commercial membranes shown in Table 8.1 [19,20]. Figure 8.6 shows the corresponding pore size distributions. For all tested commercial membranes (Gelman and Millipore), it was found that the effective porosity (ε_e) determined by Coulter porosimetry was greater than the value obtained from the gas permeation test. For Millipore membranes, the pore sizes determined by Coulter porosimetry were larger than those measured by gas permeation test and wet/dry flow method. These inaccuracies may be due to the type of the wetting liquid used. In fact, the bubble point and related methods (commonly called liquid displacement techniques) have reached the status of recommended standards ASTM F316 [21] and ASTM E1294 [22]. However, these techniques are criticizable because the mechanisms of gas transport through the membrane pores are not considered.

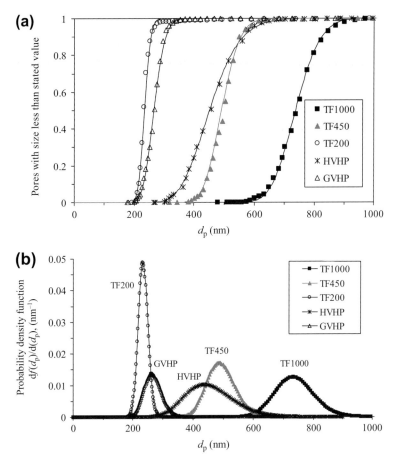

FIGURE 8.5 Cumulative pore size (a) and probability density (b) curves of various commercial membranes used in MD determined by the wet/dry flow method following an improved procedure developed by Khayet et al. [17] (d_p is the pore size). *Source: Original figures were published in [5] (www.inderscience.com)*

Mercury Porosimetry

This method of membrane pore size characterization is based on the same principles as the bubble pressure method [23]. However, in this case mercury is the employed non-wetting liquid for porous materials. Mercury is forced to penetrate into dry membrane pores and the volume of mercury that entered into the pores is determined at each applied pressure. Normally, mercury intrusion is carried out using a mercury porosimeter such as Quanta-Chrome 33000. In this instrument, a weighed amount of membrane is placed in the porosimeter chamber and evacuated to a certain pressure for a certain period. Subsequently, mercury is brought into the chamber and the pressure is increased to fill the pores of the membrane with mercury. First, the largest pores are filled with mercury at a minimum pressure. As the pressure increases, smaller pores are filled until all pores will be filled corresponding to the maximum mercury intrusion volume. The pore size distribution can be determined by registering the cumulative volume of pore-intruded mercury as a function of the operating pressure. Figure 8.7 shows

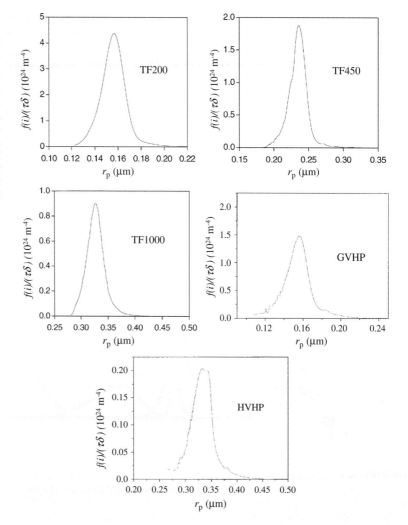

FIGURE 8.6 Pore size distributions of commercial membranes used in MD determined using coulter porosimetry ($f(i) = \frac{n(i)}{r(i-1)-r(i)}$, $i = 1 \ldots m$; n is the number of pores per m² with radius between $r(i)$ and $r(i-1)$, m is the number of classes in the pore size distribution, δ is the membrane thickness and τ is the pore tortuosity). *Source: Reprinted from [19]. Copyright 2002, with kind permission from Elsevier. Reprinted from [20]. Copyright 2003, with kind permission from Elsevier.*

a typical curve from which the mean pore size and the pore size distribution can be determined by means of Laplace equation (Eq. (8.13)), assuming the cylindrical pore shape. It is worth quoting that this technique was mainly used for the characterization of macroporous structures. Because of its simplicity, this method is used generally to characterize ceramic membranes.

One of the disadvantages of this technique is the high pressures required for membranes of small pore sizes, membrane compaction caused by the high pressures applied for mercury intrusion as well as the formation of microfractures with a consequent alteration of pore size and structure. Moreover, this method also includes dead-end pores resulting in an overestimation of the membrane pore size. All these inconveniences can be avoided by using the wet/dry flow method.

Electron Microscopy

This method includes several electronic microscopy techniques that permit the direct visualization of the morphological structure of porous membranes (top and bottom surfaces for flat sheet membranes, internal and external surfaces of capillaries and hollow fibre membranes as well as membrane cross-section), such as scanning electron microscopy (SEM), transmission electron microscopy (TEM), field emission scanning electron microscopy (FESEM), etc. Computerized image analysis of the corresponding micrographs is frequently used to obtain pore size distributions and porosities.

SEM was used in different MD studies to determine the mean pore sizes of the membrane surfaces [24–42]. Before the SEM test, the membrane samples must be first frozen in liquid nitrogen and then broken to obtain small fragments. However, one of the limitations of this technique is the heavy metal coating, required for membrane sample preparation, which gives some artifacts and tends to damage the membrane surface. Figure 8.8 shows some SEM images of commercial and laboratory prepared membranes for MD applications.

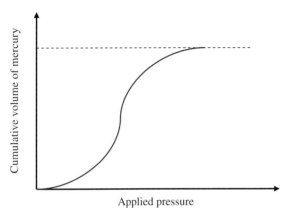

FIGURE 8.7 Cumulative volume of mercury *versus* applied pressure.

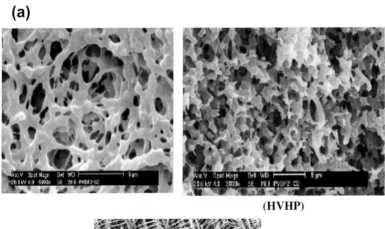

(HVHP)

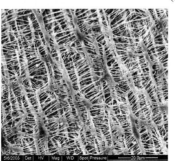

Surface (PTFE)

FIGURE 8.8 SEM images of commercial membranes (a) and laboratory prepared membranes (b) for MD applications. *Sources: (a) Reprinted from [86]. Copyright 2004, with kind permission from Elsevier. Reprinted from [87]. Copyright 2006, with kind permission from Elsevier. (b) Reprinted from [9]. Copyright 2002, with kind permission from Elsevier. Reprinted from [88]. Copyright 2002, with kind permission from Elsevier.*

FIGURE 8.8 (continued).

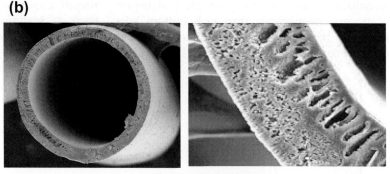
Cross-section (PVDF hollow-fibre)

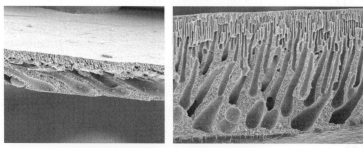
Cross-section (SMM modified membranes)

Surface (PVDF electro-spun membrane)

As can be seen in Fig. 8.8, normally various structures and shapes of pores (elliptical, circular, slit, etc.) are observed and therefore determination of the pore size from SEM images is not an easy task. Phattaranawik et al. [43] used the FESEM (Hitachi model S-900) to image the commercial membranes GVHP and HVHP (Millipore, Table 8.1) and to determine the mean pore size and the pore size distribution using the software 'AnalySIS' provided by Soft Imaging System. The pore area was measured by encircling the observed membrane pores. A total of 900−1300 pores were taken into consideration for each membrane image. The equivalent pore size (d_p) was determined by:

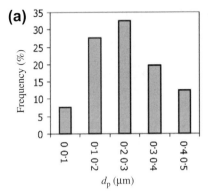
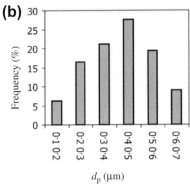

FIGURE 8.9 Pore size distributions obtained by FESEM of the commercial membranes (a) GVHP and (b) HVHP commonly used in MD. *Source: Reprinted from [43]. Copyright 2003, with kind permission from Elsevier.*

$$d_p = 2\sqrt{\frac{S_{pore}}{\pi}} \qquad (8.23)$$

where S_{pore} is the surface area of the pore that is supposed to be circular.

The obtained mean pore sizes were 251 nm for GVHP membrane and 414 nm for HVHP membrane. These values are lower than those obtained from the gas permeation test, the wet/dry flow method and Coulter porosimetry (Table 8.1). As stated previously, this difference may be attributed to contraction of the pores during metal coating of the membrane sample, which is required for the FESEM technique. Figure 8.9 shows typical pore size distributions obtained by the FESEM. The bar graph in Fig. 8.9 can be fitted to a log-normal distribution and the mean pore size together with the standard deviation can be obtained. In this case, the standard deviation of the commercial membrane GVHP was found to be 1.037 and that of the membrane HVHP was 0.636. These values are lower than those obtained by the wet/dry flow method indicating that the pore size distribution determined by the electron microscopy is narrower around the mean pore size.

Atomic Force Microscopy

Atomic Force Microscopy (AFM) is a newly developed high-resolution technique to study the surface morphology of the membrane surfaces down to the scale of nanometers. It was developed by Binning et al. [44] and its main advantage over the electron microscopy is that no previous sample preparation is needed. It is used to obtain directly the three-dimensional topographical images of the membrane surfaces (i.e. top and bottom surfaces of flat sheet membranes, internal and external surfaces of capillaries and hollow fibre membranes) as well as their cross-sections up to atomic level resolution in air or in liquid by scanning a sharp tip, situated at the end of a microscopic cantilever, over a surface [45]. More details can be found in the two recently published books [46,47].

AFM was first applied to polymeric membrane surface by Albrecht and Quate [48] in 1988 soon after it was invented by Binning et al. [44]. The technique has been applied, since then, extensively for studying various types of membranes and materials [49,50], including microfiltration (MF) membranes [51,52], ultrafiltration (UF) membranes [53–55], nanofiltration (NF) membranes [56,57], reverse osmosis (RO) membranes [58,59], gas separation membranes [60,61] and MD membranes [62,63], giving useful information about surface morphology, pore size, nodule size, pore density, porosity and roughness.

Three-dimensional images of the MD membrane surface can be obtained directly without special sample preparation. These images can be obtained over different locations of each membrane sample at room temperature using, e.g., Nanoscope III equipped with 1553D scanner (Digital Instruments Inc.). Only small pieces of approximately 0.5 cm × 0.5 cm in area are required. For an adequate comparison, it is advisable to use the same tip to scan all membrane surfaces and all captured images should be treated in the same way.

A surface structure of a MD membrane can be observed more accurately by AFM as shown in Fig. 8.10. In addition, the AFM may be used in a number of different modes: contact mode, non-contact mode and tapping or intermittent mode. The mean pore size, the pore size distribution, the surface porosity as well as its roughness parameters (mean roughness, root mean-square, average difference in height) can be determined as follows.

To determine the pore sizes and nodule sizes, cross-sectional line profiles are selected to traverse the obtained AFM images and the diameters of nodules (i.e. high peaks) or pores (i.e. low valleys) are measured by means of a pair of cursors along the reference line as shown in Fig. 8.11. The horizontal distance between each pair of cursors is taken as the diameter of the nodule or pore. The AFM software program allows quantitative determination of pores or nodules by using the images in conjunction with digitally stored line profiles.

To determine the mean pore size and pore size distribution, a minimum of 30 pores on each membrane sample should be measured by inspecting line profiles on the AFM images at different locations of a membrane surface. Subsequently, the measured pore sizes are arranged in ascending order and the corresponding median ranks (50%) are calculated using the following equation:

$$\chi = \left(\frac{i - 0.3}{n + 0.4}\right) \times 100 \quad (8.24)$$

where i is the order number of the measured pore size arranged in ascending order and n is the total number of the measured pores.

If the obtained pore sizes fit to a log-normal distribution, median ranks plotted on the ordinate against pore sizes arranged in an ascending order on the abscissa will yield a straight line on log-normal probability paper as presented in Fig. 8.12. If this condition is satisfied, from the log-normal plot, the mean pore sizes and the corresponding geometric standard deviations can be calculated. The mean values will correspond to 50% of the cumulative number of pores and the geometric standard deviation can be calculated from the ratio of 84.13% of the cumulative number of pores to that of 50%. From the mean values and the corresponding geometric standard deviations, the pore size distribution can be expressed by the probability density function described in Eq. (8.19). Typical cumulative pore size distributions together with the probability density function curves can be generated as illustrated in Fig. 8.12.

In addition, the surface pore density of each membrane, which is the number of pores per unit area, N, can be obtained directly from the AFM analysis software program. The corresponding surface porosity (ε_s) may be then determined from Eq. (8.20) using the pore size distribution.

The surfaces of the MD membranes can be compared using various roughness parameters, such as the mean roughness, R_a, the root mean square of Z data, R_q, and the mean difference in the height between the five highest peaks and the five lowest valleys, R_z. These parameters are defined later on and can be quantitatively determined by an AFM analysis software program. Different locations should be chosen for each membrane sample and the average values are calculated. The roughness parameters depend on the curvature and size of the AFM tip, the scanned area and the treatment of the data of the captured surfaces such as plane fitting, flattening, filtering, etc.

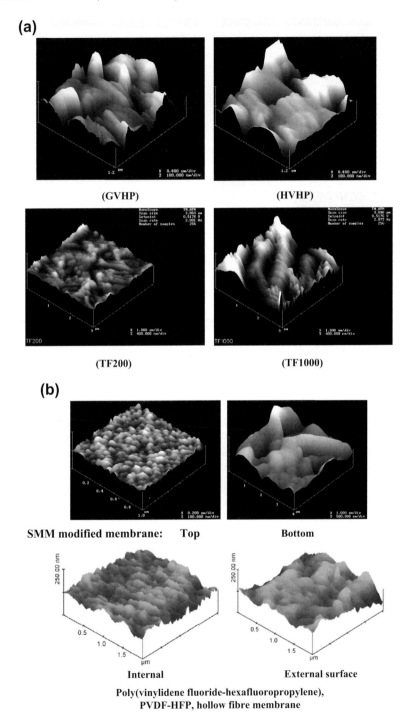

FIGURE 8.10 Three-dimensional AFM images of commercial membranes used in MD (a) and laboratory prepared MD membranes (b). *Sources: (a) Original figures were published in [5] (www.inderscience.com). (b) Reprinted from [54]. Copyright 2003, with kind permission from Elsevier. Reprinted from [40]. Copyright 2010, with kind permission from Elsevier.*

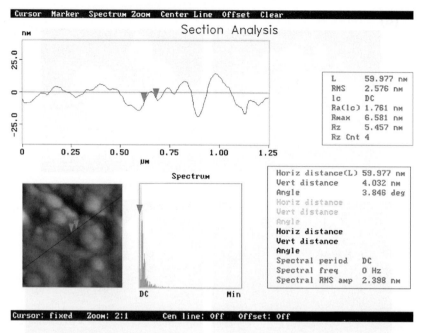

FIGURE 8.11 Section analysis of an AFM image of MD membrane. *Source: Reprinted from [62]. Copyright 2004, with kind permission from Elsevier.*

Therefore, the roughness parameters should not be considered as absolute roughness values. For comparison, the same tip should be used to produce the AFM images of the MD membranes and all captured surfaces should be treated in the same way.

The mean roughness, R_a, represents the mean value of the surface relative to the center plane for which the volumes enclosed by the images above and below this plane are equal. This parameter is expressed as:

$$R_a = \frac{1}{L_x L_y} \int_0^{L_x} \int_0^{L_y} |f(x,y)| dx dy \quad (8.25)$$

where $f(x, y)$ is the surface profile relative to the center plane and L_x and L_y are the dimensions of the surface in the x and y directions, respectively.

The root mean square roughness, R_q, is the standard deviation of the Z values within the specific area and is calculated using the following equation:

$$R_q = \sqrt{\frac{\sum (Z_i - Z_m)^2}{N_p}} \quad (8.26)$$

where Z_i is the current Z value, Z_m the average of the Z values and N_p is the number of points within a given area.

The average difference in height, R_z, between the five highest peaks and the five lowest valleys is calculated relative to the mean plane, which is a plane about which the image data has a minimum variance.

It is worth quoting that in general the roughness parameters of the membranes become larger with an increase in pore size and nodule size. This may be explained since the roughness parameters depend on the Z values. In fact, when the surface consists of deep depressions that characterize pores and high peaks that

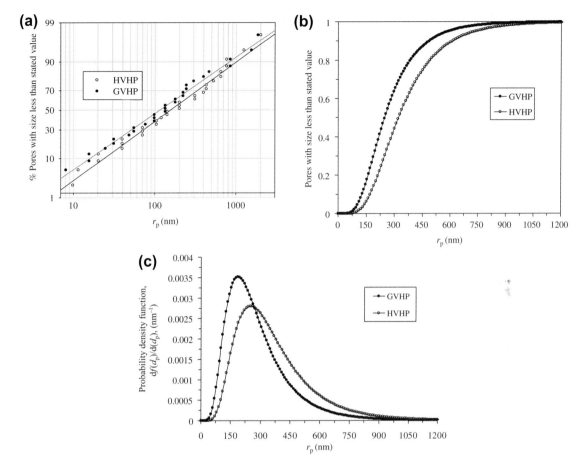

FIGURE 8.12 Log-normal pore size distribution (a), cumulative pore size distribution (b) and probability density function curves (c) obtained from AFM images of two MD membranes.

correspond to nodules, high roughness parameters are expected.

Khayet et al. [62,63] were the first to characterize MD membranes by AFM. The mean pore size, pore size distribution, surface porosity, pore density and roughness parameters of commercial membranes used commonly in MD as well as membranes prepared in their laboratory for MD were determined. The results of some commercial membranes are given in Table 8.2. It was observed that, generally, the mean pore sizes determined by AFM were 1.2 to 2.1 times larger than those determined from the gas permeation test. This was attributed to the fact that the pores measured by AFM had maximum openings at the surface entrance and a few small pores could easily be misinterpreted as one large pore when they were amalgamated, resulting in an overestimation of the pore sizes of the MD membranes. In Fig. 8.13, a comparison between the pore size distribution obtained from the wet/dry flow method and the AFM

TABLE 8.2 Characteristics of Flat Sheet Commercial Membranes Commonly used in MD Determined by AFM Technique

Membrane trade name	Manufacturer	Material	AFM technique				
			d_p (μm)	σ_p	ε_s (%)	τ	ν
TF200	Gelman	PTFE/PP	250.72	1.23	33.97	2.02	92.6
TF450			538.67	1.29	30.86	2.08	91.3
TF1000			1185.34	1.27	34.24	1.96	103.4
GVHP	Millipore	PVDF	495.70	1.69	25.28	2.77	299.7
HVHP			644.67	1.65	24.63	2.90	332.5

mean pore size, d_P; surface porosity, ε_S; geometric standard deviation, σ_p; pore tortuosity, τ and average nodule size average, ν
Source: Khayet et al. [5]; www.inderscience.com

technique is presented for the membranes GVHP and TF450 as an example. For all these membranes the pore sizes determined from the AFM technique were found to be greater than those obtained from the wet/dry flow method. Moreover, the mean pore sizes of the Millipore membranes (GVHP and HVHP) determined by AFM were 1.97 and 1.56 times larger than those obtained by the electron microscopy [43]. As reported earlier, this may be due to pore contraction during metal coating of the membrane sample, which is required to produce the electron microscopic images. In addition, as can be seen in Table 8.2, the values of the geometrical standard deviation, σ_p, of the membranes GVHP and HVHP obtained by AFM analysis were higher than those obtained from the electron microscopy.

The surfaces of the commercial membranes GVHP and HVHP contain nodule aggregates

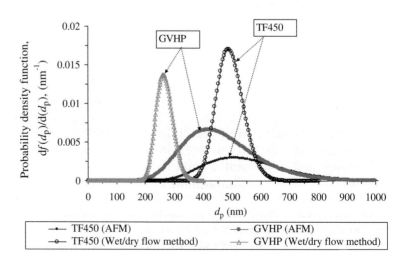

FIGURE 8.13 Pore size distribution of two commercial membranes (GVHP, Millipore) and (TF450, Gelman) obtained by AFM analysis and wet/dry flow method. *Source: Original figures were published in [5] (www.inderscience.com).*

and super nodular aggregates, while the nodule sizes of the commercial membranes (TF200, TF450 and TF1000) are smaller (Fig. 8.10). This result may be attributed to the different techniques and polymers used for membrane preparation. Khayet et al. [62] observed that the nodule sizes of both Millipore membranes (GVHP and HVHP) are greater than the nodule sizes of the PVDF membranes prepared by the phase inversion method for MD using different water concentrations in the casting solution. This may be due to the difference in polymer type or in membrane preparation conditions.

The roughness parameters of the MD membranes were also determined by the method stated above. The membranes GVHP and HVHP were found to be rougher than the membranes TF200, TF450 and TF1000. For example, the average values of R_a, R_q and R_z of the commercial membrane GVHP were 81.7 nm, 113.8 nm and 1052.0 nm, respectively; whereas, those of the commercial membrane TF200 were 44.2 nm, 59.6 nm and 441.8 nm, respectively, much lower than the values for GVHP. This may be attributed partly to the larger nodules sizes of the Millipore membranes.

POROSITY AND EFFECTIVE POROSITY: TECHNIQUES USED

The void volume fraction or the membrane porosity (ε) is defined as the volume of the pores divided by the total volume of the membrane. For MD membranes, it can be determined by measuring the density of the membrane material (ρ_{pol}) using isopropyl alcohol (IPA), which penetrates inside the pores of the membrane, and the density of the membrane (ρ_m) using pure water, which does not enter the pores because of the hydrophobic character of the MD membrane. In this method, a pycnometer and a balance are employed and the following equation can be used as suggested by Smolders and Franken [2]:

$$\varepsilon = 1 - \frac{\rho_m}{\rho_{pol}} \qquad (8.27)$$

This method can be used for flat sheet, capillaries and hollow fibre porous and hydrophobic membranes. For example, Fig. 8.14 shows the measured void volume fraction of MD hollow fibre membranes as a function of the polymer

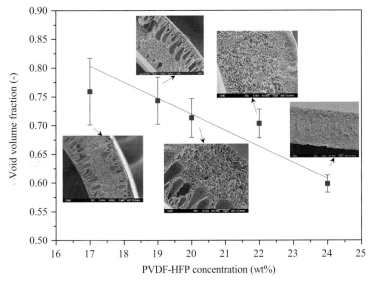

FIGURE 8.14 Measured void volume fraction and SEM cross-section images of hollow fibre membranes prepared by dry/wet spinning technique using different concentrations of the copolymer poly(vinylidene fluoride-hexafluoropropylene) (PVDF-HFP). *Source: Reprinted from [40]. Copyright 2010, with kind permission from Elsevier.*

TABLE 8.3 Measured Parameters of Flat Sheet Commercial Membranes Commonly used in MD

Membrane trade name	Manufacturer	Material	Measured values			
			δ (μm)	ε (%)	LEP_w (kPa)	k_m (W/m K)
TF200	Gelman	PTFE/PP	54.8	68.7	276	0.043
TF450			60.0	64.3	149	—
TF1000			58.4	67.1	58	—
GVHP	Millipore	PVDF	117.7	70.1	204	0.041
HVHP			115.8	71.3	105	0.040

membrane thickness, δ; void volume fraction, ε; liquid entry pressure of water, LEP_w; thermal conductivity, k_m
Source: Khayet et al. [5]; www.inderscience.com

concentration in the spinning solution. It was found that the void volume fraction decreased gradually with the increase of the polymer content in the spinning solution. This was further related to the morphology of the hollow fibre membrane, which has changed from a finger-like structure to a complete sponge-like structure.

Table 8.3 summarizes the measured void volume fraction of some commercial membranes commonly used in MD applications. The measured values are lower than the values given by the manufacturers shown in Table 8.1. This may be attributed to the different measurement techniques used. The manufacturers do not supply the technique they have used. In fact, it should be pointed out that even when using the same method (Eq. (8.27)) for the same membranes, the values obtained by Khayet and Matsuura [41] were found to be higher than the ones given by Izquierdo-Gil et al. [67,77] ($62 \pm 2\%$ for GVHP membrane and $66 \pm 2\%$ for HVHP membrane). Therefore, the proposed method should be carried out carefully. The standard deviations of the measured values are lower than 8% ($70.1 \pm 2.7\%$ for the membrane GVHP, $71.3 \pm 3.4\%$ for the membrane HVHP and $68.7 \pm 5.4\%$ for the membrane TF200).

Bonyadi and Chung [37] used a 33% LIX54 kerosene solution to measure the porosity of hollow fibre membranes prepared for MD. The hollow fibre membrane sample was first weighed (w_d) and then immersed in the solution for 1 week. In this case, it is assumed that all void volume fraction of the membrane sample is filled with the solution. Subsequently, the impregnated hollow fibre membrane sample was removed from kerosene solution and wiped in order to remove the excess of liquid present in the lumen and on the outer surfaces. Finally, the hollow fibre membrane sample was weighed again (w_w). The empty volume (V) was calculated as follows:

$$V = \frac{w_w - w_d}{\rho_K} \quad (8.28)$$

where ρ_k is the kerosene density. The porosity of the hollow fibre membrane is estimated by calculating the ratio of the empty voids (V) to the total volume of the membrane samples.

The effective porosity (ε_e) of MD membranes takes into consideration the porosity (ε) and the effective pore length (L_p) as defined in Eq. (8.10). The use of ε_e is adequate for theoretical modeling of MD process as it includes the pore tortuosity, which is considered as an adjustment factor in the models. As stated earlier, this parameter can be obtained from the gas permeation test (Eq. (8.12)) and Coulter porosimetry. The obtained values of some commercial membranes are presented in Table 8.1. It can be seen that the membranes supplied by Gelman (TF200, TF450, TF1000) exhibit higher effective porosities than the membranes supplied by Millipore (GVHP, HVHP). This result is also confirmed by Coulter porosimetry technique and may be attributed to the higher thickness of the membranes GVHP and HVHP inducing lower effective membrane porosities.

The surface porosity (ε_s) can also be determined by AFM. The pore density is obtained by counting the number of pores on the AFM images and the surface porosity is then calculated from Eq. (8.20) as stated in the previous section using the pore size distribution. Both the pore density and surface porosity data were found to be higher for the commercial membranes Gelman than for Millipore membranes (Table 8.2). This is in accordance with the porosity (ε) and the effective porosity (ε_e) determined for the same membranes. From the wet/dry flow method, the obtained effective porosity is also found to be higher for the Gelman membranes compared to Millipore membranes (Table 8.1). When comparing surface porosity determined from wet/dry flow method and AFM technique, the one obtained from AFM is lower for all commercial membranes. Moreover, it is worth noting that the surface porosity values are lower than the void volume even though the membranes are the same.

EVALUATION OF MEMBRANE PORE TORTUOSITY

It must be pointed out that if the effective membrane porosity (ε_e), the membrane thickness (δ) and porosity (ε) are known, the tortuosity of the membrane pores can be estimated as suggested in [17,62]:

$$\tau = \frac{\varepsilon}{\delta \varepsilon_e} \qquad (8.29)$$

A mean tortuosity value is obtained from this equation assuming that the membrane pore size does not change from one side to the other side of the membrane.

In MD studies, a value of 2 is frequently assumed for the tortuosity factor to predict the MD fluxes although a value of the tortuosity factor as high as 3.9 has been reported [43,64]. The calculated tortuosity factors using the wet/dry flow method are summarized in Table 8.1 and those obtained from the AFM technique are shown in Table 8.2. Due to the fact that the membranes (TF200, TF450 and TF1000) are supported by a backing polypropylene net, the membrane layer was peeled off from the support and the thickness was measured. In general, the tortuosity factor values of the Millipore membranes were found to be higher than those of the Gelman membranes. This may be attributed partly to the higher membrane thickness of Millipore membranes. Moreover, it can be observed that for the same membrane the tortuosity value determined using AFM analysis is higher than that obtained from the wet/dry flow method. This may be attributed partly to the surface porosity, which is found to be lower using AFM technique.

PENETRATION PRESSURE DETERMINATION

To avoid membrane pore wetting, at least one layer of the MD membrane must be prepared with a hydrophobic material. However pore wetting may occur and produced water quality may be deteriorated if the applied transmembrane hydrostatic pressure exceeds the liquid entry pressure (*LEP*) of the aqueous feed

FIGURE 8.15 Schema of the systems used to measure *LEP* of flat sheet (a) *(Source: Reprinted from [8]. Copyright 2001, with kind permission from American Chemical Society)* and capillaries or hollow fibre membranes (b) *(Source: Reprinted from [40]. Copyright 2010, with kind permission from Elsevier).*

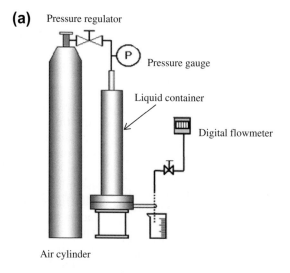

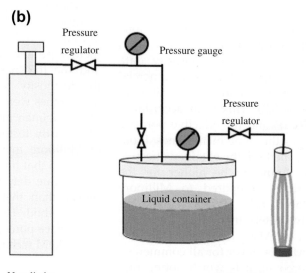

solution. *LEP* is the minimum required pressure for the aqueous solution to go into dry membrane pores. For example, LEP_w is the pressure that must be applied onto pure water before it penetrates into the pores. Figure 8.15 shows schematically a typical system for the measurement of the *LEP* of flat sheet, capillaries and hollow fibre membranes.

The capillary or hollow fibre membranes are first mounted in tubular stainless steel modules, whereas the flat sheet membrane is placed in a stainless steel cell. For the flat sheet, the membrane is placed between the upper chamber (Fig. 8.15(a)), the feed side, which is filled with the liquid solution to be treated and the lower chamber, the permeate side, which is connected to a digital capillary flowmeter. Some LEP systems do not use flowmeter on the permeate side. First, a slight pressure of about 0.3×10^5 Pa (gauge) is applied to the feed solution for at least 10 min; then the pressure is increased stepwise with a small increment until a first drop of the feed solution appears on the permeate side or until a continuous permeate flow starts to occur. The corresponding applied pressure is the membrane LEP. This method has been described by Smolders and Franken [2] in detail and has been used by various researchers for both flat sheet [8,17,28,65–69] and hollow fibre membranes [9,40].

The LEP decreases with the increase of the maximum pore size of the membrane and/or the decrease of the contact angle of the feed solution to be treated at the membrane surface.

The LEP_w of some commercial flat sheet membranes prepared from different materials are listed in Table 8.3 [5,17]. The standard deviations from the shown average values are lower than 4%. The values given by the manufacturer (Gelman) are presented in Table 8.1. It can be observed that the measured values of the membranes TF450 and TF1000 are slightly higher (i.e. 7.4% and 17.2%) than those given by the manufacturers. On the other hand, for the membrane TF200, the measured value is slightly lower than the manufacturer's value (i.e. 2.2%). In addition, as it is expected, among the membranes of the same manufacturer (Gelman or Millipore), the LEP_w is lower for membranes having greater pore size. Comparing TF200 (Gelman, polytetrafluoroethylene, PTFE) and GVHP (Millipore), both having similar pore sizes, LEP_w is higher for the former membrane since the contact angle of PTFE is higher than PVDF, and hence PTFE material is more hydrophobic than PVDF.

Khayet and Matsuura [65] found that the LEP_w values of both the modified hydrophilic polyetherimide (PEI) membranes by surface modifying macromolecules (SMMs) and the unmodified ones increased with the increase of the PEI concentration in the casting solution used for the preparation of the membranes, and for the same PEI concentration the LEP_w of the SMM-modified membrane was higher than that of the unmodified membrane. These results were attributed to the higher hydrophobicity of the SMM-modified PEI membranes and to the smaller pore size of the membranes prepared with the higher PEI concentration in the casting solution.

Tomaszewska [28] studied the variation of the LEP_w of flat sheet PVDF membranes with the lithium chloride (LiCl) concentration in the polymer casting solution. It was observed that the PVDF membranes prepared with higher amounts of LiCl exhibited lower values of LEP_w due to their larger pore sizes. Similar results were obtained by Khayet and Matsuura [8] for the supported and unsupported PVDF membranes prepared with the same solvent dimethylacetamide (DMAC) and water as nonsolvent additive.

Gostoli and Sarti [66] measured the LEP of ethanol–water mixtures for the membrane TF200 and observed a lineal decrease of the LEP with ethanol concentration. This is attributed to the decrease of the surface tension of the aqueous mixture with the increase of ethanol concentration.

Franken et al. [68] established wetting criteria for the application of MD. Various membrane types and liquids including ethanol, acids (formic acid, acetic acid, propionic acid and butyric acid), solvents (dimethylacetamide, dimethylsulfoxide and dimethylformamide), acetone and 1-4-dioxane. For each membrane, different LEP values were obtained. It was concluded that the maximum allowable concentration of organic compound in water cannot be calculated but has to be measured.

García-Payo et al. [69] studied of the effects of alcohol type, alcohol concentration, temperature and type of membrane on the LEP value. It was observed that the LEP decreased with the

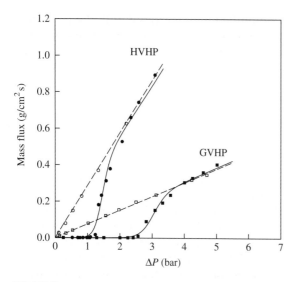

FIGURE 8.16 Water permeate flux *versus* the applied transmembrane pressure for GVHP and HVHP membranes at 25 °C. *Source: Reprinted from [69]. Copyright 2000, with kind permission from Elsevier.*

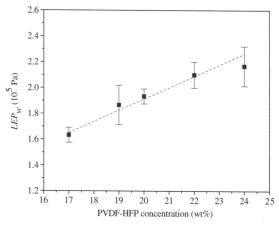

FIGURE 8.17 LEP_w of hollow fibre membranes prepared by the dry/wet spinning technique using different concentrations of the copolymer poly(vinylidene fluoride-hexafluoropropylene) (PVDF-HFP). *Source: Reprinted from [40]. Copyright 2010, with kind permission from Elsevier.*

increase of the concentration of alcohol (i.e. methanol, ethanol, isopropanol) in water, the length of the alcohol hydrocarbon chain, the temperature, which was varied in the range 25–50 °C and the membrane pore size. Figure 8.16 presents as an example a non-linear increase of the mass flux with the applied hydrostatic transmembrane pressure on distilled water at 25 °C for GVHP and HVHP membranes, after the *LEP* was surpassed. However, when the transmembrane hydrostatic pressure was decreased linear relationships were found for both membranes.

Figure 8.17 shows as an example the measured LEP_w of MD hollow fibre membranes as a function of the polymer concentration in the spinning solution. The observed gradual increase of the LEP_w with the increase of the polymer concentration is due to the decrease of the maximum pore size, because the same polymer material was used for the preparation of the hollow fibre membranes by the dry/wet spinning technique.

Some researchers measured only the water contact angle and the LEP_w, while MD was carried out with aqueous solutions containing lower surface tension compounds than water, without paying attention to the possible pore flooding of their membranes. It should be emphasized that a difference exists between the LEP_w of water and that of the feed and permeate solutions that are brought into direct contact with the membrane surface. Care must also be taken when organic compounds are removed from aqueous solutions by direct contact membrane distillation (DCMD). Since the organic compounds are more volatile than water, the permeate becomes more concentrated in organics during the DCMD process, increasing the risk of membrane pore wetting by the permeate aqueous solution.

It is obvious from the Laplace equation (Eq. (8.13)) that the membrane wetting depends on the contact angle between the liquid and the membrane surface. It would be desirable to measure both advancing and receding water contact angles on both sides of the MD membranes. To perform correct measurement, the contact angles should be measured at different spots of the membrane sample and the effect of the membrane pore size and roughness should also be taken into consideration [70–73]. For example, liquid contact angles were measured at room temperature using different systems for flat sheet [25,29,30,65,68,69,74] and capillaries or hollow fibre membranes used in MD [37]. It was observed that the measured advancing contact angles of the PTFE membranes (Gelman TF200, $113.6 \pm 2.7°$) were higher than those of the PVDF membranes (Millipore GVHP, $110.2 \pm 3.3°$), while the receding contact angles were $100.7 \pm 3.6°$ and $95.9 \pm 2.5°$, for the PTFE and PVDF membrane, respectively.

Wu et al. [25] measured the advancing water contact angles on the top surface of modified membranes by plasma grafting of polystyrene at the surface of the cellulose acetate membranes. The water contact angles were found to increase with the radiation time and exhibited a minimum for the discharge power.

Feng et al. [29,30] also measured the advancing water contact angles of membranes prepared from PVDF and the copolymer poly(vinylidene fluoride-co-tetrafluoroethylene) (F2.4) under the same conditions using the phase inversion technique. Both asymmetric porous and symmetric dense membranes were considered for each polymer. It was found that the F2.4 membrane's pore sizes were larger than the PVDF membranes and the water contact angles of the porous membranes were lower than those of the dense membranes. This may be attributed to the pore effect.

Khayet and Matsuura [65] measured both advancing and the receding water contact angles of the SMM-modified and unmodified PEI membranes. Both contact angles increased when blending SMM with the polymer PEI solution, indicating that the SMM-modified PEI membranes were more hydrophobic than the unmodified ones prepared with the same PEI polymer concentration in the casting solution. This was due to the hydrophobic character of the fluorohydrocarbon tails of the SMM molecule. Moreover, the increase in the receding contact angle of the SMM-modified membranes indicated that hydrophobic segments of the SMM remained at the membrane surface even after the surface was brought into contact with water.

Hoffmann et al. [74] measured the contact angles between PTFE foils and binary mixtures of methanol/water, ethanol/water and n-propanol/water of different concentrations. It was observed the contact angle decreased with an increase in alcohol concentration and in the length of the hydrocarbon chain of the alcohol.

García-Payo et al. [69] measured the advancing contact angle for commercial membranes of different pore sizes and materials (PVDF, PTFE) using distilled water and different alcohol–water mixtures. For the same membrane material, the water contact angle was found to be independent of the membrane pore size and the membrane support had no effect. Mean values for PVDF ($111 \pm 3°$) and PTFE ($123 \pm 2°$) were reported. Not only accurate measurement of contact angles but also correction of measured values by considering the effects of membrane roughness is necessary for MD membranes.

The water contact angles of PVDF electrospun membranes were measured at 25 °C on an optical contact angle meter CAM 100 equipped with a CCD camera, frame grabber and image analysis software. Water contact angles were obtained by placing the tip of the syringe near the sample surface and depressing the syringe to produce a constant water drop volume of about 2 μl. Five to six drops per sample and twenty reading per each drop can be carried out within the settling time of 1 s. The contact angle values are determined by a computer software using the Laplace equation (Eq. (8.13)). More than fifteen readings were obtained for each membrane sample and a mean value was calculated. Figure 8.18

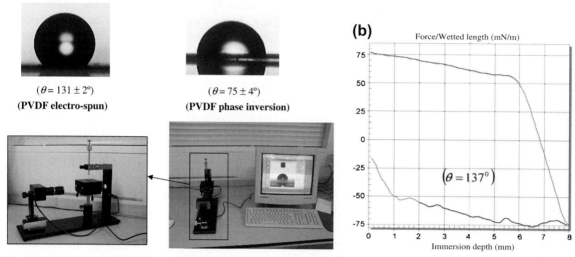

FIGURE 8.18 Measured water contact angles of flat sheet electro-spun PVDF membrane and phase inversion PVDF membrane (a) and measured force per unit length of the fibre *versus* the immersion depth used to determine the water contact angle of hollow fibre membrane prepared for MD (b). *Source: Reprinted from [37]. Copyright 2007, with kind permission from Elsevier.*

shows images of the measured water contact angles of electro-spun and phase inversion PVDF membranes prepared using the same polymer solution. The water contact angles of electro-spun membranes are higher than those of phase inversion membranes.

Bonyadi and Chung [37] used a Sigma 701 Tensiometer from KSV instruments Ltd. to measure the water contact angle of hollow fibre membranes prepared for MD. This tensiometric method determines the advancing contact angle by measuring the force when a hollow fibre membrane sample is brought into vertical contact with the liquid. The principle of this measurement is based on the forces of interaction, geometry of the solid and the surface tension of the liquid. In this case, a computer software is employed. At least ten readings should be carried out. An example of the obtained forces and the calculated values of the water contact angles together with their images are shown in Fig. 8.18. It is important to mention that in the case of hydrophilic/hydrophobic composite hollow fibre membranes, the lumen side of the hollow fibre should be blocked by inserting a metallic bar to prevent the effect of the inner hydrophilic surface on the water contact angle of the external surface.

THERMAL STABILITY TESTS

The MD membranes should exhibit good thermal stability up to temperatures as high as the boiling point of the aqueous solutions to be treated, which is usually lower than 100 °C. In general, all the used polymeric and ceramic materials have good thermal stability at temperatures even higher than 100 °C. To examine the thermal stability of a MD membrane, the membrane can be exposed to high temperatures ($\approx 95°C$) in water or in air for a predetermined time and then its *LEP* should be measured and compared with the initial value obtained before heat treatment.

Thermal degradation measurements of MD membranes can be performed using thermogravimetric analysis (TGA). Systems such as Mettler Toledo thermogravimetric analyser (TGA, model SDTA 851) or (TGA, Texas Instruments) can be employed. Various temperature programs can be run from 25 °C to 1000 °C at different heating rates, e.g. 10 °C/min in oxygen, nitrogen or air atmosphere. For example, in order to test the thermal stability of fluorosilanes (1H, 1H, 2H, 2H-perfluorodecyltriethoxysilane, C8 compound) grafted commercial ceramic membranes (zirconia layer on the microporous alumina support) prepared for desalination by air gap membrane distillation (AGMD), TGA analysis was performed using a TGA (Texas Instruments) apparatus in air atmosphere and in the temperature range of 25 °C to 1000 °C [75,76]. Figure 8.19 shows typical TGA curves obtained for the grafted zirconia powder at different C8 grafting time. The weight loss increases when increasing the grafting time and the decomposition of the fluoro-chain occurs at temperatures higher than 230 °C proving that the used modified material is thermally stable up to 200 °C.

The thermal conductivity of MD membranes should be low. In fact, heat transfer by conduction through the membrane cross-section is heat loss in MD because no mass transfer is associated with it. Therefore, the thermal conductivity of the MD membranes should be measured. The thermal conductivity of various commercial membranes commonly used in MD was measured by Izquierdo-Gil et al. [67,77] and García-Payo and Izquierdo-Gil [78] using a slightly modified Less method using a system similar to the one shown in Fig. 8.20. Superposition of three or more membranes was needed to conduct their experiments, which led to large errors due to the presence of air layer between membranes. However, the authors [78] claimed that the procedure was simple and accurate for determination of the thermal conductivity. In this method, the thermal conductivity of the MD membrane was deduced from a linear fit of the thermal resistance *versus* the number of membranes. Dependence of the thermal resistance on the thickness of the membrane samples was observed and better results were obtained when the air layer effect was considered in the linear fit. It will be interesting to apply the same system using membranes prepared in the same conditions but with different thicknesses.

The measured thermal conductivity of some commercial membranes commonly used in MD is presented in Table 8.3. The calculated values are obtained from both membrane matrix (k_p) and gas (k_g) present in the membrane pores following the Isostrain model,

$$k_m = \varepsilon k_g + (1 - \varepsilon) k_p \qquad (8.30)$$

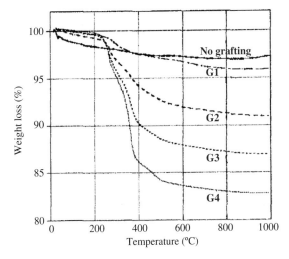

FIGURE 8.19 TGA curves of fluorosilane-grafted zirconia powder used for preparation of MD membranes at different grafting times (G1: 4 h, G2: 14 h, G3: 25 h, G4: 75 h). *Source: Reprinted from [76]. Copyright 2004, with kind permission from Elsevier.*

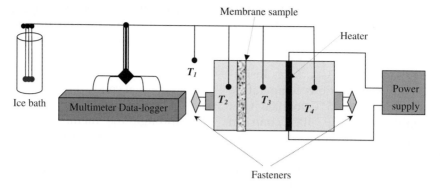

FIGURE 8.20 Schematic of the system used for thermal conductivity measurements of flat sheet membranes.

the Isostress model,

$$k_m = \left[\frac{\varepsilon}{k_g} + \frac{(1-\varepsilon)}{k_p}\right]^{-1} \quad (8.31)$$

or the flux law model:

$$k_m = k_g\left[\frac{1+(1-\varepsilon)\beta}{1+(1-\varepsilon)\beta}\right] \quad (8.32)$$

where $\beta = \dfrac{k_p/k_g - 1}{k_p/k_g + 2}$

Other models can also be found in [78]. The reported thermal conductivities of PVDF material are 0.17–0.19 W/m K at 296 K and 0.21 W/m K at 348 K. Slightly lower values were reported for polypropylene (PP) (0.11–0.16 W/m K at 296 K and 0.2 W/m K at 348 K), while those of PTFE are higher (0.25–0.27 W/m K at 296 K and 0.29 W/m K at 348 K) [79]. In fact, both the void volume and the structure of the pore affect the thermal conductivity prediction of MD membranes. Phattaranawik et al. [79] found that the measured values were close to those calculated using the Isostress model and the flux law model rather than the usually used Isostrain model in most MD studies. After testing various models based on empirical or theoretical correlations, García-Payo and Izquierdo-Gil [78] recommended the use of Maxwell equation (lower bound model or type I) to calculate the thermal conductivity of MD membranes with porosities higher than 60%:

$$k_m = k_g\left[\frac{1+2\beta(1-\varepsilon)}{1-\beta(1-\varepsilon)}\right] \quad (8.33)$$

where $\beta = \dfrac{k_p - k_g}{k_p + 2k_g}$

MECHANICAL STABILITY

The MD membranes are not required to be mechanically resistant like the membranes used in the pressure-driven membrane separation processes. Normally, pressures near atmospheric pressure are applied in MD. For industrial applications of the MD process, the flat sheet membranes should be assembled in large modules like spiral wound modules, whereas the capillaries and hollow fibre membranes should be packed in tubular modules. Therefore, the MD membranes should exhibit adequate mechanical strength for each MD configuration. In vacuum membrane distillation (VMD) applications, a low pressure is applied on the permeate side and the membrane should have better mechanical properties than the MD membranes used in the other

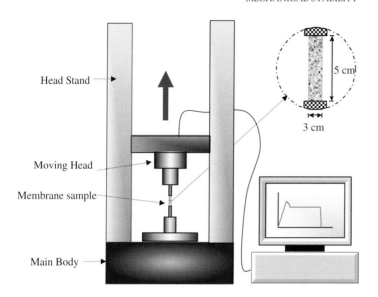

FIGURE 8.21 System used to determine mechanical properties of MD membranes.

configurations (DCMD, AGMD, sweeping gas membrane distillation (SGMD)).

The mechanical properties of MD membranes can be determined by tensile testing at room temperature on an Instron dynamometer similar to the one schematically shown in Fig. 8.21, according to ASTM D 638 M (standards). Various models exist in the market. Tests are carried out with a crosshead speed of 50 ml/min at break. The dynamic mechanical properties of the membrane samples (25 × 4 × 0.5 mm) can be determined using a dynamic mechanical thermoanalyzer. Tests are carried out in torsion deformation mode, at a frequency of 5 Hz and the temperature programs are run from −80 °C to 30 °C, at a heating rate of 2 °C/min, under a controlled sinusoidal strain in a flow of nitrogen. At least three measurements for each membrane sample should be tested. Tensile stress–strain properties are measured according to ISO 37-1977, with type 2 test specimens. Tear strength is measured following the ISO 816-1983. Rebound resilience measurements can be carried out on a Schob pendulum according to ISO 4662-1978. Shore hardness can be measured by using a Bareiss Rockwell tester according to ASTM D-2240.

Tomaszewska [28] studied the mechanical properties of PVDF membranes prepared for MD by phase inversion technique. The used strain rate of Instron test was 5 mm/min. The tensile strength was calculated dividing the force at break. It was observed that the addition of lithium chloride (LiCl) in the casting PVDF solution drastically decreased the strength at break of the prepared MD membranes and this result is in accordance with the morphology of the membranes. The presence of big cavities in the cross-section of the membranes reduced the mechanical resistance of the membranes.

Feng et al. [29,30] studied the mechanical properties of phase inversion membranes prepared from PVDF and the copolymer poly(vinylidene fluoride-co-tetrafluoroethylene) (F2.4) under the same conditions using an Instron1121 test. Parameters such as the maximal load during tensile, the stretching strength, stretching strain, elastic tensile modulus, the load at break, etc. were determined. Typical stress–strain curves are shown in Fig. 8.22. Better mechanical properties can be observed for the F2.4 membrane

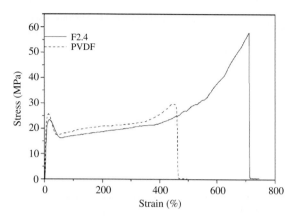

FIGURE 8.22 Stress—strain curves of two membranes prepared for MD using two different polymers (stretching speed, 200 mm/min at 20 °C). *Source: Reprinted from [29]. Copyright 2004, with kind permission from Elsevier.*

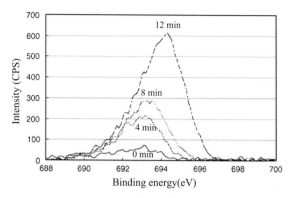

FIGURE 8.23 XPS spectra (fitted F1s) of different MD composite membranes prepared with different solvent evaporation times. *Source: Reprinted from [81]. Copyright 2006, with kind permission from Elsevier.*

compared to PVDF membrane. In other words, stretching strength of F2.4 membrane is stronger than that of PVDF membrane. The stress-at-break and stretching strength of F2.4 membrane were found to be 3.42 and 3.45 MPa; while those of PVDF membrane were 2.89 and 3.02 MPa, respectively. The elongation percentage at break of F2.4 membrane (330%) was approximately eight-fold higher than that of PVDF porous membrane (44%).

CHEMICAL STABILITY

One of the MD applications is the separation of organic compounds from water. In this case, VMD, SGMD and AGMD are used as reported in Chapters 11–13. Although the concentration of the organic compounds in the feed aqueous solutions is maintained small in order to prevent liquid penetration into the membrane pores, the MD membranes should exhibit long-term chemical stability in order to maintain the hydrophobic character of the membranes, especially when modified membranes are used. One of the characterization techniques used is the X-ray photoelectron spectroscopy (XPS). This permits to analyze the compositional gradient of different species near the membrane surface when porous composite membranes are considered for MD applications [32,33,80]. The relative atomic percentages of fluorine, nitrogen, oxygen and carbon can be determined as a function of the depth into the membrane using the variable photoelectron take-off angle method. Figure 8.23 shows as an example the XPS spectra of MD membranes. Multilab ESCA 3000 instrument (Thermo VG Scientific, East Grinstead, UK) was employed under a high vacuum of 6.894757×10^{-6} Pa to 6.894757×10^{-7} Pa. It can be observed that the intensity corresponding to the binding energy of F1s increases as the solvent evaporation time increases. In fact, fluorine is associated with the surface-modifying macromolecules (SMMs) used for membrane surface modification and SMMs migrate to the membrane surface during solvent evaporation rendering it more hydrophobic. By XPS analysis, it was proved that fluorohydrocarbon tails of SMMS were oriented perpendicular to the membrane surface.

The surface chemical properties of MD membranes can also be evaluated by its hydrophilicity, which can be measured by water contact systems. This was discussed previously.

OTHER CHARACTERIZATION TECHNIQUES

Thickness of Bilayer Hydrophobic/hydrophilic Composite MD Membranes

Khayet et al. [82] presented a method to determine the thickness of the hydrophobic layer of composite porous hydrophobic/hydrophilic membranes used in MD. The results of different DCMD experiments were shown together with a theoretical model involving the structural characteristics of the membranes and the heat and mass transfer mechanism. The proposed method can be used to measure the layer thickness of porous composite hydrophobic/hydrophilic membranes with random uncertainties less than 5% and maximum deviation of each individual thickness from the corresponding average value less than 9%. The DCMD parameters were found to exert no effect on the calculated thickness.

Mean Pore Size and Pore Size Distribution

Rather than the techniques presented in Section 'Determination of pore size (mean pore size, pore size distribution) by different physical methods' to determine the mean pore size and pore size distribution of MD membranes, attempts can be made to use other characterization techniques such as permporometry, thermoporometry, liquid permeation and solute transport. Detailed explanations of other characterization techniques are given in [3,9,53,54]. These techniques are used generally to characterize UF and NF membranes. However, when the prepared MD membranes possess small pore sizes — like the porous composite hydrophobic/hydrophilic membranes — these methods can be used for the sake of comparison [54,83]. Khayet et al. [84] compared the surface and bulk pore size of various laboratory made flat sheet and hollow fibre membranes obtained by different methods, including the gas permeation test, the solute transport method and the AFM analysis. The pore size of both the internal and external surfaces of the hollow fibre membranes as well as the pore size of the top and bottom surfaces of the flat sheet membranes were evaluated using AFM analysis. All techniques showed the same tendency in the pore size change as the membrane preparation conditions were changed. The pore size at the membrane surface was found to be larger than in the bulk of the membrane. The pores at the bottom surfaces of flat sheet membranes were 3.7–9.8 times larger than those at the top membrane surface, while the pores at the top surface were 2.1 times larger than the bulk pore sizes determined from the gas permeation test. For the PVDF hollow fibre membranes, the pore size at the outer surface was 1.7 times larger than the bulk pore size determined from the gas permeation test and the internal pore size was 1.1–1.4 times larger than the external pore size.

Optical Techniques

Optical techniques using high magnification camera or microscope combined with camera or video recorder can be employed to observe in situ microscopic views of possible MD membrane fouling or particle deposition near the membrane surface. The photos and/or the video images can be monitored on-line during MD process. These optical techniques have been applied in pressure-driven membrane separation processes such as UF and require special design of membrane modules [85].

Other techniques, such as X-Ray diffraction (XRD), Fourier transform infrared (FTIR) spectroscopic analysis, nuclear magnetic resonance (NMR) and differential scanning calorimetry (DSC), can be employed for characterization of special membranes prepared for MD process such as modified membranes and also after long-term MD process. XRD and FTIR techniques were used for investigations of MD membrane degradation [86]. It was claimed that hydrophilic groups (i.e. hydroxyl or carbonyl groups) were formed on the polymer (polypropylene, PP) surface during membrane contact with air and MD process conditions (presence of water, enhanced temperature and scaling) accelerated PP degradation. During long-term MD, the used PP membranes indicated an increase of crystallinity of polymer determined by XRD.

To conclude this chapter, it is important to keep in mind that the characterization technique should be chosen in such a way that the medium of characterization and final application are similar. Since the final application is MD, which is water vapor transport, the medium of characterization should be gas and vapor transport method.

References

[1] A.C.M. Franken, S. Ripperger, Terminology for membrane distillation, Eur. Soc. Membr. Sci. Technol (January 1988).

[2] C.A. Smolders, A.C.M. Franken, Terminology for membrane distillation, Desalination 72 (1989) 249−262.

[3] F.P. Cuperus, C.A. Smolders, Characterization of UF membranes: Membrane characteristics and characterization techniques, Adv. Colloid Interface Sci. 34 (1991) 135−173.

[4] S.I. Nakao, Determination of pore size and pore size distribution. 3. Filtration membranes: Review, J. Membr. Sci. 96 (1994) 131−165.

[5] M. Khayet, J.I. Mengual, G. Zakrzewska-Trznadel, Direct contact membrane distillation for nuclear desalination. Part I: Review of membranes used in membrane distillation and methods for their characterisation, Int. J. Nuclear Desalination 1 (2005) 435−449.

[6] S.P. Deshmuck, K. Li, Effect of ethanol composition in water coagulation bath on morphology of PVDF hollow fiber membranes, J. Membr. Sci. 150 (1998) 75−85.

[7] D. Wang, K. Li, W.K. Teo, Preparation and characterization of polyvinylidene fluoride (PVDF) hollow fiber membranes, J. Membr. Sci. 163 (1999) 211−220.

[8] M. Khayet, T. Matsuura, Preparation and characterization of polyvinylidene fluoride membranes for membrane distillation, Ind. Eng. Chem. Res. 40 (2001) 5710−5718.

[9] M. Khayet, C.Y. Feng, K.C. Khulbe, T. Matsuura, Preparation and characterization of polyvinylidene fluoride hollow fiber membranes for ultrafiltration, Polymer 43 (2002) 3879−3890.

[10] H. Yasuda, J.T. Tsai, Pore size of microporous polymer membranes, J. Appl. Polym. Sci. 18 (1974) 805−819.

[11] R.D. Present, Kinetic theory of gases, McGraw-Hill, New York, 1958.

[12] P.C. Carman, Flow of gases through porous media, Butterworth Publications, London, 1956.

[13] H.C. Shih, Y.S. Yeh, H. Yasuda, Morphology of microporous poly(vinylidene fluoride) membranes studied by gas permeation and scanning microscopy, J. Membr. Sci. 50 (1990) 299−317.

[14] K. Li, J.F. Kong, D. Wang, W.K. Teo, Taylor-made asymmetric PVDF hollow fibers for soluble gas removal, AIChE J. 45 (1999) 1211−1219.

[15] A. Bottino, G. Capannelli, S. Munari, A. Turturro, High performance ultrafiltration membranes cast from LiCl doped solutions, Desalination 68 (1988) 167−177.

[16] J. Kong, K. Li, An improved gas permeation method for characterizing and predicting the performance of microporous asymmetric hollow fiber membranes used in gas adsorption, J. Membr. Sci. 182 (2001) 271−281.

[17] M. Khayet, A. Velázquez, J.I. Mengual, Modelling mass transport through a porous partition: Effect of pore size distribution, J. Non-Equilib. Thermodyn. 29 (2004) 279−299.

[18] R.E. Kesting, Synthetic polymer membranes, McGraw Hill, New York, 1972.

[19] L. Martínez, F.J. Florido-Díaz, A. Hernández, P. Prádanos, Characterization of three hydrophobic porous membranes used in membrane distillation: Modelling and evaluation of their water vapor permeabilities, J. Membr. Sci. 203 (2002) 15−27.

[20] L. Martínez, F.J. Florido-Díaz, A. Hernández, P. Prádanos, Estimation of vapor transfer coefficient of hydrophobic porous membranes for applications in membrane distillation, Separ. Purif. Tech. 33 (2003) 45−55.

REFERENCES

[21] ASTM F316, Standard Test Method for Pore Size Characteristics of Membrane Filters by Bubble Point and Mean Flow Pore Test.

[22] ASTM E1294, Standard Test Methods for Pore Size Characteristics of Membrane Filters using Automated Liquid Porosimetry.

[23] E. Honold, E.L. Skau, Application of mercury intrusion method for determination of pore-size distribution to membrane filters, Science 120 (1954) 805–806.

[24] K. Sakai, T. Koyano, T. Muroi, M. Tamura, Effects of temperature and concentration polarization on water vapour permeability for blood in membrane distillation, Chem. Eng. J. 38 (1988) B33–B39.

[25] Y. Wu, Y. Kong, X. Lin, W. Liu, J. Xu, Surface-modified hydrophilic membranes in membrane distillation, J. Membr. Sci. 72 (1992) 189–196.

[26] Y. Fujii, S. Kigoshi, H. Iwatani, M. Aoyama, Y. Fusaoka, Selectivity and characteristics of direct contact membrane distillation type experiment: II. Membrane treatment and selectivity increase, J. Membr. Sci. 72 (1992) 73–89.

[27] J.M. Ortiz de Zárate, L. Peña, J.I. Mengual, Characterization of membrane distillation membranes prepared by phase inversion, Desalination 100 (1995) 139–148.

[28] M. Tomaszewska, Preparation and properties of flat-sheet membranes from polyvinylidene fluoride for membrane distillation, Desalination 104 (1996) 1–11.

[29] C. Feng, B. Shi, G. Li, Y. Wu, Preliminary research on microporous membrane from F2.4 for membrane distillation, Separ. Purif. Tech. 39 (2004) 221–228.

[30] C. Feng, B. Shi, G. Li, Y. Wu, Preparation and properties of microporous membrane from poly(vinylidene fluoride-co-tetrafluoroethylene) (F2.4) for membrane distillation, J. Membr. Sci. 237 (2004) 15–24.

[31] C.M. Tun, A.G. Fane, J.T. Matheickal, R. Sheikholeslami, Membrane distillation crystallization of concentrated salts-flux and crystal formation, J. Membr. Sci. 257 (2005) 144–155.

[32] M. Qtaishat, M. Khayet, T. Matsuura, Novel porous composite hydrophobic/hydrophilic polysulfone membranes for desalination by direct contact membrane distillation, J. Membr. Sci. 341 (2009) 139–148.

[33] M. Qtaishat, D. Rana, M. Khayet, T. Matsuura, Preparation and characterization of novel hydrophobic/hydrophilic polyetherimide composite membranes for desalination by direct contact membrane distillation, J. Membr. Sci. 327 (2009) 264–273.

[34] C.Y. Feng, K.C. Khulbe, T. Matsuura, R. Gopal, S. Kaur, S. Ramakrishna, M. Khayet, Production of drinking water from saline water by air-gap membrane distillation using polyvinylidene fluoride nanofiber membrane, J. Membr. Sci. 311 (2008) 1–6.

[35] Z.D. Hendren, J. Brant, M.R. Wiesner, Surface modification of nanostructured ceramic membranes for direct contact membrane distillation, J. Membr. Sci. 331 (2009) 1–10.

[36] M. Gryta, Long-term performance of membrane distillation process, J. Membr. Sci. 265 (2005) 153–159.

[37] S. Bonyadi, T.S. Chung, Flux enhancement in membrane distillation by fabrication of dual layer hydrophilic-hydrophobic hollow fiber membranes, J. Membr. Sci. 306 (2007) 134–146.

[38] K.Y. Wang, T.S. Chung, M. Gryta, Hydrophobic PVDF hollow fiber membranes with narrow pore size distribution and ultra-thin skin for the fresh water production through membrane distillation, Chem. Eng. Sci. 63 (2008) 2587–2594.

[39] S. Bonyadi, T.S. Chung, Highly porous and macro-void-free PVDF hollow fiber membranes for membrane distillation by a solvent-dope solution co-extrusion approach, J. Membr. Sci. 331 (2009) 66–74.

[40] M.C. García-Payo, M. Essalhi, M. Khayet, Effects of PVDF-HFP concentration on membrane distillation performance and structural morphology of hollow fiber membranes, J. Membr. Sci. 347 (2010) 209–219.

[41] Z. Jin, D.L. Yang, S.H. Zhang, X.G. Jian, Hydrophobic modification of poly(phthalazinone ether sulfone ketone) hollow fiber membrane for vacuum membrane distillation, J. Membr. Sci. 310 (2008) 20–27.

[42] B. Wu, X. Tan, K. Li, W.K. Teo, Removal of 1,1,1-trichloroethane from water using a polyvinylidene fluoride hollow fiber membrane module: Vacuum membrane distillation, Separ. Purif. Tech. 52 (2006) 301–309.

[43] J. Phattaranawik, R. Jiraratananon, A.G. Fane, Effect of pore size distribution and air flux on mass transport in direct contact membrane distillation, J. Membr. Sci. 215 (2003) 75–85.

[44] G. Binning, C.F. Quate, C. Gerber, Atomic force microscope, Phys. Rev. Lett. 56 (1986) 930–933.

[45] K.C. Khulbe, C. Feng, T. Matsuura, M. Khayet, AFM images of the cross-section of polyetherimide hollow fibers, Desalination 201 (2006) 130–137.

[46] K.C. Khulbe, C.Y. Feng, T. Matsuura, Synthetic polymeric membranes: characterization by atomic force microscopy, Springer, Heidelberg, Germany, 2008.

[47] W.R. Bowen, N. Hilal, Atomic force microscopy in process engineering: an introduction to afm for improved processes and products, Elsevier B.H., Oxford, 2009.

[48] T.R. Albrecht, C.F. Quate, Atomic resolution with the atomic force microscope on conductors and non-conductors, J. Vac. Sci. Technol. A 6 (1988) 271–275.

[49] K.C. Khulbe, T. Matsuura, Characterization of synthetic membranes by Raman spectroscopy, electron spin resonance, and atomic force microscopy: a review, Polymer 41 (2000) 1917–1935.

[50] W.R. Bowen, N. Hilal, R.W. Lovitt, P.M. Williams, C.J. Wright, in: T.S. Sorensen (Ed.), Surface chemistry and electrochemistry of membrane surfaces, surfactant science series, Marcel Dekker, New York, 1999, pp. 1–37. Chapter 1.

[51] P. Dietz, P.K. Hansma, O. Inacker, H.D. Lehmann, K.H. Herrmann, Surface pore structure of micro- and ultrafiltration membranes imaged with the atomic force microscope, J. Membr. Sci. 65 (1992) 101–111.

[52] A. Chahboun, R. Coratger, F. Ajustron, J. Beauvillain, A. Aimar, V. Sánchez, Comparative study of micro- and ultrafiltration membranes using STM, AFM, and SEM techniques, Ultramicroscopy 41 (1992) 235–244.

[53] S. Singh, K.C. Khulbe, T. Matsuura, P. Ramamurthy, Membrane characterization by solute transport and atomic force microscopy, J. Membr. Sci. 142 (1998) 111–127.

[54] M. Khayet, C.Y. Feng, T. Matsuura, Morphological study of fluorinated asymmetric polyetherimide ultrafiltration membranes by surface modifying macromolecules, J. Membr. Sci. 213 (2003) 159–180.

[55] W.R. Bowen, N. Hilal, R.W. Lovitt, P.M. Williams, Visualisation of an ultrafiltration membrane by non-contact atomic force microscopy at single pore resolution, J. Membr. Sci. 110 (1996) 229–232.

[56] W.R. Bowen, A.W. Mohammad, N. Hilal, Characterization of nanofiltration membrane for predictive purposes-use of salts, uncharged solutes and atomic force microscopy, J. Membr. Sci. 126 (1997) 91–105.

[57] A.W. Mohammad, N. Hilal, M.N. Abu Seman, A study on producing composite nanofiltration membranes with optimized properties, Desalination 158 (2003) 73–78.

[58] M. Hirose, H. Ito, Y. Kamiyama, Effect of skin layer surface structures on the flux behavior of RO membranes, J. Membr. Sci 121 (1996) 209–215.

[59] D.F. Stamatialis, C.R. Dias, M.N. de Pinho, Atomic force microscopy of dense and asymmetric cellulose-based membranes, J. Membr. Sci. 160 (1999) 235–242.

[60] K.C. Khulbe, B. Kruczek, G. Chowdhury, S. Gagné, T. Matsuura, S.P. Versma, Characterization of membranes prepared from PPO by Raman scattering and atomic force microscopy, J. Membr. Sci. 111 (1996) 57–70.

[61] J.M.A. Tan, T. Matsuura, Effect of non-solvent additive on the surface morphology and the gas separation performance of poly(2,6-dimethyl-1,4-phenylene) oxide membranes, J. Membr. Sci. 160 (1999) 7–16.

[62] M. Khayet, K.C. Khulbe, T. Matsuura, Characterization of membranes for membrane distillation by atomic force microscopy and estimation of their water vapor transfer coefficients in vacuum membrane distillation process, J. Membr. Sci. 238 (2004) 199–211.

[63] M. Khayet, Characterization of membrane distillation membranes by tapping mode atomic force microscopy, in: A. Méndez-Vilas (Ed.), Recent advances in multidisciplinary applied physics, Elsevier, Oxford, 2005, pp. 141–148.

[64] C. Fernández-Pineda, M.A. Izquierdo-Gil, M.C. García-Payo, Gas permeation and direct contact membrane distillation experiments and their analysis using different models, J. Membr. Sci. 198 (2002) 33–49.

[65] M. Khayet, T. Matsuura, Application of surface modifying macromolecules for the preparation of membranes for membrane distillation, Desalination 158 (2003) 51–56.

[66] C. Gostoli, G.C. Sarti, Separation of liquid mixtures by membrane distillation, J. Membr. Sci. 41 (1989) 211–224.

[67] M.A. Izquierdo-Gil, M.C. García-Payo, C. Fernández-Pineda, Direct contact membrane distillation of sugar aqueous solutions, Separ. Sci. Tech. 34 (1999) 1773–1801.

[68] A.C.M. Franken, J.A.M. Nolten, M.H.V. Mulder, D. Bargeman, C.A. Smolders, Wetting criteria for the applicability of membrane distillation, J. Membr. Sci. 33 (1987) 315–328.

[69] M.C. García-Payo, M.A. Izquierdo-Gil, C. Fernández-Pineda, Wetting study of hydrophobic membranes via liquid entry pressure measurements with aqueous alcohol solutions, J. Colloid Interface Sci. 230 (2000) 420–431.

[70] K.L. Mittal, Contact angle, wettability and adhesion, VSP, Utecht, The Netherlands, 1993.

[71] S. Wu, Polymer interface and adhesion, Marcel Dekker, New York, 1982.

[72] E. Matijevic, Surface and colloid science, Wiley, New York, 1989.

[73] M. Khayet, M.V. Álvarez, K.C. Khulbe, T. Matsuura, Preferential surface segregation of homopolymer and copolymer blend films, Surf. Sci. 601 (2007) 885–895.

[74] E. Hoffman, D.M. Pfenning, E. Phillippsen, P. Schwahn, M. Sieber, R. When, D. Woermann, Evaporation of alcohol/water mixtures through hydrophobic porous membranes, J. Membr. Sci. 34 (1987) 199–206.

[75] S.R. Krajewski, W. Kujawski, M. Bukowska, C. Picard, A. Larbot, Application of fluoroalkylsilanes (FAS)

grafted ceramic membranes in membrane distillation process of NaCl solutions, J. Membr. Sci. 281 (2006) 253—259.

[76] A. Larbot, L. Gazagnes, S. Krajewski, M. Bukowska, W. Kujawski, Water desalination using ceramic membrane distillation, Desalination 168 (2004) 367—372.

[77] M.A. Izquierdo-Gil, M.C. García-Payo, C. Fernández-Pineda, Air gap membrane distillation of sucrose aqueous solutions, J. Membr. Sci. 155 (1999) 291—307.

[78] M.C. García-Payo, M.A. Izquierdo-Gil, Thermal resistance technique for measuring the thermal conductivity of thin microporous membranes, J. Phys. D: Appl. Phys. 37 (2004) 3008—3016.

[79] J. Phattaranawik, R. Jiraratananon, A.G. Fane, Heat transport and membrane distillation coefficients in direct contact membrane distillation, J. Membr. Sci. 212 (2003) 177—193.

[80] M. Khayet, D.E. Suk, R.M. Narbaitz, J.P. Santerre, T. Matsuura, Study on surface modification by surface-modifying macromolecules and its applications in membrane-separation processes, J. Appl. Polym. Sci. 89 (2003) 2902—2916.

[81] D.E. Suk, T. Matsuura, H.B. Park, Y.M. Lee, Synthesis of a new type of surface modifying macromolecules (nSMM) and characterization and testing of nSMM blended membranes for membrane distillation, J. Membr. Sci. 277 (2006) 177—185.

[82] M. Khayet, T. Matsuura, J.I. Mengual, Porous hydrophobic/hydrophilic composite membranes: Estimation of the hydrophobic-layer thickness, J. Membr. Sci. 266 (2005) 68—79.

[83] M. Khayet, J.I. Mengual, T. Matsuura, Porous hydrophobic/hydrophilic composite membranes: Application in desalination using direct contact membrane distillation, J. Membr. Sci. 252 (2005) 101—113.

[84] M. Khayet, T. Matsuura, Determination of surface and bulk pore sizes of flat-sheet and hollow-fiber membranes by atomic force microscopy, gas permeation and solute transport methods, Desalination 158 (2003) 57—64.

[85] V. Chen, H. Li, A.G. Fane, Non-invasive observation of synthetic membrane processes: a review of methods, J. Membr. Sci. 241 (2004) 23—44.

[86] J.B. Xu, S. Lange, J.P. Bartley, R.A. Johnson, Alginate-coated microporous PTFE membranes for use in the osmotic distillation of oily feeds, J. Membr. Sci. 240 (2004) 81—89.

[87] M. Nasef, N.A. Zubir, A.F. Ismail, M. Khayet, K.Z.M. Dahlan, H. Saidi, R. Rohani, T.I.S. Ngah, N.A. Sulaiman, PSSA pore-filled PVDF membranes by simultaneous electron beam irradiation: Preparation and transport characteristics of protons and methanol, J. Membr. Sci. 268 (2006) 96—108.

[88] M. Khayet, C.Y. Feng, K.C. Khulbe, T. Matsuura, Study on the effect of a non-solvent additive on the morphology and performance of ultrafiltration hollow-fiber membranes, Desalination 148 (2002) 321—327.

CHAPTER 9

MD Membrane Modules

OUTLINE

Background	227	Spiral Wound MD Membrane Modules	241
Plate-and-Frame MD Membrane Modules	230	Design of MD Modules	242
Tubular Capillary and Hollow Fibre MD Membrane Modules	236		

BACKGROUND

As explained in the previous chapters, there are different membrane distillation (MD) configurations that can be applied in order to establish the driving force (i.e. transmembrane vapour pressure) through different membrane types (flat sheet, capillary or hollow fibre). This chapter shows that different types of membrane modules have been designed and used in MD. Most of laboratory-scale membrane modules are plate-and-frame modules designed for use with flat sheet membranes due to their versatility and simplicity in fabrication, as compared to the spiral wound or tubular (capillary, hollow fibre) counterparts. In fact, flat sheet membranes can easily be removed from a plate-and-frame module for their examination, cleaning or replacement, and the same module can be used to test different membranes. On the other hand, membranes are permanently installed in the spiral wound and tubular module. Being considered as an integral part of the module, the membranes are not easily replaceable.

Among the membranes studied in MD are commercial module units, formed in plate-and-frame, shell-and-tube or spiral wound configurations. In fact, two membrane types have been tested so far for MD (see Chapters 2–4): (i) flat sheet membranes and (ii) capillary or hollow fibre membranes. These membrane types have been packed in a large variety of membrane module configurations and tested in MD systems. Flat sheet membranes in plate-and-frame modules or spiral wound modules and capillary membranes in tubular modules were used in various MD studies. Capillary membranes were also assembled in plate-and-frame modules.

Since the early form of MD system was presented by Bodell in a 1963 U.S. Patent application [1], improvements in the MD module design occurred simultaneously with membrane developments (see Chapters 2–7). In the successive patents filed by Rodgers in 1972 and 1974 [2,3], a multiple effect direct contact membrane distillation (DCMD) desalination system was presented using a stack of plate-and-frame modules containing flat sheet membranes separated by non-permeable corrugated heat transfer films. In the early of 1980s, when novel membranes with better characteristics became available, various membrane modules were fabricated for MD [4–14]. Gore-Tex membrane (i.e. expanded polytetrafluoroethylene (PTFE) membrane with a thickness of 50 μm and a pore size of 0.5 μm) was proposed by Gore & Associated Co. under the name 'Gore-Tex Membrane Distillation' for MD application in a spiral wound-type module using the air gap membrane distillation (AGMD) or the liquid gap DCMD configuration [4,5]. Other types of MD membranes and modules have been proposed by Cheng and Wiersma in a series of patents [6–9]. The Swedish National Development Co. (Svenska Utvecklings AB, Akersberga, Sweden) presented an AGMD plate-and-frame membrane module for 'SU Membrane Distillation' [10,11]. Enka AG, a German company, developed a tubular module with polypropylene (PP) hollow fibre membranes [12]. The module was used in DCMD process with heat recovery, but the work was more academically orientated than industrially [13,14]. More MD module presentations were made at the Second World Congress on Desalination and Water Reuse (Ref. Proc. of the 1985 International Desalination Association Annual Conference, Bermuda, 1985). This renewed interest was a result of the development of various porous hydrophobic membranes and modules used in different MD configurations [15–19]. Over 100-fold increase in permeate flux was achieved compared to the membrane modules used by Weyl and Findley in the 1960s [20,21]. Although the results seemed very promising at the development stage of laboratory MD membrane modules, several attempts of commercialization have failed due to difficulties in engineering aspects. Until now, practically all membrane modules were designed for academic purposes in the research laboratories rather than for industrial use. Non-availability of the industrial MD module imposes a serious limitation for the implementation of the industrial MD process. In fact, most of the MD modules so far used were prepared commercially for other separation processes (see Table 2.2) [22]. The magnitude of the permeate flux obtained in the MD process should be affected significantly by the module design, the MD configuration and its operating conditions. This effect is probably a reason for the observed large discrepancies in the results recorded so far in the MD literature [22–27].

The Swedish companies XZero AB and Scarab Development AB have spent several years to develop the MD technology for applications in the semiconductor industry, process industries, desalination and drinking water purification. Some engineering difficulties faced in MD systems have been resolved by Scarab Development AB in cooperation with leading engineering companies, ABB and Electrolux among others [28]. Scarab Development AB was founded in 1973 in order to exploit low-temperature distillation technology concepts developed by a research group at the Royal Institute of Technology in Stockholm (Sweden). These developments lead to the invention of AGMD module(s), which was patented for the first time by Scarab Development AB in 1981. The Swedish company XZero AB acquired the licence to use Scarab's technology in semiconductor industry for making ultrapure water systems with zero liquid discharge [29]. The technology has been evaluated by different institutions including Sandia National Laboratory in the United States [30]. Since 2003, the Department of Energy Technology at KTH (Kungliga

Tekniska Högskolan) has also collaborated with XZero AB. One AGMD membrane module (63 cm long, 17.5 cm wide and 73 cm high) consisting of 10 cassettes was constructed from PTFE membrane having a porosity of 80%, a thickness of 0.2 mm, an air gap thickness of 2 mm and a total membrane area of around 2.3 m^2. It was reported that a pilot plant comprised of one module (20 membranes) produced 300–500 l/day of ultrapure water [30]. During the year 2005, XZero BA ordered the construction of a five-module AGMD pilot plant, to be manufactured and assembled by Uddevalla Finmekanik AB (UFAB), and was capable of producing 1–2 m^3/day of purified water [31]. It was found that although connection of several modules in series reduced the electrical energy consumption needed for recirculation, less permeate flux than expected was obtained [31]. This was attributed to hydrodynamic conditions causing high pressure drop, which directly affected the energy consumption. Membrane scaling also caused flow maldistribution, which resulted in the inefficient use of the membrane area and hence a relatively significant permeate flux reduction. It was also stated that scale formation at the feed inlets of the membrane cassettes led to reduction of the feed flow and to total or partial clogging of the membrane pores. It was concluded that much careful design is needed to optimize the AGMD system, to develop more efficient and inexpensive modules and to solve the higher estimated product water cost (i.e. 10–20 SEK/m^3, Swedish Krona) compared to that of reverse osmosis (RO).

Recently, XZero AB has commercialized small-scale AGMD membrane module (similar to the one presented by Andersson et al. [11]) for manufacture of ultrapure water for semiconductor and for small-scale drinking water production [30]. It is claimed that each module of 2.8 m^2 is able to produce a permeate flux (temperature dependent) of 7–20 l/m^2·h (500 l/24 h per module). The module is composed of two stainless steel (AISI 316, SS 2343) frame support and all gaskets are made of ethylene propylene diene monomer (EPDM FDA) or silicone materials, whereas frames are made of PP. However, the actual price of each membrane module is far more to compete with the price of RO membrane modules.

It is worth quoting that the Sandia National Laboratories (USA) and the Bureau for Reclamation (USA) evaluated vacuum membrane distillation (VMD) configuration for further development [32]. In their report on the VMD process [33], porous hydrophobic hollow fibre PP membranes (with and without ultrathin silicone and fluoropolymer coating layer) were used for desalination. The number of fibres in a module varied between 78 and 6000 with a membrane area of 0.023–0.96 m^2. A permeate flux as high as 15 l/m^2·h was reported [33].

Sponsored by the Bureau for Reclamation (USA) [34], an AGMD desalination system of brackish and saline waters was investigated using the commercially available AGMD module manufactured by Scarab with 2.94 m^2 membrane area. Feed and permeate temperatures were controlled by the El-Paso solar pond and high-quality distillate could be produced with a permeate flux ranging from 1 l/m^2·h to 6 l/m^2·h [34].

The Memstill technology developed by the research and technology organization TNO (Netherlands Organization for Applied Scientific Research) is composed of an AGMD module, in which a cold saline water flows through a condenser with non-permeable walls, increasing its temperature due to the condensing permeate, and then passes through a heat exchanger where additional heat is added before entering in direct contact with the membrane [35]. Some information about the constructed plants using different generations of Memstill modules were reported in [35].

It is to be noted that the choice of a membrane module for each MD configuration is usually determined by both economic and operative conditions. Some of the important criteria are high membrane module performance (i.e. high permeability and high separation factor), high membrane surface area to module volume ratio (i.e. high membrane packing density), low temperature and pressure drop along the membrane module, high heat transfer coefficients in both feed and permeate, high membrane liquid entry pressure (*LEP*) of water, good sealing and housing with good thermal and chemical resistances, and low heat transfer by conduction through the membrane material. Overall, a membrane module with efficient control of temperature and concentration polarization effects as well as low membrane fouling provides the best technical solution for specific applications.

PLATE-AND-FRAME MD MEMBRANE MODULES

It is observed throughout the MD literature that majority of the laboratory-scale modules are designed for use with flat sheet membranes (i.e. plate-and-frame modules). Schematic of typical laboratory-scale MD modules are presented in Chapter 1 (Fig. 1.1). In all MD configurations, the feed liquid is maintained in direct contact with one side of the membrane. There are several different possibilities on the permeate side to apply a driving force: (i) a liquid, mainly distilled water, or a sweep gas comes into direct contact with the membrane (Fig. 1.1(a,b)), (ii) permeate is removed and condensed externally under vacuum (Fig. 1.1(c)) or (iii) permeate is condensed onto a cooled wall (Fig. 1.1(d)). In these systems, tangential fluid flows in direct contact with the membrane are applied by means of circulating pumps.

In these MD membrane modules, the sandwiched flat sheet membranes can be easily replaced, changed, examined or cleaned. The only inconvenience of using flat sheet membranes in plate-and-frame modules is requirement of supports to hold the membrane, especially when the membrane surface area exposed to the flow is large. Appropriate supports with enough strength to prevent rupture and deflection of the membrane and with low heat and mass transfer resistances should be selected. The membranes and the support plates together with the spacers are assembled forming different cassettes, which are stacked together between two end plates that are placed in appropriate housings. The packing density can vary between $100 \, m^2/m^3$ and $400 \, m^2/m^3$ depending on the number of membrane sheets.

Different plate-and-frame modules were used in different MD configurations by various authors [36–44]. The Spanish groups [40–43] used the module for DCMD shown in Fig. 9.1(a). The module is composed of two symmetrical rectangular channels between which the flat sheet membrane is sandwiched. Each compartment is made of nine channels. The dimensions of the channel are 55 mm long, 7 mm wide and 0.4 mm deep.

Khayet et al. [44] used a more complicated module for sweeping gas membrane distillation (SGMD) presented in Fig. 9.1(b). Ohta et al. [37,38] and Liu et al. [39] used plate-and-frame AGMD modules resembling that depicted in Fig. 9.1(b). Different dimensions have been considered. The dimension of the flow channel in [37] is 840 mm long, 390 mm wide, 80 mm high; in [38] it is 840 mm long, 390 mm wide, 57 mm high; and in [39] it is 200 mm long, 100 mm wide, 10 mm high.

Cylindrical AGMD modules, like the one shown in Fig. 9.1(c), were used by Kimura and Nakao [19] and Hsu et al. [45]. The dimensions of these modules were different depending on the experiment. Typically, the cylinder had a length varying from 12 cm to 15 cm and a diameter from 5 cm to 8 cm [19,45].

To avoid the use of membrane supports, Lawson and Lloyd [46,47] used in both VMD and DCMD, a small laboratory-scale module with a very small square cross-sectional area of

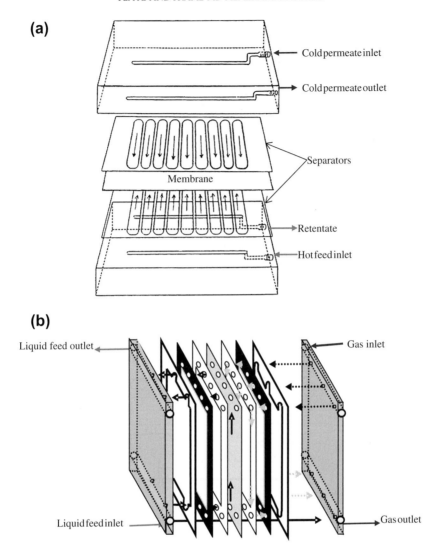

FIGURE 9.1 Schematics of some laboratory-scale modules used in MD. (a) Channelled DCMD module (*Source: Reprinted from [62]. Copyright 1998, with kind permission from Elsevier*), (b) SGMD module, (c) cylindrical AGMD module (*Source: Reprinted from [19]. Copyright 1987, with kind permission from Elsevier*), (d) thin channel module (*Source: Reprinted from [48]. Copyright 1987, with kind permission from Elsevier*).

(0.63 cm × 0.63 cm). Its total membrane area was 9.7 cm^2 and the module was designed to achieve very high Reynolds numbers up to 25000 for liquid flows.

A slightly different membrane module with tangential flow channels and a thin channel device is shown in Fig. 9.1(d). It was used for DCMD by Schofield et al. [48].

As mentioned earlier, different membrane modules have been developed by Scarab Development AB [28] and XZero AB [29] based on AGMD technology. Each of them has about

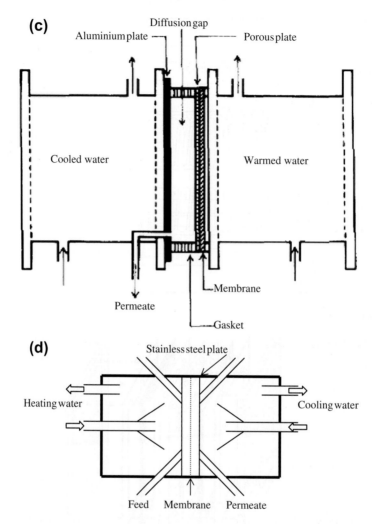

FIGURE 9.1 *(continued).*

2.3 m² membrane area. A picture of one of the modules is shown in Fig. 9.2. These modules resemble the one presented by Andersson et al. [11] in 1985 for desalination having an effective membrane area of 0.3 m² per cassette. The module is composed of plastic cassettes designed so that they can be stacked together to form modules of varying sizes. Each cassette consists of injection-moulded plastic frames containing two parallel membranes, feed and exit channels for the warm water and two condensing walls. By interconnecting the cassettes, channels for the cooling water are formed between the condensing walls of adjacent cassettes.

It must be pointed out that XZero AB has commercialized its technology for small-scale drinking water production and is interested in finding other application areas. During the year 2005, XZero AB ordered the construction

(a)

(b)

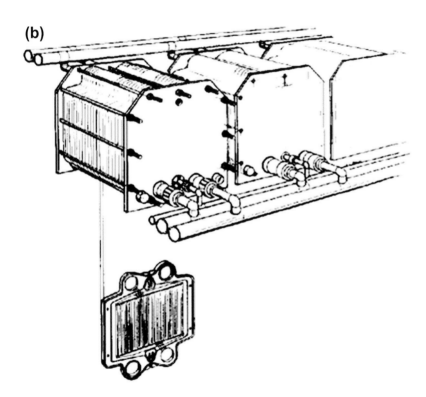

FIGURE 9.2 Picture of the AGMD module presented by Scarab AB and XZero AB [28–31]. (a) *Source: Reprinted from [31]. Copyright 2011, with kind permission from Elsevier. Other pictures and characteristics of this type of modules may be found in [28–30]. Similar design was presented previously in [11].* (b) Details of the module. *Source: Reprinted from [11]. Copyright 1985, with kind permission from Elsevier.*

of a five-module AGMD pilot plant [30,31]. The plant is capable of producing $1-2\ m^3$/day purified water, depending upon the heat source. Some problems were detected for the module, such as poor hydrodynamic conditions that increase the pressure drop throughout the membrane module and low permeate flux compared to the recirculation rate that causes a negative impact on the energy consumption [30,31]. To overcome the problems, some

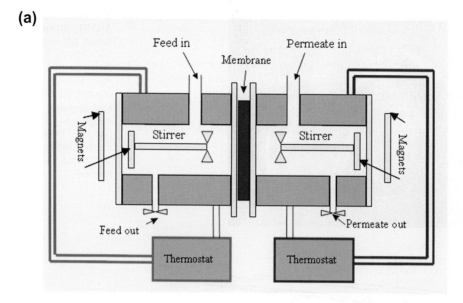
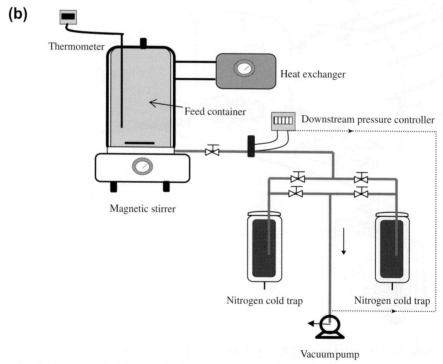

FIGURE 9.3 Schematics of other laboratory-scale modules used in MD: (a) Lewis cell used in DCMD; (b) cell used in VMD.

proposals were made; e.g. an appropriate redesign of the module taking into account the enhancement of the mass transfer and the increase of the membrane area per module volume [30,31].

Other laboratory MD systems, called Lewis cells (Fig. 9.3(a)), have also been employed for DCMD studies, in which both feed and permeate liquids are stirred inside the cells by graduated magnetic stirrers without the need of circulation pumps [49–56]. In this type of modules, the membrane is placed between two cylindrical chambers (i.e. perpendicular to the cylinder axis). The Lewis cell is only suitable for laboratory-scale tests and has been used in the majority of the MD studies to characterize flat sheet membranes for DCMD. Khayet and Matsuura [57] conducted VMD experiments in a device where the feed solution was also stirred inside the cell by a magnetic stirrer while vacuum was applied on the permeate side (Fig. 9.3(b)).

It must be mentioned that capillary membranes were also assembled in plate-and-frame membrane modules in cross-flow mode to reduce the temperature polarization effect by increasing the heat transfer coefficients [58–61]. In this case, use of support is not necessary. Different face boxes and face plates were assembled with a rectangular membrane module channel to constitute the whole device as can be seen in Fig. 9.4. The modules have different surface areas varying between 113 cm^2 and 0.66 m^2 (based on internal diameters of the

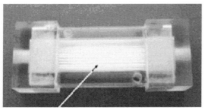

Hollow fibres

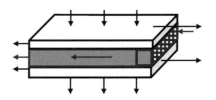

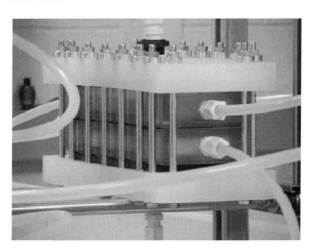

FIGURE 9.4 Rectangular cross-flow unit and connections of two of them in a DCMD module. *Sources: Reprinted from [59]. Copyright 2005, with kind permission from Elsevier. Reprinted from [61]. Copyright 2008, with kind permission from Elsevier.*

fibres). The number of modified PP fibres assembled in one unit was between 180 and 2652, so that the packing density changed between 0.12 and 0.22. The internal dimensions of each unit, module frame, are as follows: length 6.4–25.4 cm, width 2.5–8.57 cm and high 1.8–4.45 cm [58–61]. Considerably enhanced water vapour flux and high module productivity in both DCMD and VMD configurations were achieved.

Spacer-filled channels of membrane modules and spiral turbulent promoters were also considered in MD to change the flow characteristics and promote regions of turbulence, which lead to a decrease in the temperature and concentration polarization effects. Some researchers considered the use of spacer-filled channels of plate-and-frame membrane modules for DCMD [62,63]. Higher DCMD permeate fluxes were observed when a screen separator was used than when an open separator was used. The spacers changed the flow characteristics and promoted regions of turbulence, leading to a decrease in the temperature polarization (i.e. boundary layers). The temperature polarization effect was found to be higher for the coarse spacer than for the fine spacer.

TUBULAR CAPILLARY AND HOLLOW FIBRE MD MEMBRANE MODULES

Tubular, capillary or hollow fibre membranes are mainly housed in stainless steel, glass or reinforced plastic shell-and-tube modules that resemble the one depicted in Fig. 9.5. This type of modules does not require supports and the membranes are an integrated part of the module and cannot be replaced easily. From an industrial standpoint, hollow fibre modules are more attractive due to their high specific surface area to module volume ratios. From a commercial standpoint, tubular membrane modules are more attractive than plate-and-frame

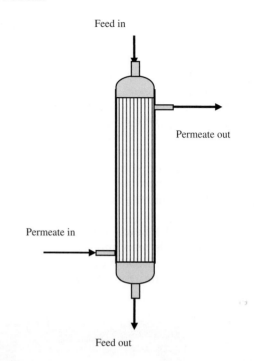

FIGURE 9.5 Shell-and-tube tangential flow module.

modules due to much higher membrane surface area to volume ratios. However, once membrane pores are wetted by liquid, the whole module becomes useless for MD applications.

The diameters of tubular membranes typically vary between 1.0 cm and 2.5 cm with packing densities of around 300 m^2/m^3 in shell-and-tube modules. These membranes are advised for high viscous fluids and when high feed flow rates are to be used. The inner diameters of capillary membranes are smaller than those of tubular ones, typically ranging from 0.2 mm to 3 mm, so that a large number of membrane capillaries can be installed in the modules with packing densities of about 600–1200 m^2/m^3. In the case of hollow fibre membranes, the inner diameters are even smaller, 50–100 μm, and thousands of hollow fibres can be packed in shell-and-tube membrane modules with very high packing densities, which may reach 3000 m^2/m^3.

It is to be noted that the first commercially available module was a shell-and-tube type produced by Enka AG (Akzo) using PP membrane [12]. Since then this module configuration has been used in various MD studies [13–15,64–69]. Similar modules with different numbers of capillaries were manufactured by Microdyn® Modulbau GmbH & Co. KG, a company founded in 1990 out of the former 'Technical Membranes' Department of the Enka-Membrana/Akzo AG. All Microdyn® modules contain hydrophobic Accurel® membranes developed by Akzo AG made of PP. These membranes are featured by a highly porous symmetrical design with a narrow pore size distribution and different inside diameters (0.6–5.5 mm). Since January 2003, the company operates under the name Microdyn-Nadir GmbH. Both commercial and laboratory-made shell-and-tube membrane modules have been applied in DCMD, SGMD and VMD configurations [15,64–69] (see Chapters 2 and 4).

As mentioned in the previous section, capillary membranes were also assembled in plate-and-frame membrane modules in a cross-flow mode to reduce the temperature polarization effect by increasing the heat transfer coefficients [58–61].

It is worth quoting that some commercial membrane modules fabricated by other companies are still not used in MD but have been applied in osmotic distillation (OD) and gas/liquid mass transfer. Some of those examples are as follows:

(i) Liqui-Cel™ Extra-Flow module was fabricated by Celgard LLC (Charlotte, NC; formerly Hoechst Celanese), with thousands of PP fibres (240 μm inner diameter, 30 μm thickness). It contains a central shell-side baffle that improves mass transfer rate. The fibres are woven into a fabric and wrapped around a central tube feeder that supplies the shell-side fluid. Thus, woven fabric allows more uniform fibre spacing. The diameters of this type of module vary from 2.5 inches to 10 inches with an effective membrane area of $1.4-130\,m^2$.

(ii) DISSO$_3$LVE™ module made with expanded PTFE membrane fibres and commercialized by WL Gore & Associates. The fibres have a diameter of 1.7 mm and a thickness of 0.5 mm and are arranged in helix geometry in order to increase mass transfer coefficients on the shell side of the membrane module. The fibres are housed in a PVDF shell and a variety of module sizes are available.

In Fig. 9.5, the feed liquid is circulated through the lumen side (i.e. tube side) of the membrane module. However, it can flow through the shell side. This type of modules is also called conventional parallel flow modules. The mass transfer coefficients can be reduced significantly if the boundary layer resistance in the shell side is high. In this case, membrane modules with cross-flow are recommended. It is known that fluid flow normal to fibre axis rather than parallel leads to higher mass transfer coefficients. However, the efficiency of the whole cross-flow membrane module is lower than that of the parallel countercurrent membrane module. One of the solutions that permit to regain the module efficiency is to use baffled modules, which combines both countercurrent and cross-flow designs. Care should be taken, however, when designing baffled modules. As the number of baffles increases, the efficiency also increases, but the pressure drop along the membrane module increases as well [70]. Furthermore, if the fibres are twisted or braided (see Fig. 9.6) instead of being arranged straight or in a fabric, more turbulent and uniform flow outside the fibres can be produced leading to an enhancement of both heat and mass transfer coefficients in the shell

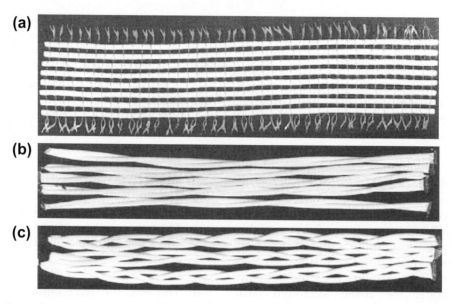

FIGURE 9.6 Membrane capillaries arranged in (a) fabric, (b) twisted and (c) braided configurations. *Source: Reprinted from [71]. Copyright 1988, with kind permission from Elsevier.*

side of the membrane [71]. Bundles with twisted or braided fibres may act like static mixers around their outer diameter.

Different hollow fibre membrane module designs with baffles, spacers and modified fibre geometries, wavy geometries such as twisted and braided, have been fabricated and tested in DCMD desalination process [72]. Increase in heat transfer coefficients together with the increase in the permeate flux (18–33% increase) was observed when baffles and spacers were incorporated in the designs. Up to 36% permeate flux enhancement was observed when twisted and braided geometries were applied. Inserting baffles in membrane modules in such a way as shown in Fig. 9.7 creates fluid instabilities in the liquid flow and the formed vortices improve the mixing between the boundary layer (i.e. temperature and concentration layers) of the membrane and the bulk liquid to a greater extent than simply generating turbulent flow augmenting the speed of the pumps.

In shell-and-tube membrane modules, the feed aqueous solution cools down when travelling along the membrane module, thereby reducing the MD driving force. To overcome this, various studies have been conducted using countercurrent flow and turbulent flow [14]. In general, the recommended flow velocity inside and outside the capillaries should exceed 0.1 m/s [71].

It must be pointed out that although one of the major advantages of hollow fibre membrane modules over others is its high packing density and compactness, the fibres are often randomly packed inside the module and care must be taken to avoid flow maldistribution caused by the polydispersity of fibre inner diameter at the lumen side and the non-uniformity of fibre packing at the shell side of the membrane module. It was reported that the effect of the randomness of fibre packing in shell side could lead to a permeate flux reduction by up to 58% corresponding to a packing fraction of 0.4 compared to the ideal

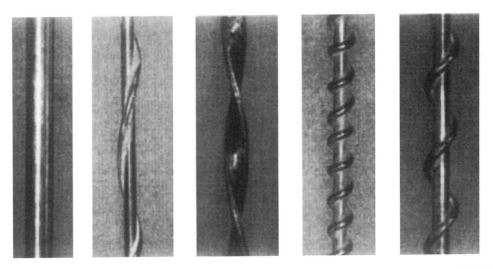

FIGURE 9.7 Images of possible baffles for capillaries or hollow fibre membrane modules. *Source: Reprinted from [98]. Copyright 2004, with kind permission from Elsevier.*

membrane module with uniform packing. By increasing the packing fraction, this effect can be minimized [73–75].

Zander et al. [76] also observed that the air pressure drop at the shell side of the hollow fibre membrane module, when used for membrane air stripping (MAS) process, was much higher than that for the packed tower. The packing density of hollow fibres in the module was thought to be the cause for this high pressure drop. Proper design of the bundle to prevent fibre blockage of

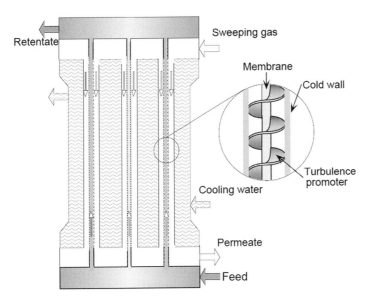

FIGURE 9.8 Schematic representation of TSGMD module with details of the turbulence promoter. *Source: Reprinted from [99]. Copyright 2002, with kind permission from Elsevier.*

the exit port and to avoid major head loss across the air ports may help reduce the pressure drop.

In SGMD, the gas temperature, the heat transfer rate and the mass transport through the membrane change during the gas progression along the membrane module. In order to overcome this problem, a novel tubular device was proposed recently by Rivier et al. [77] for SGMD, in which the increase of the sweeping gas temperature along the membrane module was minimized by using a cold wall in the permeate chamber. A schematic representation of the module is shown in Fig. 9.8. It must be mentioned here that a part of the permeating vapour condenses inside the module depending on the operating conditions, while the rest is collected in an external condenser. The liquid feed was pumped through the inner side of the membrane tubes and air through their outer side. The membrane used to construct tubes of 5 mm internal diameters and 0.2 m length was flat sheet membrane TF200 (Table 2.1). These tubes were subsequently thermo-welded to connections of the module, which were made of PP. A spiral turbulent promoter was placed around the tubular membranes in order to reduce the polarization effect in the gas side.

TNO, the patent holder and the main R&D performer, and a consortium of other eight institutions have developed a modified counter current AGMD module for desalination of seawater, combining multi-stage flash (MSF) and multi-effect distillation modes into one membrane module, called 'Memstill® technology' [78,79]. Figure 9.9 shows the principle of this technology [78–84]. Compared to both large- and small-scale RO, MSF and multi-effect distillers (MEDs), this Memstill® process claims to be a low-cost alternative solution (i.e. low energy consumption, simple construction based on prefabricated modules, lower total cost price) and exhibits a potential of very high salt separation factors, limited

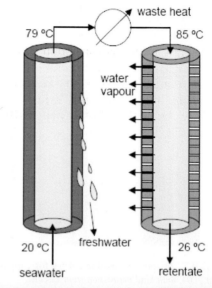

FIGURE 9.9 Principle of Memstill® process and membrane modules for pilot plant. *Source: Reprinted from [78,79]. Copyright 2006, with kind permission from Elsevier.*

corrosion and easy maintenance [78]. Feed seawater flows through a condenser, a heat exchanger, into the membrane evaporator, which consists of a tubular microporous hydrophobic membrane. The condenser and the membrane can also be flat sheets with spacers between them. The condensing surface is cooled by the entering feed flow that is preheated progressively along the condenser. It was reported that two pilots engineered by Keppel Seghers with a mean capacity of 1–2 m^3/h are under construction [79]. It is worth quoting that one of the inconveniences of AGMD configuration is the resistance to mass transfer of the air gap, which can be reduced by decreasing the air gap width, the air gap pressure or both.

SPIRAL WOUND MD MEMBRANE MODULES

The use of spiral wound modules in MD has been first communicated (more than 20 years ago) by Gore & Associated Co. [5] and by Hanbury and Hodgkiess [18] (see Table 2.1).

Flat sheet membranes are assembled in spiral wound modules as shown in Fig. 9.10. The feed

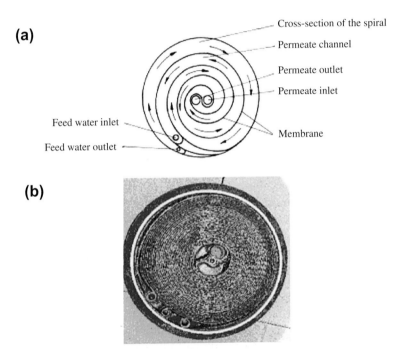

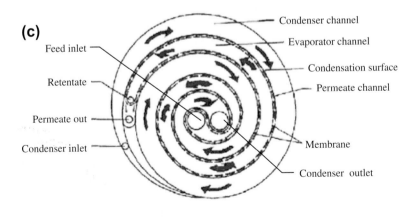

FIGURE 9.10 Spiral wound modules used in (a) DCMD (*Source: Reprinted from [85]. Copyright 1999, with kind permission from Elsevier*) and (b) AGMD (*Source: Reprinted from [90]. Copyright 2008, with kind permission from Elsevier*).

and permeate flow channel spacers, the membrane and the supports are enveloped and rolled around a perforated central collection tube. The membrane packing density normally ranges between $300\,m^2/m^3$ and $1000\,m^2/m^3$, depending on the channel height. Different commercial spiral wound modules were tested in DCMD (Fig. 9.10(a)) and AGMD (Fig. 9.10(b)) experiments [85–91]. The spiral wound AGMD module with integrated heat recovery has been used for the design of solar-powered desalination plants [86–91]. The modules (Fig. 9.10(b)) were developed and manufactured by Fraunhofer ISE (Institute for Solar Energy Systems, Germany) using a PTFE membrane with $0.2\,\mu m$ pore diameter, $35\,\mu m$ thickness, 80% porosity, 450–800 mm height, 300–400 mm diameter, $7-12\,m^2$ effective membrane area, $100-200\,kWh/m^3$ specific thermal energy consumption and a distillate output of $10-30\,l/h$ [86,88,89,91]. All components of this module are made of polymer materials (PP, polyvinyl chloride (PVC), polyethylene (PE) and synthetic resin).

It must be pointed out that in 1995 Bier and Plantikow [92] presented a similar AGMD membrane module and a prototype for an autonomous solar AGMD system for the first time, with an integrated latent hear recovery, which was installed and operated in the Island of Ibiza (Spain) since 1993. However, as reported previously, the additional mass transfer resistance created by the air gap resulted in a large reduction in the permeate production. A distillate production of $40-85\,l/h$ was reported for feed flow rates of $0.8-1.7\,m^3/h$ [92].

DESIGN OF MD MODULES

In general, well-designed membrane modules should provide high mass and heat transfer rates with low temperature and concentration polarization effects as well as less fouling in order to maintain high membrane permeability. Unfortunately, adequate MD modules of industrial scale are currently not available, which is one of the limitations for MD process implementation. Most of the shell-and-tube modules as well as the spiral wound modules prepared for commercial purposes were designed for other separation processes rather than for MD. Reliability of the membrane module is still a serious issue and each configuration imposes certain fluid dynamic conditions on both feed and permeate sides. The question is then: What are the principal features of a module to be used ideally in MD?

A membrane module to be used in MD should meet the following requirements in addition to those previously given for the MD membrane (see Chapter 2, Section 'MD membrane engineering and membrane material selection for MD'): the module should provide a high packing density (i.e. high membrane surface area), high feed and permeate flow rate tangential to the membrane, in cross-flow mode or in transverse-flow mode with high turbulence in order to reduce the temperature and concentration polarization effects, and high heat and mass transfer rate between the bulk solution and the solution at the membrane interface. The mass transfer rate at the feed boundary layer must be sufficiently high to prevent excessive concentration polarization, which can cause membrane wetting and scaling by building up salt crystals or minerals on the membrane surface. Additionally, uniform temperatures of the feed and permeate solutions must be maintained along the module length, which can be accomplished with high heat transfer coefficients. Moreover, housing should exhibit high resistance to pressure, temperature and chemicals. The membranes should be well assembled in potting resins, free of cracks and with a good adhesion. The module should permit its drying in case of membrane wetting problem as well as easy inspection and defects repair. In fact, membranes have a finite life; therefore, it is necessary to consider the possibility of membrane replacement. Care must be taken in membrane packing by taking into consideration the thermal expansion effects like in 'pure' heat exchangers (i.e. good thermal stability). Last but not least, the module should guarantee uniform flows throughout the whole membrane module, avoiding dead corners and channel formation.

In other words, the MD module should satisfy low pressure drop along the membrane module length to prevent excessive high transmembrane hydrostatic pressure that may cause flooding of membrane pores. The MD module should not only provide good flow conditions, but also guarantee low heat loss to the environment and, if possible, act as a good heat recovery system (i.e. internal heat exchanger). The possibility of using plastic equipment also reduces or avoids erosion problems.

It is to be noted that the packing density of the hollow fibre in modules may be one of the causes for the high pressure drop. For shell-and-tube membrane modules, proper design of bundle(s) to prevent fibre blockage and to avoid major head loss across the air ports may help reducing the pressure drop. For example, pumping cost decreases with increasing fibre diameter, but membrane cost increases due to the increase of membrane area. This indicates that pumping costs dominate at small diameters, while membrane costs dominate at large ones.

Some researchers filled the channels with spacers on both feed and permeate side of the plate-and-frame membrane module. This increases the heat transfer coefficients but may, at the same time, decreases the membrane surface area and, consequently, the MD module productivity. A membrane support must be strong enough to prevent deflection or rupture of the membrane. On the other hand, it should not significantly increase the heat and mass transfer resistance. It is worth quoting that temperature polarization is significantly increased by the presence of transversal filaments, which are however necessary to increase the mechanical strength of the spacer.

Computational fluid dynamics (CFD) techniques can be used to study the effects of spacer and channel geometries of membrane modules as well as to design an adequate spacer shape in order to reduce temperature polarization, concentration polarization and fouling. Preliminary results of thermo-fluid dynamics of a spiral wound MD module show how spacers can significantly affect temperature gradients within the channel of a membrane module, permitting to design an optimal spacer for the MD module [93]. The study has been conducted only for the feed MD channel to determine the velocities, pressure and temperature distributions and investigate the effect of the spacer geometry on these distributions.

It is possible to increase membrane plant capacities simply by increasing the module size. However, care must be taken when scaling up the membrane module in order to avoid the loss in module efficiency. For example, in shell-and-tube membrane modules, by passing fluid on the shell side is not a problem for laboratory scale but may become a serious issue upon scaling up to larger modules and results in a loss of efficiency. Furthermore, the optimum diameter of the fibre increases with increasing the hollow fibre length, but it may also result in an increase in the pumping cost.

Several contactor design examples have been suggested by different authors [70,94–97]. All these designs are of interest for the construction of MD systems and should be considered as starting points of the MD module design.

References

[1] Bodell BR. Silicone rubber vapor diffusion in saline water distillation, United States Patent Serial No. 285,032 (1963).

[2] Rodgers FA. Stacked microporous vapor permeable membrane distillation system, United States Patent Serial No. 3,650,905 (1972).

[3] Rodgers FA. Compact multiple effect still having stacked impervious and previous membranes, United States Patent Serial No. Re.27,982 (1974); original No. 3,497,423.

[4] K. Esato, B. Eiseman, Experimental evaluation of Gore-Tex membrane oxygenator, J. Thoracic. Cardiovasc. Surg. 69 (1975) 690—697.

[5] Gore DW. Gore-Tex membrane distillation, Proceedings of the 10th Annual Convention of the Water Supply Improvement Association, Honolulu, USA, July 25—29, 1982.

[6] Cheng DY. Method and apparatus for distillation, United States Patent Serial No. 4,265,713 (1981).

[7] Cheng DY, Wiersma SJ. Composite membrane for a membrane distillation system, United States Patent Serial No. 4,316,772 (1982).

[8] Cheng DY, Wiersma SJ., Composite membrane for a membrane distillation system, United States Patent Serial No. 4,419,242 (1983).

[9] Cheng DY, Wiersma SJ. Apparatus and method for thermal membrane distillation, United States Patent Serial No. 4,419,187 (1983).

[10] L. Carlsson, The new generation in sea water desalination: SU membrane distillation system, Desalination 45 (1983) 221—222.

[11] S.I. Andersson, N. Kjellander, B. Rodesjo, Design and field tests of a new membrane distillation desalination process, Desalination 56 (1985) 345—354.

[12] Catalogue of Enka AG presented at Europe-Japan Joint Congress on Membranes and Membrane Processes, Stresa, Italy, June, 1984.

[13] K. Schneider, T.J. van Gassel, Membrandestillation, Chem. Ing. Tech. 56 (1984) 514—521 (in German).

[14] T.J. van Gassel, K. Schneider, An energy-efficient membrane distillation process, in: E. Drioli, M. Nagaki (Eds.), Membranes and membrane processes, Plenum Press, New York, 1986, pp. 343—348.

[15] E. Drioli, Y. Wu, Membrane distillation: An experimental study, Desalination 53 (1985) 339—346.

[16] G.C. Sarti, C. Gostoli, S. Matulli, Low energy cost desalination processes using hydrophobic membranes, Desalination 56 (1985) 277—286.

[17] A.S. Jonsson, R. Wimmerstedt, A.C. Harrysson, Membrane distillation: A theoretical study of evaporation through microporous membranes, Desalination 56 (1985) 237—249.

[18] W.T. Hanbury, T. Hodgkiess, Membrane distillation: An assessment, Desalination 56 (1985) 287—297.

[19] S. Kimura, S. Nakao, Transport phenomena in membrane distillation, J. Membr. Sci. 33 (1987) 285—298.

[20] Weyl PK. Recovery of demineralized water from saline waters, United States Patent Serial No. 3,340,186 (1967).

[21] M.E. Findley, Vaporization through porous membranes, Ind. Eng. Chem. Process Des. Dev. 6 (1967) 226—237.

[22] M. Khayet, N.N. Li, A.G. Fane, W.S.W. Ho, T. Matsuura, Membrane distillation, in: Advanced membrane technology and applications, John Wiley & Sons, Inc,, New York, NY, USA, 2008, pp. 297—370.

[23] K.W. Lawson, D.R. Lloyd, Membrane distillation: Review, J. Membr. Sci. 124 (1997) 1—25.

[24] M.S. El-Bourawi, Z. Ding, R. Ma, M. Khayet, A framework for better understanding membrane distillation separation process, J. Membr. Sci. 285 (2006) 4—29.

[25] E. Curcio, E. Drioli, Membrane distillation and related operations — A review, Sep. Purif. Rev. 34 (2005) 35—86.

[26] A. Burgoyne, M.M. Vahdati, Direct contact membrane distillation: Review, Sep. Sci. Technol. 35 (2000) 1257—1284.

[27] A.M. Alklaibi, N. Lior, Membrane-distillation desalination: Status and potential, Desalination 171 (2004) 111—131.

[28] www.scarab.se

[29] www.xzero.se

[30] C. Liu, Polygeneration of electricity, heat and ultrapure water for the semiconductor industry, Master of Science Thesis, Heat & Power Technology, Department of Energy Technology, Royal Institute of Technology, Stockholm, Sweden, 2004.

[31] A. Kullab, A. Martin, Membrane distillation and applications for water purification in thermal cogeneration — Pilot plant trials, M06-611, Värmeforsk Service AB, Stockholm, Sweden, 2007.

[32] Sandia National Laboratories, USA (2003) Desalination and water purification technology roadmap. A report of the executive committee, Desalination and water purification research and development program, Report No. 95.

[33] Sirkar KK, Qin Y.(2001) Novel membrane and device for direct contact membrane distillation-based desalination process. US Department of the Interior, Bureau of Reclamation, Agreement No 99-FC-810—180, Program Report No. 87.

[34] Walton J, Lu H, Turner C, Solis S., Hein H. Solar and waste heat desalination by membrane distillation, U.S. Department of the Interior, Bureau of Reclamation, Denver Office, Technical Service Center, Environmental Services Division, Water Treatment Engineering and Research Group, Agreement No. 98-FC-81—0048, Desalination and water purification research and development program, Report No. 81, April, 2004.

[35] C. Dotremont, B. Kregersman, S. Puttemans, P. Ho, J. Hanemaaijer, Memstill: A near-future technology for sea water desalination, ICOM, Honolulu, Hawaii, USA (www.icom2008.org).

[36] C. Gostoli, G.C. Sarti, Separation of liquid mixtures by membrane distillation, J. Membr. Sci. 41 (1989) 211–224.

[37] K. Ohta, K. Kikuchi, I. Hayano, T. Okabe, T. Goto, S. Kimura, H. Ohya, Experiments on sea water desalination by membrane distillation, Desalination 78 (1990) 177–185.

[38] K. Ohta, I. Hayano, T. Okabe, T. Goto, S. Kimura, H. Ohya, Membrane distillation with fluorocarbon membranes, Desalination 81 (1991) 107–115.

[39] G.L. Liu, C. Zhu, C.S. Cheung, C.W. Leung, Theoretical and experimental studies on air gap membrane distillation, Heat Mass Transfer 34 (1998) 329–335.

[40] J.M. Ortiz de Zárate, C. Rincón, J.I. Mengual, Concentration of bovine serum albumin aqueous solutions by membrane distillation, Sep. Sci. Technol 33 (1998) 283–296.

[41] C. Rincón, J.M. Ortiz de Zárate, J.I. Mengual, Separation of water and glycols by direct contact membrane distillation, J. Membr. Sci. 158 (1999) 155–165.

[42] L. Martinez-Díez, F.J. Florido-Díaz, M.I. Vázquez-Gonzalez, Study of evaporation efficiency in membrane distillation, Desalination 126 (1999) 193–198.

[43] L. Martínez-Díez, F.J. Florido-Díaz, Theoretical and experimental studies on membrane distillation, Desalination 139 (2001) 373–379.

[44] M. Khayet, M.P. Godino, J.I. Mengual, Thermal boundary layers in sweeping gas membrane distillation processes, AIChE J. 48 (2002) 1488–1497.

[45] S.T. Hsu, K.T. Cheng, J.S. Chiou, Seawater desalination by contact membrane distillation, Desalination 143 (2002) 279–287.

[46] K.W. Lawson, D.R. Lloyd, Membrane distillation. I. Module design and performance evaluation using vacuum membrane distillation, J. Membr. Sci. 120 (1996) 111–121.

[47] K.W. Lawson, D.R. Lloyd, Membrane distillation. II. Direct contact MD, J. Membr. Sci. 120 (1996) 123–133.

[48] R.W. Schofield, A.G. Fane, C.J.D. Fell, Heat and mass transfer in membrane distillation, J. Membr. Sci. 33 (1987) 299–313.

[49] Honda Z, Komada H, Okamoto K, Kai M. Nonisothermal mass transport of organic aqueous solution in hydrophobic porous membrane, Proceedings of Membranes and Membrane Process, Stresa, Italy, 1986, pp. 587–594.

[50] K. Sakai, T. Koyano, T. Muroi, M. Tamura, Effects of temperature and concentration polarization on water vapor permeability for blood in membrane distillation, Chem. Eng. J 38 (1988) B33–B39.

[51] J.M. Ortiz de Zárate, A. Velázquez, L. Peña, J.I. Mengual, Influence of temperature polarization on separation by membrane distillation, Sep. Sci. Technol. 28 (1993) 1421–1436.

[52] L. Peña, J.M. Ortiz de Zárate, J.I. Mengual, Steady state in membrane distillation: Influence of membrane wetting, J. Chem. Soc., Faraday Trans. 89 (1993) 4333–4338.

[53] M.I. Vázquez-González, L. Martinez, Nonisothermal water transport through hydrophobic membranes in a stirred cell, Sep. Sci. Technol 29 (1994) 1957–1966.

[54] J.I. Mengual, L. Peña, Membrane distillation, Colloid Interface Sci. 1 (1997) 17–29.

[55] M. Sudoh, K. Takuwa, H. Iizuka, K. Nagamatsuya, Effects of thermal and concentration boundary layers on vapor permeation in membrane distillation of aqueous lithium bromide solution, J. Membr. Sci. 131 (1997) 1–7.

[56] M. Khayet, J.I. Mengual, T. Matsuura, Porous hydrophobic/hydrophilic composite membranes: Application in desalination using direct contact membrane distillation, J. Membr. Sci. 252 (2005) 101–113.

[57] M. Khayet, T. Matsuura, Preparation and characterization of polyvinylidene fluoride membranes for membrane distillation, Ind. Eng. Chem. Res. 40 (2001) 5710–5718.

[58] B. Li, K.K. Sirkar, Novel membrane and device for direct contact membrane distillation-based desalination process, Ind. Eng. Chem. Res. 43 (2004) 5300–5309.

[59] B. Li, K.K. Sirkar, Novel membrane and device for vacuum membrane distillation-based desalination process, J. Membr. Sci. 257 (2005) 60–75.

[60] L. Song, B. Li, K.K. Sirkar, J.L. Gilron, Direct contact membrane distillation-based desalination: novel membranes, devices, larger-scale studies, and a model, Ind. Eng. Chem. Res. 46 (2007) 2307–2323.

[61] L. Song, Z. Ma, X. Liao, P.B. Kosaraju, J.R. Irish, K.K. Sirkar, Pilot plant studies of novel membranes and devices for direct contact membrane distillation-based desalination, J. Membr. Sci. 323 (2008) 257–270.

[62] L. Martínez-Díez, M.I. Vázquez-González, F.J. Florido-Díaz, Study of membrane distillation using channel spacers, J. Membr. Sci. 144 (1998) 45–56.

[63] J. Phattaranawik, R. Jiraratananon, A.G. Fane, C. Halim, Mass flux enhancement using spacer filled channels in direct contact membrane distillation, J. Membr. Sci. 187 (2001) 193–201.

[64] M. Khayet, M.P. Godino, J.I. Mengual, Possibility of nuclear desalination through various membrane distillation configurations: a comparative study, Int. J. Nuclear Desalination 1 (2003) 30–47.

[65] M. Tomaszewska, M. Gryta, A.W. Morawski, Study on the concentration of acids by membrane distillation, J. Membr. Sci. 102 (1995) 113–122.

[66] M. Gryta, M. Tomaszewska, J. Grzechulska, A.W. Morawski, Membrane distillation of NaCl solution containing natural organic matter, J. Membr. Sci. 181 (2001) 279–287.

[67] F. Laganà, G. Barbieri, E. Drioli, Direct contact membrane distillation: modelling and concentration experiments, J. Membr. Sci. 166 (2000) 1–11.

[68] F.A. Banat, J. Simandl, Removal of benzene traces from contaminated water by vacuum membrane distillation, Chem. Eng. Sci. 51 (1996) 1257–1265.

[69] S. Al-Obaidani, E. Curcio, F. Macedonio, G.D. Profio, H. Al-Hinai, E. Drioli, Potential of membrane distillation in seawater desalination: thermal efficiency, sensitivity study and cost estimation, J. Membr. Sci. 323 (2008) 85–98.

[70] K.L. Wang, E.L. Cussler, Baffled membrane modules made with hollow fiber fabric, J. Membr. Sci. 85 (1993) 265–278.

[71] K. Schneider, W. Hölz, R. Wollbeck, Membranes and modules for transmembrane distillation, J. Membr. Sci. 39 (1988) 25–42.

[72] M.M. Teoh, S. Bonyadi, T.S. Chung, Investigation of different hollow fiber module designs for flux enhancement in the membrane distillation process, J. Membr. Sci. 311 (2008) 371–379.

[73] D. Zhongwei, L. Liying, M. Runyu, Study on the effect of flow maldistribution on the performance of the hollow fiber modules used in membrane distillation, J. Membr. Sci. 215 (2003) 11–23.

[74] J. Zheng, Y. Xu, Z. Xu, Flow distribution in a randomly packed hollow fiber membrane module, J. Membr. Sci. 211 (2003) 263–269.

[75] J. Zheng, Z. Xu, J. Li, S. Wang, Y. Xu, Influence of random arrangement of hollow fiber membranes on shell side mass transfer performance: a novel model prediction, J. Membr. Sci. 236 (2004) 145–151.

[76] A.K. Zander, M.J. Semmens, R.M. Narbaitz, Removing VOCs by membrane stripping, J. Am. Water Works Assoc. 81 (1989) 76–81.

[77] C.A. Rivier, M.C. García-Payo, I.W. Marison, U. von Stockar, Separation of binary mixtures by thermostatic sweeping gas membrane distillation: I. Theory and simulations, J. Membr. Sci. 201 (2002) 1–16.

[78] G.W. Meindersma, C.M. Guijt, A.B. Haan, Desalination and water recycling by air gap membrane distillation, Desalination 187 (2006) 291–301.

[79] J.H. Hanemaaijer, J. van Medevoort, A.E. Jansen, C. Dotremont, E. van Sonsbeek, T. Yuan, L.D. Ryck, Memstill membrane distillation – A future desalination technology, Desalination 199 (2006) 175–176.

[80] J.H. Hanemaaijer, Memstill®-low cost membrane distillation technology for seawater desalination, Desalination 168 (2004) 355.

[81] C.M. Guijt, I.G. Rácz, J.W. van Heuven, T. Reith, A.B. Haan, Modelling of a transmembrane evaporation module for desalination of seawater, Desalination 126 (1999) 119–125.

[82] C.M. Guijt, I.G. Rácz, T. Reith, A.B. Haan, Determination of membrane properties for use in the modelling of a membrane distillation module, Desalination 132 (2000) 255–261.

[83] C.M. Guijt, G.W. Meindersma, T. Reith, A.B. Haan, Air gap membrane distillation: 1. Modelling and mass transport properties for hollow fiber membranes, Sep. Purif. Tech. 43 (2005) 233–244.

[84] C.M. Guijt, G.W. Meindersma, T. Reith, A.B. Haan, Air gap membrane distillation: 2. Model validation and hollow fibre module performance analysis, Sep. Purif. Tech. 43 (2005) 245–255.

[85] G. Zakrewska-Trznadel, M. Harasimowicz, A.G. Chmielewski, Concentration of radioactive components in liquid low-level radioactive waste by membrane distillation, J. Membr. Sci. 163 (1999) 257–264.

[86] J. Koschikowski, M. Wieghaus, M. Rommel, Solar Thermal-driven desalination plants based on membrane distillation, Desalination 156 (2003) 295–304.

[87] J. Koschikowski, M. Rommel, M. Wieghaus, Solar thermal membrane distillation for small scale desalination plants, ISES World Congress, Orlando, August, 2005. 6–12.

[88] F. Banat, N. Jwaied, M. Rommel, J. Koschikowski, M. Wieghaus, Desalination by a "compact SMADES" autonomous solar-powered membrane distillation unit, Desalination 217 (2007) 29–37.

[89] F. Banat, N. Jwaied, M. Rommel, J. Koschikowski, M. Wieghaus, Performance evaluation of the "large SMADES" autonomous desalination solar-driven membrane distillation plant in Aqaba, Jordan, Desalination 217 (2007) 17–28.

[90] H.E.S. Fath, S.M. Elsherbiny, A.A. Hassan, M. Rommel, M. Wieghaus, J. Koschikowski, M. Vatansever, PV and thermally driven small-scale, stand-alone solar desalination systems with very low maintenance needs, Desalination 225 (2008) 58–69.

[91] J. Koschikowski, M. Wieghaus, M. Rommel, V.S. Ortin, B.P. Suarez, J.R.B. Rodriguez, Experimental

investigations on solar driven stand-alone membrane distillation systems for remote areas, Desalination 248 (2009) 125–131.

[92] C. Bier, U. Plantikow, Solar powered desalination by membrane distillation, IDA World Congress on Desalination and Water Science, Abu Dhabi, 1995, pp. 397–410.

[93] A. Cipollina, A.D. Miceli, J. Koschikowski, G. Micale, L. Rizzuti, CFD simulation of a membrane distillation module channel, Desalination and Water Treatment 6 (2009) 177–183.

[94] A. Gabelman, S.T. Hwang, Hollow fiber membrane contactors, J. Membr. Sci. 159 (1999) 61–106.

[95] S.R. Wickramasinghe, M.J. Semmens, E.L. Cussler, Hollow fiber modules made with hollow fiber fabric, J. Membr. Sci. 84 (1993) 1–14.

[96] R. Prasad, K.K. Sirkar, Membrane-based solvent extraction, in: W.S.W. Ho, K.K. Sirkar (Eds.), Membrane handbook, Chapman & Hall, New York, 1992, pp. 727–763.

[97] B.W. Reed, M.J. Semmens, E.L. Cussler, Membrane contactors, in: R.D. Noble, S.A. Stern (Eds.), Membrane separations technology. principles and applications, Elsevier, Amsterdam, 1995, p. 474.

[98] A.L. Ahmad, A. Mariadas, Baffled microfiltration membrane and its fouling control for feed water of desalination, Desalination 168 (2004) 223–230.

[99] M.C. García-Payo, C.A. Rivier, I.W. Marison, U. von Stockar, Separation of binary mixtures by thermostatic sweeping gas membrane distillation: II. Experimental results with aqueous formic acid solutions, J. Membr. Sci. 198 (2002) 197–210.

CHAPTER 10

Direct Contact Membrane Distillation

OUTLINE

Introduction	249	Osmotic Distillation (OD)	275
Theoretical Models	254	DCMD Applications	278
Mass Transfer Through the Membrane Pores	254	DCMD Technology in Desalination and Crystallization	278
Heat Transfer Through the Membrane and the Boundary Layers	261	DCMD Technology in Concentration of Fruit Juices	284
Experimental: Effects of DCMD Process Conditions	268	DCMD Technology for Treatment of Wastewaters	285
Effects of Feed Temperature	268	Other Applications of DCMD Technology	286
Effect of Permeate Temperature	269		
Effects of Feed and Permeate Flow Rates	271		
Effects of Solute Concentration in the Feed Aqueous Solution	272		

INTRODUCTION

Most of membrane transport processes are isothermal and their driving force is the transmembrane chemical potential difference of the transported species (i.e. hydrostatic pressure, concentration, electrical potential, etc.). For example, the well-known reverse osmosis (RO) used specially in desalination of seawater or brackish waters is an isothermal process. Less membrane processes are non-isothermal requiring a thermal driving force to establish the necessary transmembrane chemical potential difference or transmembrane partial vapour pressure difference. Among these processes, one can find direct contact membrane distillation (DCMD), which is the most used configuration of MD process (i.e. about 63.3% of the papers published up to December 2010 were on DCMD), especially for desalination of seawater and brackish waters. This is due to the condensation step that can be carried out inside the membrane module enabling a simple MD operation mode. However, the heat transferred by conduction through the membrane — considered heat lost in MD — is higher than in the other MD configurations.

The term DCMD comes from its similarity to conventional distillation (i.e. simple and multi-effect distillation) because evaporation and condensation take place at the liquid—vapour interfaces formed at the pore entrances on the feed and the permeate side, respectively. The hydrophobic porous membrane is maintained in direct contact with the feed aqueous solution to be treated and the permeate aqueous solution that is usually drinkable water or distilled water. Furthermore, both DCMD and conventional distillation technologies require latent heat of evaporation to be supplied to the aqueous feed solution in order to produce mass flux(es).

It is worth quoting that desalination was first performed with DCMD process by Weyl [1]. Permeate fluxes up to $1 \, kg/m^2 \cdot h$ were reported. These DCMD permeate fluxes were found to be in good agreement with the theoretical calculation. However, a productivity of $1 \, kg/m^2 h$ was far below the permeate fluxes obtained in RO systems (i.e. $20-75 \, kg/m^2 \cdot h$). The first paper on DCMD was published by Findley et al. [2]. Both experimental and theoretical calculations were reported. The study involved heat and mass transfer of water vapour from a hot salt aqueous solution through a hydrophobic porous membrane to a cooled water condensate. The thermal conductivity of the membrane and the heat transfer coefficients of the boundary layers on the feed and permeate sides have been considered in the theoretical model. Their experimental studies indicated that the major factor affecting the rates of heat and mass transfer was the diffusion through the stagnant gas (i.e. air) in the membrane pores. An empirical correction related to the possible internal condensation and diffusion along the membrane pores has been considered to perform the calculations.

As shown in Fig. 10.1, in DCMD process the hot feed solution and the cold permeate solution are maintained in direct contact with the feed side and the permeate side of the membrane, respectively. The feed and permeate aqueous solutions are circulated tangentially to the membrane surfaces by means of circulating pumps or stirred inside the membrane cell by means of magnetic stirrers. The hydrostatic transmembrane pressure difference must be lower than the membrane liquid entry pressure (i.e. breakthrough pressure, LEP). The hydrophobic nature of the membrane prevents liquid solutions from entering the pores due to the surface tension forces. As a result, liquid—vapour interfaces are formed at the entrances of the membrane pores. In this case the transmembrane temperature difference induces the required vapour pressure difference. Under these conditions, volatile molecules evaporate at the hot vapour—liquid interface, flow across the membrane pores in vapour phase, and finally condense at the cold liquid—vapour interface inside the membrane module, which may be a shell and tube or plate and frame configuration working under cross flow or longitudinal flow. A typical laboratory DCMD system used for flat sheet, capillary or hollow fibre membranes is presented in Fig. 10.2. Various membrane modules can also be associated simultaneously (i.e. array of DCMD modules) [3,4].

In DCMD, the temperature of the feed solution is below its boiling point. It can vary between few degrees over the ambient temperature (about 30 °C) to 90 °C. The temperature of the feed and permeate solutions is measured at both the inlet and outlet of the membrane module. The feed and permeate pressures are close to the atmospheric pressure. To avoid membrane pore wetting, the pressure is controlled continuously with manometers also placed at the inlet and outlet of the membrane module. Moreover, the absence of pore wetting can be checked by determining the permeate concentration and/or by measuring the electrical conductivity on the permeate side using aqueous salt solution on the feed side. In all MD configurations the permeate flux is

INTRODUCTION

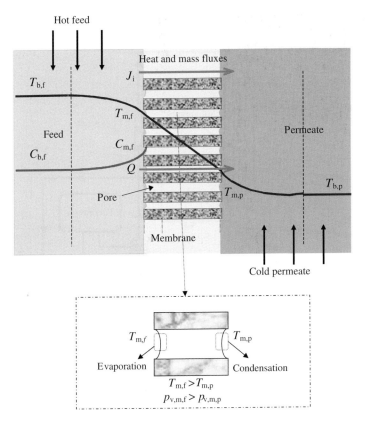

FIGURE 10.1 Schema of heat and mass transfer through a single layer hydrophobic membrane used in DCMD process for the treatment of a feed aqueous solution containing non-volatile solutes.

calculated by measuring the condensate collected on the permeate side of the membrane module for a pre-determined period of time.

It is worth quoting that DCMD is a process mainly suited for applications in which water is the major component of the feed solution containing non-volatile solutes such as salts, colloids, proteins, etc. In this case, the concentrations of both permeate and feed solutions are determined at a temperature of 20 °C, by a previously calibrated conductivity meter (i.e. electrical conductivity *vs.* solute concentration). The separation factor, α, is calculated using the following expression:

$$\alpha = \left(1 - \frac{C_{b,p}}{C_{b,f}}\right) \times 100 \quad (10.1)$$

where $C_{b,p}$ and $C_{b,f}$ are the solute concentration in the permeate and in the bulk feed solution, respectively.

The hydrophobic character of the membrane as well as the proper use of the MD configuration permits to achieve very high salt separation factors. Close to 100% separation factors, α, were obtained when aqueous solutions of salts were employed as feed.

Care must be taken when volatile solutes are present in the feed solution because of the risk of pore wetting from the permeate side. The risk will be high when the concentration of the volatile solute in the permeate increases. Moreover, wastewater containing contaminants of low surface tension such as alcohols, detergents, etc. cannot be treated directly in DCMD. Adequate pre-treatments must be made and

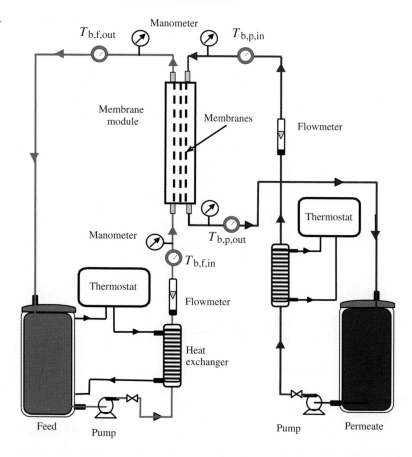

FIGURE 10.2 Typical experimental DCMD laboratory system.

rigorous studies on liquid entry pressure (*LEP*) of the membrane must be carried out before the membrane is used in order to avoid membrane pore clogging.

Porous compositae hydrophobic/hydrophilic membranes with higher *LEP* values than the commonly used commercial membranes have been proposed for DCMD applications [5–11]. In this case the feed liquid is brought into contact with the hydrophobic side of the membrane. It was found that these membranes exhibit higher desalination performance than the single hydrophobic layer membranes.

For capillaries and hollow fibre membranes, two types of DCMD experiments can be carried out: (i) feed liquid circulating along the inner surface of the membranes and (ii) feed liquid circulating along the outer surface of the membranes.

It must be pointed out that 40.6% of the publications in DCMD (up to December 2010) are concerned with theoretical models. Most of DCMD experimental studies were performed to investigate the effects of the operating conditions and very few research groups have investigated the effects of membrane parameters on DCMD performance. Compared to the other MD configurations, it was noticed that most of the fabricated membranes were tested in DCMD. The highest membrane permeate fluxes obtained so far in DCMD are 145.8 kg/m^2·h

INTRODUCTION

TABLE 10.1 Reported DCMD Permeate Flux (J_w) of Different Types of Commercial Flat Sheet Membranes Shown in Table 2.1

Membrane trade name	J (10^{-3} kg/m²s)	Observation	Reference
GVHP	13.52	Distilled water as feed; $T_{b,f}$ = 90.7 °C; $T_{b,p}$ = 19.7 °C.	[44]
	9.00	Distilled water as feed; $T_{b,f}$ = 70 °C; $T_{b,p}$ = 20 °C.	[30]
	7.00	Distilled water as feed; $T_{b,f}$ = 70 °C; $T_{b,p}$ = 20 °C.	[49]
	1.38	Distilled water as feed; $T_{b,f}$ = 40 °C; $T_{b,p}$ = 20 °C.	[112]
	0.89	0.05 M NaCl; ΔT_b = 10 °C; $T_{b,p}$ = 51.9 °C.	[66]
	0.83	1.1 M NaCl; ΔT_b = 10 °C; $T_{b,p}$ = 52.7 °C.	
HVHP	18.61 16.39 11.11	Deareation DCMD; $T_{b,f}$ = 80 °C; $T_{b,p}$ = 21 °C Distilled water NaCl (14 wt%) NaCl (25 wt%).	[23]
	10.80	Distilled water as feed; $T_{b,f}$ = 70 °C; $T_{b,p}$ = 20 °C.	[30]
TF200	18.69	Distilled water as feed; $T_{b,f}$ = 80.1 °C; $T_{b,p}$ = 20 °C.	[44]
	2.90	0.032 M NaCl; ΔT_b = 10 °C; $T_{b,p}$ = 52.2 °C.	[66]
	2.23	1.1 M NaCl; ΔT_b = 10 °C; $T_{b,p}$ = 52.7 °C.	
	2.13	Distilled water as feed; $T_{b,f}$ = 40 °C; $T_{b,p}$ = 20 °C.	[112]
PTFE Sartorious	14.00	Distilled water as feed; $T_{b,f}$ = 70 °C; $T_{b,p}$ = 20 °C.	[30]
TS22	21.67	0.6 g/l NaCl; $T_{b,f}$ = 60 °C; $T_{b,p}$ = 20 °C.	[54]
TS45	22.22	0.6 g/l NaCl; $T_{b,f}$ = 60 °C; $T_{b,p}$ = 20 °C.	
FGLP	8.56	Distilled water as feed; $T_{b,f}$ = 57.2 °C; $T_{b,p}$ = 20 °C.	[33]
PP22	7.78	0.6 g/l NaCl; $T_{b,f}$ = 60 °C; $T_{b,p}$ = 20 °C.	[54]
3MA	25.2 22.5 19.8	$T_{b,f}$ = 74 °C; $T_{b,p}$ = 20 °C; distilled water as feed 0.6 mol% 1.3 mol%.	[12]
3MB	21.6	Distilled water as feed; $T_{b,f}$ = 70 °C; $T_{b,p}$ = 20 °C.	
3MC	37.8	Distilled water as feed; $T_{b,f}$ = 80 °C; $T_{b,p}$ = 20 °C.	
3MD	27	Distilled water as feed; $T_{b,f}$ = 70 °C; $T_{b,p}$ = 20 °C.	
3ME	40.5 32.4	Distilled water; $T_{b,f}$ = 80 °C; $T_{b,p}$ = 20 °C. 1.3 mol% NaCl; $T_{b,f}$ = 74 °C; $T_{b,p}$ = 20 °C.	

and 116.6 kg/m²·h using distilled water and 1.3 mol% aqueous sodium chloride (NaCl) solution, respectively — as feed at a temperature of 74 °C and a permeate temperature of 20 °C. A commercial membrane — 3ME supplied by 3M Corporation — was used [12]. This membrane was made of polypropylene (PP) and had a thickness of 79 µm, a maximum pore size of 0.73 µm and a porosity of 85% (Table 2.1).

The potential applications of DCMD are production of high-purity water, concentration of ionic, colloidal and other non-volatile aqueous solutions. Various applications (desalination, water-reuse, food processing, medical application, etc.) are involved in DCMD. The lower operating temperatures than the conventional distillation, the lower operating hydrostatic pressures than the pressure-driven processes (i.e. reverse osmosis, RO; nanofiltration, NF; ultrafiltration, UF; microfiltration, MF), the less demanding membrane mechanical properties and the high rejection factor achieved when treating solutions containing non-volatile solutes (salts, colloids, etc.) make MD more attractive than other popular separation processes. In addition, the possibility of using waste heat and renewable energy sources enable the DCMD technique to cooperate in conjunction with other processes.

THEORETICAL MODELS

In general, in all MD configurations, simultaneous heat and mass transfer phenomena take place through the porous membrane. Any developed models for the MD process must be able to describe — among others — the mechanism(s) of mass transport through the membrane, the effects of the temperature and concentration boundary layers at the membrane surfaces (i.e. temperature and concentration polarization, see Fig. 10.1), the production rate and the solute retention or selectivity of the membrane.

The following three assumptions are made in theoretical modelling:

(i) the kinetic effects at the liquid–vapour interfaces formed at both ends of the membrane pores are negligible.
(ii) the vapour and liquid phases are in equilibrium corresponding to the temperature at each side of the pores (i.e. feed and permeate sides).
(iii) compared to a flat interface, the curvature of the liquid–vapour interface is assumed to have a negligible effect on the equilibrium. If necessary, this effect can be estimated by means of the Kelvin equation [14]:

$$p_c = p \exp\left(\frac{2\,\gamma_L}{r\,c\,R\,T}\right) \qquad (10.2)$$

where p_c and p are the liquid saturation pressure above a convex liquid surface and a flat surface, respectively; r is the radius of curvature, γ_L is the liquid surface tension, c is the liquid molar density, R is the gas constant and T is the absolute temperature.

Since DCMD configuration is — as mentioned earlier — most suited for feed solutions containing non-volatile solutes, mass transfer through the membrane will be discussed only for water vapour transfer through the membrane.

It is to be noted that transport of gases and vapours through porous media has extensively been studied and theoretical models have been developed based on the Kinetic Theory of Gases to predict the MD performance of the membranes depending on the MD configuration used. In general, the membrane pores are considered cylindrical in shape. In what follows mass and heat transfer analysis in DCMD process is addressed.

Mass Transfer Through the Membrane Pores

In the DCMD process, the water vapour transfer from the feed to the permeate is due to the difference in its chemical potential between both sides of the membrane ($\Delta\mu_w$). This depends on temperature, pressure and concentration. The DCMD permeate flux (J_w) is dependent on the membrane characteristics and the established driving force, which, in this case, is the transmembrane water vapour pressure difference (Δp_w) and can be expressed as [3,4,13–15]:

$$J_w = B_w \Delta p_w = B_w(p^0_{w,f} a_{w,f} - p^0_{w,p} a_{w,p})$$
$$= B_w(p^0_{w,f} \gamma_{w,f} x_{w,f} - p^0_{w,p} \gamma_{w,p} x_{w,p}) \quad (10.3)$$

where B_w, a_w, γ_w and x_w are the membrane DCMD coefficient (i.e. membrane permeability), activity, activity coefficient and mole fraction of water, respectively. The subscripts w, f and p refer to water, feed and permeate, respectively. As will be described more in detail, B_w is a function of the applied temperature and the membrane properties such as pore size, porosity, thickness and tortuosity of the pores.

In Eq. (10.3), the following expression has been considered:

$$p_w(x, T) = p^0_w(T) a_w(x) \quad (10.4)$$

where p_w is the partial water vapour pressure in the solution, x is the non-volatile solute mole fraction, T is the absolute temperature (K) and p^0_w (Pa) is the vapour pressure of pure water determined by means of the Antoine equation:

$$p^0_w(T) = \exp\left(23.1964 - \frac{3816.44}{T - 46.13}\right) \quad (10.5)$$

The diffusion of non-condensable gases from the aqueous feed solution across the membrane can be neglected because it is very small compared to the mass of the transported volatile molecules through the membrane pores. For example, the solubility of air in water is about 10 ppm. This indicates that the transmembrane flux of air is many orders of magnitude lower than that of water.

If distilled water is obtained in the permeate membrane side, J_w can be rewritten as:

$$J_w = B_w(p^0_{w,f} \gamma_{w,f} x_{w,f} - p^0_{w,p}) \quad (10.6)$$

As mentioned previously, DCMD desalination has been studied extensively. For an aqueous solution of sodium chloride (NaCl), an empirical correlation between γ_w and the molar fraction of the solute x_{NaCl} is often used [14]:

$$\gamma_w = 1 - 0.5 x_{NaCl} - 10 x^2_{NaCl} \quad (10.7)$$

For an aqueous solution of calcium chloride ($CaCl_2$), the following empirical correlation for the water activity a_w was used when the range of mass fraction (w_{CaCl_2}) in the solution is between 32.2% and 46.2% [15]:

$$a_w = 1.6941 - 0.041 w_{CaCl_2} + 2.4 \cdot 10^{-4} w^2_{CaCl_2} \quad (10.8)$$

where w_{CaCl_2} is the mass fraction of $CaCl_2$ in the aqueous solution.

For aqueous sugar solution at 25 °C, the water activity can be evaluated using the following equations [15]:

$$a_w = -0.27 w^3 - 0.08 w^2 - 0.09 w + 1 \quad (10.9)$$

where w is sugar weight fraction in the aqueous solution.

For dilute aqueous solution of salts, the following approximation is often considered:

$$p_w(x_w, T) = (1 - x_s) p^0_w(T) \quad (10.10)$$

where x_s is the non-volatile solute mole fraction.

For a low transmembrane bulk temperature difference ($T_{b,f} - T_{b,p} \leq 10$ K), the permeate water vapour flux (J_w) is linearly related to its partial pressure difference across the membrane pores [16,17]:

$$J_w = B_w(p_{m,f} - p_{m,p})$$
$$= B_w \left(\frac{dP}{dT}\right)_{T_m} (T_{m,f} - T_{m,p}) \quad (10.11)$$

where $p_{m,f}$ and $p_{m,p}$ are the partial vapour pressures of water at the membrane surface in the feed and the permeate side of the membrane, respectively; T_m is the mean temperature in the membrane pores and (dp/dT) can be evaluated from the Clausius–Clapeyron equation, combined with the Antoine equation, to calculate the vapour pressure:

$$\left(\frac{dp}{dT}\right)_{T_m} = \frac{\Delta H_{v,w}}{RT_m^2} \exp\left(23.238 - \frac{3841}{T_m - 45}\right) \quad (10.12)$$

where $\Delta H_{v,w}$ is the heat of vapourization of water that can be evaluated using the following equation [14]:

$$\Delta H_{v,w} = 1.7535\, T + 2024.3 \quad (10.13)$$

where T is the absolute temperature in K and $\Delta H_{v,w}$ is in kJ/kg.

For feed aqueous solutions containing non-volatile solutes, the permeate water vapour flux, J_w, can be rewritten as [14,18]:

$$J_w = B_w \left(\frac{dP}{dT}\right)_{T_m} \times \left[\left(T_{m,f} - T_{m,p}\right) - \Delta T_0\right](1 - x_{s,m,f}) \quad (10.14)$$

where $\Delta T_0 = \dfrac{RT_m^2}{\Delta H_{v,w}}\left(\dfrac{x_{s,b,f} - x_{s,b,p}}{1 - x_{s,m,f}}\right)$

and the subscripts s, m, b, f and p refer to solute, membrane surface, bulk solution, feed and permeate, respectively.

As can be seen, a threshold temperature difference, ΔT_0, must be overcome in order to produce water vapour transport towards the permeate side. A negative DCMD flux (i.e. flux from permeate side to feed side) can occur if a negative driving force is applied in DCMD process $((T_{m,f} - T_{m,p}) < \Delta T_0)$. This reversed driving force is caused mainly by the osmotic pressure of the feed aqueous solution, when it is lower than the osmotic pressure of distilled water present on the permeate side. For typical DCMD operating conditions, the estimated values of ΔT_0 is generally less than 1 °C.

To determine the water vapour permeability of a porous membrane used in DCMD, the Kinetic Theory of Gases is considered [19,20]. The mechanisms proposed for the mass transport through DCMD membranes are the Knudsen flow model and the ordinary molecular diffusion model [21,22]. Due to the fact that in DCMD configuration both feed and permeate aqueous solutions are brought into contact with the membrane under atmospheric pressure, the total pressure is maintained constant at $\approx 10^5$ Pa resulting in negligible viscous flow [21,22]. Schofield et al. [17,23–25] stated that the air flow across the membrane is extremely small relative to the flux of water, and the viscous flow can be neglected unless the solutions are degassed.

The governing quantity which provides a guideline in determining the operative mechanism in a given membrane pore under a given experimental condition is the Knudsen number (Kn) defined as:

$$Kn = \lambda/d_p \quad (10.15)$$

where d_p is the membrane pore size.

The mean free path, λ_w, of water molecules in vapour phase can be calculated using the following expression [14,26]:

$$\lambda_w = \frac{k_B T}{\sqrt{2}\pi\, P_m\, (2.641\, 10^{-10})^2} \quad (10.16)$$

where k_B is the Boltzmann constant, P_m is the mean pressure within the membrane pores, T is the absolute temperature and the value 2.641 Å is the collision diameter of water molecules, σ_w.

As an example for DCMD application, the calculated mean free path for water vapour at 50 °C under atmospheric pressure is approximately 0.14 μm, which is similar to the pore size of the membranes commonly used in MD.

If the mean free path of the transported water molecules in vapour phase is greater than the membrane pore size (i.e. $Kn > 10$ or $d_p < 0.1\lambda_w$), the molecule-pore wall collisions are dominant over the molecule–molecule collisions and Knudsen type of flow is responsible for the mass transfer through the membrane pore as schematically presented in Fig. 10.3. It was considered that Knudsen type flow is predominant when the ratio of the pore radius to the mean free path (i.e. r_p/λ_w) is lower than 0.05 [26].

In this case the permeability through each pore in Knudsen region is expressed as follows [22,26–28]:

$$B_w^K = \frac{2\pi}{3} \frac{1}{RT} \left(\frac{8RT}{\pi M_w}\right)^{1/2} \frac{r_k^3}{\tau\delta} \qquad (10.17)$$

where r_k is the pore radius in Knudsen region, M_w is the molecular weight of water, R is the gas constant and δ is the membrane thickness.

If $Kn < 0.01$ (i.e. $d_p > 100\lambda_w$), molecular diffusion is used to describe the mass transport in continuum region caused by the stagnant air trapped within each membrane pore due to the low solubility of air in water (Fig. 10.3). The following relationship is used to determine the permeability through a pore having an area of πr_D^2 in the ordinary diffusion region [22,29].

$$B_w^D = \frac{\pi}{RT} \frac{PD_w}{p_a} \frac{r_D^2}{\tau\delta} \qquad (10.18)$$

where D_w is the diffusion coefficient, P is the total pressure inside the membrane pore and p_a is the air pressure in the membrane pore.

For water/air, PD_w (Pa m²/s) can be calculated from the following equation [22,30].

$$PD_w = 1.895\ 10^{-5} T^{2.072} \qquad (10.19)$$

As mentioned above, the calculated λ_w values are similar to the pore size of the membranes used in DCMD. In transition region, $0.01 < Kn < 10$ (i.e. $0.1\lambda_w < d_p < 100\lambda_w$), the water vapour transport takes place via a combined Knudsen/ordinary diffusion mechanism. In this case, the following model is used to determine the water vapour permeability of the membrane [22].

$$B_w^C = \frac{\pi}{RT} \frac{1}{\tau\delta} \left[\left(\frac{2}{3}\left(\frac{8RT}{\pi M_w}\right)^{1/2} r_t^3\right)^{-1} + \left(\frac{PD_w}{p_a} r_t^2\right)^{-1} \right]^{-1} \qquad (10.20)$$

where r_t is the pore radius in the transition region.

Khayet et al. [21] investigated the mechanism of mass transport in DCMD process. An uniform pore size (i.e. mean pore size) was assumed for the membranes TF200, TF100,

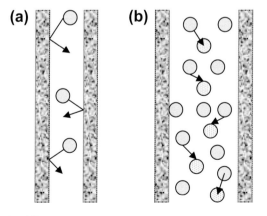

FIGURE 10.3 Transport mechanism through a pore of a membrane used in DCMD: (a) Knudsen type of flow and (b) molecular diffusion type of flow.

FGLP and FHLP listed in Table 2.1. The DCMD experiments were carried out under different operation conditions. It was claimed that the mass transport mechanism takes place via the combined Knudsen/molecular diffusion mechanism for the studied membranes having mean pore sizes between 0.2 and 1 μm. In their model the pore tortuosity was taken into account as an adjustment parameter. Since then, most authors consider this combined mechanism of transport in their theoretical models for the DCMD process [30–36].

It must be mentioned here that for given experimental conditions, the calculated DCMD flux based on the Knudsen mechanism is higher than that based on the combined Knudsen/molecular diffusion mechanism. This indicates that when the pore size is near the critical pore size (i.e. $\approx 0.1\lambda_w$), the DCMD flux will not necessarily increase with the increase of the pore size. Therefore, under some operating conditions, it would be better to use membranes with smaller pore sizes than $0.1\lambda_w$ so that Knudsen flow will take place, leading to a higher DCMD flux than membranes having larger pore sizes, where the combined Knudsen/molecular diffusion flux is responsible for mass transfer. As a consequence, it is advisable to choose the appropriate membrane with a specific pore size, taking into account the value of the mean free path of the transported molecules, and try to use the membrane under Knudsen type flow conditions. These facts must be considered when selecting or designing a membrane for DCMD applications [3,5–7].

As shown in Chapter 8, the membranes commonly used in MD exhibit distributions of pores sizes. Therefore, more than one mechanism of mass transport can occur simultaneously through the entire membrane. The following equation was proposed to determine the total DCMD water vapour permeability of the membrane [22]:

$$B_w^m = \frac{N}{\delta}\left[\sum_{i=1}^{m(r=0.05\lambda_w)} G_w^K f_i r_i^3 \right.$$
$$+ \sum_{i=m(r=0.05\lambda_w)}^{p(r=50\lambda_w)} \left(\frac{1}{G_w^K r_i} + \frac{1}{G_w^D}\right)^{-1} f_i r_i^2$$
$$\left. + \sum_{i=p(r=50\lambda_w)}^{n(r=r_{max})} G_w^D f_i r_i^2 \right] \quad (10.21)$$

where

$$G_w^K = \left(\frac{32\pi}{9M_w RT}\right)^{1/2} \quad (10.22)$$

$$G_w^D = \frac{\pi}{RT}\frac{PD_w}{p_a} \quad (10.23)$$

and f_i is the fraction of pores with pore radius r_i, N is the total number of pores per unit area, m is the largest pore group in Knudsen region, p is the largest pore group in the transition region and r_{max} is the maximum pore radius. It is to be noted that in Eq. (10.21) the upper summation limit changes depending on the maximum pore radius (r_{max}) of the membrane. If ($r_{max} < 0.05\lambda_w$), only Knudsen mechanism can be applied and if ($r_{max} > 50\lambda_w$) all mechanisms are operative simultaneously.

When an uniform pore radius, $\bar{r}$, is assumed for a given membrane and the Knudsen model is applied ($\bar{r} < 0.05\lambda_w$), the following equations is used:

$$B_w^K = \frac{2}{3RT}\frac{\varepsilon \bar{r}}{\tau \delta}\left(\frac{8RT}{\pi M_w}\right)^{1/2} \quad (10.24)$$

When the transition flow dominates ($0.5\lambda_w < \bar{r} < 50\lambda_w$), the following equation corresponding to the combined Knudsen/ordinary diffusion mechanism is used:

$$B_w^m = \frac{1}{RT\delta}\left(\frac{3\tau}{2\varepsilon \bar{r}}\left(\frac{\pi M_w}{8RT}\right)^{1/2} + \frac{p_a \tau}{\varepsilon PD}\right)^{-1} \quad (10.25)$$

It was found that the permeate flux of commercial porous membranes — calculated assuming all pores having the same size equal to the mean pore size and the permeate flux calculated with a Gaussian (symmetric) distribution function of the pore size — were similar and the predicted permeate fluxes were lower than the experimental ones [37]. When using the commercial membranes (GVHP and HVHP in Table 2.1) it was concluded that the influence of pore size distribution on the predicted DCMD permeate flux was insignificant [30]. However, as stated previously, an adjustment factor (i.e. pore tortuosity) was considered in the calculations. Khayet et al. [22] found that a slightly higher membrane permeability was obtained when taking into consideration the pore size distribution than the value obtained using the mean pore size alone. It was explained that the small effect was due to the small geometric standard deviation in the pore size distribution of the commercial membrane. A larger discrepancy may be detected if laboratory made membranes with broader pore size distributions are used. By using their theoretical model, Khayet et al. [22] found an increase in the predicted water vapour permeability with an increase of the geometric standard deviation of the pore size distribution. Theoretical modelling with and without pore size distribution has also been studied for DCMD configuration by other authors [38,39].

It is worth quoting that the presence of air within the membrane pores between the feed and the permeate liquid–vapour interfaces reduces the DCMD permeate flux. Deaerated DCMD systems were proposed by some authors [23,24,40,41]. It was found that for membranes having small pore sizes, Knudsen flow is predominant and the removal of air results only in a small increase of the DCMD permeate flux. However, for membranes having greater pore sizes than the mean free path of the volatile components, a substantial enhancement of the DCMD flux can be achieved by deareation. Deareation can be carried out by lowering the pressure of the liquid streams controlling the maximum pressure of gas within the membrane pores. As a result, a transmembrane hydrostatic pressure difference causes Poiseuille (i.e. viscous) flow, whereas the ordinary molecular diffusion model becomes negligible. It is to be noted that when the molecule–molecule collision surpasses the molecule–pore wall collisions, viscous flow becomes responsible for the mass transfer through the membrane pore. Therefore, in this special case, i.e. deaerated DCMD, the water vapour transport occurs in the Knudsen/Poiseuille transition region as will be explained in Chapter 12 for vacuum membrane distillation (VMD) configuration.

In absence of air effects (deaerated DCMD systems), the following correlation was proposed [24]:

$$J_w = a\xi^b \Delta p_w \quad (10.26)$$

where ξ is a dimensionless pressure (P/P_{ref}), a is the membrane permeability at the reference pressure P_{ref} and b is a parameter that can vary from 0 for fully Knudsen type of flow and 1 for fully Poiseuille (viscous) type of flow. The parameters a and b depend on the molecular weight of the transported components as well as on the membrane type. When the membrane is deaerated the molecular diffusion resistance, which makes Knudsen flow dominant, is decreased. It was observed that as the air partial pressure in the membrane increases (0–100 kPa) the DCMD permeate flux decreases [24,25].

In the case of aerated DCMD systems (i.e. air trapped within membrane pores), the following expression was taken into consideration [23,24]:

$$J_w = \left(\frac{1}{a\xi^b} + \frac{p_a}{d}\right)^{-1} \Delta p_w \quad (10.27)$$

where p_a is the average pressure of air within the membrane pores and d takes into consideration both the membrane characteristics and the molecular properties.

A more complete model – known as dusty gas model – was adopted in the DCMD process to predict the permeate flux [12,14,19]. As shown in Fig. 10.4, this model combines all transport mechanisms through the membranes: Knudsen diffusion, ordinary molecular diffusion and viscous or Poiseuille type of flow except for surface diffusion.

The following set of equations has been considered [12,14,19]:

$$\frac{J_w^D}{D_{we}^k} + \sum_{i=1 \neq w}^{n} \frac{p_i J_w^D - p_w J_i^D}{D_{wie}^0} = -\frac{1}{RT}\nabla p_w \quad (10.28)$$

$$J_w^v = -\frac{\varepsilon r^2 p_w}{8RT\tau\mu}\nabla P \quad (10.29)$$

$$D_{we}^k = \frac{2\varepsilon r}{3\tau}\sqrt{\frac{8RT}{\pi M_w}} \quad (10.30)$$

$$D_{wie}^0 = \frac{\varepsilon}{\tau} P D_{wi}^0 \quad (10.31)$$

$$J_w = J_w^D + J_w^v \quad (10.32)$$

where J^D is the diffusive type of flux, J^v is the viscous type of flux, D^k is the Knudsen diffusion coefficient, D^0 is the ordinary diffusion coefficient, P is the total pressure, μ is the viscosity of the gas mixture, r is the membrane radius, ε is the membrane porosity, M is the molecular weight, τ is the pore tortuosity and the subscripts e, i and w refer to the effective diffusion coefficients, the compound i that can be in this case air molecules and the transported water vapour molecules, respectively.

It is to be noted that the dusty gas model was originally developed for isothermal systems. However, it was successfully applied in MD assuming an average temperature across the membrane.

Eqs. (10.28–10.32) were reduced to the Knudsen/ordinary molecular diffusion transition form and the following integrated equation was derived for the DCMD process [12]:

$$J_w = \frac{D_{wie}^0}{\delta RT_m}\ln\left(\frac{p_{pi}D_{we}^k + D_{wie}^0}{p_{fi}D_{we}^k + D_{wie}^0}\right) \quad (10.33)$$

where T_m is the average temperature across the membrane, p_{pi} and p_{fi} are the partial pressure of air on the permeate and feed side, respectively.

It must be mentioned that the empirical models proposed as Eqs. (10.26) and (10.27) are nearly equivalent to the dusty gas model. However, the later equations are somewhat inconvenient because a and b are dependent on the tested gases and also because the value of the dimensionless pressure (ξ) is near unity.

In general, the theoretical models published in the DCMD literature assume 'cylindrical' pores in the membrane ignoring pore space interconnectivity. Monte Carlo (MC) simulation model has been developed to study both heat and mass transfer in DCMD configuration considering inter-connected pores [42–44]. The MC models were designed without using any adjustable parameters and the membrane pore space was described by a three-dimensional network of inter-connected cylindrical pores (bonds) with size distribution and nodes (sites). The models can simultaneously simulate the DCMD permeate flux and temperatures at the membrane surfaces permitting the evaluation

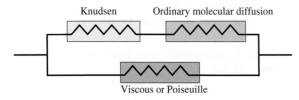

FIGURE 10.4 Electrical analogy circuit presenting the different transport mechanisms in Dusty gas model.

of the temperature polarization coefficient. The bonds of the network, which represent the membrane pore throats, were assumed to be cylindrical capillaries of a given pore size and length, and mass transfer in each capillary was based on the Kinetic Theory of Gases (Knudsen model, Poiseuille or viscous model and ordinary molecular diffusion model). Comparisons between the simulated results and the experimental ones demonstrated that the MC model was in excellent qualitative and quantitative agreement with experiments for different membranes with different characteristics. The MC simulation model may play supportive role to experimental studies, optimizing membrane structure and MD process module design.

Heat Transfer Through the Membrane and the Boundary Layers

In DCMD process, the heat transfer is described by three steps as illustrated schematically in Fig. 10.5:

(i) heat transfer through the feed boundary layer
(ii) heat transfer through the membrane
(iii) heat transfer through the permeate boundary layer.

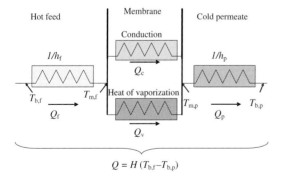

FIGURE 10.5 Heat transfer resistances in DCMD process.

The total heat flux through the membrane, Q_m, is due to two mechanisms (Fig. 10.5):

(i) conduction across the membrane material and its gas-filled pores (Q_c)
(ii) latent heat associated with the vapourized molecules (Q_v).

Therefore, the balance of energy is expressed as:

$$Q_m = Q_c + Q_v \qquad (10.34)$$

The heat transfer due to mass transfer is written as:

$$Q_v = J_w \Delta H_{v,w} \qquad (10.35)$$

The heat transfer by conduction through the membrane is given by the following integrated equation assuming a linear temperature distribution between $T_{m,f}$ and $T_{m,p}$ (Fig. 10.5):

$$Q_c = -k_m \frac{dT}{dx} = \frac{k_m}{\delta}(T_{m,f} - T_{m,p}) \qquad (10.36)$$

where k_m is the thermal conductivity of the membrane, which can be calculated by one of the following equations (Eq. (10.37) or Eq. (10.38)), x is the distance from the membrane surface (facing the feed), δ is the membrane thickness and $T_{m,f}$ and $T_{m,p}$ are the temperatures at the feed−membrane interface and at the membrane−permeate interface, respectively.

Various models have been used to evaluate the thermal conductivity of the membrane but the following equation is often used.

$$k_m = \varepsilon\, k_g + (1 - \varepsilon)k_p \qquad (10.37)$$

where k_p is the thermal conductivity of the membrane material and k_g is the thermal conductivity of the gas filling the membrane pores.

It was found that the Isostress model, defined by means of Eq. (10.38), was better than the one defined in Eq. (10.37) and is currently used in most MD studies [45].

$$k_m = \left[\frac{\varepsilon}{k_g} + \frac{(1-\varepsilon)}{k_p}\right]^{-1} \quad (10.38)$$

The reported thermal conductivities of PVDF material are 0.17–0.19 W/m K at 296 K and 0.21 W/m K at 348 K. Those reported for PP are 0.11–0.16 W/m K at 296 K and 0.2 W/m K at 348 K, while those of PTFE are 0.25–0.27 W/m K at 296 K and 0.29 W/m K at 348 K.

The calculated thermal conductivity of the PVDF membranes – GVHP and HVHP (Table 2.1) – are respectively 0.041 W/m K and 0.040 W/m K, while those of the PTFE membranes supplied by Gore (Table 2.1) are 0.043 W/m K for the supported PTFE/PP membrane, 0.031 W/m K and 0.027 W/m K for the unsupported PTFE membranes with 0.2 μm (TF200) and 0.45 μm (TF450) pore sizes, respectively. It was observed that the values of the membrane thermal conductivity calculated by Eq. (10.37) agree with the measured ones within 10% [24].

The heat transfer coefficient of the whole membrane (i.e. material plus gas-filled pores), h_m, can be written as:

$$h_m = k_m/\delta \quad (10.39)$$

The vapour heat transfer coefficient in the membrane is defined as:

$$h_v = \frac{J_w \Delta H_{v,w}}{T_{m,f} - T_{m,p}} \quad (10.40)$$

It is worth quoting that between 50% and 80% of energy is consumed as latent heat for water vapour production, while the remainder is lost by thermal conduction (Q_c). In fact, the heat used effectively in DCMD is the latent heat of evaporation associated with the mass flux (J_w), whilst the heat transferred by conduction across the membrane is considered as heat loss (Q_c). This heat loss becomes less significant at higher operating feed temperatures and the thermal efficiency, η, of a DCMD process defined by the following equation becomes higher.

$$\eta\,(\%) = \frac{Q_v}{Q_v + Q_c} \times 100 \quad (10.41)$$

Figures 10.1 and 10.5 show the temperature profile and heat transfer resistances involved in DCMD. The temperatures in the bulk aqueous solutions differ from the corresponding temperatures at the membrane–solution interfaces where vapour–liquid interfaces are formed. This is due to the presence of boundary layers adjoining the membrane surfaces at both feed and permeate sides.

On the feed side:

$$Q_f = h_f S(T_{b,f} - T_{m,f}) \quad (10.42)$$

On the permeate side:

$$Q_p = h_p S(T_{m,p} - T_{b,p}) \quad (10.43)$$

where h_f and h_p are the heat transfer coefficients in the feed and permeate boundary layers, respectively.

At steady state conditions:

$$Q_f = Q_m = Q_p \quad (10.44)$$

Thus, the heat transfer through the DCMD system 'feed aqueous boundary layer/membrane/permeate boundary layer' is summarized by the following equation:

$$h_f(T_{b,f} - T_{m,f}) = \frac{k_m}{\delta}(T_{m,f} - T_{m,p}) + J_w \Delta H_{v,w}$$
$$= h_p(T_{m,p} - T_{b,p})$$
$$= H(T_{b,f} - T_{b,p})$$
$$(10.45)$$

where H is the global heat transfer coefficient of the DCMD process. Therefore, the total heat flux (Q) can be written as a function of the total mass flux (J_w):

$$Q = \left[\frac{1}{h_f} + \frac{1}{k_m/\delta + J_w \Delta H_{v,w}/\Delta T_m} + \frac{1}{h_p}\right]^{-1} \Delta T$$
$$= H \Delta T$$
$$(10.46)$$

where ΔT is the bulk temperature difference between the feed and permeate ($T_{b,f} - T_{b,p}$) and ΔT_m is the transmembrane temperature difference ($T_{m,f} - T_{m,p}$).

The heat transfer through the boundary layers is recognized as the limiting factor of the DCMD efficiency. The temperature polarization coefficient (θ) is generally used to quantify the magnitude of the boundary layer resistances over the total heat transfer resistance. In other words, this coefficient reflects the reduction in the driving force (i.e. vapour pressure difference, Δp_w), which has a negative influence on the DCMD process productivity. It is defined as (Fig. 10.1 and 10.5):

$$\theta = \frac{T_{m,f} - T_{m,p}}{T_{b,f} - T_{b,p}} \qquad (10.47)$$

In the ideal case, θ should be equal to unity. However, usually it is lower than unity. Thus, θ indicates whether a DCMD module is poorly or well designed. If θ value is less than 0.2, the DCMD process is heat transfer limited and the DCMD module design is poor. If θ value is higher than 0.6, the DCMD process is mass transfer limited with a low membrane permeability.

If the heat transfer through both the feed and the permeate boundary layers is very high, the temperatures at the membrane surfaces approach those of the bulk phases. This means that θ approaches unity. In this case, the temperature polarization effect is negligible and the mass transfer resistance of the membrane controls the DCMD process.

If the heat transfer through both the feed and the permeate boundary layers is very small and the membrane permeability is large, the difference between the temperatures at the membrane surfaces and those at the bulk phases is large, and the transmembrane temperature difference is low. This means that the temperature polarization coefficient approaches zero. In this case, the temperature polarization effect is very significant and the heat transfer resistances of the boundary layers control the DCMD process.

Generally, for satisfactory DCMD modules, θ values ranges between 0.4 and 0.7. This means that between 30% and 60% of the applied temperature difference is dissipated in the thermal boundary layers. Enhancement of θ was observed when both the feed and permeate flow rates were increased, and the temperature, especially the feed temperature, was decreased. θ also is strongly dependent on membrane characteristics [4,6,16,46–48].

Several methods have been adopted to minimize the heat transfer resistances of the boundary layers, thus increasing the film heat transfer coefficients h_f and h_p. Those are the use of spacers, turbulence promoters or simply application of high feed and permeate flow rates as well as high stirring rates in order to generate turbulent flow regime in the DMCD membrane module. High flow rates through the membrane modules, resulting in increased Reynolds number (Re), improve heat transfer transport efficiency of the DCMD operation. Spacers can have benefits in DCMD, since the flow is destabilized creating eddy currents in the laminar regime and enhancing momentum, heat and mass transfer. The use of screen separators or net spacers in both the feed and permeate channels in DCMD systems reduce the polarization effects due to the appearance of turbulences and the formation of eddies and wakes near the membrane surfaces leading to an enhancement of the DCMD flux [45,49,50]. Generally, in DCMD, the film heat transfer coefficients are estimated from appropriate dimensionless empirical correlations that were developed for heat exchangers [14,15,31,33,45,49–52].

By using Eqs. (10.45) and (10.46), the temperature polarization coefficient can be written as a function of the heat transfer coefficients:

$$\theta = \frac{1}{1 + \dfrac{H}{h}} = 1 - \frac{H}{h} \qquad (10.48)$$

where

$$\frac{1}{h} = \frac{1}{h_f} + \frac{1}{h_p} \quad (10.49)$$

Due to water vapour transport though the membrane pores, the concentration of the non-volatile solutes at the membrane surface ($C_{m,f}$) becomes higher than the concentration in the bulk feed aqueous solution ($C_{b,f}$). This is also schematically illustrated in Fig. 10.1. This phenomenon is called concentration polarization and results in a reduction of both the driving force (i.e. water vapour pressure difference) and the permeate DCMD flux.

The concentration polarization coefficient, ξ_s, of a given non-volatile solute s present in the aqueous feed solution is generally defined as:

$$\xi_s = \frac{x_{s,m,f}}{x_{s,b,f}} \quad (10.50)$$

where x_s is the mole fraction of the solute s in the aqueous feed solution.

The concentration polarization coefficient is defined also as a function of the solute concentration as:

$$\xi_s = \frac{C_{m,f}}{C_{b,f}} \quad (10.51)$$

where $C_{b,f}$ and $C_{m,f}$ are the solute concentration at the bulk feed solution and at the feed–membrane interface, respectively.

The retained non-volatile compounds accumulate in proximity of the membrane surface and the concentration gradient between the membrane interface ($C_{m,f}$) and the bulk of the feed solution ($C_{b,f}$) leads to a diffusive flow of the solutes from the membrane surface to the bulk phase. Steady-state concentration profile is established when the convective transport of solutes to the membrane surface is counterbalanced by a diffusive flux of the retained compounds back to the bulk solution. Nernst film model that neglects the eddy and thermal diffusions in relation to the ordinary diffusion is frequently used in DCMD to relate the solute concentration at the membrane surface to that at the bulk solution if the non-volatile solutes are retained by the membrane:

$$\xi_s = \frac{C_{m,f}}{C_{b,f}} = \exp(J_w/k_s) \quad (10.52)$$

where k_s is the solute mass transfer coefficient for the diffusive mass transfer through the boundary layer ($k_s = D/\delta$), D is the molecular diffusivity and δ is the thickness of the boundary layer. The mass transfer coefficient k_s can be estimated from Sherwood number (Sh) via an appropriate dimensionless empirical correlation for mass transfer derived employing an analogy with an empirical correlation for heat transfer [14,15]. A number of empirical correlations have been developed and used in DCMD applications. These correlations are generally expressed in the form:

$$Sh = f(Re, Sc) = a Re^b Sc^c \left(\frac{d}{L}\right)^d \quad (10.53)$$

where L is the length of the feed channel, Sh is Sherwood number ($Sh = (k\,d)/D$, d: hydraulic diameter), Re is Reynolds number (Re $= (\rho\,u\,d)/\mu$, ρ: density, μ: viscosity, u: feed velocity) and Sc is Schmidt number ($Sc = \rho/(\mu D)$). For example, in laminar feed flow regime $b = c = d = 0.33$.

In general, compared to the temperature polarization, the concentration boundary layers adjoining the membrane result in a smaller contribution of concentration polarization to the overall mass transfer resistance [16].

The interfacial temperatures and concentrations cannot be measured directly in DCMD membrane modules. However, if the heat and mass transfer coefficients are known, these temperatures can be evaluated. For heat transfer coefficient, the following semi-empirical equation can be used [14–16,22,23,51,52].

$$Nu = f(\mathrm{Re}, \mathrm{Pr}) = a \mathrm{Re}^b \mathrm{Pr}^c \left(\frac{d}{L}\right)^d \quad (10.54)$$

where Nu is Nusselt number ($Nu = (h\, d)/k$, k: thermal conductivity) and Pr is Prandtl number ($\mathrm{Pr} = (\mu\, C_p)/k$).

The concentration at the membrane surface can be obtained from Eq. (10.51), whereas the temperatures at the feed–membrane and membrane–permeate interfaces can be determined from the following equations obtained from Eq. (10.45):

$$T_{m,f} = \frac{\frac{k_m}{\delta}\left(T_{b,p} + \frac{h_f}{h_p}T_{b,f}\right) + h_f T_{b,f} - J_w \Delta H_{v,w}}{\frac{k_m}{\delta} + h_f\left(1 + \frac{k_m}{\delta h_p}\right)}$$

$$(10.55)$$

$$T_{m,p} = \frac{\frac{k_m}{\delta}\left(T_{b,f} + \frac{h_p}{h_f}T_{b,p}\right) + h_p T_{b,p} + J_w \Delta H_{v,w}}{\frac{k_m}{\delta} + h_p\left(1 + \frac{k_m}{\delta h_f}\right)}$$

$$(10.56)$$

Both the temperature polarization and concentration polarization produce a decrease of the driving force (i.e. vapour pressure difference). Therefore, both coefficients θ and ξ_s can be combined in one coefficient termed vapour pressure polarization coefficient (ψ) that is defined as:

$$\psi = \frac{\Delta p_w}{\Delta p_{w,b}} = \frac{p_{m,f} - p_{m,p}}{p_{b,f} - p_{b,p}} \quad (10.57)$$

where $\Delta p_{w,b}$ is the externally applied driving force (i.e. bulk water vapour pressure difference) related to the temperature difference.

It is to be noted that there is a temperature drop in the tangential flow membrane modules. This means that the inlet feed temperature is different from the outlet feed temperature depending on the applied feed flow rate. Therefore, a local permeate flux and local temperature and concentration polarization coefficients must be considered. However, when a permeation cell with stirring is used in laboratory scale, no change in the bulk solution temperature from cell inlet to cell outlet was detected.

By using Eqs. (10.3) and (10.57) the DCMD permeate flux can be given as a function of the bulk vapour pressure difference:

$$J_w = B'_w \Delta p_{w,b} = B'_w \psi^{-1} \Delta p_w = B_w \Delta p_w$$

$$(10.58)$$

where B'_w is the global mass transfer coefficient of the membrane taking into consideration the effects of the feed and permeate boundary layers (i.e. feed and permeate mass transfer resistances), the DCMD operating conditions and the characteristics of the used membrane.

Based on Eqs. (10.24), (10.25), (10.18–10.32), the porosity (void volume fraction open to DCMD vapour flux, ε) of the membrane must be as high as possible in order to enhance the DCMD membrane permeability. Membranes with higher porosity can provide larger spaces for evaporation. Therefore, it is generally agreed upon that the higher membrane porosity results in higher permeate flux. In most cases, ε lies between 30 and 85%. Another advantage of using membranes with high porosities is the low conductive heat loss (Eqs. (10.37, 10.38)) since the thermal conductivity of the gases entrapped within the membrane pores is an order of magnitude smaller than that of the hydrophobic polymer used for membrane preparation.

The permeate flux is inversely proportional to the membrane thickness. This factor plays a significant role on the water vapour transfer and the membrane must be as thin as possible in order to obtain a high DCMD permeability. However, when a membrane is thin, a large amount of heat will be transferred by conduction through the membrane leading to low heat efficiency of the DCMD process, as explained previously based on Eq. (10.41). Therefore, a compromise should be made between the mass and the heat transfer, by properly adjusting the membrane thickness. In general, the thickness of the commercial membranes is not optimized for the use in MD. One advantage of using multi-layered membrane (i.e. hydrophobic and hydrophilic layers) is that a high mass transport is enabled by making the hydrophobic layer as thin as possible, while a low heat transfer is enabled by making the overall membrane thickness as thick as possible. This was confirmed by the use of porous compositae hydrophobic/hydrophilic membranes with a very thin hydrophobic layer that was as low as 5 μm [7].

The tortuosity factor is the measure of the deviation of the pore structure from straight cylindrical pores normal to the surface. This factor gives the average length of the pores compared to the membrane thickness, and must be small because it is inversely proportional to the permeability of the membrane (Eq. (10.17—10.25)). Generally, the membrane pores do not go straight across the membrane and the diffusing water molecules in vapour phase must move along tortuous paths reducing the membrane permeability. It is worth quoting that in order to predict the DCMD permeate flux, a value of 2 is frequently assumed for the membrane tortuosity factor. In various theoretical models the tortuosity factor was used as an adjusting parameter to obtain better agreement between the simulated and the experimental permeate fluxes. In fact, it is difficult to measure the pore tortuosity. However, this factor was estimated based on the gas permeation test combined with the measurement of the porosity (see Chapter 8) [22].

It is generally agreed upon that the DCMD permeate flux increases with an increase in the membrane pore size but with some limitations, i.e. the pore size should allow a sufficiently high liquid entry pressure (LEP) in order to avoid wetting of membrane pores. In other words, the maximum pore size of the membrane must be as small as possible in order to guarantee a high LEP value, which is characteristic of each membrane (see Chapter 8). Again, the MD membrane should exhibit an optimum pore size.

Based on the above cited Eq. (10.21), the change of the DCMD membrane permeability with membrane pore size is related to the change in mechanism of water vapour transport from the Knudsen mechanism for small pore sizes to the combined mechanism for larger pore sizes and then to the ordinary molecular diffusion mechanism. Moreover, membranes may exhibit pore size distributions rather than uniform pore sizes and more than one mechanism can take place simultaneously depending on the pore sizes and the DCMD operating conditions. The effect of pore size distribution on the DCMD flux has been studied [22,30,37]. It was observed that the predicted DCMD flux assuming uniform pore size was quite similar to that calculated using the pore size distribution when commercial membranes with narrow pore size distributions are considered. However, when using laboratory made membranes having broader pore size distributions the difference between the calculated permeate fluxes is larger. It is to be noted that the effect of the pore size distribution on the DCMD flux has not yet been clarified. Further studies are needed to clarify this effect using different types of membranes.

It is to be mentioned here that the DCMD process has been applied effectively to determine the permeability (B_w) of hydrophobic porous membranes with a single layer and compositae hydrophobic/hydrophilic porous membranes, when distilled water was used as feed, transmembrane bulk temperature difference was low ($T_{b,f} - T_{b,p} \leq 10\,K$) and the mean temperature $\left(T_m = \dfrac{T_{b,f} + T_{b,p}}{2}\right)$ ranged from 20 °C to 55 °C [16,17,22]. As stated in Eq. (10.11) the permeate water vapour flux (J_w) is linearly related to its partial pressure difference across the membrane pores. From Eqs. (10.11–10.13, 10.37, 10.45, 10.49), the following equation can be derived:

$$\dfrac{T_{b,f} - T_{b,p}}{J_w \Delta H_{v,w}} = \dfrac{1 + \dfrac{k_m}{\delta h}}{B_w \Delta H_{v,w} \left(\dfrac{dp}{dT}\right)_{T_m}} + \dfrac{1}{h} \quad (10.59)$$

The above equation can be used for the analysis of the DCMD experimental results for which the bulk feed and permeate temperatures together with the DCMD permeate flux are known. As shown in Fig. 10.6, a fit to a linear function of $(T_{b,f} - T_{b,p})/(J_w \Delta H_{v,w})$ versus $[\Delta H_{v,w} (dp/dT)_{Tm}]^{-1}$ should yield an intercept of $1/h$ and a slope of $(1/J_w \Delta H_{v,w})(1+k_m/\delta \cdot h)$, from which B_w can be obtained if the thermal conductivity k_m is known or calculated using Eqs. (10.37, 10.38). The obtained B_w values were $(2.9 \pm 0.1) \times 10^{-7}$ kg m^{-2}s^{-1}Pa^{-1} for the membrane

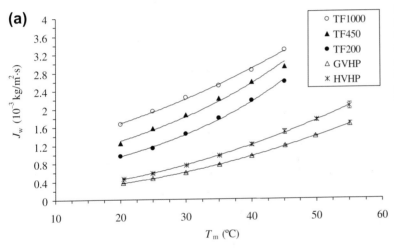

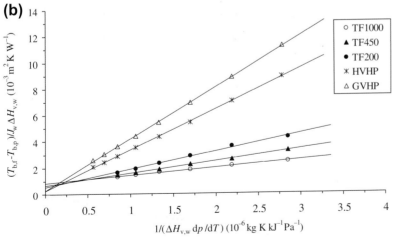

FIGURE 10.6 DCMD flux (J_w) versus mean temperature (T_m) maintaining the transmembrane temperature difference fixed at 10 K (a) and corresponding plots of Eq. (10.59) of these experimental data for the commercial membranes GVHP, HVHP, TF200, TF450 and TF100 (Table 2.1).

GVHP, $(3.6 \pm 0.1) \times 10^{-7}$ kg m^{-2}s^{-1}Pa^{-1} for the membrane HVHP, $(10.7 \pm 0.3) \times 10^{-7}$ kg m^{-2}s^{-1}Pa^{-1} for the membrane TF200, $(13.6 \pm 0.4) \times 10^{-7}$ kg m^{-2}s^{-1}Pa^{-1} for the membrane TF450 and $(23.9 \pm 0.5) \times 10^{-7}$ kg m^{-2}s^{-1}Pa^{-1} for the membrane TF1000. The obtained heat transfer coefficient h was 3522.4 W/m^2 K for the membrane GVHP, 4153.7 W/m^2 K for the membrane HVHP, 1937.2 W/m^2 K for the membrane TF200, 1561.8 W/m^2 K for the membrane TF450 and 1188.9 W/m^2 K for the membrane TF1000.

The DCMD process has also been proposed as an experimental method to estimate the thickness of the hydrophobic layer and the hydrophilic layer of porous compositae hydrophobic/hydrophilic membranes [7].

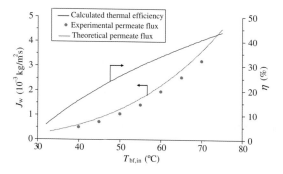

FIGURE 10.7 Effect of the feed inlet temperature ($T_{b,f,in}$) on the DCMD permeate flux (J_w) and thermal efficiency (η) of the commercial membrane module Accurel® S6/2 MD020CP2N (see Table 2.2). Distilled water was used as feed with flow rate 293 l/h, permeate flow rate 245 l/h and permeate inlet temperature $T_{b,p,in} = 20\,°C$.

EXPERIMENTAL: EFFECTS OF DCMD PROCESS CONDITIONS

Effects of Feed Temperature

In DCMD process the applied feed temperature ($T_{b,f}$) commonly ranges between 20 °C and 90 °C. The highest temperature used is below the boiling point of the feed aqueous solution. The effect of the feed temperature on the permeate flux has widely been investigated for different DCMD modules and systems [3,4,14]. This operating parameter is the most important factor affecting the DCMD permeate flux. As illustrated in Fig. 10.7, e.g., an exponential increase of the permeate flux was observed with the increase of the feed temperature maintaining the constant permeate temperature.

Based on Eqs. (10.3–10.5), the observed trend is due to the exponential increase of the vapour pressure of the feed aqueous solution with temperature, which enhances the driving force. For example, increasing the feed temperature from 50 °C to 70 °C can enhance the DCMD permeate flux by more than 3-fold. By using Eqs. (10.25), (10.54–10.56) reported previously, reasonably good agreement was obtained between the predicted and the experimental DCMD permeate flux. It is to be noted that no feed temperature effect was detected on the non-volatile solute rejection factor. If the temperature of the feed stream is measured at the entrance and exit of the membrane module, the temperature difference (ΔT_{module}) can be determined, which allows to evaluate the total heat transferred from the feed to the permeate. As well, the amount of heat consumed by the DCMD process (Q_v) can be estimated from the permeate flux. The difference corresponds to the heat loss by conduction. It was observed that the heat loss was very sensitive to the measured ΔT_{module} and depended on the DCMD permeate flux [12]. The ratio of the heat loss by conduction (Q_c) to the total heat transferred from the feed to the permeate (Q_m) decreases linearly with the permeate flux. When the production rate of distilled water is kept constant, the heat loss is higher for greater ΔT_{module}. It was reported that for a permeate flux of 97.2 kg/m^2·h, 22% of heat was lost by conduction at ΔT_{module} of

0.3 °C and it increased to 42% at ΔT_{module} of 0.4 °C.

In general, in the DCMD process the temperature polarization effect is increased at higher feed temperatures, reducing the permeate flux considerably. As can be seen in Fig. 10.8, the temperature polarization coefficient, θ, is reduced with increasing the feed temperature. However, as shown in Fig. 10.7 the internal evaporation efficiency or thermal efficiency, η, defined in Eq. (10.41) is higher at higher feed temperatures. Therefore, it is advisable to work under high feed temperatures although the effect of the temperature polarization is enhanced.

Effect of Permeate Temperature

In general, in DCMD studies the permeate temperature ($T_{b,p}$) varies between 10 °C and 40 °C. As may be expected from Eq. (10.3), an increase in permeate temperature results in a reduction of the transmembrane driving force, leading to a decrease of the permeate flux. This can be expected from Eq. (10.6). Both exponential decrease and linear decrease of the DCMD permeate flux with the increase of the permeate temperature was observed experimentally. These two different tendencies may be attributed to the temperature variation along the membrane module as well as to the characteristics of the membrane module. Figure 10.9 shows an example for the decrease

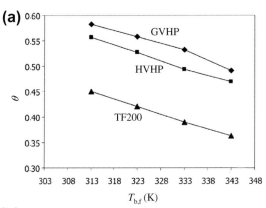

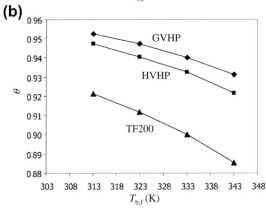

FIGURE 10.8 Effect of the feed temperature ($T_{b,f}$) on the temperature polarization, θ, for laminar, $h = 1498.9$ W/m²·K (a) and turbulent $h = 21423.4$ W/m²·K (b) flow conditions in DCMD using the membranes GVHP, HVHP and TF200 (Table 2.1). *Source: Reprinted from [45]. Copyright 2003, with kind permission from Elsevier.*

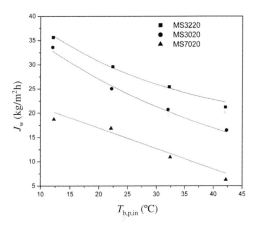

FIGURE 10.9 Effect of the permeate inlet temperature ($T_{b,p,in}$) on the DCMD permeate flux (J_w) of three flat sheet commercial membranes MS3220, MS3020 and MS7020 (Membrane Solutions, Shanghai, China). Arsenic aqueous solution of concentration 394 ppb was used as feed with flow rate 0.028 m/s, permeate flow rate 0.052 m/s and feed inlet temperature $T_{b,f,in} = 60-61$ °C. *Source: Reprinted from [53]. Copyright 2010, with kind permission from Elsevier.*

of the DCMD permeate flux with the increase of the permeate temperature. If the feed temperature is kept constant, while varying the permeate temperature, it is more likely to have an increase of the permeate flux tending to approach an asymptotic value at high temperature differences.

It is to be noted that ambient permeate temperature is commonly used in DCMD applications. However, the permeate temperature increases along the membrane module length from the inlet to the outlet. This variation depends strongly on the flow rate of the permeate and the design of the membrane module affecting the driving force and consequently the permeate flux. Therefore, cooler(s) must be used as shown in Fig. 10.2.

For the same temperature difference between the feed and the permeate, the effect of the permeate temperature on the permeate flux is more than 2 fold smaller than the effect of the feed temperature.

When the temperature difference is maintained constant, the change of the DCMD permeate flux with the change in the mean temperature (T_m) becomes as follows:

$$J_w = \frac{A}{T_m} \exp\left(-\frac{\Delta H_{v,w}}{RT_m}\right) \quad (10.60)$$

where A is a fitting parameter that depends on the membrane module and the DCMD operating conditions. Figure 10.10 illustrates as an example the variation of the DCMD permeate flux of different flat-sheet membranes with the mean temperature together with the fitting of experimental data to Eq. (10.60). As can be seen there is a very good fitting of the experimental data to Eq. (10.60).

When the mean temperature (T_m) is maintained constant, it may be expected that the DCMD permeate flux increases linearly with the applied temperature difference (see Eq. (10.11)). However, in this case, there is also the asymmetric effect of the temperature polarization (i.e. the thickness of the feed and permeate boundary layers adjoining the membrane

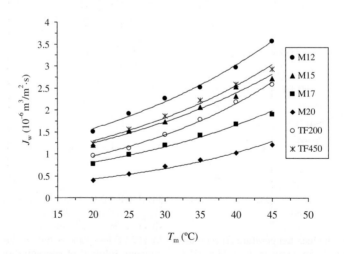

FIGURE 10.10 Effect of the mean temperature (T_m) on the DCMD permeate flux (J_w) of two flat sheet commercial membranes TF200 and TF450 (see Table 2.1) and laboratory made porous hydrophobic/hydrophilic membranes (M12, M15, M17 and M20). Distilled water was used as feed with a stirring rate of 500 rpm (Lewis cell, see Fig. 9.3(a)) and at a bulk temperature difference of 10 K. *Source: Reprinted from [6]. Copyright 2005, with kind permission from Elsevier.*

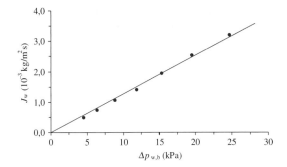

FIGURE 10.11 DCMD permeate flux (J_w) as a function of the driving force, vapor pressure difference ($\Delta p_{w,b}$), of the commercial membrane module Accurel® S6/2 MD020CP2N (see Table 2.2). Distilled water was used as feed with flow rate 293 l/h and permeate flow rate 245 l/h.

surfaces are different) [16]. In fact, there was a monotonic increase of the DCMD permeate flux with the increase of the vapour pressure difference as depicted in Fig. 10.11. The slope of the observed straight line permits the evaluation of the global mass transfer coefficient of the membrane module B'_w (see Eq. (10.58)). If the temperature polarization coefficient is determined by means of Eqs. (10.55) and (10.56), the DCMD coefficient of the membrane (i.e. membrane permeability, B_w) can be evaluated (Eq. (10.58)).

It is worth quoting that the temperature polarization coefficient decreases with increasing feed, permeate and mean temperature in the DCMD process. Moreover, if large differences exist between the temperatures of the feed and permeate aqueous solutions at the inlet and the outlet of the membrane module, the use of the logarithmic mean vapour pressure difference is recommended:

$$\Delta p_{w,b} = \frac{\Delta p_{w,1} - \Delta p_{w,2}}{\ln\left(\dfrac{\Delta p_{w,1}}{\Delta p_{w,2}}\right)} \quad (10.61)$$

where $\Delta p_{w,1} = p_w(T_{b,f,out}) - p_w(T_{b,p,in})$ and $\Delta p_{w,2} = p_w(T_{b,f,in}) - p_w(T_{b,p,out})$ are the water vapour pressure difference between the feed and the permeate at both sides of the membrane module as indicated in Fig. 2.

Effects of Feed and Permeate Flow Rates

To reduce both temperature and concentration polarization effects, the feed and permeate flow rates (i.e. feed circulation velocity or feed stirring rate in case of Lewis cell) must be increased. In this case, the heat transfer coefficient of the membrane module, H in Eq. (10.46), increases resulting in higher DCMD permeate fluxes. In fact, the heat transfer coefficients in the feed and/or permeate boundary layers (h_f, h_p) are enhanced. When the flow rate is increased the temperature and non-volatile solute concentration at the membrane surface become closer to the corresponding bulk temperature and bulk concentration, resulting in higher transmembrane temperature difference and greater DCMD permeate flux. However, the flow rate must be varied with due precautions in order to avoid membrane pore wetting as the transmembrane hydrostatic pressure must be lower than the LEP and, at the same time, to assure working under turbulent flow regime in order to obtain high productivity. Figure 10.12 shows the effects of the feed and permeate flow rates (i.e. stirring rate in Lewis cell) on the temperature polarization coefficient and on the DCMD productivity of various types of membranes. Asymptotic values are reached at high flow rates indicating that the flow regime at high stirring rates is turbulent.

The trends presented in Fig. 10.12 depend on the membrane module and may differ from one DCMD system to another. In some studies, a practically linear increase of the DCMD flux with the flow rate was observed [31,54]. Probably, the DCMD process was operated under laminar or transitional (laminar/turbulent) flow regimes. In other studies, an only slight enhancement was observed between the DCMD permeate flux and the feed and permeate flow rates.

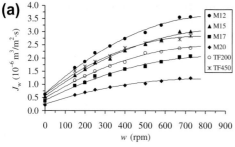

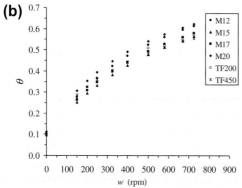

FIGURE 10.12 Effect of feed and permeate stirring rate (w) on the DCMD permeate flux (J_w) (a) and on the corresponding temperature polarization coefficient (θ) of two flat sheet commercial membranes TF200 and TF450 (see Table 2.1) and laboratory made porous hydrophobic/hydrophilic membranes (M12, M15, M17 and M20). Distilled water was used as feed with a stirring rate of 500 rpm (Lewis cell) and a bulk temperature difference of 10 K. *Source: Reprinted from [6]. Copyright 2005, with kind permission from Elsevier.*

Effects of Solute Concentration in the Feed Aqueous Solution

The increase of the non-volatile solute concentration in the feed aqueous solution results in a reduction of the DCMD permeate flux. This behaviour is attributed to the decrease of the water vapour pressure, the driving force, with the addition of non-volatile solute in water due to the decrease in water activity in the feed as shown by Eqs. (10.3–10.10). Furthermore, there is also the contribution of the concentration polarization effect as reported previously by means of Eqs. (10.50–10.52).

However, the latter contribution is very small compared to the temperature polarization effect. Figure 10.13 shows, as an example, the effects of salt (sodium chloride, NaCl) and sucrose concentrations on the DCMD permeate flux as well as on the driving force ($\Delta p_{w,b}$). It is reported that when the sucrose concentration

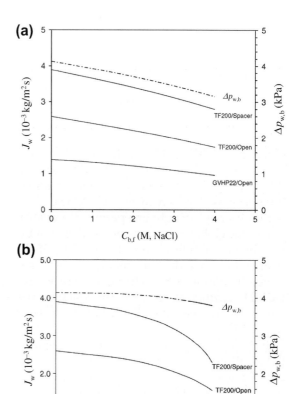

FIGURE 10.13 Effect of the feed concentration ($C_{b,f}$) on the DCMD permeate flux (J_w) and on the driving force ($\Delta p_{w,b}$) when using as non-volatile solute NaCl (a) and sucrose (b), two commercial membranes GVHP and TF200 (see Table 2.1), open and spacer for the membrane TF200, a bulk feed temperature 36 °C and a bulk permeate temperature 16 °C. *Source: Reprinted from [55]. Copyright 2007, with kind permission from Elsevier.*

rises from 0% to 47% (by weight) in water, the decrease of $\Delta p_{w,b}$ is lower than that of NaCl when the concentration of NaCl in water rises from 0 to 4 M.

It is to be noted that also the thermal efficiency (η) as well as the temperature polarization coefficient (θ) are affected by the variation of the feed concentration [55]. A reduction of both η and θ were observed with the increase of the NaCl concentration in the feed aqueous solutions and only a slight decrease of η was observed when sucrose aqueous solutions with different concentrations were used. This was attributed to the different increase of the viscosity of the feed solution with the concentration affecting the heat transfer coefficient in the feed boundary layer. In other words, a large decrease of the DCMD permeate flux with the increase of sucrose concentration is due to the feed resistance caused by the increase of the viscosity of the feed aqueous solution. In the case of NaCl aqueous solutions, the decrease of the DCMD permeate flux is mainly due to the decrease of the driving force as mentioned earlier.

It was reported that a salt concentration increase by more than 5-fold decreased the permeate flux by only 1.15 fold. In some studies it was found that the DCMD permeate flux decreased in an approximately linear dependence with the salt concentration [56]. This linear behaviour may depend on the range of the salt concentration, the other DCMD operating conditions and on the membrane module.

A DCMD permeate flux decline from 5.83 kg/m^2·h to 1.66 kg/m^2·h was observed during the first 3 days of DCMD operation with 0.58 wt% NaCl aqueous solution at a feed temperature of 50 °C and a permeate temperature of 20 °C using the membrane TF450 (see Table 2.1) [57]. By employing three different feed solutions such as raw seawater, raw water pretreated by MF and 3% NaCl aqueous solution, it was found that the pretreatment of the raw water increased the permeate flux by about 25% and the DCMD permeate flux obtained from the 3% NaCl feed solution was twice as high as that obtained from raw seawater. The test was carried out for a period of 160 h applying a feed temperature of 45 °C and a permeate temperature of 20 °C. The obtained concentration of the permeate water was 140 ppm, 8.4 ppm and 1.7 ppm when using feed concentrations of 50,000 ppm, 30,000 ppm and 20,000 ppm, respectively.

It is worth quoting that there is a considerable change in performance when seawater is treated. Scaling potential due to sparingly dissolved soluble salts is a critical problem in the DCMD process, particularly when the process is operated at high recovery factors.

DCMD can be used for the treatment of highly concentrated aqueous solutions of salts without suffering large drop in productivity compared to other membrane separation processes such as RO commonly used in industrial desalination [3,4,12,14,54]. However, when DCMD is applied for the treatment of solutions of very high salt concentration, the behaviour is very different from that observed for dilute aqueous solutions of the salt. The DCMD permeate flux also decreases as the salt concentration of the feed solution increases and at very high salt concentrations crystallization occurs [58,59,60,61].

Membrane crystallization technology was proposed for concentration of aqueous salt solutions above their saturation limit. In this case, a supersaturated environment is attained and crystals nucleate and grow. Figure 10.14 illustrates the presence of crystals with long needle shape structure of mixed calcite and gypsum grown on the surface of polypropylene membrane (Accurel PP hollow fibre, Table 2.2). A progressive decrease of the DCMD productivity was observed (i.e. 45% reduction after 35 h of uninterrupted operation) [61]. Two-step cleaning procedure with aqueous solutions of citric acid and sodium hydroxide — each for 20 min — was proposed to completely restore the membrane and the DCMD permeate flux

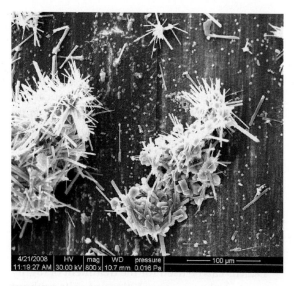

FIGURE 10.14 Scanning electron microscopy (SEM) image of mixed calcite and gypsum crystals grown on the surface of Accurel PP hollow fibre membrane (see Table 2.2). *Source: Reprinted from [61]. Copyright 2010, with kind permission from Elsevier.*

to its initial value of $2.5 \, l/m^2 \cdot h$ obtained with 40 °C feed temperature and 20 °C permeate temperature.

In the case of DCMD desalination, the permeate flux continues to decrease until supersaturation is reached and crystals start to precipitate. Scale formation fouling or crystallization fouling results from deposition or growth of crystals on the membrane surface, especially when treating aqueous solutions of highly concentrated salt. This fouling induces clogging of some membrane pores and reduction of the permeate flux [62–65].

To confirm that scaling occurs only on the membrane surface and it is a reversible fouling, membrane permeability should be measured after washing the membrane with distilled water and the results must be compared to the initial membrane permeability. In some cases, no significant difference was observed between the initial membrane permeability and the permeability after the membrane was washed, confirming that scaling was reversible and it was taking place only on the membrane surface.

In general, compared to other membrane separation processes, fouling is less studied in MD. Undesirable layers may be formed on the feed membrane surface and/or membrane pores by suspended particles, corrosion products, biological growth and variety of crystalline deposits increasing the risk of membrane pore wetting and/or plugging of the pore entrances causing decay of MD production and quality. Fouling in MD may be different from that encountered in the pressure-driven membrane processes (i.e. RO, NF, UF, etc.). More details about fouling in MD may be found in Chapter 14.

It is to be noted that a large part of the reported studies on DCMD focused on the conventional method of experimentation, which involves changing one of the independent parameters while maintaining the others constant as shown in Figs. 7–13. This 'classical' method of experimentation requires many experimental runs, which are time consuming, ignores interaction effects between the operating parameters of the DCMD process and leads to a low efficiency in optimization. Response surface methodology (RSM) that involves statistical design of experiments (DoE), in which all factors are varied simultaneously over a set of experimental runs, has been applied in the DCMD process in order to study the interactions between the operating parameters [66]. RSM is a collection of mathematical and statistical techniques useful for developing, improving and optimizing processes, and can be used to evaluate the relative significance of several affecting factors even in the presence of complex interactions. Furthermore, the developed RSM models can be used for optimization of the DCMD performance as well as optimization of energy efficiency. The central compositae design of orthogonal type has been considered in desalination by DCMD using different membrane types (both laboratory fabricated and

TABLE 10.2 Reported DCMD Permeate Flux (J_w) of Different Types of Capillary and Hollow Fibre Commercial Membranes Shown in Table 2.2

Membrane trade name	J (10^{-3} kg/m²s)	Observation	Reference
PP Accurel® S6/2 MD020CP2N	8.56 – 6.36 9.25	$T_{b,f}$ = 80 °C; $T_{b,p}$ = 20 °C; distilled water (reduction of permeate flux) $T_{b,f}$ = 85 °C; $T_{b,p}$ = 20 °C; tap water (605–650 μs/cm).	[138]
	8.1 – 6.36	3 years DCMD operation; $T_{b,f}$ = 80 °C; $T_{b,p}$ = 20 °C; distilled water and tap water $CaCO_3$ (cleaning procedures and reduction of permeate flux).	[65]
	3	Distilled water as feed in lumen side; $T_{b,f}$ = 70 °C; $T_{b,p}$ = 20 °C.	[139]
	3.61	35 g/l NaCl; $T_{b,f}$ = 70 °C; $T_{b,p}$ = 15 °C.	[36]
	3.47	5 wt% NaCl; $T_{b,f}$ = 85 °C; $T_{b,p}$ = 20 °C.	[140]
	3.30 2.08 1.74	Distilled water as feed; $T_{b,f}$ = 80 °C; $T_{b,p}$ = 20 °C. 10 wt% NaCl; $T_{b,f}$ = 70 °C; $T_{b,p}$ = 20 °C. 20 wt% NaCl; $T_{b,f}$ = 70 °C; $T_{b,p}$ = 20 °C.	[51]
	2.65 2.51 2.36	$T_{b,f}$ = 62 °C; $T_{b,p}$ = 22 °C. distilled water NaCl aqueous solution (78 μs/cm) NaCl aqueous solution (135 μs/cm).	[16]
MD080CO2N	0.83	35 g/l NaCl; $T_{b,f}$ = 55 °C; $T_{b,p}$ = 15 °C.	[36]
MD020TP2N	0.97	35 g/l NaCl; $T_{b,f}$ = 70 °C; $T_{b,p}$ = 15 °C.	
PP capillary Accurel® Enka d_p = 0.43 μm	4.63	0.05 M NaCl; $T_{b,f}$ = 42.5 °C; $T_{b,p}$ = 22.5 °C.	[81]

commercial membranes: TF200, TF450 and GVHP, see Table 2.1). It was observed that the interactions between parameters affect the DCMD permeate flux. Figure 10.15 shows an example of such interactions. Canonical analysis was employed for optimization. It was found that the optimum conditions of the DCMD process depended on the type of the used membrane and the predicted permeate fluxes by RSM were confirmed experimentally.

Optimization of different DCMD installations and systems is still to be done in order to study further the interaction effects between parameters, to increase DCMD performance and to decrease energy consumption. This can be carried out applying RSM. These mathematical–statistical tools are also a promising alternative for final fabrication of industrial DCMD membrane modules and DCMD pilot plants.

OSMOTIC DISTILLATION (OD)

Osmotic distillation (OD) is an isothermal process patented at the end of 1990s [67]. In this process the driving force is also the vapour pressure difference between both sides of

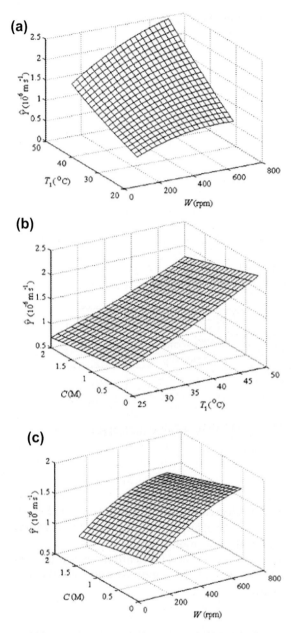

FIGURE 10.15 Response surface plots of the DCMD permeate flux of the membrane TF200 (see Table 2.1) as a function of: (a) stirring rate (w in Lewis cell) and temperature of the feed solution ($T_{b,f}$) at NaCl concentration $C_{b,f} = 1.1$ M, (b) $T_{b,f}$ and $C_{b,f}$ at $w = 437.5$ rpm, and (c) w and $C_{b,f}$ at $T_{b,f} = 37.5\,°C$. Source: Reprinted from [66]. Copyright 2007, with kind permission from American Chemical Society.

a non-wetted porous hydrophobic membrane. The same membrane modules as in DCMD can be employed in OD. The difference between DCMD and OD is the way to generate the driving force. In OD the vapour pressure gradient results from a concentration gradient from the feed to the permeate side of the membrane — generated using an extracting solution on the permeate side of the module. The examples of the extracting solutions are hypertonic solutions or solutions containing osmotic pressure agents. Both water vapour and the volatile components present in the aqueous feed solution are transferred through the membrane to the permeate side where a liquid capable of absorbing those components is circulating. The salts used as osmotic pressure agents are in general sodium chloride (NaCl because of its low cost), magnesium chloride ($MgCl_2$), calcium chloride ($CaCl_2$) or magnesium sulphate ($MgSO_4$). Among many salts, potassium salts of ortho- and pyrophosphoric acid offer several advantages, including low equivalent weight, high water solubility, steep positive temperature coefficients of solubility and safe use in foods and pharmaceuticals [68,69].

OD can be performed at ambient temperature, which is particularly attractive for heat sensitive products such as those present in food processing (e.g. concentration of fruit and vegetables juices) and pharmaceutical industries (e.g. concentration of miscible solvent extracts of intracellular products from fermentation broths and selective removal of solvent to concentrate antibiotics, vaccines, hormones and other heat-sensitive biochemicals). It is also applied for the treatment of dilute aqueous solution of non-volatile solutes such as sugars, polysaccharides, amino acids, proteins or carboxylic acid salts. The objective is to concentrate the solution by removing the solvent water.

OD can also be used to concentrate aqueous solutions of volatile compounds by selective removal of water, leaving a concentrate of flavour and fragrance compounds. This can be achieved because at low operating temperatures the vapour pressure of these compounds is depressed relative to that of water and also because many flavour/fragrance volatiles are hydrophobic, rendering them poor solubility in concentrated saline solutions [68]. Because of the nature of the driving force used in OD process (i.e. ambient temperature), flavour and fragrance compounds can be conveniently preserved. Moreover, flavour and fragrance have high molecular weights and a low diffusive permeabilities of these compounds through the membrane are expected. Similarly, OD can be used to reduce alcohol content in wines produced from grapes with high sugar.

As it was reported previously, the addition of non-volatile solutes in water reduces the water vapour pressure and the decrease in vapour pressure can be enhanced by using higher solute concentrations. If the temperature of both feed and permeate solutions that are in direct contact with the membrane is maintained constant (e.g. at ambient temperature) and the water vapour pressure is reduced on the permeate side by highly concentrated non-volatile solutes, water will evaporate at the liquid–vapour interface formed at the feed side of membranes pores, cross the pores as vapour and finally condense at the liquid–vapour interface formed on the permeate side (Fig. 10.16). It is to be noted that

OD is not a purely mass transfer operation. Both evaporation and condensation are taking place inside the membrane module. Therefore, a temperature difference at the membrane interfaces is created, even if the bulk temperatures of the two liquids are maintained equal [70].

A pilot plant for the concentration of fruit juice and vegetable juice by OD was built in Australia with a capacity of 50 l/h (maximum capacity 100 l/h) and 65–70 wt% product [68].

Osmotic membrane distillation (OMD) or osmotic direct contact membrane distillation is also a variant of DCMD process. In this case, the driving force includes both the transmembrane temperature gradient and the concentration gradient that are generated employing an hypertonic solution on the permeate side of the membrane [13,71,72]. A concentrated aqueous salt solution (brine) is employed on the permeate side of the membrane to increase further the vapour pressure driving force. As a consequence, it is expected that the mass transport through the membrane in OMD is higher than in DCMD or OD alone.

The temperature and activity (i.e. concentration) gradients can act either in a synergistic way or in an antagonistic way to each other [73]. Moreover, both OMD and OD are strongly affected by the concentration polarization — occurring particularly on the permeate side. Appropriate hydrodynamic conditions as well as membrane module design are necessary to decrease the concentration polarization effects.

The set of equations derived for the DCMD process, which were shown in the theoretical section of this chapter, have successfully been applied for the description of the OD and OMD processes [13,70,73–75].

It is to be noted that, besides the terms OMD and OD, other terms are used such as osmotic evaporation (OE) [74,76], membrane osmotic distillation (MOD) [77,78] and isothermal membrane distillation (IMD) [79] in the membrane literature.

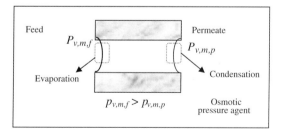

FIGURE 10.16 Schema of mass transfer through a single layer hydrophobic membrane used in OD process for the concentration of feed aqueous solutions.

DCMD APPLICATIONS

In general, DCMD technology is applied in the following:

- Desalination and production of high purity water from brackish water, seawater and brines.
- Crystallization.
- Nuclear industry for concentration of radioactive solutions, nuclear desalination and pure water production.
- Textile industries (removal of dyes, wastewater treatments).
- Pharmaceutical and biomedical industries (removal of water from blood and protein solutions, wastewater treatments).
- Food industry (concentration of fruit juices and milk processing) and in areas where high temperature applications lead to degradation of process fluids.
- Chemical industry (concentration of acids, removal of trace volatile organic compounds (VOCs) from wastewater, separation of azeotropic aqueous mixtures such as alcohol/water).
- Wastewater reclamation in space using integrated DCMD and *OD* processes.
- Concentration of extract of traditional Chinese medicine (TCM).
- Boron and arsenic removal from aqueous solutions.
- Treatment of coolant liquids (concentration of ethylene glycol−water mixtures).
- Concentration of olive mill wastewater for polyphenols recovery.

DCMD Technology in Desalination and Crystallization

As mentioned earlier, DCMD is a technology applicable in desalination of brackish waters, seawater, brines and saline waters in general. Different types of membranes under distinct DCMD operating conditions and configurations have been utilized. The majority of applications is for production of high purity water [1,12,23,47,56,57,80−90]. Near 100% rejection of non-volatile electrolytes (i.e. NaCl, KCl, LiBr, etc.) was achieved. As the permeate product is very pure, it is suitable for use in the medical, pharmaceutical and semiconductor sectors. Schneider et al. [82] reported a quality water as low as 0.4 µs/cm when using Accurel® PP capillary membranes (75% porosity, 5.5 mm inner diameter and 8.6 mm outer diameter) for feed temperatures varying from 58 °C to 100 °C, permeate temperatures from 42 °C to 86 °C and NaCl salt concentrations varying from 0.05% to 20% in water. Karakulski et al. [87] reported slightly higher electrical conductivity values of produced water, 0.8 µs/cm with 0.6 ppm TDS (total dissolved solids), when using Accurel® S6/2 membrane module (Table 2.2). Tables 10.1−10.4 summarize the DCMD performance of commercial and fabricated flat sheet and hollow fibre membranes. It must be pointed out that other than the permeate flux, the advantage of DCMD desalination is the observed high rejection factor, which cannot be accomplished by RO at high permeate fluxes. It was observed that when using commercial membranes in DCMD, different values of the permeate flux were reported for the same membranes working under the same operating conditions. The discrepancies may depend on the membrane modules used.

Weyl [1] was the first in conducting desalination by DCMD process, but the obtained permeate fluxes were lower than $1\,kg/m^2\cdot h$. Nowadays the DCMD permeate fluxes are two orders of magnitude greater. As can be seen in Tables 10.1−10.4, the highest DCMD permeate fluxes of $145.8\,kg/m^2\cdot h$ and $116.6\,kg/m^2\cdot h$ were achieved using distilled water as feed and 1.3 mol% of NaCl feed aqueous solution, respectively. The feed temperature was 74 °C, the permeate temperature was 20 °C and the membrane was 3ME (Table 2.1) [12]. In addition,

TABLE 10.3 Reported Permeate Flux (J_w) of Different Types of Fabricated and Modified Flat Sheet Membranes for DCMD Applications

Membrane type	J (10^{-3} kg/m^{2}s)	Observation	Reference
PVDF unsupported	2.70	1–2 wt% NaCl; $T_{b,f}$ = 60 °C; $T_{b,p}$ = 20 °C.	[141]
	1.86	0.3 M NaCl; $T_{b,f}$ = 55 °C; $T_{b,p}$ = 25 °C.	[142]
Copolymer F2.4[a]	2.03	0.3 M NaCl; $T_{b,f}$ = 55 °C; $T_{b,p}$ = 25 °C.	[142]
Modified CN[b]	8.33	0.5 M NaCl; $T_{b,f}$ = 60 °C; $T_{b,p}$ = 25 °C (99% salt retention).	[143]
Modified CA[c]	0.38	0.3 M NaCl; $T_{b,f}$ = 50 °C; $T_{b,p}$ = 20 °C (99.1% salt retention).	[143]
SMM/PEI (M12)[d]	4.1 3.5	$T_{b,f}$ = 50 °C; $T_{b,p}$ = 40 °C Distilled water as feed 0.5 M NaCl (99.9% salt retention).	[6]
SMM/PEI (M1)[d]	7.5 5.8	$T_{b,f}$ = 55 °C; $T_{b,p}$ = 15 °C Distilled water as feed 0.5 M NaCl (>99% salt retention).	[10]
SMM/PEI (M2)[d]	5.1 3.9	$T_{b,f}$ = 65 °C; $T_{b,p}$ = 15 °C Distilled water as feed 0.5 M NaCl (99.9% salt retention).	[11]
SMM/PS (M1)[e]	2.65 2.30	$T_{b,f}$ = 50 °C; $T_{b,p}$ = 40 °C Distilled water as feed 0.5 M NaCl (99.9% salt retention).	[144]
SMM/PES (M1)[f]	3.0 2.6	$T_{b,f}$ = 50 °C; $T_{b,p}$ = 40 °C Distilled water as feed 0.5 M NaCl (99.9% salt retention).	[145]
PFS/anodiscs[g]	4.78	0.1 M NaCl; $T_{b,f}$ = 53 °C; $T_{b,p}$ = 18 °C (93%–99% salt retention).	[146]
PVA/PEG/PVDF[h]	6.72 6.53	$T_{b,f}$ = 70 °C; $T_{b,p}$ = 22 °C Distilled water as feed 3.5 wt% NaCl (>99% salt retention).	[147]

[a] F2.4: Poly(vinylidene fluoride-co-tetrafluoroethylene).
[b] Modified cellulose nitrate by plasma polymerization using vinyltrimethylsilicon (VTMS)/carbon tetrafluoride (CF$_4$).
[c] Modified cellulose acetate by radiation polystyrene grafting using styrene (St)- pyridine (Pyd)-carbon tetrachloride (CCl$_4$).
[d] polyetherimidie (PEI) fabricated using surface modifying macromolecules (SMMs).
[e] polysulfone (PS) fabricated using surface modifying macromolecules (SMMs).
[f] polyethersulfone (PES) fabricated using surface modifying macromolecules (SMMs).
[g] Modified alumina anodisc^{TM} membrane of pore size 200 nm by surface treatment using 1H, 1H, 2H, 2H-perfluorodecyltriethoxysilane (PFS).
[h] Hydrophoilic modified PVDF (GVSP, Millipore) membranes using polyvinyl alcohol (PVA) blended with polyethylene glycol (PEG) and cross-linked by aldehydes and sodium acetate.

TABLE 10.4 Reported Permeate Flux (J_w) of Fabricated and Modified Hollow Fibre Membranes for DCMD Applications

Membrane trade name	J (10^{-3} kg/m²s)	Observation	Reference
PVDF 93-2[a]	0.876 0.148 0.728	$T_{b,f}$ = 52 °C; $T_{b,p}$ = 19 °C; Ethanol/water initial concentration (5.02 wt% in feed; 5.07 wt% in permeate); final concentration in permeate 8.80 wt% Total permeate flux Ethanol permeate flux Water permeate flux.	[133]
PVDF 43-1[b] Uncoated Si-LTV coated	 0.311 0.067 0.244 0.146 0.067 0.079	$T_{b,f}$ = 47.8 °C; $T_{b,p}$ = 16.3 °C; Ethanol/water initial concentration (5.1 wt% in feed & in permeate); Total permeate flux Ethanol permeate flux (ethanol selectivity 5.1) Water permeate flux. Total permeate flux; Ethanol permeate flux (ethanol selectivity 5.1) Water permeate flux.	[134]
PP[c]	0.09	35 g/l NaCl; $T_{b,f}$ = 60.5 °C; $T_{b,p}$ = not mentioned.	[148]
PE[c]	0.24	35 g/l NaCl; $T_{b,f}$ = 60.5 °C; $T_{b,p}$ = not mentioned.	[148]
PVDF[d]	11.53	3.5 wt% NaCl; $T_{b,f}$ = 79.3 °C; $T_{b,p}$ = 17.5 °C (99.99% salt retention).	[149]
PVDF mixed matrix[e]	22	3.5 wt% NaCl; $T_{b,f}$ = 81.3 °C; $T_{b,p}$ = 17.5 °C (100% salt retention).	[150]
PVDF/PTFE[f]	11.22	3.5 wt% NaCl; $T_{b,f}$ = 79.5 °C; $T_{b,p}$ = 17.5 °C (99.8% salt retention).	[151]
CO17[g]	0.36	Distilled water as feed $T_{b,f}$ = 45 °C; $T_{b,p}$ = 20 °C.	[152]
Modified ceramic Al_2O_3[h]	1.626 1.499	$T_{b,f}$ = 95 °C; $T_{b,p}$ = 5 °C 0.001 M NaCl (96% salt retention) 1 M NaCl (100% salt retention).	[153]
Modified TiO_2 (Ti5)[h]	0.231	0.5 M NaCl; $T_{b,f}$ = 95 °C; $T_{b,p}$ = 5 °C; (99.1% salt retention).	[154]
Modified ceramic ZrO_2[h]	2.341 1.918	$T_{b,f}$ = 95 °C; $T_{b,p}$ = 5 °C; 0.1 M NaCl (99.5% salt retention) 1 M NaCl (98.5% salt retention).	[153]
Modified ZrO_2 (Zr50)[h]	1.099	0.5 M NaCl; $T_{b,f}$ = 95 °C; $T_{b,p}$ = 5 °C; (99.8% salt retention).	[154]
Modified PP MXFR 3[i]	21.94	1 wt% NaCl; $T_{b,f}$ = 90 °C; $T_{b,p}$ = 15 °C–17 °C (100% salt retention).	[89]

TABLE 10.4 Reported Permeate Flux (J_w) of Fabricated and Modified Hollow Fibre Membranes for DCMD Applications—Cont'd

Membrane trade name	J (10^{-3} kg/m^{2}s)	Observation	Reference
Dual layer PVDF[j]	15.3	3.5 wt% NaCl; $T_{b,f}$ = 90 °C; $T_{b,p}$ = 16.5 °C (99.8% salt retention).	[155]
PVDF (M4)[k]	11.25	3.5 wt% NaCl; $T_{b,f}$ = 81.8 °C; $T_{b,p}$ = 20 °C (99.99% salt retention).	[156]

[a] Poly(vinylidene fluoride) (PVDF 93-2: d_i/d_o = 0.844/1.024 mm; ε = 58%, d_p = 8.14 nm; τ = 3.06) hollow fibre membranes prepared by dry/jet wet spinning technique.

[b] Silicone (Si-LTV, silicone rubber) inside coated PVDF hollow fibre membranes (PVDF 13–1: d_i/d_o = 1.042/0.761 mm; d_p = 43.8 nm; τ = 2.93).

[c] Polypropylene (PP: d_i/d_o = 342.5/442.5 µm; ε = 53.3%, d_p = 7.4 10^{-2} µm) and polyethylene (PE: d_i/d_o = 267.5/367.5 µm; ε = 66.3%, d_p = 8.7 10^{-2} µm) hollow fibre membranes prepared by melt extruded/cold-stretched.

[d] PVDF hollow fibre membrane (d_i/d_o = 0.6/0.82 mm; d_p = 0.16 µm) prepared by dry/jet wet phase inversion method.

[e] Mixed matrix PVDF hollow fibre membrane (Kureha20A-C; d_i/d_o = 1.02/1.38 mm; $d_{p,inner\ surface}$ = 1 µm; $d_{p,\ outer\ surface}$ = 50 nm; ε = 86.7%) prepared by dry/jet wet phase inversion method.

[f] PVDF/PTFE (50 wt% PTFE particle with size <1 m) hollow fibre membranes (PVDF-T50-C; d_i/d_o = 0.53/0.75 mm; d_p = 0.248 µm; ε = 80%) prepared by dry/jet wet phase inversion method.

[g] Poly(vinylidene fluoride-hexafluoropropylene), PVDF-HFP hollow fibre membranes (CO17, d_i/d_o = 1.53/1.64 mm; $d_{p,inner\ surface}$ = 77.83 nm; $d_{p,\ outer\ surface}$ = 114.17 nm; ε = 76.3%; LEP_w = 1.65 10^5 Pa) prepared by dry/wet spinning technique.

[h] grafted ceramic hollow fibre membranes by 1H,1H,2H,2H-perfluorodecyltriethoxysilane (Zirconia, ZrO_2: d_p = 50 nm); (Alumina, Al_2O_3: d_p = 200 nm); (Titania, TiO_2: d_p = 5 nm).

[i] Polypropylene (PP) porous hollow fibre membrane (Accurel Membrana: PP 150/330; d_i/d_o = 0.33/0.63 mm; ε = 65%; maximum pore size = >0.2 µm) coated by plasma polymerization sing silicone fluoropolymer.

[j] Hydrophobic/hydrophilic PVDF hollow fibre membrane fabricated by co-extrusion dry/jet wet spinning method (PVDF/hydrophobic cloisite NA$^+$ particles and PVDF/PAN/hydrophilic cloisite 15A particles: d_i/d_o = 0.52/1.2 mm, d_p = 0.41 µm).

[k] PVDF hollow fibre membrane fabricated by phase inversion method (d_i/d_o = 0.94/1.2 mm, d_p = 0.25–0.3 µm; ε = 79.7%).

DCMD permeate fluxes as high as those of RO systems were obtained by Schneider et al. [82], Schofield et al. [23] and during last years in number of studies (see Tables 10.1—10.4).

Drioli and Wu [57] and Drioli et al. [80] applied DCMD for the treatment of NaCl at different concentrations and temperature gradients using both flat sheet (PTFE) and capillary PP and PVDF membranes. Rejection factor was shown to increase with the decrease of membrane pore size from 96.5% to 100% and membranes having nominal pore sizes near 0.2 µm exhibited rejection factors higher than 99% with permeate fluxes lower than 2.5 kg/m^2·h. Ohta et al. [83], by performing experimental studies on seawater desalination using a plate and frame module containing 3.12 m^2 of non-porous silicone—polysulfone compositae membrane, obtained DCMD fluxes lower than 2.5 kg/m^2·h. Martínez and Florido-Díaz [86] were able to predict permeate fluxes about 65 kg/m^2·h for the DCMD desalination of seawater by HVHP and GVHP (Table 2.1) membranes at high feed temperature (i.e. 85 °C). In addition, Schofield et al. [23] stated that DCMD of salt aqueous solutions (up to 25%) was able to give permeate fluxes that were 60% to 70% of those obtained for feed distilled water. Their analysis revealed that the flux reduction for salt NaCl aqueous solutions was largely due to vapour pressure reduction with small effect of increased viscosity.

It is worth quoting that, in some cases, the DCMD process was carried out for several months without detection of permeate flux decay; while in other studies, permeate flux declined down to more than 70% in the first days of operation. This phenomenon is not well understood and more studies must be conducted on the long-term DCMD performance. Only a few studies have been conducted for DCMD over a period of several weeks [65,82,87,90]. For example, in 3 days of DCMD operation a permeate flux decline of 71.5% was observed for the membrane TF450 (see Table 2.1) when using 0.58 wt% NaCl aqueous solution with a feed temperature of 50 °C and a permeate temperature of 20 °C [57]. However, when a DCMD study was performed over 3 years for production of pure water using the membrane module Accurel PP hollow fibre (Table 2.2) under a feed temperature of 80 °C (i.e. lumen side) and a permeate temperature of 20 °C (i.e. shell side), it was claimed that the used membranes were thermally stable, remained unchanged based on SEM analysis and good separation factors were maintained throughout the whole period of experiments [65]. The measured electrical conductivity of the permeate was increased slightly from 0.9 μs/cm to 2.5 μs/cm. The corresponding decrease of the permeate flux was 21.4% (i.e. decrease from 7000 l/m²·day to 5500 l/m²·day). Moreover, when tap water was used as feed, precipitation of calcium carbonate ($CaCO_3$) was detected on the membrane surface. Scaling was significantly limited by the acidification of the feed water at pH 4. The membrane was cleaned by using 2—5 wt% HCl aqueous solutions, rinsing with pure water and final drying. Figure 10.17 shows the DCMD permeate flux and the electrical conductivity variation with the operating time as well as the nature of the feed aqueous solution.

It is to be advised that the deposit that may be formed on the membrane surface can cause partial clogging of the membrane pores adjacent to the deposit leading to a decline of both the quantity and quality of the produced water as well as to the efficiency of the whole DCMD process. Compared to the hydrostatic pressure-driven processes such as RO, much lower operating hydrostatic pressures are needed in DCMD (i.e. near atmospheric pressure). Therefore, it may be expected that membrane fouling and scaling effects are of less problem in DCMD. Moreover, the observed decline of the permeate flux is due also to the decrease of the driving force.

After only 35 h of uninterrupted DCMD operation using a semi-pilot plant with Accurel PP fibres (Table 2.2) — at a feed temperature of 40 °C and a permeate temperature of 20 °C — it was observed that the permeate flux was reduced by 45% from the

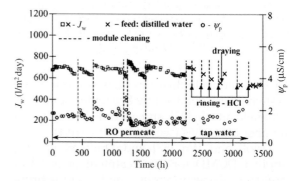

FIGURE 10.17 Influence of operating time on DCMD performance of Accurel PP hollow fibre (Table 2.2), $T_{b,f} = 80\,°C$ and $T_{b,p} = 20\,°C$ using first the permeate of RO plant and then tap water as feed. *Source: Reprinted from [65]. Copyright 2005, with kind permission from Elsevier.*

initial permeate flux of $2.05 \, l/m^2 \cdot h$. $CaCO_3$ scaling significantly reduced the permeate flux by 33% [61]. The membrane was restored by cleaning with aqueous citric acid ($C_6H_8O_7$) solution for 20 min and with sodium hydroxide (NaOH) aqueous solution for another 20 min.

Membrane scaling by $CaCO_3$, $CaSO_4$ and their mixtures as well as the effects of different antiscalants types on the DCMD performance of fluorosilicone-modified PP hollow fibre membranes were studied recently by He et al. [62–64]. It was shown that DCMD is more resistant to scaling than other conventional thermal processes. It was found that precipitation of $CaSO_4$ from supersaturated solutions having high saturation index (SI) values did not reduce the water vapour flux or lead to pore wetting. Moreover, based on various $CaCO_3$ scaling experiments with tap water over a wide range of SI values (10–64) at high temperatures ranging from 70 °C to 80 °C, it was concluded that no significant drop of the DCMD permeate flux was detected and the increase of distillate electrical conductivity was attributed to CO_2 transport through the membrane. Furthermore, it was reported that for the DCMD process with scalants such as $CaCO_3$ and the mixture $CaCO_3$–$CaSO_4$, the concentration polarization effect was more important than the temperature polarization. Fluorosilicone coating could however eliminate $CaSO_4$ scaling in the DCMD process using feed solutions of at high SI values. When antiscalants were added in the feed solutions, it was observed – as occurred in other membrane processes such as NF and RO – that antiscalants slowed down the precipitation rate of crystals, and no sign of DCMD permeate flux reduction nor any increase of the electrical conductivity were detected. This also proves that the used antiscalants did not cause any membrane pore wetting during DCMD operation.

Currently, about 48% of all desalination plants discharge their concentrate waste stream into surface waters or oceans. It was reported in the previous section that membrane crystallization was proposed as one of the most interesting and promising extension of the DCMD concept to recover solid substances present in the concentrated streams of the desalination plants and improve the efficiency of desalination processes. The main objectives are to recover water and to generate the desired supersaturation in the crystallizer where product crystals (i.e. valuable salts) can be precipitated. When concentrating feed solutions above their saturation limit, a supersaturated environment will be attained where crystals may nucleate and grow. The presence of a polymeric membrane increases the probability of nucleation with respect to other locations in the system (heterogeneous nucleation). This application is also termed membrane distillation crystallization (MDC) [59,60,91]. Figure 10.18 shows a typical MDC setup, which is very similar to DCMD systems (Fig. 10.2). The only difference is, instead of a feed container, there is a batch-type crystallizer with the retentate stream from the DCMD module returned to the crystallizer and the mother liquor from the crystallizer recirculated as feed solution by the peristaltic pump to the DCMD module. The crystallizer has a stirrer and temperature controller.

Wu and Drioli [92] reported for the first time that at a sufficiently high concentration crystallization of solutes may occur leading to the possibility of MDC. Later on Wu et al. [59] carried out pharmaceutical wastewater treatment for taurine production by MDC concluding that crystallization occurred after reduction of the DCMD permeate flux to practically zero. Gryta [90] performed experiments using an integrated DCMD and crystallization to concentrate NaCl solution in both batch and continuous modes. A production of $100 \, kg/m^2 \cdot day$ NaCl was reported with a feed temperature of 85 °C, a permeate temperature of 20 °C and a permeate

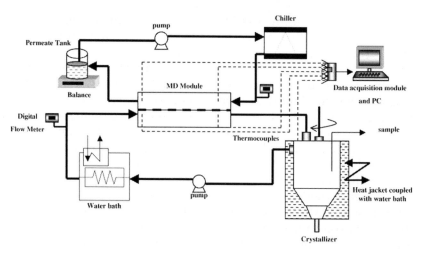

FIGURE 10.18 Experimental MDC setup [91]. *Source: Reprinted from [91]. Copyright 2005, with kind permission from Elsevier.*

flux of $23\,kg/m^2 \cdot h$. However, at the applied feed temperature, the risk of membrane pore wetting is high. Concentration of sulphuric acid solutions up to 40% was performed by Tomaszewska [93] in order to recover lanthane compounds. A 25% lanthane recovery could be achieved from the precipitate obtained from the concentrated solution. Curcio et al. [58] have demonstrated the production of NaCl crystals by MDC process at a temperature of 29 °C while crystallization was carried out at 25 °C. Recently, Tun et al. [91] showed that anhydrous sodium sulphate (Na_2SO_4) crystals could be produced by MDC with a relatively narrow crystal size distribution and 60–80 μm average crystal size. It was concluded that DCMD can operate with aqueous salt solutions of extremely high concentrations at reasonably high permeate fluxes. Up to a critical supersaturation of the salts feed solution, a gradual decrease of the DCMD permeate flux was observed; however, a drastic and rapid permeate flux decline occurred beyond the critical supersaturation due to the rapid growth of crystal deposition on the membrane surface.

It is worth quoting that for aqueous salt solutions with positive solubility/temperature coefficients, this means lower solubility at lower temperature (e.g. NaCl) — the combined temperature and concentration polarization effects in DCMD process could encourage crystal formation on the membrane surface. In this case, DCMD operation at low feed temperatures is advisable. In contrast, for aqueous salts solutions with negative solubility/temperature coefficients, e.g. Na_2SO_4, higher temperatures are more favourable.

It is worth mentioning here that more progresses in DCMD process have to be accomplished in order to overcome the limits suffered by desalination technology such as low desalted water costs, high recovery factors, scaling and brine disposal.

DCMD Technology in Concentration of Fruit Juices

The potential of DCMD process as concentration technique to remove water from fruit juices is a novel and interesting technology in agro-food industry together with OD and OMD process because they can be performed at relatively low feed temperatures in order to maintain the properties of thermo-labile solutes.

These processes can be used as single units or integrated with other systems to increase the performance and efficiency of the whole plant, even at high osmotic pressures.

Remember that one of the advantages of the DCMD process is that it is not limited by high osmotic pressure constraints and fouling as the membrane works at atmospheric pressure and does not require elevated temperatures. Therefore, the heat can be supplied by solar collectors and other forms of low-grade waste heat from industries or geothermal energy.

It has been shown that DCMD is effective in the concentration of various types of juices including orange juice [29,94,95], apple juice [37,96,97], sugarcane juice [98], etc. For example, PP Accurel® MD020CP2N membrane module (Table 2.2) was used to produce a highly concentrated apple juice (64° Brix) [37]. DCMD was also used to concentrate clarified cane sugar solution (20° Brix) obtained from sugar mill [98].

Fruit juices have a high content of solids and pectins. For this reason, when directly applied in DCMD, low permeate fluxes were observed. Moreover, there is a limiting concentration of juices in OMD process associated with their very high viscosity ($>0.2\,N/m$) at concentrations of sucrose exceeding 68° Brix [13]. For feed concentrations lower than 40 wt% of sugar, the process is controlled by the membrane resistance and therefore the permeate flux varies in accordance with the membrane permeability (B_w); however, at higher feed concentrations (50–68 wt%), the mass transfer resistance due to the feed boundary layer is dominant and, in this case, using membranes with high B_w values is not necessarily advantageous [13]. Schofield et al. [23] stated that DCMD of sucrose aqueous solutions (up to 30%) was able to give permeate fluxes of 60–70% of that of feed distilled pure water. Their analysis revealed that for the aqueous sucrose solutions the viscosity was the major factor in reducing the DCMD permeate flux. Near 100% rejection of non-electrolyte solutes (i.e. glucose, sucrose, fructose, etc.) in aqueous solutions was achieved [23,55,99].

Contrary to DCMD process, the performance of OMD process was evaluated on larger scale pilot plants [100,101]. Other details have been reported previously in section 10.4.

DCMD technology applied to the concentration of fruit juices has important benefits over other processes such as conventional evaporation mainly in terms of improved product quality, low energy consumption and easy scale up. However, the risk of fouling and wetting of membrane pores compromises the durability of the membranes limiting their applications on industrial scale. Nowadays, extensive studies aiming to enhance the performance of the membrane modules as well as long-term applications are in progress to improve the DCMD, OD and OMD concentration of fruit juices.

DCMD Technology for Treatment of Wastewaters

DCMD process has been applied successfully to wastewater treatment. The obtained permeate is less hazardous to the environment, whereas the obtained retentate is more concentrated in valuable chemicals. More specifically, among different studies carried out, DCMD has been applied to textile wastewater contaminated with dyes [102]. Dyes are non-volatile substances — therefore are completely separated by DCMD process. However, the treatment of this wastewater is reasonable only if it is focused on water recovery rather than on concentration of dyes because generally a mixture of dyes is concentrated. Integration of photocatalytic membrane reactors and DCMD for removal of mono- and poly-azo dyes from water has been proposed [103,104]. Significant improvements have been reported including absence of fouling of the catalyst-based TiO_2. A high product quality with an electrical conductivity lower than 3 µs/cm and pH above 5.2 has been achieved using Accurel® S6/2 and different process parameters (see Table 2.2). In this field, the DCMD process is still under development while the pressure-driven membrane processes such as MF, UF and NF are preferably applied. In fact, DCMD application may be more advantageous when a significant fouling is observed in other processes.

DCMD process was proposed for the treatment of radioactive waste water solutions [105–108]. Zakrewska-Trznadel et al. [105,106] proposed DCMD to clean low-level radioactive wastes from a nuclear centre at throughput of about $0.05 \, m^3/h$. A spiral wound module equipped with a PTFE membrane of an effective surface area of $4 \, m^2$ was used (Table 2.1). The pilot plant laboratory experiments were able to produce very pure water with about 30 mg/l total solute concentration and activity at the level of natural background, and permeate fluxes between 30 and 50 l/h (i.e. 180–300 $l/m^2 \cdot day$) using feed inlet temperature of 45–80 °C, permeate inlet temperature of 5–20 °C, and feed and distillate flow rates of up to 1500 l/h. Moreover, it was proved that DCMD is feasible to process low- and medium-level radioactive wastes giving high decontamination factor in only one stage [105–108].

DCMD process was also applied for the treatment of pharmaceutical wastewater containing taurine [59], wastewater contaminated with heavy metals [109], wastewater reclamation in space in a combined direct osmosis system [110], oil–water emulsions [111] and olive mill wastewaters (OMW) [112]. In this last application the objective was to obtain pure water and a concentrate containing high amounts of polyphenols, which can be much easier to extract later. A separation coefficient of 99% was observed for the membrane TF200 (Table 2.1) after 9 h DCMD operation with OMW concentration factor of 1.72, whereas those of the membrane GVHP (Table 2.1) were lower, 89% OMW separation factor and 1.4 concentration factor.

Other Applications of DCMD Technology

Other applications of DCMD technology include separation of water and ethylene glycol for possible treatment of coolant liquids observing complete rejection of ethylene glycol [113], treatment of humic acid aqueous solutions [61,114–116], sulphuric acid solutions rich in specific compounds such as lanthane [93,117–122], removal of arsenic from contaminated groundwater [53,123,124], boron removal from aqueous solutions [125] and concentration of extract of traditional Chinese medicine (TCM) [126,127]. When concentrating TCM by DCMD process, temperatures up to 60 °C were used. In this case the observed permeate flux decline was attributed to membrane fouling, decrease of the driving force in the presence of TCM and to the increased mass transfer resistance in the feed boundary layer. It was reported that no considerable membrane wetting due to TCM deposition on the membrane surface was detected [127].

By using PVDF hollow fibre membranes with a pore size 0.15 µm and a porosity of 80%, the observed removal efficiencies of Arsenite, As(III), and arsenate, As(V), were above 99.95% and in the produced permeate both contaminants were below their maximum permitted limit (10 µg/l) when the concentration of the feed aqueous solutions were varied from 40 mg/l to 2 g/l [123]. Furthermore, after 250 h DCMD operation, no arsenic was detected in the permeate. DCMD permeate flux up to $95 \, kg/m^2 \cdot h$ with almost 100% rejection factor was reported when using arsenic contaminated groundwater (396 ppb) with a feed temperature of 60 °C, a permeate temperature of 20–22 °C and a flat sheet PVDF membrane, supplied from Sepro membranes, with an average pore size of 0.13 μm, a porosity of 70–75% and a thickness of 150 μm [124].

DCMD process exhibits high boron rejection factors (> 99.8%), which were found to be independent of the feed pH and salt concentrations [125]. When boron concentration in feed aqueous solution was as high as 750 mg/l, the boron concentration in the produced permeate was below the maximum permitted level when a fabricated PVDF hollow fibre membrane (0.15 μm pore size, 79.5% porosity, 150 μm thickness) was used, while for natural groundwater with a boron concentration of 12.7 mg/l, the boron concentration in the permeate was less than 20 μg/l with an electrical conductivity below 5 μs/cm. This result was maintained whether the feed solution was acidified or not, although pre-acidification was found to be advisable to maintain stability in the DCMD permeate flux [125].

Various biological solutions have been concentrated by DCMD. Sakai et al. [128,129] have applied the process to the concentration of bovine plasma and bovine blood using PTFE membranes of different pore sizes and thicknesses. It was found that the permeate flux was directly proportional to the vapour pressure difference and exhibited only a slight decrease during blood treatment concluding that PTFE membranes, which possess outstanding properties of biocompatibility, were suitable for stable removal of solute-free water from blood with a haematocrit of 45% by DCMD process. Ortiz de Zárate [130] looked at the concentration of protein (0.4% and 1% bovine serum albumin at pH 7.4) aqueous solutions by DCMD at low temperatures and found that fouling effects were practically absent, while the limiting factor of the process was the temperature polarization. Capuano et al. [131] and Criscuoli et al. [132] suggested DCMD as innovative tool for ameliorate treatment of uraemia by allowing purification of the blood ultrafiltrate and the re-injection of the purified water to the patients. Tests have been made on artificial solutions and plasma ultrafiltrate obtained from a haemofiltration unit. The researchers proposed the clinical application of DCMD to patients' treatment by operating online with haemofiltration or haemodialysis.

It is worth quoting that few papers appeared on the DCMD treatment of aqueous solutions containing volatile organic compounds (VOCs). As it is reported previously, DCMD cannot be used for VOCs removal from water because of the risk of pore wetting from the membrane permeate side. Fujii et al. [133,134] studied the removal of low concentration organics (ethanol, acetone, acetonitrile, n-butanol) from water using various coated and uncoated hollow fibre fine porous membranes. It was found that the selectivity varied depending on the properties of the polymer used, the membrane characteristics and the DCMD operation conditions. It must be pointed out that in DCMD, the permeate must not wet the membranes pores, while in the other MD configurations concentration of the condensed permeate is not a concern as it does not come into contact with the membrane.

DCMD has also potential applications in biotechnology, for the removal of toxic products from culture broths. Udriot et al. [135] described the application of DCMD unit connected to a laboratory bioreactor for the selective recovery of ethanol from the culture medium. The experiments were run at a constant temperature of 38 °C on an anaerobic culture of *fragilis* using a PTFE membrane. The researchers found that continuous extraction of ethanol using DCMD resulted in an increase in ethanol productivity by 87%. Gryta et al. [136] also investigated the batch fermentation combined with the removal of ethanol from the broth by means of DCMD, where an increase in productivity and rate of conversion of sugar to ethanol was observed.

The application of DCMD for breaking azeotropic mixtures was first proposed by Udriot et al. [137], who looked at separating hydrochloric acid–water and propionic acid–water azeotrope mixtures. In fact, the azeotrope mixtures are impossible to separate by simple distillation. Retention selectivities of the solute between 0.6 and 0.8 were achieved instead of unity as implied by vapour liquid equilibrium (VLE). In this case DCMD may be used to shift the selectivity above or below the one obtained by the VLE.

It is known that DCMD technology exhibits various advantages. For example, it is not limited by high osmotic pressure constraints and fouling as the membrane works at atmospheric pressure. High purity water can be obtained at high concentrations of feed aqueous solutions and does not require elevated temperatures. Therefore different proposals were made to integrate DCMD units to other conventional processes such as distillation and pressure-driven membrane processes such as RO and NF. As well, proposals were made to combine DCMD with high heat recovery systems and with alternative energy sources such as solar and geothermal energy, and also to build DCMD units around nuclear installations where waste heat can be recovered. These lead to various important benefits such as high product quality, plant compactness, environmental impact and energetic aspects. These applications are commented in more details in Chapters 14 and 15.

References

[1] Weyl PK. Recovery of demineralized water from saline waters, United States Patent 3,340,186 (1967).
[2] M.E. Findley, V.V. Tanna, Y.B. Rao, C.L. Yeh, Mass and heat transfer relations in evaporation through porous membranes, AIChE J. 15 (1969) 483–489.
[3] M. Khayet, Membrane distillation, in: N.N. Li, A.G. Fane, W.S.W. Ho, T. Matsuura (Eds.), Advanced membrane technology and applications, John Wiley & Sons, Inc, New York, NY, USA, 2008, pp. 297–370.
[4] M.S. El-Bourawi, Z. Ding, R. Ma, M. Khayet, A framework for better understanding membrane distillation separation process, J. Membr. Sci. 285 (2006) 4–29.
[5] M. Khayet, T. Matsuura, Application of surface modifying macromolecules for the preparation of membranes for membrane distillation, Desalination 158 (2003) 51–56.
[6] M. Khayet, J.I. Mengual, T. Matsuura, Porous hydrophobic/hydrophilic composite membranes: Application in desalination using direct contact membrane distillation, J. Membr. Sci. 252 (2005) 101–113.
[7] M. Khayet, T. Matsuura, J.I. Mengual, Porous hydrophobic/hydrophilic composite membranes: Estimation of the hydrophobic-layer thickness, J. Membr. Sci. 266 (2005) 68–79.
[8] D.E. Suk, T. Matsuura, H.B. Park, Y.M. Lee, Synthesis of a new type of surface modifying macromolecules (nSMM) and characterization and testing of nSMM blended membranes for membrane distillation, J. Membr. Sci. 277 (2006) 177–185.
[9] M. Qtaishat, M. Khayet, T. Matsuura, Guidelines for preparation of higher flux hydrophobic/hydrophilic composite membranes for membrane distillation, J. Membr. Sci. 329 (2009) 193–200.
[10] M. Qtaishat, D. Rana, M. Khayet, T. Matsuura, Preparation and characterization of novel hydrophobic/hydrophilic polyetherimide composite membranes for desalination by direct contact membrane distillation, J. Membr. Sci. 327 (2009) 264–273.
[11] M. Qtaishat, D. Rana, T. Matsuura, M. Khayet, Effect of surface modifying macromolecules stoichiometric ratio on composite hydrophobic/hydrophilic membranes characteristics and performance in direct contact membrane distillation, AIChE J. 55 (2009) 3145–3151.
[12] K.W. Lawson, D.R. Lloyd, Membrane distillation: II. Direct contact MD, J. Membr. Sci. 120 (1996) 123–133.
[13] M. Gryta, Osmotic MD and other membrane distillation variants, J. Membr. Sci. 246 (2005) 145–156.
[14] K.W. Lawson, D.R. Lloyd, Review: membrane distillation, J. Membr. Sci. 124 (1997) 1–25.
[15] E. Curcio, E. Drioli, Membrane distillation and related operations: A review, Sep. Purif. Rev. 34 (2005) 35–86.
[16] M. Khayet, M.P. Godino, J.I. Mengual, Study of asymmetric polarization in direct contact membrane distillation, Sep. Sci. Tech. 39 (2004) 125–147.
[17] R.W. Schofield, A.G. Fane, C.J.D. Fell, Heat and mass transfer in membrane distillation, J. Membr. Sci. 33 (1987) 299–313.
[18] M.E. Findley, Vaporization through porous membranes, Ind. Eng. Chem. Process Des. Dev. 6 (1967) 226–237.
[19] E.A. Mason, A.P. Malinauskas, Gas transport in porous media: The Dusty-Gas model, Elsevier, Amsterdam, 1983.

[20] R.D. Present, Kinetic theory of gases, McGraw-Hill, New York, 1958.

[21] M. Khayet, M.P. Godino, J.I. Mengual, Modelling transport mechanism through a porous partition, J. Non-Equilb. Thermodyn 26 (2001) 1–14.

[22] M. Khayet, A. Velázquez, J.I. Mengual, Modelling mass transport through a porous partition: Effect of pore size distribution, J. Non-Equilib. Thermodyn. 29 (2004) 279–299.

[23] R.W. Schofield, A.G. Fane, C.J.D. Fell, R. Macoun, Factors affecting flux in membrane distillation, Desalination 77 (1990) 279–294.

[24] R.W. Schofield, A.G. Fane, C.J.D. Fell, Gas and vapour transport through microporous membranes, I. Knudsen-Poiseuille transition, J. Membr. Sci. 53 (1990) 159–171.

[25] R.W. Schofield, A.G. Fane, C.J.D. Fell, R, Gas and vapour transport through microporous membranes, II. Membrane Distillation, J. Membr. Sci. 53 (1990) 173–185.

[26] T. Matsuura, Synthetic membranes and membrane separation processes, CRC Press, Boca Raton, USA, 1994.

[27] M. Khayet, K.C. Khulbe, T. Matsuura, Characterization of membranes for membrane distillation by atomic force microscopy and estimation of their water vapour transfer coefficients in vacuum membrane distillation process, J. Membr. Sci. 238 (2004) 199–211.

[28] M. Khayet, T. Matsuura, Pervaporation and vacuum membrane distillation processes: Modeling and experiments, AIChE J. 50 (2004) 1697–1712.

[29] S. Kimura, S. Nakao, Transport phenomena in membrane distillation, J. Membr. Sci. 33 (1987) 285–298.

[30] J. Phattaranawik, R. Jiraratananon, A.G. Fane, Effect of pore size distribution and air flux on mass transport in direct contact membrane distillation, J. Membr. Sci. 215 (2003) 75–85.

[31] L. Martínez, J.M. Rodríguez-Maroto, Characterization of membrane distillation modules and analysis of mass flux enhancement by channel spacers, J. Membr. Sci. 274 (2006) 123–137.

[32] S. Srisurichan, R. Jiraratananon, A.G. Fane, Mass transfer mechanisms and transport resistances in direct contact membrane distillation process, J. Membr. Sci. 277 (2006) 186–194.

[33] M. Qtaishat, T. Matsuura, B. Kruczek, M. Khayet, Heat and mass transfer analysis in direct contact membrane distillation, Desalination 219 (2008) 272–292.

[34] M.M. Teoh, S. Bonyadi, T.S. Chung, Investigation of different hollow fiber module designs for flux enhancement in the membrane distillation process, J. Membr. Sci. 311 (2008) 371–379.

[35] M.A. Izquierdo-Gil, C. Fernández-Pineda, M.G. Lorenz, Flow rate influence on direct contact membrane distillation experiments: Different empirical correlations for Nusselt number, J. Membr. Sci. 321 (2008) 356–363.

[36] S. Al-Obaidani, E. Curcio, F. Macedonio, G.D. Profio, H. Al-Hinai, E. Drioli, Potential of membrane distillation in seawater desalination: thermal efficiency, sensitivity study and cost estimation, J. Membr. Sci. 323 (2008) 85–98.

[37] F. Laganà, G. Barbieri, E. Drioli, Direct contact membrane distillation: Modelling and concentration experiments, J. Membr. Sci. 166 (2000) 1–11.

[38] L. Martínez, F.J. Florido-Díaz, A. Hernández, P. Prádanos, Characterization of three hydrophobic porous membranes used in membrane distillation: Modelling and evaluation of their water vapour permeabilities, J. Membr. Sci. 203 (2002) 15–27.

[39] L. Martínez, F.J. Florido-Díaz, A. Hernández, P. Prádanos, Estimation of vapour transfer coefficient of hydrophobic porous membranes for applications in membrane distillation, Sep. Purif. Tech. 33 (2003) 45–55.

[40] K. Schneider, T.J. van Gassel, Membrandestillation, Chem. Ing. Tech. 56 (1984) 514–521 (in German).

[41] A.G. Fane, R.W. Schofield, C.J.D. Fell, The efficient use of energy in membrane distillation, Desalination 64 (1987) 231–243.

[42] A.O. Imdakm, T. Matsuura, A Monte Carlo simulation model for membrane distillation processes: direct contact (MD), J. Membr. Sci. 237 (2004) 51–59.

[43] A.O. Imdakm, T. Matsuura, Simulation of heat and mass transfer in direct contact membrane distillation (MD): The effect of membrane physical properties, J. Membr. Sci. 262 (2005) 117–128.

[44] M. Khayet, A.O. Imdakm, T. Matsuura, Monte Carlo simulation and experimental heat and mass transfer in direct contact membrane distillation, Int. J. Heat Mass Transfer 53 (2010) 1249–1259.

[45] J. Phattaranawik, R. Jiraratananon, A.G. Fane, Heat transport and membrane distillation coefficients in direct contact membrane distillation, J. Membr. Sci. 212 (2003) 177–193.

[46] J.M. Rodríguez-Maroto, L. Martínez, Bulk and measured temperatures in direct contact membrane distillation, J. Membr. Sci. 250 (2005) 141–149.

[47] J.M. Ortiz-Zárate, F. García López, J.I. Mengual, Non-isothermal water transport through PTFE membranes, J. Membr. Sci. 56 (1991) 181–194.

[48] A. Velázquez, J.I. Mengual, Temperature polarization coefficients in membrane distillation, Ind. Eng. Chem. Res. 34 (1995) 585–590.

[49] J. Phattaranawik, R. Jiraratananon, A.G. Fane, Effects of net-type spacers on heat and mass transfer in direct contact membrane distillation and comparison with ultrafiltration studies, J. Membr. Sci. 217 (2003) 193–206.

[50] L. Martínez-Díez, M.I. Vázquez-González, F.J. Florido-Díaz, Study of membrane distillation using channel spacers, J. Membr. Sci. 144 (1998) 45–56.

[51] M. Gryta, M. Tomaszewska, Heat transport in the membrane distillation process, J. Membr. Sci. 144 (1998) 211–222.

[52] M. Gryta, M. tomaszewska, A.W. Morawski, Membrane distillation with laminar flow, Sep. Purif. Tech. 11 (1997) 93–101.

[53] P. Pal, A.K. Manna, Removal of arsenic from contaminated groundwater by solar-driven membrane distillation using three different commercial membranes, Water Res. 44 (2010) 5750–5760.

[54] T. Cath, V.D. Adams, A.E. Childress, Experimental study of desalination using direct contact membrane distillation: A new approach to flux enhancement, J. Membr. Sci. 228 (2004) 5–16.

[55] L. Martínez, J.M. Rodríguez-Maroto, On transport resistances in direct contact membrane distillation, J. Membr. Sci. 295 (2007) 28–39.

[56] L. Martínez, F.J. Florido-Díaz, Desalination of brines by membrane distillation, Desalination 137 (2001) 267–273.

[57] E. Drioli, Y. Wu, Membrane distillation: an experimental study, Desalination 53 (1985) 339–346.

[58] E. Curcio, A. Criscuoli, E. Drioli, Membrane crystallizers, Ind. & Eng. Chem. Res. 40 (2001) 2679–2684.

[59] Y. Wu, Y. Kong, J. Liu, J. Zhang, J. Xu, An experimental study on membrane distillation-crystallization for treating waste water in taurine production, Desalination 80 (1991) 235–242.

[60] L. Mariah, C.A. Buckley, C.J. Brouckaert, E. Curcio, E. Drioli, D. Jaganyi, D. Ramjugernath, Membrane distillation of concentrated brines – Role of water activities in the evaluation of driving force, J. Membr. Sci. 280 (2006) 937–947.

[61] E. Curcio, X. Ji, G.D. Profio, A.O. Sulaiman, E. Fontananova, E. Drioli, Membrane distillation operated at high seawater concentration factors: Role of the membrane on $CaCO_3$ scaling in presence of humic acid, J. Membr. Sci. 346 (2010) 263–269.

[62] F. He, J. Gilron, H. Lee, L. Song, K.K. Sirkar, Potential for scaling by sparingly soluble salts in crossflow DCMD, J. Membr. Sci. 311 (2008) 68–80.

[63] F. He, K. Sirkar, J. Gilron, Effects of antiscalants to mitigate membrane scaling by direct contact membrane distillation, J. Membr. Sci. 345 (2009) 53–58.

[64] F. He, K. Sirkar, J. Gilron, Studies on scaling of membranes in desalination by direct contact membrane distillation: $CaCO_3$ and mixed $CaCO_3$/$CaSO_4$ systems, Chemical Eng. Sci. 64 (2009) 1844–1859.

[65] M. Gryta, Long-term performance of membrane distillation process, J. Membr. Sci. 265 (2005) 153–159.

[66] M. Khayet, C. Cojocaru, M.C. García-Payo, Application of response surface methodology and experimental design in direct contact membrane distillation, Ind. Eng. Chem. Res. 46 (2007) 5673–5685.

[67] Lefebvre MSM. Method of performing osmotic distillation, United States Patent 4,781,837 (1988).

[68] P.A. Hogan, R.P. Canning, P.A. Peterson, R.A. Johnson, A.S. Michaels, A new option: osmotic distillation, Chem. Eng. Prog. (July 1998) 49–61.

[69] A. Gabelman, S.T. Hwang, Hollow fiber membrane contactors, J. Membr. Sci. 159 (1999) 61–106.

[70] C. Gostoli, Thermal effects in osmotic distillation, J. Membr. Sci. 163 (1999) 75–91.

[71] B.R. Babu, N.K. Rastogi, K.S.M.S. Raghavarao, Concentration and temperature polarization effects during osmotic membrane distillation, J. Membr. Sci. 322 (2008) 146–153.

[72] K. Bélafi-Bakó, B. Koroknai, Enhanced water flux in fruit juice concentration: Coupled operation of osmotic evaporation and membrane distillation, J. Membr. Sci. 269 (2006) 187–193.

[73] J.I. Mengual, J.M. Ortiz de Zárate, L. Peña, A. Velázquez, Osmotic distillation through porous hydrophobic membranes, J. Membr. Sci. 82 (1993) 129–140.

[74] W. Kunz, A. Benhabiles, R. Ben-Aïm, Osmotic evaporation through macroporous hydrophobic membranes: a survey of current research and applications, J. Membr. Sci. 121 (1996) 25–36.

[75] M. Courel, M.M. Dornier, G.M. Rios, M. Reynes, Modelling of water transport in osmotic distillation using asymmetric membrane, J. Membr. Sci. 173 (2000) 107–122.

[76] V.D. Alves, I.M. Coelhoso, Mass transfer in osmotic evaporation: effect of process parameters, J. Membr. Sci. 208 (2002) 171–179.

[77] Z. Wang, F. Zheng, S. Wang, Experimental study of membrane distillation with brine circulated in the cold side, J. Membr. Sci. 183 (2001) 171–179.

[78] Z. Wang, F. Zheng, Y. Wu, S. Wang, Membrane osmotic distillation and its mathematical simulation, Desalination 139 (2001) 423–428.

[79] A.J. Costello, P.A. Hogan, A.G. Fane, Performance of helically wound hollow fibre modules and their application to isothermal membrane distillation. in: Euromembrane '97, 23–27 June, University of Twente, The Netherlands, 1997 (book of abstracts)403–405.

[80] E. Drioli, V. Calabró, Y. Wu, Microporous membranes in membrane distillation, Pure Appl. Chem. 58 (1986) 1657–1662.

[81] E. Drioli, Y. Wu, V. Calabró, Membrane distillation in the treatment of aqueous solutions, J. Membr. Sci. 33 (1987) 277–284.

[82] K. Schneider, W. Holz, R. Wollbeck, Membranes and modules for transmembrane distillation, J. Membr. Sci. 39 (1988) 25–42.

[83] K. Ohta, K. Kikuchi, I. Hayano, T. Okabe, T. Goto, S. Kimura, H. Ohya, Experiments on sea water desalination by membrane distillation, Desalination 78 (1990) 177–185.

[84] J.M. Ortiz de Zárate, A. Velázquez, L. Peña, F. García-López, J.I. Mengual, Non-isothermal solute transport through PTFE membranes, J. Membr. Sci. 69 (1992) 169–178.

[85] M. Sudoh, K. Takuwa, H. Iizuka, K. Nagamatsuya, Effects of thermal and concentration boundary layers on vapour permeation in membrane distillation of aqueous lithium bromide solution, J. Membr. Sci. 131 (1997) 1–7.

[86] L. Martínez, F.J. Florido-Díaz, Theoretical and experimental studies on desalination using membrane distillation, Desalination 139 (2001) 373–379.

[87] K. Karakulski, M. Gryta, A. Morawski, Membrane processes used for potable water quality improvement, Desalination 145 (2002) 315–319.

[88] S.T. Hsu, K.T. Cheng, J.S. Chiou, Seawater desalination by direct contact membrane distillation, Desalination 143 (2002) 279–287.

[89] B. Li, K.K. Sirkar, Novel membrane and device for direct contact membrane distillation-based desalination process, Ind. Eng. Chem. Res. 43 (2004) 5300–5309.

[90] M. Gryta, Concentration of NaCl solution by membrane distillation integrated with crystallization, Sep. Sci. Tech. 37 (2002) 3535–3558.

[91] C.M. Tun, A.G. Fane, J.T. Matheickal, R. Sheikholeslami, Membrane distillation crystallization of concentrated salts-flux and crystal formation, J. Membr. Sci. 257 (2005) 144–155.

[92] Y. Wu, E. Drioli, The behaviour of membrane distillation of concentrated aqueous solution, Water Treat. 4 (1989) 399–415 (1989).

[93] M. Tomaszewska, Concentration of the extraction fluid from sulphuric acid treatment of phosphogypsum by membrane distillation, J. Membr. Sci. 78 (1993) 277–282.

[94] E. Drioli, B.L. Jiao, V. Calabró, The preliminary study on the concentration of orange juice by membrane distillation, Proc. Int. Soc. Citriculture 3 (1992) 1140–1144.

[95] V. Calabró, B.L. Jiao, E. Drioli, Theoretical and experimental study on membrane distillation in the concentration of orange juice, Ind. Eng. Chem. Res. 33 (1994) 1803–1808.

[96] S. Gunko, S. Verbych, M. Bryk, N. Hilal, Concentration of apple juice using direct contact membrane distillation, Desalination 190 (2006) 117–124.

[97] O.S. Lukanin, S.M. Gunko, M.T. Bryk, R.R. Nigmatullin, The effect of content of apple juice biopolymers on the concentration by membrane distillation, J. Food Eng. 60 (2003) 275–280.

[98] S. Nene, S. Kaur, K.K. Sumod, B. Joshi, K.S.M.S. Raghavarao, Membrane distillation for the concentration of raw-cane sugar syrup and membrane clarified sugarcane juice, Desalination 147 (2002) 157–160.

[99] M.A. Izquierdo-Gil, M.C. García-Payo, C. Fernández-Pineda, Direct contact membrane distillation of sugar aqueous solutions, Sep. Sci. Tech. 34 (1999) 1773–1801.

[100] F. Ali, M. Dornier, A. Duquenoy, M. Reynes, Evaluating transfer of aroma compounds during the concentration of sucrose solutions by osmotic distillation in batch-type pilot plant, J. Food. Eng. 60 (2003) 1–8.

[101] F. Vaillant, E. Jeanton, M. Dornier, G.M. O'Brien, M. Reynes, M. Decloux, Concentration of passion fruit juice on an industrial pilot scale using osmotic evaporation, J. Food. Eng. 47 (2001) 195–202.

[102] V. Calabró, E. Drioli, F. Matera, Membrane distillation in the textile wastewater treatment, Desalination 83 (1991) 209–224.

[103] S. Mozia, A.W. Morawski, M. Toyoda, T. Tsumura, Effect of process parameters on photodegradation of Acid Yellow 36 in a hybrid photocatalysis-membrane distillation system, J. Chem. Eng. 150 (2009) 152–159.

[104] S. Mozia, A.W. Morawski, M. Toyoda, T. Tsumura, Integration of photocatalysis and membrane distillation for removal of mono- and poly-azo dyes from water, Desalination 250 (2010) 666–672.

[105] G. Zakrzewska-Trznadel, A.G. Chmielewski, N.R. Miljevic, Separation of protium/deuterium and oxygen-16/oxygen-18 by membrane distillation, J. Membr. Sci. 113 (1996) 337–342.

[106] G. Zakrzewska-Trznadel, M. Harasimowicz, A.G. Chmielewski, Concentration of radioactive components in liquid low-level radioactive waste by membrane distillation, J. Membr. Sci. 163 (1999) 257–264.

[107] G. Zakrzewska-Trznadel, M. Harasimowicz, A.G. Chmielewski, Membrane processes in nuclear technology-application for liquid radioactive waste treatment, Sep. Purif. Tech. 22-23 (2001) 617–625.

[108] M. Khayet, J.I. Mengual, G. Zakrzewska-Trznadel, Direct contact membrane distillation for nuclear desalination. Part II. Experiments with radioactive solutions, Int. J. Nuclear Desalination 2 (2006) 56–73.

[109] P.P. Zolotarev, V.V. Urgozov, I.B. Volkina, V.N. Nikulin, Treatment of waste water for removing heavy metals by membrane distillation, J. Hazard. Mater. 37 (1994) 77–82.

[110] T.Y. Cath, D. Adams, A.E. Childress, Membrane contactor processes for wastewater reclamation in space, II. Combined direct osmosis, osmotic distillation, and membrane distillation for treatment of metabolic wastewater, J. Membr. Sci. 257 (2005) 111–119.

[111] M. Gryta, K. Karakulski, The application of membrane distillation for the concentration of oil-water emulsions, Desalination 121 (1999) 23–29.

[112] A. El-Abbassi, A. Hafidi, M.C. García-Payo, M. Khayet, Concentration of olive mill wastewater by membrane distillation for polyphenols recovery, Desalination 245 (2009) 670–674.

[113] C. Rincón, J.M. Ortiz de Zárate, J.I. Mengual, Separation of water and glycols by direct contact membrane distillation, J. Membr. Sci. 158 (1999) 155–165.

[114] M. Khayet, A. Velázquez, J.I. Mengual, Direct contact membrane distillation of humic acid solutions, J. Membr. Sci. 240 (2004) 123–128.

[115] M. Khayet, J.I. Mengual, Effect of salt concentration during the treatment of humic acid solutions by membrane distillation, Desalination 168 (2004) 373–381.

[116] S. Srisurichan, R. Jiraratananon, A.G. Fane, Humic acid fouling in the membrane distillation process, Desalination 174 (2005) 63–72.

[117] M. Tomaszewska, M. Gryta, A.W. Morawski, Study on the concentration of acids by membrane distillation, J. Membr. Sci. 102 (1995) 113–122.

[118] M. Tomaszewska, M. Gryta, A.W. Morawski, The influence of salt in solutions on hydrochloric acid recovery by membrane distillation, Sep. Purif. Tech. 14 (1998) 183–188.

[119] M. Tomaszewska, Concentration and purification of fluosilicic acid by membrane distillation, End. Eng. Chem. Res. 39 (2000) 3038–3041.

[120] M. Tomaszewska, M. Gryta, A.W. Morawski, Recovery of hydrochloric acid from metal pickling solutions by membrane distillation, Sep. Purif. Tech. 22-23 (2001) 591–600.

[121] M. Tomaszewska, Mass transfer of HCl and H_2O across the hydrophobic membrane during membrane distillation, J. Membr. Sci. 166 (2000) 149–157.

[122] V.V. Ugrosov, I.B. Elkina, Concentration of binary aqueous solutions by the method of membrane distillation, Theor. Found. Chem. Eng. 32 (1998) 97–100.

[123] D. Qu, J. Wang, D. Hou, Z. Luan, B. Fan, C. Zhao, Experimental study of arsenic removal by direct contact membrane distillation, J. Hazard. Mater. 163 (2009) 874–879.

[124] A.K. Manna, M. Sen, A.R. Martin, P. Pal, Removal of arsenic from contaminated groundwater by solar-driven membrane distillation, Environ. Pollut. 158 (2010) 805–811.

[125] D. Hou, J. Wang, X. Sun, Z. Luan, C. Zhao, X. Ren, Boron removal from aqueous solution by direct contact membrane distillation, J. Hazard. Mater. 177 (2010) 613–619.

[126] Z. Ding, L. Liu, J. Yu, R. Ma, Z. Yang, Concentrating the extract of traditional Chinese medicine, J. Membr. Sci. 310 (2008) 539–549.

[127] Z. Ding, L. Liu, Z. Liu, R. Ma, Fouling resistance in concentrating TCM extract by direct contact membrane distillation, J. Membr. Sci. 362 (2010) 317–325.

[128] K. Sakai, T. Koyano, T. Muroi, M. Tamura, Effects of temperature and concentration polarization on water vapour permeability for blood in membrane distillation, Chem. Eng. J. 38 (1988) B33–B39.

[129] K. Sakai, T. Muroi, K. Ozawa, S. Takesawa, M. Tamura, T. Nakane, Extraction of solute-free water from blood by MD, Am. Soc. Artif. Intern. Organs 32 (1986) 397–400.

[130] J.M. Ortiz de Zárate, C. Rincón, J.I. Mengual, Concentration of bovine serum albumin aqueous solutions by membrane distillation, Sep. Sci. Tech. 33 (1998) 283–296.

[131] A. Capuano, B. Memoli, V.E. Andreucci, A. Criscuoli, E. Drioli, Membrane distillation of human plasma ultrafiltrate and its theoretical applications to haemodialysis techniques, Int. J. Artif. Organs 23 (2000) 415–422.

[132] A. Criscuoli, E. Drioli, A. Capuano, B. Memoli, V.E. Andreucci, Human plasma ultrafiltrate purification by membrane distillation: process optimisation and evaluation of its possible application on-line, Desalination 147 (2002) 147–148.

[133] Y. Fujii, S. Kigoshi, H. Iwatani, M. Aoyama, Selectivity and characteristics of direct contact membrane distillation type experiment: I. Permeability and selectivity through dried hydrophobic fine porous membranes, J. Membr. Sci. 72 (1992) 53–72.

[134] Y. Fujii, S. Kigoshi, H. Iwatani, M. Aoyama, Y. Fusaoka, Selectivity and characteristics of direct contact membrane distillation type experiment: II. Membrane treatment and selectivity increase, J. Membr. Sci. 72 (1992) 73–89.

[135] H. Udriot, S. Ampuero, I.W. Marison, U. von Stokar, Extractive fermentation of ethanol using membrane distillation, Biotechnol. Lett. 11 (1989) 509–514.

[136] M. Gryta, A.W. Morawski, M. Tomaszewska, Ethanol production in membrane distillation bioreactor, Catal. Today 56 (2000) 159–165.

[137] H. Udriot, A. Araque, U. von Stockar, Azeotropic mixtures may be broken by membrane distillation, Cheng. Eng. J. 54 (1994) 87–93.

[138] M. Gryta, Influence of polypropylene membrane surface porosity on the performance of membrane distillation process, J. Membr. Sci. 287 (2007) 67–78.

[139] M. Khayet, M.P. Godino, J.I. Mengual, Possibility of nuclear desalination through various membrane distillation configurations: a comparative study, Int. J. Nuclear Desalination 1 (2003) 30–46.

[140] M. Gryta, M. Tomaszewska, J. Grzechulska, A.W. Morawski, Membrane distillation of NaCl solution containing organic matter, J. Membr. Sci. 181 (2001) 279–287.

[141] M. Tomaszewska, Preparation and properties of flat-sheet membranes from polyvinylidene fluoride for membrane distillation, Desalination 104 (1996) 1–11.

[142] C. Feng, B. Shi, G. Li, Y. Wu, Preparation and properties of microporous membrane from poly(vinylidene fluoride-co-tetrafluoroethylene) (F2.4) for membrane distillation, J. Membr. Sci. 237 (2004) 15–24.

[143] Y. Wu, Y. Kong, X. Lin, W. Liu, J. Xu, Surface-modified hydrophilic membranes in membrane distillation, J. Membr. Sci. 72 (1992) 189–196.

[144] M. Qtaishat, M. Khayet, T. Matsuura, Novel porous composite hydrophobic/hydrophilic polysulfone membranes for desalination by direct contact membrane distillation, J. Membr. Sci. 341 (2009) 139–148.

[145] M. Qtaishat, T. Matsuura, M. Khayet, K.C. Khulbe, Comparing the desalination performance of SMM blended polyethersulfone to SMM blended polyetherimide membranes by direct contact membrane distillation, Desalination and Water Treat. 5 (2009) 91–98.

[146] Z.D. Hendren, J. Brant, M.R. Wiesner, Surface modification of nanostructured ceramic membranes for direct contact membrane distillation, J. Membr. Sci. 331 (2009) 1–10.

[147] P. Peng, A.G. Fane, X. Li, Desalination by membrane distillation adopting a hydrophilic membrane, Desalination 173 (2005) 45–54.

[148] J. Li, Z. Xu, Z. Liu, W. Yuan, H. Xiang, S. Wang, Y. Xu, Microporous polypropylene and polyethylene hollow fiber membranes: Part 3. Experimental studies on membrane distillation for desalination, Desalination 155 (2003) 153–156.

[149] K.Y. Wang, T.S. Chung, M. Gryta, Hydrophobic PVDF hollow fiber membranes with narrow pore size distribution and ultra-skin for the fresh water production through membrane distillation, Chem. Eng. Sci. 63 (2008) 2587–2594.

[150] K.Y. Wang, S.W. Foo, T.S. Chung, Mixed matrix PVDF hollow fiber membranes with nanoscale pores for desalination through direct contact membrane distillation, Ind. Eng. Chem. Res. 48 (2009) 4474–4483.

[151] M.M. Teoh, T.S. Chung, Membrane distillation with hydrophobic macrovoid-free PVDF-PTFE hollow fiber membranes, Sep. Purif. Tech. 66 (2009) 229–236.

[152] M.C. García-Payo, M. Essalhi, M. Khayet, Effects of PVDF-HFP concentration on membrane distillation performance and structural morphology of hollow fiber membranes, J. Membr. Sci. 347 (2010) 209–219.

[153] A. Larbot, L. Gazagnes, S. Krajewski, M. Bukowska, W. Kujawski, Water desalination using ceramic membrane distillation, Desalination 168 (2004) 367–372.

[154] S. Cerneaux, I. Stuzynska, W.M. Kujawski, M. Persin, A. Larbot, Comparison of various membrane distillation methods for desalination using hydrophobic ceramic membranes, J. Membr. Sci. 337 (2009) 55–60.

[155] S. Bonyadi, T.S. Chung, Flux enhancement in membrane distillation by fabrication of dual layer hydrophilic-hydrophobic hollow fiber membranes, J. Membr. Sci. 306 (2007) 134–146.

[156] D. Hou, J. Wang, D. Qu, Z. Luan, X. Ren, Fabrication and characterization of hydrophobic PVDF hollow fiber membranes for desalination through direct contact membrane distillation, Sep. Purif. Tech. 69 (2009) 78–86.

CHAPTER 11

Sweeping Gas Membrane Distillation

OUTLINE

Introduction	295	Effect of Gas Temperature	313
Theoretical Models	300	Effects of Feed Flow Rate	314
Experimental: Effects of Process Conditions	309	Gas Flow Rate	315
Effects of Feed Temperature	309	SGMD Applications	318
Effects of Solutes in Feed Aqueous Solution	311		

INTRODUCTION

In Chapter 1, the different membrane distillation (MD) configurations and their possible applications were outlined. The feed side in all MD modes is similar, whereas changes are made on the permeate side in order to establish the driving force, which is the vapour pressure difference between the feed and permeate sides of the porous hydrophobic membrane module.

Sweeping gas membrane distillation (SGMD), which is rarely studied (i.e. about 4.5% of the papers published up to December 2010 in refereed journals), consists of a gas that sweeps the permeate side of the membrane carrying the vapourous permeate away from the permeate side of the membrane pore. Condensation of the vapour takes place outside the membrane module. Therefore, external condensers are required to collect the vapour in the permeate side stream, complicating in this way the system design and increasing its cost. The schematics of the SGMD process and a system typically used in the laboratory are presented in Fig. 11.1. As can be seen, SGMD involves evaporation of water and volatile molecules at the hot feed side, transport of these molecules through dry pores of hydrophobic membranes driven by the transmembrane vapour pressure difference, collection of the permeating molecules by an inert cold sweeping gas and finally condensation outside the membrane module.

In SGMD, the gas temperature, the heat transfer rate and the mass transport through the membrane change considerably during the gas circulation along the membrane module. The increase of the gas temperature from the inlet

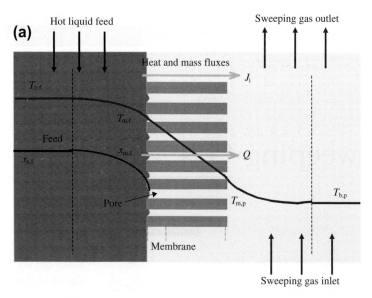
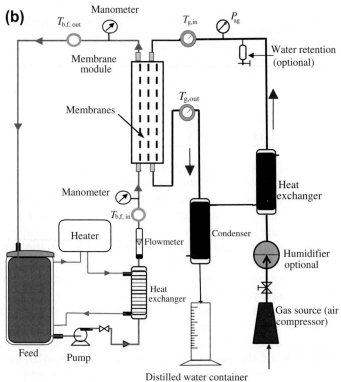

FIGURE 11.1 Schematics of SGMD process when volatile organic solutes are present in the feed (a) and a typical experimental laboratory system (b).

INTRODUCTION

to the outlet of the membrane module can be minimized using a cold surface in the permeate side. In this case, both air gap membrane distillation (AGMD) and SGMD can be combined in another variant of MD termed thermostatic sweeping gas membrane distillation (TSGMD). Figure 11.2 presents this MD mode. It must be mentioned here that a part of the permeating vapour condenses inside the module depending on the operating conditions, while the rest is collected in an external condenser.

It is to be noted that SGMD configuration has a great perspective for the future, because it combines a relatively low conductive heat loss (HL) through the membrane with a reduced mass transfer resistance. In AGMD configuration, there is a gas barrier, which results in reduction in HL by conduction through the membrane. However, the gas in SGMD is not stationary and sweeps the membrane, resulting in higher mass transfer coefficients leading to higher permeate fluxes than in AGMD. Khayet et al. [1], by comparing SGMD and DCMD configuration using the same membrane module and the same feed operating conditions, found higher fluxes (1.4 times) and lower internal HL by conduction in SGMD.

Like DCMD process, SGMD can also be applied for production of high-purity water and concentration of ionic, colloid or other nonvolatile aqueous solutions. However, SGMD is more suitable than DCMD for removal of organic compounds from water because there is no risk of membrane pore wetting from the permeate side.

It is worth quoting that SGMD configuration was used for the first time by Bodell in 1963, describing a system and a method to convert undrinkable aqueous solutions to potable water using a parallel array of tubular silicone membranes having 0.30 mm inner diameter and 0.64 mm outer diameter [2,3]. Air was circulated through the lumen side of the tubular membranes and condensation has been carried out in an external condenser. The recommended water vapour pressure in the air side of the SGMD system was at least 4 kPa below the feed aqueous solution. Moreover, Bodell suggested, for the first time, an alternative means of providing low water vapour pressure in the tubes by applying vacuum, leading to the actual known vacuum membrane distillation (VMD) configuration [2,3]. SGMD process was then considered by Henderyckx and Van Haute in 1967 [4,5]. The authors claimed that the SGMD system can utilize waste hot water and can be employed for solar distillation. However, the system has not been further developed. Since then very few studies have been carried out on SGMD process [6–15].

SGMD technology was applied successfully for desalination of aqueous solutions, first by

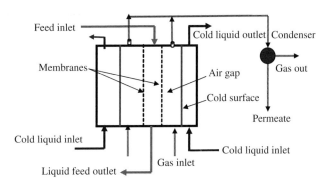

FIGURE 11.2 Schema of TSGMD process.

Khayet et al. [1,11]. Salt rejection factors of practically 100% were achieved. SGMD was applied by Lee and Hong [9] for separation of isopropanol/water mixtures and by Boi et al. [12] for removal of organics (ethanol, acetone) from wastewater [9,12]. Attempts have been made for ammonia removal from water using SGMD by Ding et al. [13] and Xie et al. [14]. The performance of SGMD was compared to that of DCMD and VMD processes [13].

Separation of binary mixtures by TSGMD process was studied theoretically by Rivier et al. [16] and then experimentally with aqueous formic acid solutions by García-Payo et al. [17].

It is worth to mention that the operation in SGMD process is similar to membrane evaporation (ME) [18,19], sweeping gas pervaporation (SPV) [20,21] and membrane air stripping (MAS) [22–26] when using porous and hydrophobic membranes. These processes were applied for removal of organic compounds (chloroform, toluene, trichlroethane, trichloroethylene, tetrachloroethylene, carbon tetrachloride, bromoform, bromodichloromethane, alcohols, etc.) from aqueous mixtures. The same devices and the same membranes (i.e. microporous hydrophobic) used in SGMD can be employed in ME, SPV and MAS instead of dense and selective membranes. Some differences exist. For example, in MAS, the pores of the membrane are permitted to be wetted by the feed aqueous solutions. By contrast, as it is indicated throughout this book for all MD configurations, in SGMD the pores must be maintained dry.

As it is indicated in Chapter 2, the number of papers published on fabrication of membranes for MD process starts growing. However, very few research groups have considered the possibility of manufacturing novel membranes and membrane modules designed specifically for SGMD [18,19]. Most of the developed membranes were tested in DCMD process. For example, the French group from the *Institut Européen des Membranes* in Montpellier [18,19] proposed the use of a surface-modified flat metallic (stainless steel) microfiltration membranes for SGMD configuration. Different structural characteristics of the metallic microfiltration membrane support have been considered (i.e. one layer with a granular structure, fibrous supported membrane and a thin separative layer of granular structure with a fibrous support from Pall Corporation). To render the metallic membrane hydrophobic, a surface treatment was applied consisting of deposition of a very thin film of a silicone compound. The deposition was carried out by immersion of the metallic membrane in a solution containing 2% w/w silicone and a cross-linking agent (3,5-dimethylhex-1-yne-3-ol) in hexane for 15 min. Subsequently, the membrane was drained to remove the excess of solution and then dried at 60 °C for 24 h. The obtained compositae membranes exhibited different characteristics (i.e. structure, porosity, pore size). Compared to the membranes commonly used in MD, these hydrophobic metallic membranes have greater pore sizes (>2.6 μm, reaching 5 μm) but lower permeate fluxes (0.12 kg/h·m^2 with a sweep air temperature of 20 °C and a feed water temperature of 25 °C) [18]. A heat and mass transport model was validated with experimental results using water as feed [19]. Unfortunately, the proposed hydrophobic metallic membranes are not used yet for MD separation of dissolved compounds from water such as desalination or for the treatment of contaminated aqueous solutions.

Table 11.1 summarizes the highest reported SGMD permeate flux (J) of different types of membranes tested under different operating conditions. In SGMD desalination, the highest permeate flux was 14.4 kg/m^2·h and 11.2 kg/m^2·h for distilled water used as feed with an electrical conductivity of 9 μS/cm and NaCl feed aqueous solution of 182 μS/cm, respectively, with a feed temperature of 65 °C, air temperature of 20 °C and PP Accurel® S6/2 MD020CP2N membrane module (Table 2.2) [11]. In this last case, the obtained permeate

INTRODUCTION

TABLE 11.1 Reported SGMD Total Permeate Flux (J) of Different Types of Membranes Shown in Tables 2.1 and 2.2

Membrane trade name	J (10^{-3} kg/m²s)	Observation	Reference
TF200	5.2	SGMD, distilled water as feed, sweeping humid air, $T_{b,f}$ = 70 °C, $T_{b,p}$ = 20 °C, v_l = 0.15 m/s, v_g = 1.5 m/s	[7,8]
	0.44	SGMD, distilled water as feed, sweeping dry nitrogen, $T_{b,f}$ = 35 °C, $T_{b,p}$ = 32 °C, ϕ_l = 200 l/h, ϕ_g = 0.63 m³/h	
	~1.32	SGMD, acetone/water mixture, sweeping dry nitrogen, $T_{b,f}$ = 35 °C, $T_{b,p}$ = 33 °C, ϕ_l = 200 l/h, ϕ_g = 0.63 m³/h	
	~1.06	Acetone concentration 10 wt%, acetone in permeate ~0.5 wt%	
		Acetone concentration 5 wt%, acetone in permeate ~0.25 wt%	[12]
	~0.56	Acetone concentration 2 wt%, acetone in permeate ~0.075 wt%	
	~0.53	SGMD, ethanol/water mixture, sweeping dry nitrogen, $T_{b,f}$ = 35 °C, $T_{b,p}$ = 33 °C, ϕ_l = 200 l/h, ϕ_g = 0.65 m³/h	
		Ethanol concentration 5 wt%, ethanol in permeate ~0.2 wt%	
	~1.06	Ethanol concentration 2 wt%, ethanol in permeate ~0.1 wt%	
	5.28	TSGMD, sweeping air, distilled water, $T_{b,f}$ = 70 °C, $T_{b,p}$ = 30 °C, v_g = 2.6 m/s	[17]
	6.94	TSGMD, sweeping air, $T_{b,f}$ = 75 °C, $T_{b,p}$ = 25 °C, v_g = 2.6 m/s. Formic acid/water (feed molar fraction 0.7), formic acid selectivity ~0.78	
TF450	6.10	SGMD, distilled water as feed, sweeping humid air, $T_{b,f}$ = 70 °C, $T_{b,p}$ = 20 °C, v_l = 0.15 m/s, v_g = 1.5 m/s	[7,8]
PTFE Hollow fibre POREFLON	0.55	SGMD, isopropanol IPA/water mixture, dry sweeping nitrogen, $T_{b,f}$ = 46 °C, inlet $T_{b,p}$ = 25 °C. IPA concentration 9.5 wt%, $T_{b,f}$ = 46 °C, $T_{b,p}$ = 25 °C, v_g = 8.7 m/s, ϕ_l = 140 kg/h. IPA selectivity ~16.5	[9]
	0.28	IPA concentration 5 wt%, ϕ_l = 12.9 kg/h. Reynolds number in permeate = 1000. IPA selectivity ~35.5	
PP Accurel® S6/2 MD020CP2N	5.0	SGMD, sweeping humid air, distilled water as feed, $T_{b,f}$ = 70 °C, $T_{b,p}$ = 20 °C, v_l = 0.8 m/s, v_g = 11.3 m/s	[11]
	4.0	SGMD, sweeping humid air, $T_{b,f}$ = 65 °C, $T_{b,p}$ = 20 °C, (air). Distilled water as feed (9 µS/cm)	
	3.1	NaCl aqueous solution (182 µS/cm). NaCl rejection factor >99.9%	

$T_{b,f}$, Bulk feed temperature; $T_{b,p}$, bulk gas temperature; ϕ_l, liquid flow rate; v_l, liquid circulation velocity; ϕ_g, gas flow rate; v_g, gas circulation velocity.

flux exhibited an electrical conductivity lower than 9 µS/cm corresponding to a salt rejection factor greater than 99.9%.

THEORETICAL MODELS

When SGMD process is used for the treatment of aqueous solutions containing non-volatile solutes such as salts, colloids and proteins, the same heat and mass transfer equations presented previously for DCMD configuration in Chapter 10 can be adopted in SGMD process in which air or any other inert gas is used as water vapour carriers prior to condensation.

The water vapour permeate flux, J_w, is also written as a function of the transmembrane water vapour pressure difference, Δp_w, and the applied bulk water vapour pressure difference, $\Delta p_{w,b}$:

$$J_w = B'_w \Delta p_{w,b} = B_w \Delta p_w$$
$$= B_w(p^0_{w,f} a_{w,f} - p_{w,p}) \quad (11.1)$$

where B_w is the SGMD coefficient of the membrane (i.e. membrane permeability), B'_w is the global mass transfer coefficient of the membrane that includes the effects of the feed and permeate boundary layers, a_w is the activity of water and p is the partial pressure of water. The superscript 0 is for pure water and the subscripts w, f and p refer to water, feed and permeate, respectively.

The partial pressure of pure water, $p^0_{w,f}$, and the activity $a_{w,f}$ depend on the temperature of the feed aqueous solution at the membrane surface ($T_{m,f}$); whereas the partial pressure at the permeate side $p_{w,p}$ depends on the temperature of the gas at the membrane surface ($T_{m,p}$) and can be written as a function of the total pressure in the permeate side (P) and the humidity ratio (w) defined for a given moist air sample as the quotient between the mass of water vapour and the mass of dry air [7,10,11]:

$$p_{w,p} = \frac{w P}{w + 0.622} \quad (11.2)$$

The humidity ratio along the membrane module, w, can be related to the gas flow rate ($\dot{m}_a$), the humidity ratio at the membrane module inlet (w_{in}), the SGMD permeate water vapor flux (J_w) and the effective membrane area (A):

$$w = w_{in} + \frac{J_w A}{\dot{m}_a} \quad (11.3)$$

Equations (11.1)–(11.3) can be rearranged in a second-degree equation for the permeate water vapour flux:

$$J_w^2 + bJ_w + c = 0 \quad (11.4)$$

where the coefficients b and c are as follows:

$$b = (w_{in} + 0.622)\frac{\dot{m}_a}{A} + B_w(P - p^0_{w,f} a_w) \quad (11.5)$$

$$c = B_w \frac{\dot{m}_a}{A}\left[P w_{in} - p^0_{w,f} a_w (w_{in} + 0.622)\right] \quad (11.6)$$

The theoretical SGMD permeate flux can be estimated from Eqs. (11.4)–(11.6) when feed aqueous solutions containing non-volatile solutes are treated by SGMD process using air as sweeping gas. In other words, only vapourous water molecules are transported through the membrane pores. Calculations were carried out using the commercial membrane module Accurel® S6/2 MD020CP2N (see Table 2.2) under different operating conditions and good agreements were found between the predicted and experimental SGMD permeate fluxes [1,11].

In the case of feed aqueous solutions containing volatile compounds, water vapour molecules together with the volatile compounds are transported through the membrane pores and the separation factor of species i can be calculated as follows:

$$\alpha_i = \frac{y_{i,p}/(1-y_{i,p})}{x_{i,f}/(1-x_{i,f})} \quad (11.7)$$

where $y_{i,p}$ and $x_{i,f}$ are the mole fractions of the volatile compound (i.e. species i) in the gas permeate (p) and liquid feed (f), respectively.

In this case, if the coupling effects are not taken into consideration, the total SGMD permeate flux, J, can be expressed as:

$$J = \sum_j B_j(p^0_{j,f}a_{j,f} - p_{j,p})$$

$$= \sum_j B_j(p^0_{j,f}\gamma_{j,f}x_{j,f} - f_{j,p}Py_{j,p}) \quad (11.8)$$

where γ_j, x_j, y_j and f_j are the activity coefficient, the mole fraction in the liquid phase, the mole fraction in the gas phase and the fugacity coefficient of the specie j, respectively. The subscript j refers to all volatile compounds including water.

As it was reported in the previous chapter, the mean free path of the volatile molecules that are transported through the membrane pores, λ_j, is the governing magnitude, providing a guideline in determining the mechanism of mass transport in a given pore under given experimental conditions. This parameter can be calculated using the following expression [27]:

$$\lambda_j = \frac{k_B T}{\sqrt{2}\pi P_m \sigma_j^2} \quad (11.9)$$

where σ_j is the collision diameter of the species j, k_B is the Boltzmann constant, P_m is the mean pressure within the membrane pores and T is the absolute temperature.

For a binary mixture (i and j) in air, the mean free path can be evaluated by the following equation:

$$\lambda_{i/j} = \frac{k_B T}{\pi P_m \left(\frac{\sigma_i + \sigma_j}{2}\right)^2} \frac{1}{\sqrt{1 + M_j/M_i}} \quad (11.10)$$

where σ_i and σ_j are the collision diameters and M_i and M_j the molecular weight of the volatile compounds i and j, respectively.

If the mean free path of the transported molecules is large in relation to the membrane pore size, the molecule–pore wall collisions are dominant over the molecule–molecule collisions and the mass transport takes place via Knudsen type of flow (see Fig. 10.3). In this case, the SGMD coefficient of the membrane, B_j in Eq. (11.8), can be calculated by means of the following expression [8]:

$$B_j^K = \frac{2}{3RT}\frac{\varepsilon \langle r \rangle}{\tau \delta}\left(\frac{8RT}{\pi M_j}\right)^{1/2} \quad (11.11)$$

where $\langle r \rangle$ is the mean pore radius of the membrane, M_j is the molecular weight of the species j, R is the gas constant, ε is the membrane porosity, T is the absolute temperature and δ the membrane thickness.

If the membrane pore size is large in relation to the mean free path of the transported molecules, molecule–molecule collisions are dominant over the molecule–pore wall collisions and the mass transport takes place via Poiseuille type of flow also called viscous type of flow as schematized in Fig. 11.3.

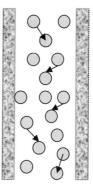

FIGURE 11.3 Poiseuille type of flow through a pore of a membrane used in SGMD process.

In this case, B_j can be calculated using the following expression [8]:

$$B_j^V = \frac{\varepsilon <r>^2}{\tau \delta} \frac{P_m}{8\eta_j RT} \quad (11.12)$$

where η_j is the viscosity of species j and P_m is the average hydrostatic pressure in the membrane pores.

If the ordinary molecular diffusion through the air molecules trapped within the membrane pores is dominant, the following relationship can be used to determine the SGMD coefficient of the membrane [8]:

$$B_j^D = \frac{1}{RT} \frac{PD_j}{p_a} \frac{\varepsilon}{\delta \tau} = \frac{1}{RT} \frac{D_j}{Y_{lm}} \frac{\varepsilon}{\delta \tau} \quad (11.13)$$

where D_j is the diffusion coefficient of the species j, P is the total pressure, p_a is the air pressure in the membrane pore and Y_{lm} is the mole fraction of air (log-mean) defined as:

$$Y_{lm} = \left(\frac{Y_{a,m,f} - Y_{a,m,p}}{\ln(Y_{a,m,f}/Y_{a,m,p})} \right) \quad (11.14)$$

If the total pressures on both sides of the membranes are similar, viscous (i.e. Poiseuille) type of flow is negligible and the mass transport will take place via a combined Knudsen/ordinary molecular diffusion mechanism. In this case, the following model was used to determine the membrane permeability following an association in series of the mass transfer resistances as schematized in Fig. 11.4 [1,8,10,11].

$$B_j^C = \left(\frac{1}{B_j^K} + \frac{1}{B_j^D} \right)^{-1} \quad (11.15)$$

In terms of mass transfer resistances:

$$R_j^C = R_j^K + R_j^D \quad (11.16)$$

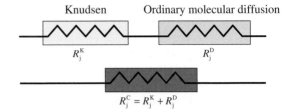

FIGURE 11.4 Electrical circuit analogue presenting the combined Knudsen/ordinary molecular diffusion transport mechanisms by means of mass transfer resistances through the membrane.

Figure 11.5 illustrates the experimental and calculated SGMD permeate fluxes using Knudsen model, Poiseuille model and combined Knudsen/ordinary molecular diffusion model for different liquid feed temperatures. Uniform pore sizes (i.e. mean pore sizes) and tortuosity factors of 1.1 were assumed for the membranes TF200 and TF450 (Table 2.1). It can be seen that under the same SGMD operating conditions the calculated permeate flux considering Knudsen model is greater than the combined Kndusen/ordinary molecular diffusion model and Poiseuille model exhibits the smallest calculated SGMD permeate flux, which is even smaller than the experimental values. Khayet et al. [8] observed this result also for other membranes and concluded that the mass transport mechanism takes place via the combined Knudsen/ordinary molecular diffusion for the studied membranes having mean pore sizes between 0.2 μm and 1 μm. Moreover, the SGMD permeate flux was found to be higher for membranes having larger pore sizes.

To be more rigorous, the contribution of each mass transfer mechanism in Eq. (11.15) is not necessarily equal. Khayet et al. [8] reported that the ordinary molecular diffusion is dominant and its contribution is greater for membrane having wider pore size. It was found that the contribution of the ordinary molecular diffusion

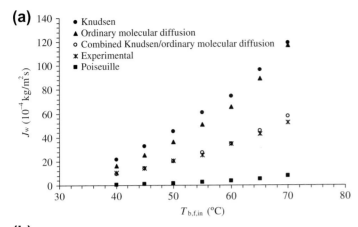
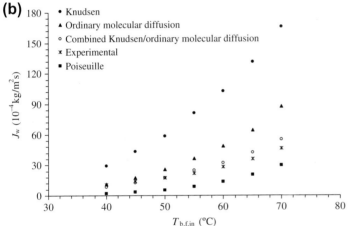
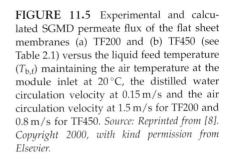

FIGURE 11.5 Experimental and calculated SGMD permeate flux of the flat sheet membranes (a) TF200 and (b) TF450 (see Table 2.1) versus the liquid feed temperature ($T_{b,f}$) maintaining the air temperature at the module inlet at 20 °C, the distilled water circulation velocity at 0.15 m/s and the air circulation velocity at 1.5 m/s for TF200 and 0.8 m/s for TF450. *Source: Reprinted from [8]. Copyright 2000, with kind permission from Elsevier.*

varies between 50% and 58% for the membrane TF200 and between 69% and 75% for the membrane TF450 having a larger pore size.

In SGMD process, the pore size distribution has not yet been considered for prediction of the permeate flux. In this case, more than one mechanism can take place in the porous membrane. Studies on the effects of pore size distribution on the SGMD permeate flux may help understanding the nature of mass transport in SGMD configuration through membranes with pore size distributions instead of a uniform pore size, e.g. the mean pore size of the membrane. The same equations proposed in Chapter 10 for DCMD process can be used in SGMD.

Most of MD authors believe that the MD membrane acts only as a barrier to hold the liquid/vapour interfaces and it is not involved in the transport phenomena and the separation performance is predominantly determined by the vapour/liquid equilibrium (VLE). The solution-diffusion flow through the membrane material is not considered in spite of the possible existence of high "affinity" between the membrane matrix and the organic compound. Rivier et al. [16] and García-Payo et al. [17] indicated that selectivity in MD depends on both the differences in volatility and diffusion rates of the components across the membrane and the gas gap in TSGMD process. Finally, it was explained that the separation performance in MD is based not only on VLE of the involved components in the feed mixture but also on the operating temperature (thermodynamic effects) and on the kinetic effects (diffusion of the components through the membrane and the gas gap). Therefore, the contribution of the membrane material in the selectivity and permeability of the membrane in SGMD should be rigorously investigated using membranes with different materials and characteristics.

The dusty gas model reported in Chapter 10 for DCMD process can also be applied to predict the permeate flux of SGMD process. This model combines all the previously cited transport mechanisms. However, the dusty gas model has not yet been applied for SGMD.

A Stefan–Maxwell-based model was applied by Rivier et al. [16] for the treatment of binary aqueous solutions by the TSGMD process, which involves multicomponent diffusion. For a system of n components, Stefan–Maxwell equations provide $(n-1)$ relations, which form a differential system written in a matrix form as follows:

$$\widehat{J} = c_t [\beta][\kappa][\xi] \Delta \widehat{p} \tag{11.17}$$

where $\widehat{J}$ is the vector permeate flux composed by all individual fluxes, c_t is the gas molar density, $[\beta]$ is the "bootstrap" matrix, $[\kappa]$ is the matrix of multicomponent mass transfer coefficients, $[\xi]$ is the matrix of high flux correction factors and $\Delta \widehat{p}$ is the vector driving force composed by all individual driving forces.

The matrix $[\beta]$ for binary aqueous solutions is written as [16]:

$$[\beta] = \begin{bmatrix} 1 + \dfrac{y_w}{y_a} & \dfrac{y_w}{y_a} \\ \dfrac{y_i}{y_a} & 1 + \dfrac{y_i}{y_a} \end{bmatrix} \tag{11.18}$$

and the matrix $[\kappa]$ as:

$$[\kappa] = \begin{bmatrix} \dfrac{k_{w,a}(y_w k_{i,a} + (1-y_w)k_{w,i})}{y_w k_{i,a} + y_i k_{w,a} + y_a k_{w,i}} & \dfrac{y_w k_{i,a}(k_{w,a} - k_{w,i})}{y_w k_{i,a} + y_i k_{w,a} + y_a k_{w,i}} \\ \dfrac{y_i k_{w,a}(k_{i,a} - k_{w,i})}{y_w k_{i,a} + y_i k_{w,a} + y_a k_{w,i}} & \dfrac{k_{i,a}(y_i k_{w,a} + (1-y_i)k_{w,i})}{y_w k_{i,a} + y_i k_{w,a} + y_a k_{w,i}} \end{bmatrix} \tag{11.19}$$

where y and k are the gaseous molar fraction and mass transfer coefficient for a binary mixture, respectively. The subscripts w, a and i refer to water, air and the volatile species i present in the feed aqueous solution, respectively.

The matrix of correction factors is expressed by [28,29]:

$$[\xi] = [\phi] \{\exp [\phi] - [I]\}^{-1} \tag{11.20}$$

where the coefficients ϕ_{ii} and ϕ_{ij} are written as:

$$\phi_{ii} = -\dfrac{RT}{P} \left(\dfrac{J_i}{k_{i,n}} + \sum_{k=1, k \neq i}^{n} \dfrac{J_k}{k_{i,k}} \right) \tag{11.21}$$

$$\phi_{ij} = -\dfrac{J_i RT}{P} \left(\dfrac{1}{k_{i,j}} - \dfrac{1}{k_{i,n}} \right) \tag{11.22}$$

In Eqs. (11.21) and (11.22), the subscripts i, j, k and n are indexes denoting the component number.

Stefan–Maxwell-based model was validated by García-Payo et al. [17], comparing the predicted and experimental permeate fluxes as well as the the selectivity of a binary aqueous mixture containing formic acid and water, which was treated by TSGMD under different operating conditions. There was good agreement between the predicted and experimental values. In their study, only one membrane was used [17]. For a more rigorous test of the Stefan–Maxwell model, prediction should be made for membranes of different materials and with different characteristics, especially of different pore sizes.

A simplified theoretical model was used by Boi et al. [12] to describe mass transfer in SGMD for the treatment of wastewaters containing volatile organic compounds (VOCs) such as acetone and ethanol. Molecular diffusion was considered as the prevailing transport mechanism through the membrane. In this model, the coupling effects between VOCs and water were assumed to be negligible and the permeate flux of an individual compound i transported through the membrane was described by the following equation:

$$J_t = k_{im} c_m \ln\left(\frac{J_i/J_t - y_{i,m,p}}{J_i/J_t - y_{i,m,f}}\right) \quad (11.23)$$

where k_{im} is the mass transfer coefficient of the compound i in the membrane, c_m is the total mole concentration in the membrane pores, y is the mole fraction in the gas phase and the subscript t refers to total permeate flux. While the total permeate flux was predicted correctly by their approach, the partial organic permeate flux was overestimated. This may be attributed partly to the technique used to analyze the permeate because loss of some VOCs may have occurred.

The heat transport in SGMD process is also described by three steps: (i) heat transport through the liquid feed boundary layer (Q_f); (ii) heat transport though the membrane (Q_m) and (iii) heat transport through the gas permeate boundary layer (Q_a).

The heat fluxes and the thermal resistances in SGMD are similar to those shown previously for DCMD process (Fig. 10.5).

As it was reported for DCMD process, the heat transfer within the membrane in SGMD takes place by two mechanisms: (i) conduction across the membrane material and its gas-filled pores (Q_c) and (ii) latent heat associated with the vapourized molecules (Q_v). Therefore, the following equation can be used in SGMD configuration:

$$Q_m = Q_c + Q_v = \frac{k_m}{\delta}(T_{m,f} - T_{m,p}) + \sum_i J_i \Delta H_{v,i}$$

(11.24)

where k_m is the thermal conductivity of the membrane, which can be estimated as it was explained in Chapter 10, $\Delta H_{v,i}$ is the evaporation enthalpy of the species i transported through the membrane pores with a transmembrane permeate flux J_i and $T_{m,f}$ and $T_{m,p}$ are the feed and permeate temperatures at the membrane surface, respectively.

The thermal efficiency, η, of an SGMD process is also defined as:

$$\eta\,(\%) = \frac{Q_v}{Q_v + Q_c} \times 100$$

$$= \frac{\sum_i J_i \Delta H_{v,i}}{k_m(T_{m,f} - T_{m,p}) + \sum_i J_i \Delta H_{v,i}} \times 100$$

(11.25)

When using the same membrane module and the same feed operating conditions, it was observed that the thermal efficiency was higher in SGMD configuration than in DCMD because the internal heat loss (HL) by conduction, Q_c, through the membrane was lower and decreased with the increase of the liquid feed temperature [1]. Operating at high temperatures increases the evaporation efficiency due to the exponential increase of the SGMD permeate flux with the feed temperature. These results indicate that the energy consumption per unit of produced distillate may be reduced appreciably at high operating feed temperatures. It was found that 9.5–28.6% of the total heat transferred in the SGMD process is lost by conduction, whereas in DCMD process, the HL is 58.9–82.3%.

The heat transfer through each individual heat transfer resistance, i.e. liquid boundary layer, membrane and gas boundary layer, should be equal to the overall heat transfer through the SGMD system and therefore

$$Q = h_f(T_{b,f} - T_{m,f})$$
$$= \frac{k_m}{\delta}(T_{m,f} - T_{m,p}) + \sum_i J_i \Delta H_{v,i}$$
$$= h_p(T_{m,p} - T_{b,p}) = H(T_{b,f} - T_{b,p}) \quad (11.26)$$

where h_f and h_p are the heat transfer coefficients on the liquid and gas boundary layers, respectively, and H is the overall heat transfer coefficient of the SGMD process.

The total heat flux, Q, transferred from the liquid feed to the gas permeate side can be expressed as:

$$Q = \left(\frac{1}{h_f} + \frac{1}{(k_m/\delta) + \sum_i J_i \Delta H_{v,i}/(T_{m,f} - T_{m,p})} + \frac{1}{h_p}\right)^{-1} (T_{b,f} - T_{b,p}) = H(T_{b,f} - T_{b,p}) \quad (11.27)$$

At steady-state conditions, when air is used as sweeping gas and no heat is lost from the membrane module to the surroundings, the heat transfer can be calculated according to:

$$Q = Q_a = \frac{\dot{m}_a(h_{a,out} - h_{a,in})}{A} \quad (11.28)$$

where Q_a is the heat transfer in the permeate and $h_{a,out}$ and $h_{a,in}$ are the specific air enthalpies at the membrane module outlet and inlet, respectively.

The enthalpy of moist air is expressed as [8]:

$$h_a = c_{h,a}T_a + w\Delta H_v^0 = (c_a + wc_{w,v})T_a + w\Delta H_v^0 \quad (11.29)$$

where $c_{h,a}$ is the specific heat of the humid air, ΔH_v^0 is the heat of vapourization of water at a temperature of 0 °C, c_a is the specific heat of the dry air, w is the humidity ratio and $c_{w,v}$ is the specific heat of the water vapour.

At steady state, when only water is transported through the membrane pores, Eqs. (11.3), (11.28) and (11.29) yield the following equation for the heat transfer on the permeate side:

$$Q_a = \frac{\dot{m}_a(c_a + w_{in}c_{w,v})(T_{a,out} - T_{a,in})}{A} + J_w(\Delta H_v^0 + c_{w,v}T_{a,out}) \quad (11.30)$$

As mentioned previously, the gas temperature along the membrane module changes considerably, resulting in a large difference in the driving force from the inlet to the outlet of the membrane module. Consequently, it is more informative to predict the local permeate flux $J_i(x)$ along the membrane module length (L) instead of a global SGMD permeate flux. Integrating along the entire module length:

$$J_i = \frac{1}{L}\int_0^L J_i(x)dx \quad (11.31)$$

Therefore, it is necessary to determine the temperature and concentration profiles along the membrane module length. Extensive two-dimensional theoretical models have been proposed by Khayet et al. [7,8,10] and Charfi et al. [15], permitting the determination of temperatures and concentrations in a plate and frame SGMD module as well as the local and global permeate fluxes. The effects of various SGMD operating parameters have been studied. Charfi et al. [15] performed a two-dimensional numerical simulation model based on Navier–Stokes equations coupled with Darcy–Brinkman–Forcheimer formulation in transient regime for SGMD process. A solver based on compact Hermitian method has been used to solve the derivative equations. The computed results obtained from numerical simulations were validated by comparison with experimental data. Figure 11.6 shows, as an example, the obtained temperature and concentration profiles in the whole membrane module. Good agreement was obtained between the simulated and experimental results.

The temperature and concentration polarization coefficients defined for DCMD in Chapter 10 can also be applied for SGMD. In the latter process, like in AGMD, the heat transfer coefficient in the permeate side is much smaller than that in the liquid feed side and therefore will dominate the overall heat transfer coefficient, H (Eq. 11.27). In other words the temperature polarization is found to be localized at the permeate side in SGMD. Hence small changes in the sweeping gas flow rate have much effects on the permeate flux. Therefore, gas flow rate is more likely to control the SGMD process [7,10].

As heat is removed from the liquid feed to vapourize the permeate, the temperature in the liquid/membrane interface decreases and the gas temperature increases. This implies a decrease in the temperature difference as driving force, leading to a decrease in the vapour pressure gradient across the membrane and a corresponding reduction in the SGMD permeate flux. If only non-volatile solutes are present in the feed aqueous solution, during SGMD process, the solute concentration at the liquid feed/membrane interface becomes greater than that in the bulk feed solution resulting in a decrease of the permeate flux. However, if the feed solution contains volatile compounds other than water, the concentrations of the volatile compounds at the liquid/membrane surface become smaller than the corresponding concentrations in the bulk feed aqueous solution, resulting in an increase of the permeate fluxes of the volatile compounds.

The temperature polarization coefficient (θ) can be split into a function of the temperature polarization coefficient of the liquid feed aqueous solution (θ_f) and that of the gas permeate side (θ_p):

$$\theta = \frac{T_{m,f} - T_{m,p}}{T_{b,f} - T_{b,p}}$$
$$= \frac{T_{m,f} - T_{b,p}}{T_{b,f} - T_{b,p}} + \frac{T_{b,f} - T_{m,p}}{T_{b,f} - T_{b,p}} - 1 = \theta_f + \theta_p - 1$$
(11.32)

If the heat transfer through the sweeping gas phase is very high, the temperatures $T_{m,p}$ and $T_{b,p}$ are very similar. This means that θ_p approaches unity. In this case, Eq. (11.32) shows that the temperature polarization effects in the liquid phase are important, and the SGMD process is controlled by the heat transfer resistance of the feed aqueous layer and the mass transfer resistance of the membrane.

If the heat transfer through the feed aqueous solution is very high, the temperatures $T_{m,f}$ and $T_{b,f}$ are very similar and θ_f approaches unity. In this case, the temperature polarization effects in the gas phase are important. Therefore, the heat transfer resistance of the gas layer and the mass transfer resistance of the membrane control the SGMD process.

The effects of different SGMD operating parameters on the temperature polarization coefficient have been studied and reported [7,10,15]. It was found that θ is temperature and flow rate dependent. It becomes more significant at higher temperatures (i.e. higher SGMD fluxes) and at lower flow rates [7,10]. The values of θ increased slightly with the feed flow rate, increased more visibly with the air circulation velocity, decreased slightly with the gas (i.e. humid air) inlet temperature and decreased more visibly with the water inlet temperature. Moreover, only a slight difference

308 11. SWEEPING GAS MEMBRANE DISTILLATION

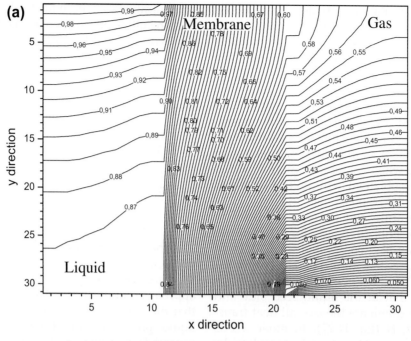

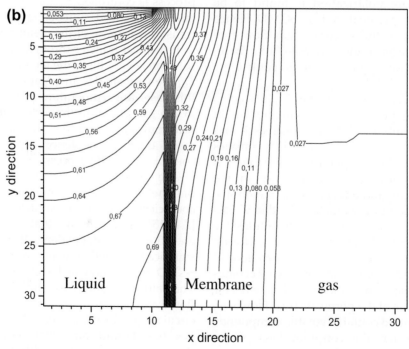

between θ values of two PTFE membranes of different pore sizes (TF200 and TF450 indicated in Table 2.1) was observed, indicating that the mass transfer is predominantly controlled by heat transfer through the boundary layers adjacent to the membrane surfaces. The effect of the gas boundary layer is stronger than that of the liquid boundary layer, due to much larger heat transfer coefficient in the liquid phase than in the gas phase.

When using plate and frame membrane module, Khayet et al. [7,10] obtained θ values lower than 0.44. These values were attributed to the values of θ_p.

As mentioned previously, in some membrane modules, the temperatures vary along the module length (x-axis). Therefore, a local temperature polarization coefficient is defined as indicated in Eq. (11.33) from which the global θ value can be calculated.

$$\theta = \frac{1}{L}\int_0^L \frac{T_{m,f}(x) - T_{m,p}(x)}{T_{b,f}(x) - T_{b,p}(x)} dx \quad (11.33)$$

It is worth quoting that an extensive study on concentration polarization effects on SGMD performance has not yet been performed. This should be done for aqueous solutions containing non-volatile solutes as well as organic volatile compounds.

EXPERIMENTAL: EFFECTS OF PROCESS CONDITIONS

Effects of Feed Temperature

In SGMD process, like in DCMD, the feed temperature ranges between 30 °C and the boiling point of the feed aqueous solution. The change in partial vapour pressure corresponding to the same temperature change becomes more as the temperature increases. For example, a temperature drop of 0.1 °C at a temperature of 30 °C results in a variation in the vapour pressure of approximately 24.4 Pa, whereas at 70 °C, the vapour pressure changes by 135 Pa.

The liquid feed temperature together with the sweeping gas flow rate was found to be the important parameters controlling the SGMD flux. In fact, the SGMD permeate flux is strongly dependent on the feed temperature because of the exponential increase of the vapour pressure with temperature. Note that, due to the additional effect of the temperature polarization, the permeate flux does not increase with the liquid feed temperature as fast as the vapour pressure curve. Figure 11.7 shows the effect of the liquid inlet temperature on both the SGMD flux and the temperature polarization coefficients for two commercial membranes of different pore sizes. As mentioned earlier, the solid lines refer to the applied combined Knudsen/ordinary molecular diffusion model (Eq. 11.15). As can be observed, the agreement between the model predictions and the experimental data is very good.

It is worth noticing that two opposite contributions are taking place simultaneously when the liquid feed temperature is increased. An increase in the liquid feed temperature implies not only an increase in the temperature polarization effect but also an increase in the SGMD permeate flux. The temperature polarization coefficient of the liquid feed aqueous solution, θ_f, is close to unity (i.e. 5% contribution to the temperature polarization coefficient θ), whereas the temperature polarization coefficient of the gas permeate side, θ_p, is much lower (<0.4).

FIGURE 11.6 Temperature (a) and concentration (b) line distributions in a plate and frame membrane module (TF200, see Table 2.1). (a) Feed inlet temperature 70 °C, air inlet temperature 20 °C, liquid circulation velocity 0.15 m/s, air circulation velocity 0.8 m/s, distilled water as feed; (b) Feed inlet temperature 50 °C, air inlet temperature 20 °C, liquid circulation velocity 0.15 m/s, air circulation velocity 0.8 m/s, salt concentration in feed 40 g/l. Source: Reprinted from [15]. Copyright 2010, with kind permission from Elsevier.

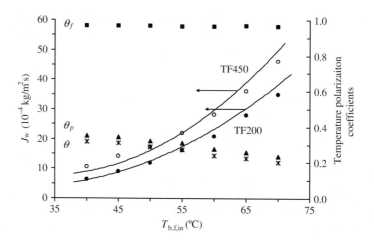

FIGURE 11.7 Effect of feed inlet temperature on experimental and predicted SGMD permeate flux of two flat sheet membranes TF200 and TF450 (see Table 2.1) as well as on the temperature polarization coefficients of the membrane TF200. Air inlet temperature 20 °C, liquid circulation velocity 0.15 m/s, air circulation velocity 1.5 m/s and distilled water used as feed. *Source: Reprinted from [10]. Copyright 2002, with kind permission from John Wiley & Sons.*

This indicates that the temperature polarization is an important factor affecting the SGMD permeate flux, and the air boundary layer exerts the greatest effect in the SGMD process. It can be seen in Fig. 11.7 that θ_f remains constant with the increase of the liquid feed temperature, while θ and θ_p decrease. Similar trends have also been observed for other membranes.

The effects of the feed temperature on the thermal efficiency, η, and the internal HL by conduction through the membrane for each kilogram of the produced permeate flux are shown in Fig. 11.8 for both DCMD and SGMD processes in which the same membrane module was operated under the same liquid feed operating conditions. The thermal efficiency increases, while HL decreases with an increase in the feed liquid temperature. The HL trend is more remarkable for DCMD. On the other hand, the thermal efficiency is higher for the SGMD configuration. This is attributed partly to the lower values of HL for SGMD compared to DCMD. These results explain the higher permeate flux obtained for SGMD, which was found to be approximately 1.4 times higher than DCMD.

When aqueous salt solutions were treated by SGMD process at different feed temperatures,

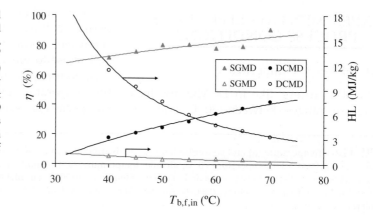

FIGURE 11.8 Effect of feed inlet temperature on the thermal efficiency (η) and the internal HL of the membrane module Accurel® S6/2 MD020CP2N (see Table 2.2) used in DCMD (permeate flow velocity 0.8 m/s, permeate inlet temperature 20 °C) and in SGMD (gas flow velocity 11.3 m/s, gas inlet temperature 20 °C) with a feed circulation velocity of 0.8 m/s.

no effect was observed on the salt rejection factor (i.e. 99.9%) [1,11].

The effects of the liquid feed temperature on the overall mass transfer coefficient of ammonia and selectivity (i.e. preferential transport of ammonia) were studied by Ding et al. [13]. It was observed that the overall mass transfer coefficient increased with the increase of the feed temperature and its value was higher for SGMD than for DCMD process. It was also reported that the solubility of ammonia in water is high and the mass transfer in gas phase dominates the overall mass transfer in both MD configurations. Furthermore, experiments showed that the ammonia selectivity of both processes decreased with the increase of the feed temperature in spite of the positive effect of the feed temperature on the mass transfer coefficient of ammonia. With the increase of the feed temperature, the mass transfer of water vapour also increased and this resulted in a decrease of the ammonia selectivity. Moreover, the ammonia selectivity was found to be greater for the DCMD process. This was attributed partly to the lower pore size of the membrane used in DCMD (0.1 μm), compared to the pore size of the membrane used in SGMD which was 0.2 μm. Xie et al. [14] also observed a decrease in ammonia selectivity by SGMD with the increase of the feed temperature and an increase of ammonia removal to a value of 97% at 75 °C for a water feed solution of 3.3 mg/l ammonia.

It is to be noted that the increase of the feed temperature not only affects the vapour pressure of water and the volatile compounds present in the feed solution, leading to an exponential increase of the permeate flux, but also their diffusivities. The selectivity in MD depends on both the differences in volatility and diffusion rates of the components across the membrane and the gas gap, if it exists, as in the case of TSGMD process. For the latter process, García-Payo et al. [17] observed a decrease of formic acid selectivity with the increase of both the temperature difference and the mean temperature. It was concluded that the thermodynamic (operating temperature) and kinetic (diffusion of the components through the membrane and the gas gap) effects are the main reasons for this reduction of selectivity. The selectivity factors were found to be lower than unity, indicating that water was enriched in the permeate more than what was expected by the VLE. It was explained that the separation performance in MD is not only based on VLE of the involved components in the feed mixture but also on the thermodynamics and kinetic effects. It is to be highlighted that that this aspect is not fully investigated yet and more experiments should be done using membranes with different pore sizes, different configurations and different materials [16,17].

For low concentrations of isopropyl alcohol (IPA) in water (<5 wt%), an increase of the selectivity (6–25) with the increase of the feed temperature was observed in SGMD process using PTFE hollow fibre membrane module supplied by Sumitomo Electric Co. (POREFLON, Table 2.2) [9]. It is to be noted that in most of the membrane processes where the solution/diffusion mechanism of mass transport through swollen polymeric membranes is dominant, e.g. pervaporation (PV), the organic selectivity decreases with the increase of the feed temperature. For higher IPA concentrations up to 9.5 wt%, the IPA selectivity showed a maximum at a temperature of around 40 °C. The selectivity then decreased. This reduction in the IPA selectivity at the high feed temperature was attributed to the effect of the viscosity. At a higher feed temperature, IPA diffusion coefficient increases; however, at the same time the self diffusion of IPA decreases due to the high viscosity of the mixtures.

Effects of Solutes in Feed Aqueous Solution

The increase of the concentration of non-volatile solutes in the feed aqueous solution results in a reduction of the water activity (i.e. water vapour pressure) at the feed/membrane surface and a decrease in the SGMD permeate flux. This can be understood from Eq. (11.1). An example is shown in Fig. 11.9 for aqueous salt (sodium chloride, NaCl) solution. As it is clear from the figure, the feed electrical conductivity was above 60 ms/cm for all cases (except for the experiments with pure water), while the permeate conductivity was below 9 µs/cm, indicating salt rejection factors more than 99.9% [11]. The predicted SGMD permeate fluxes were found to be greater than the experimental ones and this result was attributed to the limitation of the efficiency of the used condensers. Furthermore, there is also the contribution of the concentration polarization effect that leads to an enhancement of the non-volatile solutes at the feed/membrane interface and to a decrease of the SGMD permeate flux. However, the latter contribution is very small compared to the temperature polarization effect.

Crystallization may occur also in SGMD process when treating feed aqueous solutions above their saturation limit similar to what was observed in DCMD process. However, SGMD crystallization, scale formation fouling or crystallization fouling has not been studied yet.

When using SGMD process for the treatment of mixtures containing volatile solutes such as IPA, it was observed that the addition of salt (magnesium chloride, $MgCl_2$) in the feed solution improved the volatile solute selectivity because of the decrease of the evaporation rate of water as explained previously, whereas the total permeate flux changed only slightly, maintaining its electrical conductivity at around 2 µs/cm [9].

The effects of the organic volatile solute concentration on the permeate flux and selectivity were studied using SGMD [9,12]. In this case, both water and the volatile solute are transported through the membrane pores. As the organic composition in the aqueous feed solution increases, the total permeate flux is also increased [9,12]. This is expected since the volatility of the organic compound increases with respect to water. However, when using TSGMD process, both the simulated and experimental total permeate fluxes decreased with increasing formic acid concentration in the feed aqueous

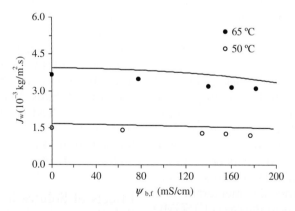

FIGURE 11.9 Effect of feed salt concentration on the predicted and measured SGMD permeate flux of the membrane module Accurel® S6/2 MD020CP2N (see Table 2.2) used at two different inlet feed temperatures of 50 °C and 65 °C, an inlet air temperature of 20 °C, air circulation velocity of 11.3 m/s and a feed circulation velocity of 0.8 m/s. *Source: Reprinted from [11]. Copyright 2003, with kind permission from Elsevier.*

solution, reached a minimum at a concentration close to the azeotropic concentration of the formic acid/water mixture and then increased [7]. Furthermore, a decrease of IPA selectivity with the increase of IPA concentration in feed aqueous solutions [9] was observed. Also, higher ethanol and acetone concentrations in the permeate were observed when their concentrations in the feed aqueous solutions were higher [12]. Care should have been taken in these experiments in order not to wet the membrane pores by increasing the volatile solute concentration in the feed solution.

The addition of ammonia in feed aqueous solution also reduces the ammonia selectivity due to the increase of the evaporation rate of water (i.e. increase of water vapour pressure mass transfer) [13]. This effect is stronger with the increase of ammonia concentration in the feed, leading to a decrease of ammonia selectivity.

Effect of Gas Temperature

The major disadvantage of the SGMD configuration is the very high increase of the gas temperature along the membrane module from the inlet to the outlet compared to the decrease of the feed liquid temperature as shown in Fig. 11.10. This variation affects the local driving force and consequently the overall SGMD permeate flux. Compared to DCMD configuration, the faster increase of the sweeping gas temperature from the inlet to the outlet of the membrane module causes a greater decrease in the transmembrane temperature drop along the membrane module. Therefore, in SGMD configuration, the effect of the gas temperature on the permeate flux is less than in DCMD.

In some SGMD studies, the gas inlet temperature was varied from 10 °C to 30 °C. Although the gas temperature increases rapidly along the membrane module length, the effect of the gas inlet temperature on the SGMD permeate flux is not negligible. The enhancement of the gas inlet temperature results in reduction of the vapour pressure gradient across the membrane (i.e. transmembrane driving force) and, consequently, in the permeate flux [7,8,10]. The results from these works are not necessarily in agreement with the one reported by Basini et al. [6] who concluded that the SGMD permeate flux was practically insensitive to the inlet gas temperature. This may be attributed partly to the length of the membrane module. As the membrane module is longer, the SGMD

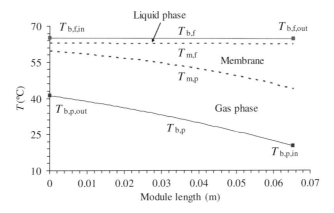

FIGURE 11.10 Temperature profile along a plate and frame SGMD membrane module using TF200 membrane (see Table 2.1), a feed liquid velocity of 0.15 m/s and an air circulation velocity of 1.5 m/s.

permeate flux will become less sensitive to the inlet gas temperature. Figure 11.11 shows an example of the negative influence of the air inlet temperature on the predicted and experimental SGMD permeate fluxes for a plate and frame membrane module of 0.065 m length. The combined Kndusen/ordinary molecular diffusion mechanism was used and the calculated values fit the experimental data quite well. Moreover, the decrease of the SGMD permeate flux with the gas inlet temperature increase is also related to the decrease in temperature polarization coefficient.

It was found that the liquid temperature polarization coefficient, θ_f, in Eq. (11.32) was not affected by the changes of the air inlet temperature, being maintained nearly equal to unity. However, a slight decrease of the air temperature polarization coefficient, θ_p, was detected with the increase of the air inlet temperature, leading to a decrease in overall temperature polarization coefficient, θ, of no more than 17%. Therefore, the SGMD permeate flux is also controlled by heat transfer, especially by the heat transfer through the gas boundary layer [8,10,15].

It may be advised that little benefit would be derived from cooling the gas inlet permeate temperature since the SGMD permeate flux is not very sensitive to this parameter compared to the feed liquid inlet temperature. Ambient temperature can be applied and it is more effective to increase the feed side temperature rather than to decrease the permeate side temperature in order to obtain higher SGMD permeate flux.

Effects of Feed Flow Rate

In SGMD, it was found that the effect of the feed flow rate on the permeate flux was practically negligible, while in the other MD configurations, the mass flux increased and in most cases tended to approach asymptotic values [7,10,27]. Particularly, in DCMD and VMD configurations, when employing the same

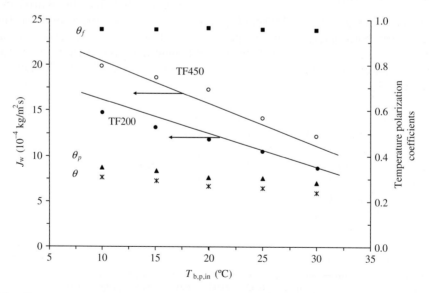

FIGURE 11.11 Effect of gas inlet temperature on the experimental and predicted SGMD permeate fluxes of two flat sheet membranes TF200 and TF450 (see Table 2.1) as well as on the temperature polarization coefficients of the membrane TF200. Feed inlet temperature 50 °C, liquid circulation velocity 0.15 m/s, air circulation velocity 0.8 m/s, distilled water as feed. *Source: Reprinted from [10]. Copyright 2002, with kind permission from John Wiley & Sons.*

membrane module and the same feed operating conditions, the permeate fluxes increased with the feed flow rate and approached asymptotic values at higher feed flow rates [1]. This is due to the reduction of the temperature polarization effect on the feed side of the membrane. It is known that the asymptotic value is observed when the turbulent flow regime is reached. With an increase in the Reynolds number, which is proportional to the fluid circulation velocity, the heat transfer coefficient of the liquid boundary layer increases and the temperature polarization effect is reduced. The negligible effect of the feed flow rate in SGMD flux is due to the fact that the temperature polarization effect is practically governed by the gas flow rate on the permeate side. It was shown in Fig. 11.11 that the temperature polarization is localized in the permeate side in SGMD and neither the temperature polarization coefficient, θ, nor the SGMD permeate flux was affected significantly by the liquid flow rate [10].

The preceding discussions on the effect of the feed flow rate were corroborated by Ding et al. [13] who used ammonia aqueous solutions for SGMD and found that the overall mass transfer coefficient of ammonia was only slightly influenced by the liquid flow rate.

The effects of the liquid feed flow rate on SGMD permeate flux and selectivity were studied by Lee and Hong [9] using PTFE hollow fibre membrane module supplied by Sumitomo Electric Co. (POREFLON, Table 2.2) under different operating conditions for the separation of IPA aqueous solutions. Figure 11.12 shows the effects of the feed flow rate on the SGMD permeate flux and selectivity for different IPA concentrations. An increase of the liquid feed flow rate led to an increase of the total permeate flux. This is due to the increase of the mass transfer coefficient and to the reduction of the resistance of the liquid boundary layer for IPA diffusion.

One should take it into account that if the liquid circulation velocity is increased, the liquid feed hydrostatic pressure would increase, and the risk of membrane wetting would become higher.

Gas Flow Rate

In SGMD, the gas flow rate must be varied with precautions so that the permeate hydrostatic pressure is kept lower than the hydrostatic pressure at the feed side and the transmembrane hydrostatic pressure must be lower than the liquid entry pressure (LEP) to avoid membrane pore wetting. Therefore, there is an optimum permeate flow rate beyond which any enhancement of the permeate flow rate may result in a decline of the SGMD permeate flux. An optimum value should be determined for each membrane module in order to obtain a maximum distilled water production.

The general effect of increasing the gas flow rate in SGMD is an increase in the permeate flux due to an increase in the Reynolds number, with which the gas flow regime changes from the laminar to the transitional and then to the turbulent flow regime. The enhancement of the gas flow rate increases the heat transfer coefficient on the permeate side by reducing the temperature polarization effect. This means that the gas temperature at the membrane surface approaches that of the bulk gas stream and consequently the driving force as well as the permeate flux increases. Figure 11.13 illustrates the increase of the SGMD permeate flux with the air circulation velocity for two types of membranes TF200 and TF450 (see Table 2.1). It seems that the SGMD permeate fluxes approach asymptotic values. A similar trend was found by simulation for a plate and frame membrane module by Charfi et al. [15] and for a shell and tube membrane module by Khayet et al. [11]. Basini et al. [6] observed an increase in the permeate flux up to asymptotic values and a slight decrease with further increase in gas velocities for both flat sheet and tubular membranes. The observed slight decrease of the permeate flux after a maximum was reached

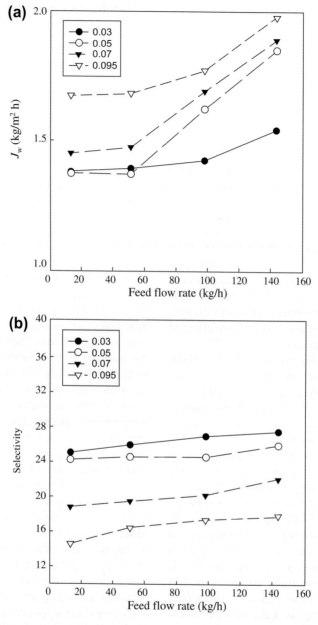

FIGURE 11.12 Effect of liquid flow rate on total SGMD permeate flux (a) and selectivity (b) for different IPA concentrations in water. Feed temperature 46 °C, sweep gas velocity 8.7 m/s, sweep gas temperature 25 °C. *Source: Reprinted from [9]. Copyright 2001, with kind permission from Elsevier.*

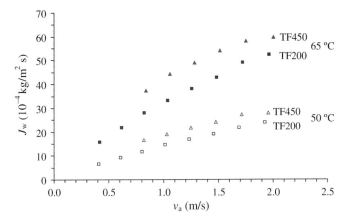

FIGURE 11.13 SGMD permeate flux of the commercial membranes TF200 and TF450 (see Table 2.1) as a function of the air circulation velocity using two liquid feed inlet temperatures of 65 °C and 50 °C, a liquid circulation velocity of 0.15 m/s and a sweep gas temperature of 20 °C. *Source: Reprinted from [7]. Copyright 2000, with kind permission from Elsevier.*

was associated with the higher gas pressure needed to enable higher gas flow rates.

Basini et al. [6] used dry air as sweeping gas to conduct SGMD experiments on both PTFE and PP flat sheet membranes as well as tubular PP membranes. A significant increase of the permeate flux was observed with an increase in air velocity, reaching a maximum value and decreasing with further increase in air velocity. Under the same operational conditions, the PTFE membrane exhibited higher SGMD performance than the PP capillary membrane due to its larger pore size and porosity. Basini and co-authors [6] did not indicate the value of the membrane area, which would enable the comparison of their results with other SGMD studies. They concluded that the mass transfer resistance in the gas phase was the rate-determining step when the gas flow was low, while at higher gas flow rates the resistance exerted by the membrane itself became dominant. An iterative theoretical model, which takes the combined Knudsen/ordinary molecular diffusion mechanism into account, was presented assuming a tortuosity factor of 3 (adjustment factor) for the PP membrane. Good agreement was obtained between the predicted SGMD fluxes and the experimental ones for the shell and tube PP capillary membrane module. However, some deviations appeared for the plate and frame PTFE membrane module and the authors attributed them to small size of the module for which no valid heat and mass transfer correlation was available.

It was found the temperature polarization coefficient in the feed side is very close to unity, whereas the temperature polarization in SGMD is primarily located in the air phase, concluding that the mass flux in the SGMD process is mostly controlled by the heat transfer through the air boundary layer [10]. The liquid temperature polarization coefficient, θ_f, hardly varies with the gas circulation velocity and the air temperature polarization coefficient, θ_p, together with the global temperature polarization coefficient, θ, increases with the increase of the gas circulation velocity [10]. This was interpreted based on the overall heat transfer coefficient, H in Eqs. (11.26) and (11.27), which was dominated by the sweeping gas contribution. In other words, as it was stated previously, the gas temperature polarization coefficient is the dominant parameter in the SGMD process.

It is worth quoting that the effect of the fluid flow rate in the permeate side is greater in SGMD than in DCMD due to the large difference in thermal conductivities of liquid and gas. For the same feed inlet temperature and feed flow rate, the permeate flux reaches higher values in the SGMD than in the DCMD configuration. This may be due in part to the smaller internal HL through the membrane in SGMD configuration.

With an increase in the sweeping gas flow rate, it was observed that both total permeate flux and ammonia removal efficiency increased but leveled off at higher gas flow rates [14]. When the sweeping gas flow rate was increased, the vapour pressure of both water and ammonia on the permeate side decreased resulting in an increase of the driving force. The boundary layer resistance decreased at the same time reaching a minimum value. It was concluded that an optimum sweeping gas flow rate existed, guaranteeing a high ammonia removal efficiency. It is to be noted that the selectivity of ammonia was found to decrease slightly with the increase of the gas flow rate for the feed temperatures of 55 °C and 65 °C but remained unchanged for the feed temperature of 75 °C. This result was attributed to the greater amount of water vapour transported through the membrane to the gas side with the increase of the sweeping gas flow rate compared to that of ammonia. However, Ding et al. [13] reported a gradual increase of ammonia selectivity with the increase of the gas flow rate and attributed this result to the increase of the overall mass transfer coefficient of ammonia for the feed temperatures of 59 °C and 65 °C.

When using organic aqueous solutions containing alcohols such as ethanol and isopropanol as well as acetone, increase in the total permeate flux was observed with an increase in the gas flow rate, reaching an asymptotic value at high sweeping gas velocities [9,12]. However, Lee and Hong [9] reported a decrease of the organic selectivity with the increase of the gas flow rate, reaching an asymptotic value at high sweeping gas velocities. Decrease of the organic concentration in the permeate with an increase in the sweeping gas flow rate was reported by Boi et al. [12].

Coupling and interaction effects between operating parameters and between the different permeating species through the membrane and its pores occur also in SGMD process. Studies on this subject as well as on optimization of different SGMD systems are needed in order to increase SGMD performance and decrease energy consumption.

SGMD APPLICATIONS

SGMD technology has been applied at laboratory scale for the treatment of aqueous solutions containing non-volatile solutes such as salts (NaCl) as well as volatile solutes such as ammonia, alcohols (ethanol, isopropanol) and acetone. However, the number of applications is much less compared to DCMD. Use of renewable energy systems to run the SGMD process has not yet been considered.

Theoretical and experimental desalination by SGMD was studied by Khayet et al. [11] using shell-and-tube PP membrane module (Accurel® S6/2 MD020CP2N, Table 2.2). Humid air was employed as the vapour carrier gas and the feed solution was circulated in countercurrent through the lumen side of the membrane module. The effects of various process parameters including salt (NaCl) concentration on the permeate flux have been investigated. Only a slight decrease of the SGMD flux with the NaCl concentration in the feed aqueous solution was detected and high purity water with an electrical conductivity less than 9 µS/cm was obtained, indicating a salt rejection of more than 99.9%. It was also found that the predicted SGMD fluxes were slightly higher than the experimental ones due partly to the limitation of the condenser efficiency. Recently, these results were confirmed theoretically by Charfi et al. [15], performing two-dimensional numerical simulations based on Navier–Stokes equations coupled with the Darcy–Forcheimer formulation in transient regime.

Lee and Hong [9] studied the effect of different operating parameters on the SGMD flux and selectivity when using dilute aqueous isopropanol (IPA) solutions and PTFE hollow fibre membrane module (Table 2.2). Nitrogen was employed as sweeping gas through the lumen side of the membrane module in a countercurrent configuration. The highest IPA concentration in feed was 10 wt% and the feed temperature was varied from 20 °C to 50 °C. The obtained IPA selectivity was ranged between 6 and 27. It was found that IPA selectivity was not affected by the feed flow rate and decreased with the increase of the sweeping gas flow rate. Moreover, the total SGMD flux increased and the IPA selectivity decreased with the IPA concentration. IPA selectivity was found to increase continuously with the feed temperature at lower IPA concentrations, whereas for higher IPA concentrations an optimum value was observed at a feed temperature close to 40 °C. This result was explained by the negative effect of the viscosity. By contrast, at lower IPA concentrations, the diffusion coefficient for IPA in water increases with an increase in temperature, leading to an increase of the mass transfer coefficient and therefore higher selectivity. The addition of the salt ($MgCl_2$) to the feed solution was found to increase IPA selectivity significantly with a slight decrease in total permeate flux. This was attributed to the reduction in water vapour pressure, leading to a decrease in the water mass transfer through the membrane.

SGMD process was applied for the treatment of wastewaters containing volatile organic compounds such as acetone and ethanol at different concentrations [12]. Experiments have been performed to study the effects of the flow rate of the sweeping gas (dry nitrogen) using flat sheet TF200 membrane (see Table 2.1) in a laboratory-scale circular cell with radial flow in both the feed and permeate chambers. A simplified mathematical model was developed to describe multicomponent mass transfer in the gas phases, in which a pseudo-binary diffusion approach was assumed. Interactions between the transported water and the organic compounds in a gaseous film were assumed negligible. Molecular diffusion was considered as the prevailing transport mechanism through the membrane. The calculated results were compared with the experimental data, obtaining a remarkable precision in the prediction of the total permeate flux but an overestimation of the organic permeate flux. In general, an increase of the permeate flux and a decrease of the organic concentration on the permeate side were observed with increasing the gas flow rate. It was concluded that the basic requirement for a successful SGMD application on an industrial scale is a good membrane module design.

Aqueous solutions containing ammonia have been treated by SGMD process [13,14]. When using aqueous solutions with low levels of ammonia (100 mg/l) at pH 11.5, a PTFE membrane with 0.45 μm pore size, 70% porosity and 100–200 μm thickness (Advantec MFS Inc.) and compressed air for sweeping gas, it was found that at the best SGMD conditions of high feed temperature and high gas flow rate, 97% removal of the ammonia with a concentration of the treated water of only 3.3 mg/l could be achieved. The liquid feed temperature, feed flow rate and gas flow rate were found to have significant influences on the efficiency of ammonia removal. By contrast, sweeping gas inlet temperature had a negligible effect on ammonia removal. It was concluded that the feed temperature is a crucial SGMD operating parameter affecting considerably both the total permeate flux and the ammonia removal from aqueous solutions. In other words, by increasing the feed temperature, the permeate flux can be increased significantly, but the ammonia selectivity is reduced [14].

The ammonia separation performance of SGMD, VMD and DCMD has been compared by Ding et al. [13] using PTFE flat sheet membranes of different pore sizes (0.1 μm and 0.2 μm), different thicknesses (60 μm and 80 μm) and 60% porosity, provided by Beijing Institute of Plastic Research. The compared parameters were the total mass transfer coefficient of ammonia and ammonia selectivity. It was found that under similar feed operating conditions VMD mode exhibited the highest mass transfer coefficient of ammonia but the lowest ammonia selectivity. DCMD mode gave the highest ammonia selectivity and moderate mass transfer coefficient of ammonia, and SGMD process has moderate ammonia selectivity and the lowest mass transfer coefficient of ammonia. It was concluded that the ammonia separation performance is greatly affected by the membrane characteristics. The membranes having larger pore size and less thickness exhibit higher mass transfer coefficient of ammonia and lower ammonia selectivity. For all the three MD configurations, it was found that higher feed temperatures resulted in higher mass transfer coefficients of ammonia but lower values of ammonia selectivity. Moreover, the pH of the aqueous ammonia solution exerted great effects on the module performance. The increase in pH improved both the mass transfer coefficient of ammonia and its selectivity. Furthermore, in the three MD configurations, a reduction of the ammonia selectivity was observed for aqueous solutions containing higher ammonia concentrations. In the case of SGMD process, by increasing the velocity of the sweeping gas (i.e. dry air), both mass transfer coefficient of ammonia and ammonia selectivity increased significantly. The obtained ammonia selectivity in SGMD ranged between 9 and 14.

One of the inconveniencies of the traditional SGMD is the steep increase of the gas temperature along the membrane module. Probably, this is one of the reasons to explain why this MD mode is still at its early stage and why only small laboratory-scale membrane modules are being used. Rivier et al. [16] and García-Payo et al. [17] proposed a TSGMD process, in which the increase of the sweeping gas temperature is minimized by using a cold wall in the permeate chamber for the treatment of aqueous formic acid solutions. A Stefan—Maxwell-based model including VLE and heat and mass transfer relations was developed. The effects of the mean temperature and the temperature difference on the temperature polarization coefficient (θ) and the energy efficiency (η) were studied for both TSGMD and DCMD configurations. It was observed that the TSGMD attained higher θ and η values than those of DCMD due to a larger heat transfer resistance of the former configuration at the permeate boundary layer. This means that temperature polarization coefficient for DCMD is significantly higher than that for TSGMD. Based on simulation results, higher membrane performance was predicted when using TSGMD configuration.

García-Payo et al. [17] applied the TSGMD process for the separation of the azeotropic mixture formic acid/water using a laboratory tubular module with TF200 flat sheet membrane (see Table 2.1) sealed in a tubular form. The effect of concentration on the entire range of formic acid mass fraction was studied and no membrane wetting was observed. It was found that the permeate flux increased with the temperature difference, the mean temperature and the sweeping air velocity, and formic acid fluxes were higher than water fluxes. The formic acid selectivity was found to be always lower than unity when the formic acid mass fraction in the feed was 0.7; by contrast, the selectivity for 0.8 mass fraction could be greater than unity depending on the applied temperature. It should be noted that a selectivity higher than 1 indicates formic acid enrichment in the permeate, while lower than 1 means water enrichment in the permeate.

Possible SGMD applications can be in the fields where DCMD process cannot be used because of the risk of membrane pore wetting from the permeate side. These can be the removal/concentration of organics from dilute organic/water mixtures such as esters, ethers, chlorinated hydrocarbons and aromatic compounds. These applications are very appropriate for environmental, chemical, petrochemical and biotechnology industries, as they need removal or recovery of organics from dilute solutions. In fact, MAS has shown potential for the removal of volatile organics (VOCs) from aqueous streams over a wide range of concentration levels and demonstrated to be usefully applied for water pollution reduction, groundwater clean-up, and organic recovery and reuse from industrial and petroleum wastewater streams [22–26]. In the case of SGMD process, as shown previously, care must be taken in order to maintain membrane pores dry.

Other interesting SGMD applications are those for which DCMD process is applicable, such as the concentration of fruit juices and concentration of aqueous sucrose solutions among others.

References

[1] M. Khayet, M.P. Godino, J.I. Mengual, Possibility of nuclear desalination through various membrane distillation configurations: a comparative study, Int. J. Nuclear Desalination 1 (2003) 30–46.

[2] B.R. Bodell Silicone rubber vapor diffusion in saline water distillation, US Patent 285,032 (1963).

[3] B.R. Bodell Distillation of saline water using silicone rubber membrane, US Patent 3,361,645 (1968).

[4] Y. Henderyckx, Diffusion doublet research, Desalination 3 (1967) 237–242.

[5] A. Van Haute, Y. Henderyckx, The permeability of membranes to water vapor, Desalination 3 (1967) 169–173.

[6] L. Basini, G. D'Angelo, M. Gobbi, G.C. Sarti, C. Gostoli, A desalination process through sweeping gas membrane distillation, Desalination 64 (1987) 245–257.

[7] M. Khayet, P. Godino, J.I. Mengual, Theory and experiments on sweeping gas membrane distillation, J. Membr. Sci. 165 (2000) 261–272.

[8] M. Khayet, P. Godino, J.I. Mengual, Nature of flow on sweeping gas membrane distillation, J. Membr. Sci. 170 (2000) 243–255.

[9] C.H. Lee, W. Hong, Effect of operating variables on the flux and selectivity in sweep gas membrane distillation for dilute aqueous isopropanol, J. Membr. Sci. 188 (2001) 79–86.

[10] M. Khayet, M.P. Godino, J.I. Mengual, Thermal boundary layers in sweeping gas membrane distillation, AIChE J. 48 (2002) 1488–1497.

[11] M. Khayet, P. Godino, J.I. Mengual, Theoretical and experimental studies on desalination using sweeping gas membrane method, Desalination 157 (2003) 297–305.

[12] C. Boi, S. Bandini, G.C. Sarti, Pollutants removal from wastewaters through membrane distillation, Desalination 183 (2005) 383–394.

[13] Z. Ding, L. Liu, Z. Li, R. Ma, Z. Yang, Experimental study of ammonia removal from water by membrane distillation (MD): The comparison of three configurations, J. Membr. Sci. 286 (2006) 93–103.

[14] Z. Xie, T. Duong, M. Hoang, C. Nguyen, B. Bolto, Ammonia removal by sweep gas membrane distillation, Water Research 43 (2009) 1693–1699.

[15] K. Charfi, M. Khayet, M.J. Safi, Numerical simulation and experimental studies on heat and mass transfer using sweeping gas membrane distillation, Desalination 259 (2010) 84–96.

[16] C.A. Rivier, M.C. García-Payo, I.W. Marison, U. von Stockar, Separation of binary mixtures by thermostatic sweeping gas membrane distillation: I. Theory and simulations, J. Membr. Sci. 201 (2002) 1–16.

[17] M.C. García-Payo, C.A. Rivier, I.W. Marison, U. von Stockar, Separation of binary mixtures by thermostatic sweeping gas membrane distillation: II. Experimental results with aqueous formic acid solutions, J. Membr. Sci. 198 (2002) 197–210.

[18] N. Hengl, A. Mourgues, E. Pomier, M.P. Belleville, D. Paolucci-Jeanjean, J. Sánchez, G. Rios, Study of a new membrane evaporator with a hydrophobic metallic membrane, J. Membr. Sci. 289 (2007) 169–177.

[19] A. Mourgues, N. Hengl, M.P. Belleville, D. Paolucci-Jeanjean, J. Sanchez, Membrane contactor with hydrophobic metallic membranes: 1. Modelling of coupled mass and heat transfers in membrane evaporation, J. Membr. Sci. 355 (2010) 112–125.

[20] R.L. Calibo, M. Matsumura, J. Takahashi, H. Kataoka, Ethanol stripping by pervaporation using porous PTFE membrane, J. Ferment. Technol. 65 (1987) 665–674.

[21] E. Korngold, E. Korin, Air sweep water pervaporation with hollow fiber membranes, Desalination 91 (1993) 187–197.

[22] H. Mahmud, A. Kumar, R.M. Narbaitz, T. Matsuura, Membrane air stripping: A process for removal of organics from aqueous solutions, Sep. Sci. Technol. 33 (1998) 2241–2255.

[23] H. Mahmud, A. Kumar, R.M. Narbaitz, T. Matsuura, A study of mass transfer in the membrane air-stripping process using microporous polypropylene hollow fibers, J. Membr. Sci. 179 (2000) 29–41.

[24] H. Mahmud, A. Kumar, R.M. Narbaitz, T. Matsuura, Mass transport in the membrane air-stripping process using microporous polypropylene hollow fibres: effect of toluene in aqueous feed, J. Membr. Sci. 209 (2002) 207–219.

[25] R.S. Juang, S.H. Lin, M.C. Yang, Mass transfer analysis on air stripping of VOCs from water in microporous hollow fibers, J. Membr. Sci. 255 (2005) 79–87.

[26] F.G. Viladomat, I. Souchon, V. Athès, M. Marin, Membrane air-stripping for aroma compounds, J. Membr. Sci. 277 (2006) 129–136.

[27] M. Khayet, Membrane distillation, In: N.N. Li, A.G. Fane, W.S.W. Ho, T. Matsuura (Eds.), Advanced membrane technology and applications, John Wiley & Sons, Inc, New York, NY, USA, 2008, pp. 297–370.

[28] F.A. Banat, F. Abu Al-Rub, R. Jumah, M. Al-Shannag, Application of Stefan-Maxwell approach to azeotropic separation by membrane distillation, Chem. Eng. J. 73 (1999) 71–75.

[29] F.A. Banat, F. Abu Al-Rub, R. Jumah, M. Shannag, Theoretical investigation of membrane distillation role in breaking the formic acid-water azeotropic point: Comparison between Fickian and Stefan-Maxwell-Based models, Int. Comm. Heat Mass Transfer 26 (1999) 879–888.

CHAPTER 12

Vacuum Membrane Distillation

OUTLINE

Introduction 323	Extraction of Volatile Organic Compounds 346
Theoretical Models 328	Treatment of Alcohol Aqueous Solutions 347
Experimental: Effects of Process Conditions 338	VMD in Concentration of Fruit Juices and Recovery of Aroma Compounds 348
Effects of Feed Temperature 338	VMD Technology in Desalination 349
Effects of Solute Concentration in the Feed Solution 340	Treatment of Textile Wastewaters 351
Effects of Feed Flow Rate 342	Other Applications of VMD Technology 352
Effects of Downstream Pressure 344	Comparison to Pervaporation 354
VMD Applications 346	

INTRODUCTION

One possible way to increase membrane permeability in membrane distillation (MD) is by removing air from its pores by deaeration (i.e. removing the effect of the partial air pressure within the membrane pores, which is the same as removing the molecular diffusion resistance) or by applying a continuous vacuum in the permeate side below the equilibrium vapour pressure.

Vacuum membrane distillation (VMD) is another variant of MD. In this configuration low pressure or vacuum is applied on the permeate side of the membrane module by means of vacuum pumps. The applied permeate pressure is lower than the saturation pressure of volatile molecules to be separated from the feed solution and condensation takes place outside the membrane module at temperatures much lower than the ambient temperature. Normally, at laboratory scale, nitrogen liquid filled condensers are used. Figure 12.1 shows schematics of heat and mass transfer through a porous hydrophobic membrane and a typical laboratory VMD system with a tangential flow cell. A VMD cell with a magnetic stirrer is shown in Fig. 9.3. In the latter cell the feed aqueous solution is stirred inside a container by a magnetic stirrer, while vacuum is applied

12. VACUUM MEMBRANE DISTILLATION

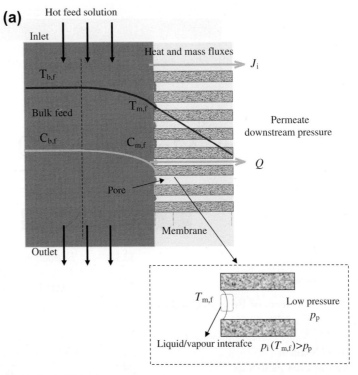
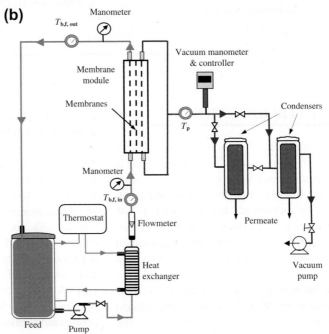

FIGURE 12.1 Schematics of heat and mass transfer through a porous hydrophobic membrane in VMD drawn for volatile solutes (a) and typical experimental VMD laboratory system with tangential flow cell (b).

on the permeate side of the membrane held horizontally. This type of cell is only used for laboratory scale studies to characterize the membranes. For the system with tangential flow cell, single or arrays of flat sheet, capillaries or hollow fibre membrane modules can be used.

In VMD, because of the hydrophobic nature of the used membranes, the feed cannot penetrate inside dried membrane pores unless a transmembrane hydrostatic pressure exceeds the 'liquid entry pressure of water (*LEP*),' which is characteristic to each membrane. This condition results in the formation of liquid/vapour interfaces at the entrances of the membrane pores and, because of the applied low pressure on the permeate side, molecules evaporate from the feed side of the membrane, cross the pores in vapour phase, and condense outside the membrane module by means of external condensers.

This configuration has the following two advantages:

(i) a very low conductive heat loss: This is due to the insulation against conductive heat loss through the membrane provided by the applied vacuum. The boundary layer in the vacuum side is negligible, which implies a decrease in the heat conducted through the membrane and enhancement of the VMD performance.
(ii) a reduced mass transfer resistance: The diffusion inside the pores of the evaporated molecules at the liquid feed/membrane interface is favoured.

In other words, the resistance to heat transfer on the permeate side and the heat transfer by conduction through the membrane can generally be neglected in VMD configuration. However, there is a drawback, i.e. due to the application of vacuum on the permeate side of the membrane module the risk of membrane pore wetting increases.

It is worth quoting that Bodell [1,2] suggested, for the first time, an alternative means of providing low water vapour pressure inside a tubular silicone membrane by applying vacuum to the lumen side of the tube, a system that is currently known as the VMD configuration.

It is to be noted that VMD process is mistakenly thought to be the same as pervaporation (PV) process. The same systems can be applied for experiments. In both the processes, the upstream side of the membrane is in contact with feed liquid while vacuum is applied on the downstream side of the membrane. The fundamental difference between them is the role that the membrane plays in the separation. VMD uses a porous and hydrophobic membrane, whereas PV requires dense and selective membranes and the separation is based on solubility and diffusivity of each feed component in the membrane material [3].

An examination of the MD literature permits to assert that VMD has attracted less attention compared with DCMD process, which is the most used configuration. As can be seen in Fig. 1.3, about 16.7% of the papers published up to December 2010 were on VMD. This is because external condensers are required to collect the distillate, complicating in this way the system design and increasing the cost.

In general, 50.4% of the MD publications dealt with theoretical models. In the case of VMD configuration, 40.3% of the published papers up to December 2010 involved theoretical models. However, very few papers have focused on the fabrication of membranes and modules for VMD applications [3–15].

Khayet and Matsuura [4] and Khayet et al. [5] used liquid water as a non-solvent additive to improve the VMD permeability of polyvinylidene fluoride (PVDF) membranes and to reduce their cost. Both supported and unsupported membranes were fabricated from 15 wt% PVDF in the solvent dimethyl acetamide (DMAC). It was observed that the porosity (26.8–79.6%) and pore size (0.02–0.7 μm) both increased with increasing water content in the casting solution, whereas the LEP_w decreased. The VMD permeate flux increased exponentially with the water content in the casting solution for both supported and unsupported membranes (1–14 kg/m$^2 \cdot$h for the supported membranes and 0.6–16 kg/m$^2 \cdot$h for the unsupported membranes). For water content in the casting solution lower than 4.25 wt%, both VMD permeate flux and mass transfer coefficient of the supported membranes were higher than those of the unsupported membranes, whereas, for higher water concentrations, the overall mass transfer coefficients were lower for the supported membranes. This result was attributed to the resistance of the used support. Furthermore, it was found that the separation factor decreased with increasing concentration of water in the PVDF casting solution and was generally lower for the supported membranes. This result was explained by the increase of the permeation rate of water through the membrane pores under the investigated range of the VMD operating conditions.

Melt-extruded/cold-stretching method was used by Li et al. [6] to prepare polyethylene (PE) and polypropylene (PP) hollow fibre membranes for desalination by VMD. Higher water permeate fluxes were obtained for the PE membranes than for the PP membranes and the result was attributed to the larger pore size of the PE membranes. Direct contact membrane distillation (DCMD) was also used and the highest permeate flux reported was 0.8 l/m$^2 \cdot$h in DCMD and about 4 l/m$^2 \cdot$h in VMD.

Li and Sirkar [7] also presented a novel hollow fibre membrane and a module for VMD desalination. The external surfaces of the commercial porous PP hollow fibres (Accurel Membrana, Wuppertal, Germany) of different dimensions and thicknesses were coated with a variety of ultrathin microporous silicone-fluoropolymer layer by means of plasma polymerization. The reason for applying the coating layer was to provide an additional porous layer having higher hydrophobicity than PP, which itself is one of the polymeric materials with very low surface energy. The fibres were arranged in a rectangular cross-flow module design for the hot feed to flow over the outside surface of the fibres and to reduce the temperature polarization effect. VMD experiments were carried out at feed temperatures ranging from 60 to 90 °C with 1% NaCl aqueous solution. Higher permeate fluxes (41–79 kg/m$^2 \cdot$h), complete absence of membrane pore wetting, higher temperature polarization coefficients (93–99% in VMD) than in other MD membrane modules and higher heat transfer coefficients on the feed side of the membrane module have been reported.

Various asymmetric microporous PVDF hollow fibre membranes with different pore sizes (i.e. 0.031–0.068 μm), effective porosities (71–1516 m^{-1}) and morphologies were fabricated for VMD by the wet spinning technique using the solvent DMAC and the non-solvent additives lithium chloride and water [8,9]. The membranes were employed to remove 1,1,1-trichloroethane (TCA) from aqueous solutions of different TCA concentrations and for toluene and benzene removal from water. It was observed that the PVDF hollow fibres post-treated by the ethanol solvent exchange method exhibited higher porosity and higher permeability. Under optimum VMD-operating parameters, particularly at a downstream pressure of 8–10.7 kPa, a feed temperature of 50 °C and a feed flow rate of 10^{-3} m^3/h up to 97% TCA removal efficiency has been achieved. For benzene and toluene removal from water by VMD, a separation factor over 99% was obtained with 8 kPa downstream pressure, 50 °C feed temperature and a feed flow rate of 10^{-3} m^3/h.

Hydrophobic/hydrophilic poly(phthalazinone ether sulfone ketone) (PPESK) hollow fibre composite membranes were prepared by coating the internal surface of the hollow fibres with silicone rubber and sol-gel polytrifluoropropylsiloxane. The membranes were used in desalination by VMD [10–12]. First, the PPESK hollow fibre ultrafiltration membranes were prepared by the dry/jet wet spinning technique. The effects of coating conditions (i.e. coating time, coating temperature and concentration of silicone rubber) on the VMD performance were studied and the membrane stability was evaluated based on long-term experiments (up to 14 days). It was found that the higher VMD permeate flux was obtained at lower coating temperature and lower concentration of silicone rubber in the coating solution. A permeate VMD flux as high as $3.5 \, l/m^2 \cdot h$ with a salt rejection factor of 99% was reached using a feed NaCl aqueous solution with a concentration of $5 \, g/l$. Feed temperature was $40\,°C$ and downstream pressure was 0.078 MPa. Hollow fibres coated with $5 \, g/l$ silicone rubber at $60\,°C$ for 9 h were used. Moreover, the studied composite hollow fibre membrane showed stable VMD performance in a long-term experiment. In the case of polytrifluoropropylsiloxane-coated hollow fibre membranes the highest permeate flux of $3.7 \, l/m^2 \cdot h$ with a NaCl rejection factor of 94.6% was observed for 30 min pre-polymerization time. It was proved for this type of membranes that the pre-polymerization time is an important factor affecting VMD performance. Longer pre-polymerization time decreased the permeability of the membrane.

Recently, comparisons of hydrophobic Zirconia (50 nm pore size) and Titania (5 nm pore size) tubular ceramic membranes used in different MD configurations (VMD, DCMD and air gap membrane distillation, AGMD) have been carried out [13]. The internal surface of the tubular membranes was chemically modified by grafting perfluoroalkylsilane molecule, $C_8F_{17}(CH_2)_2Si(OC_2H_5)_3$, achieving coating layers of 10 μm for zirconia and 5 μm for titania. Salt rejection factors higher than 99% have been obtained for all tested MD configurations when using salt (NaCl) aqueous solutions of concentrations 0.5 and 1 M. The highest permeate fluxes obtained using zirconia-modified membrane are 180, 95 and $113 \, l/m^2 \cdot day$ in desalination by VMD, DCMD and AGMD of a salt NaCl aqueous solution of 0.5 M, respectively. The corresponding permeate fluxes of Titania-modified membrane are $146 \, l/m^2 \cdot day$ for VMD and $20 \, l/m^2 \cdot day$ for DCMD and AGMD.

Surface-modified polyethersulfone flat sheet membranes using surface modifying macromolecules (SMMs: nSMM) have been used in VMD for the treatment of ethanol aqueous solution (1000 ppm ethanol in water) [14]. The membranes were prepared by the phase-inversion technique. During membrane formation the nSMM macromolecules migrate to the top surface (i.e. polymer/air interface) rendering it hydrophobic and changing its characteristics. The results showed that the mean pore size as well as the surface hydrophobicity of the membrane increased with the increase of the solvent evaporation time when the casting solution was blended with nSMM. However, the surface hydrophobicity decreased with the increase in gelation bath temperature. Moreover, when polyvinyl pyrrolidone was added to the casting solution containing nSMM, the mean pore size decreased but, at the same time, the contact angle decreased. The contact angles depended on the conditions of membrane preparation but a contact angle as high as $120°$ was achieved. The highest VMD permeate flux for the feed pure water was lower than $5.5 \times 10^{-4} \, kg/m^2 \cdot s$ for a feed temperature of $26\,°C$ and a downstream pressure in the range 0.4–5.3 kPa, and decreased with the increase of the downstream pressure. As expected, when 1000 ppm ethanol aqueous solution was used as feed at a temperature of $26\,°C$, the total permeate flux became higher (1.27×10^{-4}–$0.9 \times 10^{-4} \, kg/m^2 \cdot s$) than the partial permeate flux of water (0.84×10^{-4}–$0.72 \times 10^{-4} \, kg/m^2 \cdot s$) at a downstream pressure of 0.6–2 kPa.

Isotactic polypropylene hydrophobic microporous flat sheet membranes were fabricated via thermally induced phase separation for desalination by VMD process [15]. These membranes exhibited a narrow pore size distribution and an asymmetric cross-sectional structure with cellular pores on the skin layer and dendritic pores throughout the cross-section. It was reported that the formation of these membranes is not very sensitive to the quench bath temperature, whereas the diluent agent (soybean oil) has a strong effect on the morphology and performance of the membranes. For a feed temperature of 70 °C and a downstream pressure of 3 kPa, a VMD permeate flux of 28.92 kg/m^2·h was obtained when pure water was used as feed and 24.81 kg/m^2·h when 0.5 mol/l NaCl aqueous solution was used as feed with a salt rejection factor of 99.9%.

Table 12.1 reviews the highest permeate fluxes observed in VMD for some commercial and laboratory-fabricated membranes. The highest membrane permeate fluxes obtained so far in VMD were 372.6, 421.2 and 576.7 kg/m^2·h using distilled water as feed with a feed temperature of 74 °C and a downstream pressure of 3 kPa for the commercial membranes 3MA, 3MB and 3MC supplied by 3M Corporation (see Table 2.1) [16].

When VMD is compared to the other MD configurations, DCMD and sweeping gas membrane distillation (SGMD), operating under the same feed conditions and using the same membrane module (shell and tube PP membrane module, Accurel® S6/2 MD020CP2N, Table 2.2), it was observed that the VMD permeate flux was about 1.4 times greater than the corresponding SGMD flux and was 2.8–3.1 times higher than the DCMD permeate flux [17]. This was attributed to the thermal efficiency, which was found to be higher in VMD process. The internal heat loss by conduction through the membrane was very low in SGMD and VMD than the corresponding heat loss in DCMD.

The potential applications of VMD are generally the extraction of volatile organic compounds (VOCs) from aqueous solutions such as chloroform, benzene, toluene, methyl tert-butyl ether (MTBE), TCA, 2,4-dichlorophenol, tetracholoroethylene, etc. [3,4,8,9,11,18–21]. Treatment of aqueous alcohol solutions by VMD is also an important application [14,16,18,22,23]. Other research areas involving aqueous solutions containing non-volatile compounds have also been considered such as desalination [6,7,10–13,15,17,24–28] for production of distilled water, concentration of aqueous sucrose solutions [29], treatment of dye solutions [30,31], concentration of ginseng extracts in aqueous solutions [32,33] and ethylene glycol from used coolant liquids [34]. Moreover, VMD process was used for ammonia removal [35] and for the concentration of fruit juices and recovery of volatile aroma compounds [36–39]. Additionally, attempts were made to couple VMD technology with renewable energy sources or with other membrane processes. For example, by coupling with reverse osmosis (RO), brines produced by RO seawater desalination can be further concentrated [26,28,40].

THEORETICAL MODELS

In VMD, the driving force is maintained by applying a continuous vacuum at the permeate side (i.e. downstream pressure) below the equilibrium vapour pressure. The feed solution is brought into contact with one side of the membrane (i.e. upstream side). In the case of mass transport of a single component, i, through the membrane, the permeate flux, J_i, is written as:

$$J_i = B_i \Delta p_i = B_i(p_{i,m} - p_{i,p}) \quad (12.1)$$

where B_i is the VMD coefficient of the membrane (i.e. membrane permeability), $p_{i,m}$ is

TABLE 12.1 Reported VMD permeate flux (J) of different types of commercial membranes shown in Tables 2.1 and 2.2 and some fabricated membranes for VMD process

Membrane trade name	J (10^{-3} kg/m²s)	Observation	Reference
TF200	5.54	5 wt% ethanol/water mixture, $\phi_l = 4.75$ l/min, $T_{b,f} = 35$ °C, $p_p = 25$ torr, 27.0 wt% ethanol in permeate, separation factor = 7.0	[22]
	2.48	5 wt% ethanol/water mixture, $\phi_l = 4.75$/min, $T_{b,f} = 35$ °C, $p_p = 40$ torr, 31.7 wt% ethanol in permeate, separation factor = 8.8	
	8.0	5 wt% acetone/water mixture, $\phi_l = 2.5$ l/min, $T_{b,f} = 35$ °C, $p_p = 30$ mbar, 20 wt% acetone in permeate	[41]
	6.0	5 wt% acetone/water mixture, $\phi_l = 2.5$ l/min, $T_{b,f} = 25$ °C, $p_p = 10$ mbar, 20 wt% acetone in permeate	
	5.7	9 wt% acetone/water mixture, $\phi_l = 2.5$ l/min, $T_{b,f} = 25$ °C, $p_p = 20$ mbar, 40 wt% acetone in permeate	
	1.0	9 wt% acetone/water mixture, $\phi_l = 2.5$ l/min, $T_{b,f} = 25$ °C, $p_p = 60$ mbar, 75 wt% acetone in permeate	
	6.9	5 wt% IPA/water mixture, $\phi_l = 2.5$ l/min, $T_{b,f} = 35$ °C, $p_p = 30$ mbar, 20 wt% IPA in permeate	
	2.2	5 wt% IPA/water mixture, $\phi_l = 2.5$ l/min, $T_{b,f} = 35$ °C, $p_p = 55$ mbar, 32 wt% IPA in permeate	
	0.5	8.7 wt% IPA/water mixture, $\phi_l = 2.5$ l/min, $T_{b,f} = 35$ °C, $p_p = 70$ mbar, 50 wt% IPA in permeate	
GVHP	1.81	Distilled water as feed, $T_{b,f} = 25$ °C, $p_p = 1.7$ kPa	[5]
PP Membrana, Germany	15.61	Cross-flow VMD, distilled water as feed, $T_{b,f} = 59.1$ °C, $p_p = 6$ kPa, $\phi_l = 200$ l/h, 1.13 kW/kgh^{-1}, 222.7 W	[48]
	13.86	Cross-flow VMD, distilled water as feed, $T_{b,f} = 59.2$ °C, $p_p = 1$ kPa, $\phi_l = 235$ l/h, 1.98 kW/kgh^{-1}, 441.2 W	
	12.13	Longitudinal VMD, distilled water, $T_{b,f} = 59.3$ °C, $p_p = 6$ kPa; $\phi_l = 200$ l/h, 1.15 kW/kgh^{-1}, 441.2 W	
3MB	117.0	Distilled water as feed, $\phi_l = 63$ cm³/s, $T_{b,f} = 74$ °C, $p_p = 3$ kPa	[16]
3MC	160.2		
3MA	103.5		
PP Accurel® S6/2	7.9	Distilled water as feed in lumen side, $T_{b,f} = 65$ °C, $p_p = 3.5$ kPa, $v_l = 0.8$ m/s	[17]
MD020CP2N	7.6	Distilled water as feed; $T_{b,f} = 65$ °C, $p_p = 4$ kPa, $v_l = 0.6$ m/s	[42]
PVDF unsupported	5.0	$T_{b,f} = 25$ °C, $p_p = 1666.5$ Pa, Distilled water as feed	[4]
	4.46	Chloroform/water mixture (1000 mg/l), separation factor = 7.89	

(Continued)

TABLE 12.1 Reported VMD permeate flux (J) of different types of commercial membranes shown in Tables 2.1 and 2.2 and some fabricated membranes for VMD process—Cont'd

Membrane trade name	J (10^{-3} kg/m²s)	Observation	Reference
PVDF Supported	4.5 3.91	$T_{b,f}$ = 25 °C, p_p = 1666.5 Pa, Distilled water as feed Chloroform/water mixture (1000 mg/l), separation factor = 6.81	
PVDF[a]	0.076 0.056	1,1,1-trichloroethane (TCA) 620 ppm in feed water $T_{b,f}$ = 50 °C, p_p = 2.67 10^3 Pa $T_{b,f}$ = 60 °C, p_p = 10.7 10^3 Pa (TCA removal up to 97%)	[8]
PVDF[b] M2 M3	0.142 0.140	TCA in water 430 ppm, $T_{b,f}$ = 50 °C, p_p = 5.3 10^3 Pa Total flux, (TCA removal up to 83%) Total flux, (TCA removal up to 85%)	[9]
PP[c]	0.83	$T_{b,f}$ = 60.5 °C, p_p = 80 10^3 Pa, 35 g/l NaCl	
PE[c]	1.11	$T_{b,f}$ = 60.5 °C, p_p = 80 10^3 Pa, 35 g/l NaCl	[6]
Modified TiO$_2$ (Ti5)[d]	1.690 1.157	$T_{b,f}$ = 40 °C, p_p = 0.3 kPa, 0.5 M NaCl, 99.5% salt retention 1 M NaCl, 99% salt retention	[13]
Modified ZrO$_2$ (Zr50)[d]	2.08	$T_{b,f}$ = 40 °C, p_p = 0.3 kPa, 0.5 M NaCl, 96.1% salt retention	
Modified PP MXFR 3[e]	19.72 19.16	$T_{b,f}$ = 85 °C, p_p = 8–8.8 10^3 Pa, Distilled water as feed 1 wt% NaCl	[7]
SMM/PES V3[f]	0.54 0.127 0.084	Pure water as feed, $T_{b,f}$ = 26 °C, p_p = 400 Pa 1000 ppm ethanol/water solution, $T_{b,f}$ = 26 °C, p_p = 650 Pa, Total permeate flux Water partial permeate flux	[14]
TIPS iPP[g]	8.03 6.89	$T_{b,f}$ = 70 °C, feed flow rate 50 l/h, p_p = 3 kPa Distilled water 0.5 mol/l NaCl, 99.9% salt rejection	[15]

[a] PVDF hollow fibre membrane (d_i/d_o = 0.52/0.90 mm; d_p = 3.24 10^{-7}m; ε/L_p = 118 m^{-1}) prepared by wet phase inversion method.

[b] PVDF hollow fibre membranes, M2 (d_i/d_o = 0.53/0.84 mm; d_p = 1.36 10^{-7} m; ε/L_p = 542 m^{-1}) and M3 (d_i/d_o = 0.52/0.82 mm; d_p = 6.2 10^{-8} m; ε/L_p = 1516 m^{-1}) prepared by wet phase inversion method.

[c] Polypropylene (PP: d_i/d_o = 342.5/442.5 μm; ε = 53.3%, d_p = 7.4 $10^{-2}\mu m$) and polyethylene (PE: d_i/d_o = 267.5/367.5 μm; ε = 66.3%, d_p = 8.7 $10^{-2}\mu m$) hollow fibre membranes prepared by melt extruded/cold-stretched.

[d] grafted ceramic hollow fibre membranes by 1H,1H,2H,2H-perfluorodecyltriethoxysilane (Zirconia, ZrO$_2$: d_p = 50 nm); (Alumina, Al$_2$O$_3$: d_p = 200 nm); (Titania, TiO$_2$: d_p = 5 nm).

[e] Polypropylene (PP) porous hollow fibre membrane (Accurel Membrana: PP 150/330; d_i/d_o = 0.33/0.63 mm; ε = 65%; maximum pore size = ≥0.2 μm) coated by plasma polymerization using silicone fluoropolymer.

[f] Surface modified flat sheet membrane by surface modifying macromolecules (SMMs) using polyethersulfone host polymer (porous hydrophobic/hydrophilic membrane).

[g] flat sheet membrane prepared via thermally induced phase separation (TIPS) using isotactic polypropylene (iPP) polymer.

$T_{b,f}$: bulk feed temperature, ϕ_l: liquid flow rate, v_l: liquid circulation velocity, p_p: downstream pressure the partial pressure of the component i at the liquid feed/membrane interface, $p_{i,p}$ is the partial pressure of the component i at the downstream membrane surface.

One possible way to increase membrane permeability in MD is to remove air from the membrane pores by deaeration or by applying a continuous vacuum on the permeate side to maintain the permeate pressure below the equilibrium vapour pressure. In such a case the ordinary molecular diffusion resistance can be neglected, since as reported in the previous chapters for DCMD and SGMD processes the diffusion depends on the partial pressure of air in the membrane pores and in VMD only traces of air are present within the membrane pores. Thus, the mass transport mechanisms through porous and hydrophobic membranes in VMD are only Knudsen flow model, viscous flow model and their combination [3,4,16,23]. As a matter of fact, a Knudsen type of diffusion is commonly considered by the majority of authors [4,9,18,19,21,22,30,36,41,42]. This is because the membranes for the VMD process should have small pore sizes in order to avoid their wetting when contacted with feed solutions, which often are aqueous solutions of organic liquids with low surface tension. Moreover, the calculated values of the mean free path of the transported molecules through the membrane pores are higher in VMD than in SGMD or DCMD configurations because of the low pressure applied on the permeate side. Remember that the mean free path is inversely proportional to the pressure. This is why different mass transport mechanisms should be applied when the same membrane is used under different MD configurations. For example, for temperatures ranging between 30 and 65 °C, the mean free path of water vapour varies between 2.8 and 3.4 µm. If the pore size of the membrane is 0.2 µm (TF200, see Table 2.1), Knudsen number, Kn, defined as the ratio of the mean free path to the pore size (Eq. (10.15)), will vary from 14 to 17.

In most of the reported theoretical VMD studies, membranes of uniform and non-interconnected cylindrical pores are assumed [3,4,16,18,19,21–23,30,36,41,42]. When Kn is higher than 10 (that is, the pore radius $\bar{r} < 0.05\lambda_i$), the molecule–pore wall collisions are dominant in comparison with the molecule–molecule collisions, and Knudsen-type diffusion of the vapour molecules through the membrane pores is applied by means of the following equations (see Chapter 10) [3–5,41–44]:

$$B_i^K = \frac{2}{3RT} \frac{\varepsilon \bar{r}}{\tau \delta} \left(\frac{8RT}{\pi M_i}\right)^{1/2} \quad (12.2)$$

where $\bar{r}$ is the mean pore radius assumed to be uniformly applied for all pores, T is the absolute temperature, M_i is the molecular weight of the component i, R is the gas constant, ε is the membrane porosity, τ is the pore tortuosity and δ is the membrane thickness.

When $0.05\lambda_i < \bar{r} < 50\lambda_i$, the transition flow dominates and the following equation corresponding to the combined Knudsen/viscous mechanism is considered [3,5,16]:

$$B_i^m = \frac{1}{RT\delta}\left(\frac{2\varepsilon\bar{r}}{3\tau}\left(\frac{8RT}{\pi M_i}\right)^{1/2} + \frac{\varepsilon\bar{r}^2}{8\tau\eta_i}\bar{p}\right) \quad (12.3)$$

where η_i is the viscosity of specie i and $\bar{p}$ is the average pressure in the membrane pores.

Lawson and Lloyd [16] reported that for VMD configuration, the more complete equations of dusty gas model, reported previously in Chapter 10 (Eqs. (10.28–10.32)), are reduced to the Knudsen/viscous transition flow (Eq. (12.3)). In this case, the contribution of the viscous flux in the total permeate flux can be estimated by the ratio of the viscous and Knudsen terms:

$$\kappa_i^{v/K} = 0.2 \frac{\bar{p}\bar{r}}{\nu\eta_i} \quad (12.4)$$

where ν is the mean molecular speed of vapour molecules.

If the mass transport is considered through each pore individually, as was described in Chapter 10, the size of each individual pore (d_p) must be compared to the mean free path of the vapour molecules, λ_i, and the following three possibilities may occur:

(1) $d_p < 0.1\lambda_i$. In this case the permeability through each pore in Knudsen region is expressed as follows [3,5]

$$B_i^K = \frac{2\pi}{3} \frac{1}{RT} \left(\frac{8RT}{\pi M_i}\right) \frac{r_k^3}{\tau\delta} \qquad (12.5)$$

where r_k is the pore radius in Knudsen region.

(2) $d_p > 100\lambda_i$. In this case, molecule-molecule collisions dominate and the permeability through a single pore in viscous (i.e. Poiseuille) region may be calculated using the following equations [3,5].

$$B_i^v = \frac{\pi r_v^4}{8\eta_i} \frac{\bar{p}}{RT} \frac{1}{\tau\delta} \qquad (12.6)$$

where r_v is the pore radius in Poiseuille region.

(3) When the pore size is between $0.1\lambda_i$ and $100\lambda_i$, both molecule–molecule and molecule–pore wall interactions have to be considered inside the pore. In this case, the pore contributes to the total mass transport by a mechanism operative in the Knudsen/viscous transition region and the following expression may be used to determine the permeability through each single pore [3,5]:

$$B_i^t = \frac{\pi}{RT\tau\delta}\left[\frac{2}{3}\left(\frac{8RT}{\pi M_i}\right)^{1/2} r_t^3 + \frac{r_t^4}{8\eta_i}\bar{p}\right] \qquad (12.7)$$

where r_t is the pore radius in the transition region.

When a membrane with a pore size distribution is used in VMD, all mechanisms can occur simultaneously, depending on the operating condition, and the total VMD permeability of the membrane can be calculated using the following equation [3,5]:

$$B_i^m = \frac{N}{\tau\delta}\left(\sum_{j=1}^{m(r=0.05\lambda_i)} G_i^k f_j r_j^3 \right.$$

$$+ \sum_{j=m(r=0.05\lambda_i)}^{p(r=50\lambda_i)} (G_i^k f_j r_j^3 + G_i^v f_j r_j^4 \bar{p})$$

$$\left. + \sum_{j=p(r=50\lambda_i)}^{n(r=r_{max})} G_i^v f_j r_j^4 \bar{p} \right) \qquad (12.8)$$

$$G_i^k = \left(\frac{32\pi}{9 M_i RT}\right)^{1/2} \qquad (12.9)$$

$$G_i^v = \frac{\pi}{8\eta_i} \frac{1}{RT} \qquad (12.10)$$

where f_j is the fraction of pores with pore radius r_j, N is the total number of pores per unit area, m is the last class of pores in Knudsen region and p is the last class of pores in the transition region.

It is to be noted that in Eq. (12.8) the upper limit of each summation may be changed by the relative values of the maximum pore radius (r_{max}). The following three cases are possible: (i) if $r_{max} \leq 0.05\lambda_i$, only Knudsen mechanism prevails; (ii) if $r_{max} \leq 50\lambda_i$, both Knudsen and transition mechanisms are applicable; and (iii) if $r_{max} > 50\lambda_i$, all mechanisms are operative simultaneously.

For a binary mixture, each component may travel by any of the transport mechanisms above described, depending on the pore size and the absolute pressure in the pore. In the Knudsen regime, transport of molecules is controlled by molecule–wall collisions and the molecules travel independently from each other. In viscous flow, the pore size is much larger than the mean free path of the molecules and molecule–molecule interactions dominate.

For vapour mixtures of two components, the mean free path and the collision diameter are different from the corresponding values for the pure component. The following relationship can be applied for binary mixtures to calculate the mean free path:

$$\sigma_{i-w} = \frac{\sigma_i + \sigma_w}{2} \quad (12.11)$$

where σ_{i-w}, σ_i and σ_w are the collision diameters of the mixture, pure component i and pure component w, respectively.

From Eqs. (12.1) and (12.8), the individual fluxes of the mixture can be written as [3]:

$$J_i = \frac{N}{\tau \delta} \left(\sum_{j=1}^{m(r=0.05\lambda_i)} G_i^k f_j r_j^3 \Delta p_i \right.$$
$$+ \sum_{j=m(r=0.05\lambda_i)}^{p(r=50\lambda_i)} (G_i^k f_j r_j^3 \Delta p_i + G_i^v f_j r_j^4 \overline{p}_i \Delta p)$$
$$\left. + \sum_{j=p(r=50\lambda_i)}^{n(r=r_{max})} G_i^v f_j r_j^4 \overline{p}_i \Delta p \right) \quad (12.12)$$

where Δp is the total pressure difference across the membrane, Δp_i is the partial pressure difference of component i and $\overline{p}_i$ is the average partial pressure of component i.

For organic–water binary mixtures, the following relationships can be used:

$$\Delta p_o = Y_{o,m} p_m - Y_{o,p} p_p \quad (12.13)$$

$$\Delta p_w = Y_{w,m} p_m - Y_{w,p} p_p \quad (12.14)$$

$$\overline{p}_o = \frac{Y_{o,m} p_m + Y_{o,p} p_p}{2} \quad (12.15)$$

$$\overline{p}_w = \frac{Y_{w,m} p_m + Y_{w,p} p_p}{2} \quad (12.16)$$

$$Y_{o,m} + Y_{w,m} = 1 \quad (12.17)$$

and

$$Y_{o,p} + Y_{w,p} = 1 \quad (12.18)$$

where Y is the mole fraction of each species in the vapour phase. The subscripts w and o refer to water and organic components, respectively, while m and p indicate feed membrane surface and permeate, respectively.

Consequently, from the above equations:

$$\Delta p = \Delta p_o + \Delta p_w = p_m - p_p \quad (12.19)$$

Moreover, the mole fractions of the organic species and water in the permeate can be written as:

$$Y_{o,p} = \frac{J_o}{J_o + J_w} \quad (12.20)$$

$$Y_{w,p} = \frac{J_w}{J_o + J_w} \quad (12.21)$$

where J_o and J_w are the VMD partial permeate flux of the organic species and water respectively.

Rewriting the above Eq. (12.20) and substituting for J_o and J_w using Eqs. (12.12–12.19), the following quadratic equation can be obtained in terms of $Y_{o,p}$ after appropriate simplifications:

$$\left[(G_w^k - G_o^k) I_1 p_p + \frac{1}{2}(G_o^k - G_w^k) I_2 (p_m - p_p) p_p \right] Y_{o,p}^2$$
$$+ \left[\begin{array}{l} G_o^k I_1 Y_{o,m} p_m + G_w^k I_1 Y_{w,m} p_m + \frac{1}{2} G_o^v I_2 (p_m - p_p)(Y_{o,m} p_m - p_p) \\ + \frac{1}{2} G_w^v I_2 (p_m - p_p)(Y_{w,m} p_m + p_p) + (G_o^k - G_w^k) I_1 p_p \end{array} \right] Y_{o,p}$$
$$- G_o^k I_1 Y_{o,m} p_m - \frac{1}{2} G_o^v I_2 (p_m - p_p) Y_{o,m} p_m = 0 \quad (12.22)$$

where

$$I_1 = \frac{N}{\tau\delta} \sum_{j=1}^{p(r=50\lambda_i)} f_j r_j^3 \quad (12.23)$$

$$I_2 = \frac{N}{\tau\delta} \sum_{j=m(r=0.05\lambda_i)}^{n(r=r_{max})} f_j r_j^4 \quad (12.24)$$

It is to be mentioned that for a dilute organic aqueous mixture, the partial pressure of the organic compound ($Y_{o,m}$ p_m) can be estimated using Henry's law, while that of water ($Y_{w,m}$ p_m) can be evaluated from Raoult's law [3,20,21].

The separation factor or membrane selectivity can be calculated using the following expression:

$$\alpha = \frac{Y_{o,p}/Y_{w,p}}{x_{o,f}/x_{w,f}} \quad (12.25)$$

where $x_{o,f}$ and $x_{w,f}$ are the mole fractions of organic and water in the bulk liquid feed, respectively.

The prediction of the separation factor together with the organic and water permeate fluxes can be performed following an iterative procedure. Initially, λ_{o-w} must be calculated using Eq. (12.11) and the summations, I_1 and I_2 are evaluated from Eqs. (12.23) and (12.24), when the characteristics (pore size distribution, porosity, thickness, etc.) of the membrane are known. G_i^k and G_i^v are then calculated at the bulk feed temperature using Eqs. (12.9) and (12.10), respectively. After these calculations, Eq. (12.22) is solved in terms of $Y_{o,p}$. Subsequently, the VMD permeate flux for each species, J_i, is calculated using Eq. (12.12) and then both temperature and composition at the membrane surface are calculated from the heat and mass transfer coefficients corresponding to the feed thermal and concentration boundary layers (i.e. temperature and concentration polarization effects) as will be explained later on. The quantities G_i^k and G_i^v are re-evaluated taking into account both the temperature and concentration polarization effects and second values of the VMD permeate flux of each specie is obtained for the second time. The procedures are repeated until the difference between two successive calculated permeate fluxes becomes less than 0.1%.

In general, in MD the mass transport through the membrane matrix (i.e. surface diffusion) is neglected due to the fact that the diffusion area of the membrane matrix is small compared with the pore area [3,44]. For hydrophobic MD membranes, the 'affinity' between water and the membrane material is very low and it is allowed to neglect the contribution of transport through the membrane matrix, especially for porous membranes with large pore sizes and high porosities. Nevertheless, when some feed components have strong interaction or 'affinity' with the membrane material, the transport mechanism through the material of the membrane may have a significant effect and must be included in the theoretical models. This is explained in this chapter in the section 'Comparison to pervaporation' when VMD process is compared with PV.

Instead of assuming cylindrical pores of the membranes, interconnections between pores have been considered using Monte Carlo (MC) simulation model for VMD configuration [45]. The membrane pore space was described by a three-dimensional network of interconnected cylindrical pores with size distribution as shown in Fig. 12.2. VMD permeate flux through the membrane pores was described by all the aforementioned vapour transport mechanisms, i.e. the MC model was developed so that it could take into account all possible transport mechanisms (Knudsen, Poiseuille, combined Knudsen/Poiseuille) and the boundary conditions, which may affect VMD process. The effects of the bulk feed temperature, downstream pressure and membrane pore size on the simulated VMD permeate flux were

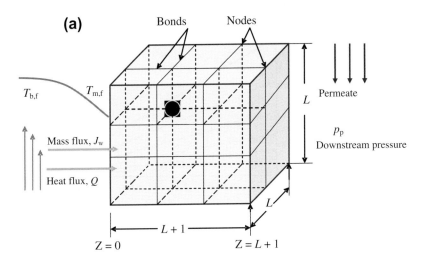
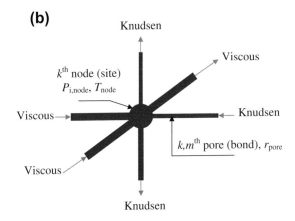

FIGURE 12.2 Network model applied for the description of membrane pore space in VMD process (a) and a schematic representation of pore level vapour flux transport mechanisms (b).

investigated. It was found that when the simulation was carried out using Knudsen type of flow alone, the obtained results were in qualitative agreement with experimental data and in accordance with the theoretical expectations; however, when viscous type of flow was considered, the simulated results indicated that the higher feed solution temperature and higher pore size do not necessarily increase the VMD permeate flux (see Fig. 12.3 as an example). These simulated results must be confirmed experimentally.

Similar to the other MD configurations, a simultaneous heat and mass transfer through the membrane occurs also in VMD process and the temperature and concentration at the vapour/liquid interface differ from the bulk conditions (Fig. 12.1). In VMD, the heat transfer within the membrane is due to the latent heat accompanying vapour flux, whereas the boundary layer resistance on the permeate side and the contribution of the heat transported by conduction through the membrane are negligible [3–5,16,41,42,44]. In such a case, a simple

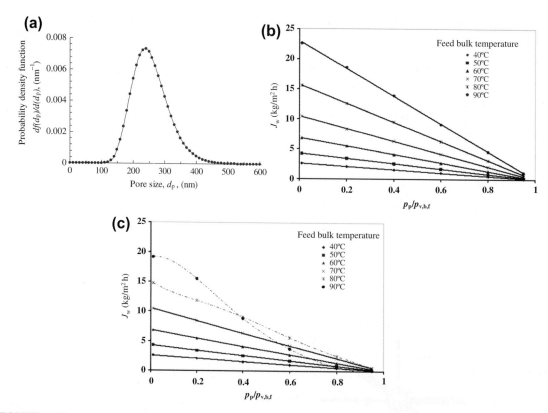

FIGURE 12.3 Simulated VMD permeate flux of a membrane with pore size distribution (a) versus the ratio of the downstream pressure (p_p) to the bulk feed pressure of water ($p_{v,b,f}$) at different bulk feed temperatures when the process is controlled by Knudsen mechanism (b) and by Knudsen and Poiseuille mechanisms (c).

enthalpy balance usually used in VMD process is [3–5,16]:

$$Q = h_f(T_{b,f} - T_{m,f}) = \sum_{i=1}^{n} J_i \Delta H_{v,i} \quad (12.26)$$

where $\Delta H_{v,i}$ is the evaporation enthalpy of specie i of the transmembrane flux J_i, n is the number of permeating species, h_f is the heat transfer coefficient in the liquid phase, $T_{b,f}$ is the feed bulk temperature and $T_{m,f}$ is the temperature at the liquid/vapour interface. For a pure component, Antoine's equation can be used to calculate the partial pressure of the liquid feed [44].

To maximize the VMD permeate flux, the temperature $T_{m,f}$, which governs the vapour pressure of liquid at the feed/membrane interface, must be increased to enhance the temperature polarization coefficient that is defined in the MD literature for VMD configuration by different expressions [3,7,16,29,30,36,41]:

$$\theta = \frac{T_{m,f}}{T_{b,f}} \quad (12.27)$$

$$\theta = \frac{T_{b,f} - T_{m,f}}{T_{b,f} - T_p} \quad (12.28)$$

$$\theta = \frac{T_{m,f} - T_p}{T_{b,f} - T_p} \quad (12.29)$$

where T_p refers to the equilibrium temperature of the liquid corresponding to the pressure at the permeate side.

Following Eq. (12.28), the membrane modules limited by heat transfer through the boundary layers will have θ values near unity, whereas the modules limited by mass transfer will exhibit θ values approaching 0. The temperature polarization coefficient was found to be as high as 0.7 for the membranes exhibiting relatively high permeabilities and independent of the feed temperature. For membranes having low permeabilities, θ increased with the feed temperature [16].

Following Eq. (12.29), it was found that the temperature polarization coefficient decreased from 0.98 to 0.67 as the membrane permeability (i.e. pore size and porosity) increased [3].

Following Eq. (12.27), a value as high as unity was reported for θ [30].

Normally, the temperature polarization coefficient is related to the heat transfer coefficient in the feed boundary layer. Such heat transfer coefficient is estimated by means of the empirical heat transfer correlations developed for different systems under different flow regimes [42,44].

$$Nu = f(Re, Pr) \quad (12.30)$$

where Nu, Re and Pr are Nusselt, Reynolds and Prandlt numbers.

When an aqueous solution containing non-volatile solutes is used as the feed, the concentration of the solutes at the feed-side membrane surface becomes greater than in the bulk due to concentration polarization, which results in reduction of both driving force and permeate flux. The same definitions and expressions used in the previous chapters, for DCMD and SGMD processes, can be used in VMD process.

The solute concentration polarization coefficient (ξ_s) is defined as [16,29,30,44]:

$$\xi_s = \frac{x_{s,m,f} - x_{s,b,f}}{x_{s,b,f}} \quad (12.31)$$

or

$$\xi_s = \frac{x_{s,m,f}}{x_{s,b,f}} = \exp(J_w/k_s) \quad (12.32)$$

where x is the mole fraction of the solute s in the aqueous feed solution, J_w is the VMD permeate flux of water and k_s is the solute mass transfer coefficient for the diffusive mass transfer through the feed boundary layer. The subscripts s, b, f and m refer to solute, bulk, feed and feed/membrane interface, respectively.

In Eq. (12.32) the mass transfer through the feed liquid phase was described using the film theory.

In Eqs. (12.31) and (12.32), the solute concentration in the feed solution (C) can be used instead of the solute molar fraction (x).

The coefficient k_s is expressed as a function of the molecular diffusivity D and the thickness of the boundary layer δ_b.

$$k_s = D/\delta_b \quad (12.33)$$

The coefficient k_s can be estimated from Sherwood number (Sh) via an appropriate dimensionless empirical correlation for mass transfer employing analogy to the empirical correlation for heat transfer (Eq. (12.30)).

$$Sh = f(Re, Sc) \quad (12.34)$$

where Sc is Schmidt number.

It is worth quoting here also that, compared with the temperature polarization, the boundary layers adjoining the membrane generally result in a smaller contribution of concentration polarization to the overall mass transfer resistance.

When aqueous solutions containing volatile solutes, such as VOCs, are treated by VMD, the mole fraction of the volatile solute at the feed/membrane interface becomes smaller than the bulk value and consequently, the mole fraction of water at the interface is larger than in the bulk. The following equation was used to define the concentration polarization coefficient of the specie i in VMD applied for the treatment of mixtures containing VOCs [3,22]:

$$\xi_i = \frac{x_{i,m,f} - x_{i,p}}{x_{i,b,f} - x_{i,p}} \quad (12.35)$$

Again, the film theory is applied for the mass transfer through the feed liquid phase in order to determine the mole fraction at the liquid/membrane interface. However, total permeate flux (i.e. permeation fluxes of both organic compounds and water) must be considered for this case instead of water flux alone [3]. Furthermore, the Wilke–Chang correlation is commonly used to calculate the value of the ordinary diffusion coefficient of organic compounds in water [46]. For dilute aqueous solutions of organic solutes, the physical properties of pure water are commonly used in theoretical models [3].

When pure water is used as feed, VMD process can be used to determine the temperature at the membrane surface ($T_{m,f}$) (Eqs. (12.1–12.3)) and therefore the boundary layer heat transfer coefficients in the membrane module can be evaluated as a function of the feed temperature and flow rate using Eq. (12.26) [16,42]. This can help selecting the adequate empirical heat transfer correlation of a given MD system, which is a complex task when developing theoretical models to determine the temperature polarization coefficients. This procedure is especially interesting for membrane modules using supports and channel spacers as well as complicated turbulent promoters.

A critical review of the most frequently used empirical heat transfer correlations in MD studies was presented by Mengual et al. [42] concluding that special care must be taken into account when the empirical correlations, developed originally for non-porous heat exchangers, are used in MD for prediction of the permeate flux and for the calculation of the temperature polarization coefficients. The characteristic constants that appeared in the empirical heat and mass transfer correlations (Eqs. (12.30) and (12.34)) must be re-evaluated for application in different MD systems in which both heat and mass transfer are taking place.

EXPERIMENTAL: EFFECTS OF PROCESS CONDITIONS

Effects of Feed Temperature

In VMD process, the effect of the feed temperature on the permeate flux is similar to the DCMD and SGMD processes. An exponential increase of the VMD permeate flux is observed with the increase of the feed temperature. The feed temperature is a very sensitive operating parameter, which significantly affects both the permeate flux and the total energy requirement.

Sensitivity of the VMD permeate flux to the process operating parameters was investigated by Banat et al. [47]. It was found that the VMD permeate flux is highly sensitive to the feed temperature especially at higher permeate pressures [47].

Figure 12.4 shows the experimental data and the model predictions of the permeate flux as a function of the inlet feed temperature in DCMD and VMD configurations, using the same membrane module under the same operating feed conditions. The theoretical predictions agree well with the experimental data and in the entire feed temperature range, the distillate flux is higher in the VMD configuration than in the DCMD configuration. The VMD flux is 2.8–3.1 times higher than the mass flux obtained by DCMD.

In Chapters 10 and 11, the thermal efficiency, η, is defined as the ratio between the heat that contributes to evaporation and the total heat transferred from the feed to the permeate. The thermal efficiency of VMD and DCMD configurations is plotted in Fig. 12.4 as a function of the feed temperature at the membrane module

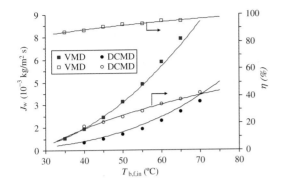

FIGURE 12.4 Effect of the feed inlet temperature ($T_{b,f,in}$) on the MD permeate flux (J_w) and thermal efficiency (η) of the commercial membrane module Accurel® S6/2 MD020CP2N (see Table 2.2) when distilled water is used as feed with 0.8 m/s circulation velocity. (DCMD) 0.8 m/s permeate circulation velocity and 20 °C permeate inlet temperature. (VMD) 3.5 kPa downstream pressure.

inlet. It is clear that operating at high temperatures increases the evaporation efficiency, especially in the DCMD configuration. This is mainly due to the exponential increase of the permeate flux with temperature. It can be seen that η is higher in VMD than in DCMD. This is related to the internal heat loss by conduction that is negligible in VMD. Khayet et al. [17] proved that the heat transferred by conduction across the membrane is 4.8–12.7% of the total heat transferred. Therefore, the heat transferred by conduction may affect the prediction of the VMD permeate flux, and if so, Eq. (12.26) should be slightly modified introducing the heat transfer by conduction through the membrane. In DCMD, only 17.7–42.1% of the total heat is consumed as latent heat, while rest of the heat is lost by internal conduction. In the SGMD configuration, 9.5–28.6% of the total heat transferred in the process is lost by conduction.

It was reported that an increase in temperature from 20 to 70 °C increased the permeate flux but the total energy requirement was also drastically increased. At a feed temperature of 70 °C more than 99% of the total energy requirement was heat energy [28]. Moreover, when the feed temperature was increased from 53.9 to 59.3 °C in a cross-flow plate and frame module with a membrane having 0.2 µm pore size and 91 µm thickness (Membrana, Germany), at a feed flow rate of 150 l/h and a downstream pressure of 10 mbar, the permeate flux increased from 29.7 to 51.5 kg/m²·h. At the same time the energy consumption increased from 354.6 to 441.8 W but the energy consumption per unit of permeate flow rate was decreased from 2.98 to 2.15 kWh/kg [48]. In general, VMD performance was found to be better than DCMD in terms of permeate flux and energy consumption per unit permeate mass and thermal efficiency.

The temperature polarization coefficient (θ) defined earlier by means of Eqs. (12.27–12.29) indicates the effects of the boundary layer heat transfer resistances on the total heat transfer resistance of the VMD system. When treating aqueous solutions containing non-volatile solutes by VMD process, it was found that the temperature polarization coefficient (θ) calculated by means of Eq. (12.27) decreased with the increase of the feed temperature [29]. However, when Eq. (12.28) is used, as shown in Fig. 12.5, θ is practically independent of the feed temperature for the membranes of high water vapour permeability, but θ increases with the feed temperature for the membrane of low permeability. Moreover, θ is higher for the membranes with higher permeability. It is to be noted that the membranes used in [16] exhibited permeate fluxes an order of magnitude greater than the commonly used membranes (see Table 12.1). The authors attributed the relatively high VMD permeate fluxes to a combination of the design of the membrane module and the membranes, achieving heat transfer coefficients on the order of 10^4 W/m²·K [16].

Since evaporation takes place at the feed liquid/membrane interface, temperature and concentration of volatile solutes at the interface are generally different and lower than the corresponding values of feed liquid. When volatile

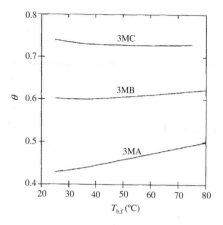

FIGURE 12.5 Effect of the feed temperature ($T_{b,f}$) on the calculated temperature polarization coefficient (θ) of three flat sheet membranes 3MA, 3MB and 3MC (see Table 2.1). Distilled water used as feed at downstream pressure of 3 kPa. *Source: Reprinted from [16]. Copyright 1996, with kind permission from Elsevier.*

compounds are present in the feed solution, for example, in the case of TCA/water binary mixtures, it was observed that the removal efficiency of TCA increased with the feed temperature and levelled off at high temperatures as can be seen in Fig. 12.6. The driving forces for the permeation of both water and TCA were increased with temperature due to the increase of the partial vapour pressures of both species with temperature. However, because water is the major component of the feed solution, the increase in the partial vapour pressure of water was much higher than that of TCA. Consequently, there is a considerable enhancement of the partial permeate flux of water at a higher feed temperature, which reduced the TCA concentration in the permeate. This is a common qualitative observation made in various VMD systems and feed solutions for a given downstream pressure. It is worth mentioning that operating at high feed temperatures is not necessarily advisable as the increase in water permeation flux may result in wetting of membrane pores and higher energy consumption with little or no improvement in TCA removal efficiency.

It is worth quoting that at higher feed temperatures there is a broader downstream pressure range at which VMD can be effective due to the higher vapour pressures of the feed aqueous solutions.

As expected, a linear correlation between the VMD permeate flux and the pressure difference across the membrane was reported, whereby the pressure difference was increased either by increasing the feed temperature while maintaining the feed flow rate and the vacuum pressure constant or by decreasing the permeate pressure while maintaining the feed flow rate and the feed temperature fixed [7,9,22,29,37]. The slope of the linear correlation depended mainly on the permeability of the used membranes (i.e. higher for membranes with greater permeability).

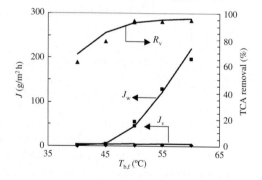

FIGURE 12.6 Effect of the bulk feed temperature on the partial permeate fluxes (water, J_w and TCA, J_v) and on the TCA removal of fabricated a PVDF hollow fibre membrane at a feed TCA concentration of 620 ppm, at a downstream pressure of 80 mmHg and a feed flow rate of 10^{-3} m^3/h. *Source: Reprinted from [8]. Copyright 2006, with kind permission from Elsevier.*

Effects of Solute Concentration in the Feed Solution

In all MD configurations, the increase in the concentration of non-volatile solutes in the aqueous feed solution results in reduction of the permeate flux due to the decrease of water vapour pressure with the addition of solutes in water (i.e. decrease of the driving force in Eq. (12.1)) as well as to the contribution of the concentration polarization effect. Very high rejection factors were observed in desalination, concentration of aqueous sucrose solutions, treatment of dye solutions, etc. [13,15,17,24, 29,31].

VMD can be used for the treatment of highly concentrated aqueous solutions of salts without suffering from a large drop in productivity compared with other membrane separation processes such as RO [25,27,28]. For example, VMD is used to concentrate RO brines from 50 to 300 g/l salt concentrations. However, when VMD is applied to the solution of very high salt concentration, the performance is very different from the treatment of dilute salt solutions. The permeate flux decreases as the salt concentration in the feed increases and this leads to MD crystallization and scaling as it was explained in Chapter 10 for DCMD process. The effect of scaling in desalination by VMD was reportedly not remarkably strong [27]. No significant difference was detected between the initial permeability of polytetrafluoroethylene (PTFE) membrane and the permeability of the membrane washed after VMD desalination, confirming that scaling is reversible and occurs only on the membrane surface.

Compared to the other MD configurations, considerably higher permeate fluxes were obtained in VMD. However, among the few studies carried out in desalination by VMD, one of the highest VMD permeate flux was found to be 80 and 69 kg/m²·h for feed distilled water and 1 wt% NaCl feed aqueous solution, respectively, at a feed temperature of 85 °C, a vacuum pressure ranging between 8×10^3 and 8.8×10^3 Pa and with a modified PP MXFR 3 membrane (PP porous hollow fibre membrane, Accurel Membrana: PP 150/330 with 0.33 mm fibre inner diameter, 0.63 mm fibre outer diameter, 65% porosity and a maximum pore size ≥ 0.2 µm, outer surface coated by plasma polymerization using a silicone fluoropolymer) [7]. These permeate fluxes are far lower than those reported for the membranes 3MA, 3MB and 3MC commercialized by 3M Corporation, 372.6, 421.2 and 576.7 kg/m²·h, respectively, using distilled water as feed at a feed temperature of 74 °C and a downstream pressure of 3 kPa [16].

When an aqueous solution containing volatile solutes is used as feed, both water and the volatile solutes are transported through the membrane and the change in the feed concentration with time allows us to determine the overall mass transfer coefficient (K) that is given by the following equation in which the mass transfer coefficients of the feed liquid boundary layer (K_f), membrane (K_m) and permeate boundary layer (K_p) are combined by the resistances in series model as follows:

$$\frac{1}{K} = \frac{1}{K_f} + \frac{1}{K_m} + \frac{1}{K_p} \quad (12.36)$$

In most of studies, the resistance term corresponding to K_m is neglected, especially when very low downstream pressures are applied.

Figure 12.7 presents an example of the change in feed chloroform concentration with time illustrating the effect of the initial chloroform concentration in the feed solution (C_0). It can be seen that the concentration of chloroform in the feed (C) decreases as an exponential function of time; the plot of $\ln(C_0/C)$ versus time becomes linear and the slope depends on the membrane regardless of the initial concentration. In other words, the dimensionless concentration of chloroform, C_0/C, is not influenced by the initial chloroform concentration in the feed solution. Accordingly, K can be determined using the following relationship [4,21]:

$$K = \frac{V}{At} \ln\left(\frac{C_0}{C}\right) \quad (12.37)$$

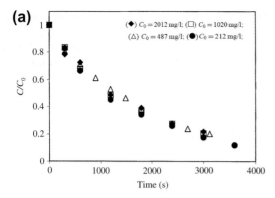

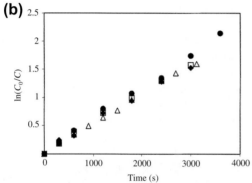

FIGURE 12.7 Effect of initial concentration of the volatile compound chloroform in the feed aqueous solution (C_0) on the dimensionless concentration C/C_0 versus time (a) and $\ln(C_0/C)$ versus time (b) with respect to a PP hollow fibre membrane module MD020TP2N (Enka-Mycrodyn, see Table 2.2) at a feed temperature 25 °C and a feed flow rate 0.23 l/min. Source: Reprinted from [21]. Copyright 2000, with kind permission from Elsevier.

where V is the initial volume of the liquid in the feed container and A is the membrane area.

In general, the total permeate flux and the partial permeate flux of the volatile solute (i.e. more volatile than water) both increase with increasing solute concentration in the feed due to the increase of the driving force for the solute mass transfer, leading to enrichment of the volatile solute in the permeate. This will depend also on the applied downstream pressure and reduction of the driving force for water permeate flux. This fact will be explained later on when analysing the effects of the downstream pressure. Theoretically, the VMD permeate flux of a given component should be zero when the downstream pressure is above the saturation vapour pressure of the component. Care should be taken in order to avoid wetting the membrane pores by increasing the volatile solute concentration in the feed solution. Moreover, the addition of ammonia (NH_3) in feed resulted in an increase of the total permeate flux but no clear trend could be observed between the ammonia separation factor and the initial ammonia concentration in the feed aqueous solutions [35].

Effects of Feed Flow Rate

As it was shown in the last two chapters, to reduce the temperature and concentration polarization effects, the feed flow rate must be increased or turbulent promoters and channel spacers must be installed. Then, the heat transfer coefficient of the feed boundary layer and VMD permeate flux increases. It should be pointed out that the increase of feed flow rate results in a significant increase in VMD permeate flux, especially in the laminar and transitional flow regimes. Further increase in the feed flow rate may result only in a marginal gain of the VMD permeate flux. As shown in Fig. 12.8, the MD permeate flux increases in DCMD and VMD configurations with increasing feed circulation velocity and approaches an asymptote value at higher feed flow rates. This is due to the reduction of the temperature polarization effect at the membrane feed side. It should be noted that the same membrane module and the same feed-operating conditions were used for both configurations. Agreement between the solid lines (theoretical) and the symbols (experimental) in Fig. 12.8 for DCMD is excellent. However, in the VMD system, agreement between the experimental and the theoretical values is better at higher feed circulation velocities ($v_f > 0.5$ m/s) than at low circulation

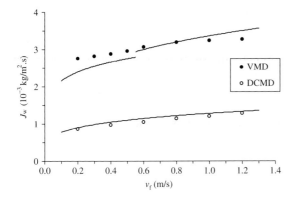

FIGURE 12.8 Effect of feed flow circulation velocity (v_f) on the permeate flux (J_w) of VMD and DCMD applied under the same feed-operating conditions using the commercial membrane module Accurel® S6/2 MD020CP2N (see Table 2.2). Distilled water was used as feed at an inlet temperature of 50 °C. In DCMD the permeate circulation velocity is 1 m/s and the permeate inlet temperature is 20 °C. In VMD the downstream pressure is 3500 Pa.

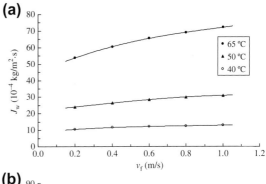

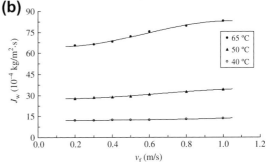

FIGURE 12.9 VMD permeate flux (J_w) versus feed flow circulation velocity (v_f) at different feed inlet temperatures: (a) water circulation on the shell side and (b) water circulation on the lumen side (3.5 kPa).

velocities. The discontinuity in the VMD line is due to the transition from laminar to transitional flow correlations. This proves one of the limitations of the empirical equations (the functional form of Eq. (12.30)) that have been developed for non-porous and rigid heat transfer exchangers. It can also be seen in Fig. 12.8 that the permeate flux for the VMD configuration is more than 2.5 times higher than the DCMD flux.

Figure 12.9 shows the experimental values of the VMD flux as a function of the feed circulation velocity (v_f) with respect to a shell and tube membrane module Accurel® S6/2 MD020CP2N (see Table 2.2) at different inlet temperatures. Experiments were done for the following two cases: (i) water circulating on the lumen side and vacuum applied on the shell side and (ii) water circulating on the shell side and vacuum applied on the lumen side. In both the cases, the VMD permeate flux increases with the feed circulation velocity and with the water inlet temperature. Moreover, the permeate flux is greater when water is circulating on the lumen side than on the shell side. This may be due to the higher Reynolds numbers achieved when water is circulating on the lumen side of the membrane module.

When the feed flow rate was increased from 150 to 235 l/h maintaining the feed inlet temperature at 59.1 °C and a downstream pressure at 60 mbar, the VMD permeate flux increased from 46.9 to 50.5 kg/m²·h but the energy consumption also increased from 199.5 to 223.9 W, whereas the energy consumption per unit of permeate flow rate was slightly increased from 1.08 to 1.10 kWh/kg [48].

In the case of feed aqueous solutions containing volatile solutes, the total and partial permeate fluxes also increase with the increase of the feed flow rate [8,41]. This is due also to the increase of the mass transfer coefficients, which results in a lower boundary layer heat and mass transfer resistances. Depending on the downstream pressure applied, the effect of the feed flow rate on the volatile solute flux may be higher than on the water flux. In fact, since dilute aqueous solutions are used to avoid pore wetting, the mole fraction of water within the liquid phase does not change appreciably suggesting that the mass transfer resistance in the liquid phase does not contribute to the change in water flux. Therefore, the observed increase in the water flux with increasing feed flow rate is to be attributed to the heat transfer resistance within the feed liquid. This is not the same for the volatile solute. Both the heat and mass transfer resistances in the liquid phase are important.

Effects of Downstream Pressure

In general, the permeate flux increases with a decrease in the permeate pressure in all VMD systems following the S-shaped or inverted S-shaped trend. In other words, both the permeate flux and the transmembrane hydrostatic pressure difference increase with the decrease of the downstream pressure. As a result, the risk of membrane pore wetting becomes very high.

Different permeate compositions are obtained depending on the downstream pressure. With respect to feed solutions containing volatile solutes, when the downstream pressure is very low, much lower than the saturation vapour pressure of water corresponding to the feed temperature, the water flux becomes much larger than the flux of the volatile solutes and the solute concentration in the permeate becomes very low. On the other hand, when the downstream pressure is high, the water flux decreases rapidly and the concentration of the volatile solutes becomes higher in the permeate. Therefore, if high separation factors are required in favour of the volatile solutes, the downstream pressure should be maintained higher than the vapour pressure of water.

It is important to mention here about the appreciable effect of the downstream pressure on the volatile solute concentration in the permeate. A concentration as high as 68% MTBE in water could be reached in the permeate when the feed MTBE concentration was only 2 wt% [41]. Similarly, 67 wt% acetone was obtained for 5 wt% acetone in aqueous feed solution. These results can be obtained when the driving force for permeate water flux is significantly reduced, whereas the driving forces for the volatile solutes are reduced only slightly by increasing the downstream pressure.

As the downstream pressure increases the VMD permeate flux decreases and correspondingly the volatile solute concentration in the permeate increases. Yet, the volatile solute concentrations in the permeate are smaller than those in vapour liquid equilibrium (VLE) with the feed solution, which are approached at higher feed pressures. These results are justified by separation due to Knudsen type of transport mechanism and polarization phenomena in the feed liquid. Since the molar mass of the organic compounds are larger than that of water molecule, Knudsen separation works unfavourably to the VMD process separation and therefore the upper limit for the composition of the permeate is dictated by the vapour/liquid equilibrium value corresponding to the feed liquid conditions. However, it is worth quoting that the relative importance of Knudsen separation and polarization effects on the VMD process performance depends upon the permeating species and operating conditions such as the downstream pressure, the hydrodynamic feed conditions and the membrane module used. Figure 12.10 shows two examples illustrating the increase of the total and water VMD permeate fluxes and/or the volatile solute flux with the decrease of the downstream pressure. The

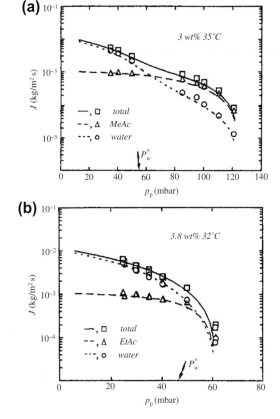

FIGURE 12.10 Effect of the downstream pressure (P_p) on the total and partial VMD permeate fluxes (J) using TF200 membrane (see Table 2.1), 3 wt% methyl acetate/water (MeAC) mixture at a temperature of 35 °C (a) and 3.8 wt% ethyl acetate/water (EtAC) mixture at a temperature of 32 °C (b) with a feed flow rate of 2.8 l/min. Lines are model predictions. *Source: Reprinted from [41]. Copyright 1997, with kind permission from John Wiley and Sons.*

decreases and the volatile solute concentration becomes higher than water.

It is important to mention that more energy is consumed to lower the downstream pressure with only a slight or no improvement of volatile solute separation factors and this suggests that VMD process is preferably to operate at moderate downstream pressures. It was found that by increasing the downstream pressure from 10 to 60 mbar (cross-flow plate and frame module (membrane pore size 0.2 μm and thickness 91 μm supplied by Membrana, Germany) operated at a feed temperature of 59.2 °C and a flow rate of 235 l/h) the energy consumption decreased from 441.2 to 223.9 W, the VMD permeate flux decreased from 56.2 to 50.5 kg/m²·h and the energy consumption per unit of permeate flow rate decreased from 1.98 to 1.10 kWh/kg [48]. Therefore, it is advisable to determine the optimum downstream pressure taking into account economic considerations.

The interactions between VMD operating parameters have to be studied to optimize the performance of VMD systems taking energy consumption into consideration. This can be done by applying response surface methodology or artificial neural network modelling. Taguchi method was applied by Mohammadi and Safavi [27] to optimize a VMD desalination system using a PP commercial membrane (0.2 μm mean pore size, 75% porosity, 163 μm thickness, purchased from Membrana, Germany). Feed temperature, vacuum pressure, feed flow rate and NaCl feed aqueous concentration were considered as variables for the surface model. Permeate flux was used as a single response. This is because distilled water was obtained for all the experimental runs. As it was expected and also well known, the results showed that the permeate flux was improved with increasing the feed temperature and decreasing the vacuum pressure. However, the permeate flux increased up to a maximum value with the variation of the feed flow rate and then decreased. This result

lower separation factors or lower volatile solute compositions are found at low downstream pressures corresponding to high total VMD permeate fluxes. At low downstream pressures the water flux is very large compared with the volatile solute permeate flux and approaches the total permeate flux. Therefore, the concentration of volatile solvent in the permeate becomes very low. On the contrary, at high downstream pressures, the water flux rapidly is not expected in VMD systems as shown previously in Fig. 12.8 and in Fig. 12.9. The optimum operating conditions were determined and confirmed experimentally. It was found that the deviation between the experimental and the predicted VMD fluxes were between 2 and 6%. However, in this study, energy consumption was not considered as a response value.

VMD APPLICATIONS

Extraction of Volatile Organic Compounds

The potential applications of VMD are generally the extraction of VOCs from dilute aqueous solutions [3,4,8,9,11,18–22,49,50].

Banat and Simandl [49] investigated the removal of benzene traces from contaminated water using tubular PP membrane module. Among the three resistances involved in the mass transport (i.e. liquid layer, membrane, gas layer), the liquid layer resistance was found to be the predominant.

Sarti et al. [18] showed that the overall flux was high, relative to the benzene flux, at low downstream pressures, resulting in a relatively poor selectivity, whereas at permeate pressures higher than the vapour pressure of water, the overall flux was found to be similar to that of benzene, resulting in a highly concentrated benzene permeate.

Couffin et al. [19] applied VMD to remove halogenated VOCs, namely chloroform, trichloroethylene and tetrachloroethylene, from drinking water at very low concentration (400 μg/l) using a VMD pilot plant designed to test both flat sheet and hollow fibre membrane module. The partial VMD permeate flux of the VOCs was found to be higher for chloroform followed by that of trichloroethylene and then tetrachloroethylene. Trichloroethylene was removed from water at very high selectivity varying from 9 to 860 depending on the feed temperature and on the downstream pressure. The total flux was decreased with increasing downstream pressure and decreasing the temperature (30–50 °C), while the selectivity was found to increase.

The same tendencies were observed by Bandini et al. [41] when using other dilute binary aqueous solutions containing acetone, ethanol, isopropanol, ethyl acetate, methyl acetate or MTBE in the concentration range 2–10 wt%. Bandini et al. [41] tested various diluted binary mixtures by VMD finding that the VOCs concentration in the permeate was lower than the corresponding equilibrium value with the feed. When both the downstream pressure and the feed temperature were low, the permeate flux was high and VOC concentration approached the equilibrium value. As it was shown in Fig. 12.10, at a low permeate pressure, the water flux is higher than the organic flux approaching the total flux and leading to a low separation factor for the organic solute. In contrast, at a permeate pressure higher than the saturation vapour pressure of water the water flux becomes lower and the organic concentration in the permeate is high.

Urtiaga and Ortiz [21] looked at removing chloroform from dilute aqueous solutions (500–2012 mg/l) using porous PP hollow fibre membranes at different VMD-operating parameters. It was concluded by means of a kinetic analysis that only under turbulent flow regime the resistance to mass transfer in the membrane affected the overall mass transfer coefficient, whereas in the laminar regime, the diffusion coefficient of chloroform in the feed liquid phase was the only parameter describing the separation performance.

Similar experiments were conducted by Khayet and Matsuura [3,4] using PVDF flat sheet membranes of different pore sizes. The chloroform selectivity was found to decrease from 72.5 to 7.9 with the increase of membrane pore size (0.01–0.2 μm), while both the chloroform and the water fluxes were increased.

PVDF hollow fibre membranes with different pore sizes and effective porosities were also tested in VMD for removal of TCA from aqueous solutions using different operating conditions and TCA initial concentrations in feed [8]. Under optimum VMD-operating parameters, particularly at a downstream pressure of 8–10.7 kPa, a feed temperature of 50 °C and a feed flow rate of 10^{-3} m^3/h, up to 97% of TCA removal efficiency has been achieved. Moreover, the same membranes have been used for the removal of toluene and benzene from water by VMD process. A TCA removal efficiency over 99% was obtained under a downstream pressure of 8 kPa, a feed temperature of 50 °C and a feed flow rate of 10^{-3} m^3/h.

Hydrophobic/hydrophilic PPESK hollow fibre composite membranes were fabricated for removal of 2,4-dicholorophenol from wastewater by VMD [11]. It was observed that the removal factor declined during VMD process because the partial vapour pressure of 2,4-dicholorophenol changes with the decrease of its concentration in the feed solution. A removal factor varying from 2 to 10 was obtained. Moreover, the removal factor was affected by the feed temperature (i.e. decreased with the increase of the feed temperature) and was not affected significantly by the variation of the applied downstream pressure. However, when treating aqueous solutions containing low concentrations of 2,4-dicholorophenol, the increase of the feed temperature was not favourable to the removal of 2,4-dicholorophenol.

Treatment of Alcohol Aqueous Solutions

VMD technology has been applied for the treatment of dilute alcohol water solutions [14,16,18,22,23,41,51].

Hoffman et al. [51], by performing VMD experiments of the binary mixtures of methanol/water, ethanol/water and *n*-propanol/water of different concentrations, found that the selectivities of the organic solutes were smaller than those calculated from the corresponding VLE data, when Gore-tex PTFE membranes with 0.45 µm nominal pore radius were used. On the contrary, maximal selectivities, higher by more than a factor three compared with those corresponding to VLE, were achieved by the PTFE membranes having lower nominal pore radius of 0.2 µm.

Bandini et al. [22] investigated VMD for the removal of ethanol using PTFE and PP membranes and found that separation was limited by concentration polarization. They also found that increases in permeate flux were usually accompanied by decreases in selectivity. It was reported that the liquid mass transfer resistance greatly affected the separation factor of dilute (5 wt%) ethanol/water mixture, which was found to be lower than 10 and increased with the mass transfer coefficient in the liquid boundary layer.

Izquierdo-Gil and Jonsson [23] studied the factors affecting VMD permeate flux and ethanol separation. It was reported that the separation factor increased with the feed circulation velocity from 5 to 7 for the PTFE and PVDF membranes, whereas it was maintained practically constant at around 7–8 for the PP membranes although the membranes have similar pore sizes (i.e. 0.1–0.2 µm).

Bandini et al. [41] studied VMD isopropyl alcohol (IPA)/water mixtures with different IPA concentrations (2.8–10 wt%) and found that IPA concentration in the permeate was higher than that in the feed aqueous solution reaching values as high as 55 wt%. The IPA concentration in the permeate was enhanced by the increase in downstream pressure and the increase in feed IPA concentration, and was higher for lower feed temperatures.

Double-layer hydrophobic/hydrophilic porous membranes have been tested in VMD for the treatment of 1000 ppm ethanol aqueous solution [14]. The membranes were fabricated using fluorinated surface-modifying macromolecules (SMMs: nSMM) and the hydrophilic polymer polyethersulfone. An increase of the ethanol concentration in the permeate from 1200 to 1900 ppm was detected when the downstream pressure was increased from 0.6 to 2 kPa at a feed temperature of 26 °C, whereas the total VMD permeate flux as well as the partial water flux both decreased with the increase of the downstream pressure.

Lawson and Lloyd [16] observed a decrease of ethanol concentration in permeate from 3.8 to about 2 mol% with the increase of the feed temperature from 35 to 75 °C when using the membrane 3MC (see Table 2.1), 2 mol% ethanol concentration in the feed aqueous solution and 3 kPa downstream pressure. The simulated results were found to be much higher than the experimental ones. They justified the deviations by the experimental errors and the errors in the simulation program resulted from the considered assumptions.

VMD of dilute aqueous ethanol solutions have been experimentally investigated under a wide range of operating conditions by Sarti et al. [18] using PP hollow fibre membranes (Celgard X-20, Table 2.21). It was claimed that at low downstream pressures the VMD permeate flux was generally dominated by Knudsen type of flow, whereas the ethanol selectivity was essentially determined by the liquid/vapour equilibrium at the feed/membrane interface conditions. Higher ethanol concentrations in the permeate and lower VMD permeate fluxes were obtained at higher downstream pressures. When using 5 wt% ethanol concentration in feed and a temperature of 50 °C, the ethanol content in the permeate was found to be lower than the calculated values based on the VLE for the feed-operating conditions, which was around 34 wt%. The experimental values were ranging from 30 to 32 wt% at the downstream pressure 60 mbar and even lower values ranging from 21 to 25 wt% were obtained at 26 mbar.

VMD in Concentration of Fruit Juices and Recovery of Aroma Compounds

VMD technology was used in food applications for concentration of fruit juices and recovery of thermally sensitive volatile aroma compounds [36–39]. One of the advantages of MD is that the organoleptic properties as well as flavours of fresh juices are better preserved in comparison with other processes such as thermal evaporation.

Bandini and Sarti [36] applied VMD process for the concentration of must (i.e. juice obtained from grape pressing) up to 50°Brix. The objective of the study was to increase the alcoholic potential of musts preserving quality and quantity of the contained aromas. In fact, musts contain sugars and a wide variety of aroma compounds. Model feed aqueous solutions were composed of glucose and typical aroma compounds (i.e. 0.246 ppm 1-hexanol, 0.344 ppm linalool, 0.176 ppm geraniol and 0.79 wt% ethanol in glucose-free solution) were used. Shell and tube PP fibre membrane module (V8/2 Akzo-Nobel) was employed. It was observed that water flux decreased with increasing glucose concentration in the feed solution and the interface temperature approached the bulk feed temperature. For lower glucose concentrations, aroma compositions at the feed/membrane interface remained higher because Reynolds number and water permeate flux were higher. The calculated mass transfer coefficients in the feed liquid phase were found to be remarkably lower at higher glucose concentrations and aroma composition at the feed/membrane interface remained lower. Moreover, at higher glucose concentrations, it was observed that mass transfer resistance in the feed liquid membrane side controlled the VMD process and the aroma rejection increased.

VMD was used by Bagger-Jorgensen et al. [37] to recover thermally sensitive volatile aroma compounds from black currant juice at low temperatures (10–45 °C) using PTFE (K150) membrane with a pore size 0.1 μm supplied by Osmonics. Seven aroma compounds have been considered, namely, methyl butanoate ($C_5H_{10}O_2$), ethyl butanoate ($C_6H_{12}O_2$), furfural ($C_5H_4O_2$), ethyl hexanoate ($C_8H_{16}O_2$), 1,8-Cineole ($C_{10}H_{18}O$), cis-3-hexene-1-ol ($C_6H_{12}O_2$) and β-damascenone ($C_{13}H_{18}O$). A linear relationship was obtained between the VMD permeate flux and the water vapour pressure difference across the membrane suggesting that Knudsen type of flow was the mechanism of transport that was taking place inside the membrane pores. The obtained concentration factors (C_p/C_f) were ranging from 5.9 to 31.4. It was observed that low feed temperatures and high feed flow rates favoured the concentration of aroma compounds. The recovered values of the highly volatile aroma compounds ranged from 68 to 83% by volume with a feed volume reduction of 5% at 10 °C and 400 l/h feed flow rate, whereas for the poorly volatile compound it was between 32 and 38% by volume.

Soni et al. [38] developed a theoretical model and analysed the recovery of volatile aroma compounds from black currant juice by VMD. The model was able to predict both the temperature and the concentration polarization effects. The model was validated by available data taken from MD literature such as Ref. [37]. The effects of various VMD operating conditions and membrane parameters have been simulated and compared to experimental data. It was found that the simulated values of the molar fractions of aroma compounds in the permeate deviated by maximum 15% from the experimental values concluding that the model was reasonably good.

Recently, recovery of 2,4-decadienoate (i.e. the main pear aroma compound) by VMD process was studied by Diban et al. [39] using tubular PP membrane module (MD020TP2N) (see Table 2.2). Model aqueous solution containing 2,4-decadienoate and ethanol were used at different VMD-operating conditions and aroma enrichment factors up to 15 were obtained. It was found that the feed temperature and the downstream pressure were the operating parameters affecting strongly the aroma enrichment factor, which was high at low feed temperature and high downstream pressure.

VMD Technology in Desalination

VMD process has been successfully applied for production of distilled water from seawater or brackish waters and for concentration of brines [6,7,10–13,15,17,24–28,52]. Among these studies, the highest reported VMD permeate flux was about 70 kg/m^2·h when treating 1 wt% sodium chloride (NaCl) feed aqueous solution with a feed temperature of 85 °C, a vacuum pressure ranging between 8 and 8.8 kPa and a modified PP MXFR 3 membrane (PP porous hollow fibre membrane, Accurel Membrana: PP 150/330 with 0.33 mm fibre inner diameter, 0.63 mm fibre outer diameter, 65% porosity and a maximum pore size ≥ 0.2 μm, outer surface coated by plasma polymerization using a silicone fluoropolymer) [7]. The feed solution was circulated in a cross-flow mode over the outer surfaces of the hollow fibres, which lead to considerable reduction of the temperature polarization effect (i.e. temperature polarization coefficient from 0.93 to 0.99). No trace of salt was detected in the permeate.

Using a salt NaCl aqueous solution of 35 g/l and different PE and PP hollow fibre membranes of different characteristics, Li et al. [6] observed that, for the same hollow fibre membrane, the permeate flux of VMD was greater than DCMD and concluded that the effects of concentration and temperature polarization on VMD permeate flux were less in VMD than in DCMD process.

A pilot VMD plant was constructed for seawater desalination on ships using a PVC module containing PP hollow fibre membranes prepared by melt-stretching technique with an inner diameter of 327 μm and a thickness of

53 μm [24]. Feed seawater was heated by the waste-heat generated from vessel engine. Different operating conditions have been investigated. It was found that the salt rejection factor of the pilot plant was up to 99.99% with a permeate flux of 5.4 kg/m^2·h at a feed temperature of 55 °C and a downstream pressure of 93 kPa. After 5 months, the high salt rejection factor and the permeate flux were maintained.

A VMD unit was coupled to 8 m^2 solar energy collector to heat the feed solution taken from underground water with an electrical conductivity higher than 230 μS/cm, in order to produce potable water [26]. A compact and simple system was constructed using a PP hollow fibre membrane module (inner diameter 371 μm, thickness 35 μm, membrane pore size 0.1 μm, 500 hollow fibres with a length of 0.14 m, total membrane area 0.09 m^2). A production of 32.2 kg/m^2·h permeate was obtained with an electrical conductivity less than 4 μS/cm. The daily water production was 173.5 kg/m^2. It was claimed that this permeate flux could be more than 170 kg/m^2 under good weather conditions and more than 50 kg/m^2 in cloudy weather conditions. However, in this study, no energy consumption analysis was performed and no photovoltaic (PV) panels were used to run the pumps.

VMD technology was applied for the treatment of aqueous solutions with high salt contents (up to 300 g/l) and RO brines [25,27,28]. Mohammadi and Safavi used aqueous NaCl solutions with concentrations up to 150 g/l and observed about 53% decrease of the permeate flux when the concentration was increased from 50 to 150 g/l. The electrical conductivity of the permeate was in the range 0.14–2.8 μS/cm. This indicates very high rejection factors of salt were achieved.

Wirth and Cabassud [25] studied VMD of aqueous NaCl solutions with concentrations varying from 15 up to 300 g/l and observed no traces of salt in the permeate. It was concluded that the VMD permeate flux was not affected too much by the increase of the salt feed concentration. A reduction by less than 30% was observed when the concentration was increased from 15 to 300 g/l. The same authors [52], based on a developed computational analysis, stated that for a membrane area 100 times larger than that of the experimental membranes and at low temperatures (25 °C), VMD can compete with RO on energy consumption (<2 kWh/m^3) with approximately the same water production.

Using RO brines from a plant installed in the Mediterranean sea with a total concentration of 50 g/l, Mericq et al. [28] reported that very few crystal deposits (calcium sulphate) were observed on the membrane surface and they did not cover the membrane pores. Moreover, it was found that when integrating VMD to RO used for seawater treatment, high recovery factors up to 89% were obtained with a brine volume reduction by a factor of 5.5 and an increase in the production by a factor more than 2. It was stated that during the concentration of RO brines from 64 to 300 g/l, the VMD permeate flux ranged from 17 to 71/m^2·h at a downstream pressure of 6 kPa, a feed temperature of 50 °C and a Reynolds number of 4000. It was also found that the temperature and concentration polarization exerted small effects on the VMD permeate flux even though the salt concentration was very high and scaling impact in VMD was very limited.

Simulations on coupled VMD and solar energy for seawater desalination have been carried out by Mericq et al. [40]. Salinity gradient solar ponds (SGSP) and solar collectors (SC) have been analysed as shown in Fig. 12.11, considering two possibilities for each solar source: (i) preheating the feed seawater by SGSP (Fig. 12.11a) or SC (Fig. 12.11c) before the VMD process, (ii) the membrane module was directly coupled with solar energy source by either submerging the membrane in SGSP (Fig. 12.11b) or integrating a SC at the surface of the membrane module (Fig. 12.11d). It was found that submerged membrane module in an

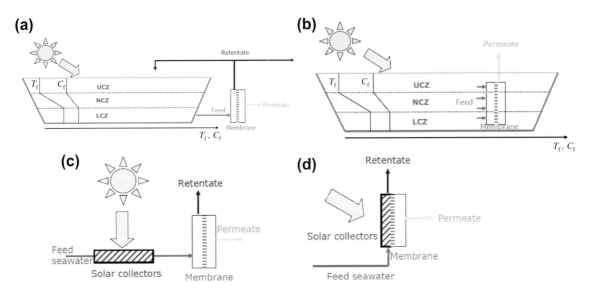

FIGURE 12.11 Different possibilities of coupling VMD membrane module and solar energy: (a) use of feed seawater taken from SGSP (UCZ: upper convective zone; NCZ: non-convective zone and LCZ: lower convective zone); (b) submerged membrane module in SGSP; (c) use of feed seawater heated by SC; and (d) SC heating seawater directly on membrane module. *Source: Reprinted from [40]. Copyright 2011, with kind permission from Elsevier.*

SGSP induced high concentration and temperature polarization effects reducing the VMD permeate flux considerably. Energy consumption, technical feasibility, maintenance and costs were qualitatively analysed in order to select the best configuration. It was concluded that the use of SC was the most interesting option simulating very high VMD permeate fluxes of $142 l/m^2 \cdot h$ (i.e. $617 l/h$) for a downstream pressure of 500 Pa and a Knudsen membrane permeability of $1.85 \ 10^{-5} s \cdot mol^{1/2} \cdot m^{-1} \cdot kg^{-1/2}$. It was reported that a semi-industrial VMD pilot plant was under testing, operating in Tunisia with a plane SC. It is to be noted that the experimental SGSP powered desalination has been considered for the first time at the University of Texas, El Paso, in 1987, using the AGMD configuration [53]. Aqueous NaCl solutions of concentrations 35 and 269.6 g/l were used as feed solutions heated by low-grade thermal energy supplied by a salt-gradient solar pond applying a heat exchanger. The obtained permeate fluxes were lower than $6 l/h \cdot m^2$. Recently, a theoretical study was made by Suarez et al. [54] on a DCMD-integrated SGSP, showing that a production of $1.6 l/d/m^2$ of SGSP can be obtained with membrane areas ranging from 1 to $1.3 m^2$.

Treatment of Textile Wastewaters

VMD was also investigated as a tool for the treatment of textile wastewater coloured with dyes [30,31]. It is to be mentioned that textile dyeing demands large amounts of water and produces high amounts of wastewaters that are harmful to the environment.

Banat et al. [30] investigated the potential of VMD for the treatment of dyed solutions using methylene blue ($C_{16}H_{18}ClN_3S.3H_2O$) as a model dye and tubular PP hollow fibre membrane module (Enka-Microdyn, see Table 2.2). The effects of the feed temperature, feed flow rate and initial dye concentration were investigated.

During VMD experiments, dye concentration in feed and the temperature polarization coefficient (Eq. (12.27)) were increased, whereas the VMD permeate flow rate and the concentration polarization coefficient (Eq. (12.32)) were decreased exponentially with time. It was concluded that the dye was concentrated in the feed container and was not detected in the permeate. For example, the dye concentration in the feed was increased from 18.5 to 32 ppm at a feed temperature of 50 °C, a Reynolds number of 1761 and a downstream pressure of 5 mmHg.

Recently, Criscuoli et al. [31] also considered VMD for the treatment of different dye aqueous solutions using PP hollow fibre membrane module (membrane pore size 0.2 μm, thickness 0.51 mm, inner diameter 1.79 mm supplied by Membrana, Germany) and found a complete rejection of dye producing pure water. The used dyes are remazol brilliant blue R ($C_{22}H_{16}N_2Na_2O_{11}$), reactive black 5 ($C_{26}H_{21}N_5Na_4O_{19}S_6$), indigo vat blue 1 ($C_{16}H_{10}N_2O_2$), acid red 4 ($C_{17}H_{14}N_2O_5SNa$) and methylene blue ($C_{16}H_{18}ClN_3S.3H_2O$) and their concentrations were varied from 25 to 500 ppm. VMD permeate flux decay was also observed during the first 30 min of VMD operation for all dyes with an increase of their concentration in the feed. Permeate flux decay was attributed to membrane fouling. However, it was confirmed that initial VMD permeate fluxes are recovered after cleaning with distilled water during prolonged period of time. Surprisingly, higher initial VMD permeate fluxes were observed when the dye solution (remazol brilliant blue R) was used as feed compared with the VMD permeate flux of water used as feed. This result was attributed to a possible swelling of the PP membrane structure when it was brought into contact with the dye solution.

Other Applications of VMD Technology

VMD configuration has been proposed for applications in other fields such as in lithium bromide (LiBr) adsorption refrigeration system (LBARS), energy storage system and the regeneration of liquid desiccant solution in temperature humidity independent control air conditioning system [55,56]. It was reported that VMD was an efficient and cheap desorption mode. Further studies are needed in order to confirm these potential applications of VMD, especially to be a novel generator or to be used as a secondary generator in desorption of aqueous LiBr solution using low temperature heat energy (i.e. low grade energy such as waste energy, solar energy, geothermal energy). In fact, in the traditional generator, LiBr is desorbed under vacuum and the applied temperature must be higher than the boiling point of the LiBr aqueous solution.

LiBr concentrations between 45 and 55 wt% were considered to simulate the absorption cooling cycle [55,56]. Wang et al. [55] used fabricated PVDF hollow fibre membrane module with an effective membrane area of 0.3 m², packing density of 41.16%, length of membrane 400 mm, membrane porosity 85%, membrane inner diameter 0.8 mm, membrane thickness 0.15 mm and membrane pore size 0.16 μm (Key Laboratory of Hollow Fibre Membrane Materials and Membrane Process of Ministry of Education, Tianjin Polytechnic University). It was observed in all experimental tests that the electrical conductivity of the obtained permeate was less than 12 μS/cm. It is worth quoting that desorption of aqueous LiBr solution by DCMD has been mentioned previously by Sudoh et al. [57] in 1997 using PTFE membrane and different LiBr concentrations up to 55 wt% in water.

Attempts have been made to use VMD process for the treatment of coolant liquids employed in automotive, air conditioning, railways, etc. [34]. Typical coolant liquid mixtures contain water and glycols with concentrations ranging from 20 to 40 wt% plus small content of chemical additives. Among the glycols used ethylene glycol is commonly employed in most of commercial coolants. Mohammadi and Akbarabadi [34] tested ethylene glycol water mixtures (20–60 wt%) in VMD using a flat sheet PP membrane (PP Accurel 2E with 0.2 μm pore size, 75% porosity, 163 μm thickness supplied by Membrana, Germany) under different temperatures (i.e. 40–60 °C) and feed flow rates, and found that increasing both parameters resulted in higher VMD permeate flux with no effect on the ethylene glycol separation factor, which was very high. The obtained permeate was almost distilled water because the vapour pressure of ethylene glycol is negligible compared with that of water.

VMD process has been applied for ammonia (NH_3) removal [35]. Ammonia in wastewater exists in two forms, as volatile ammonia and ammonium ions (NH_4^+). Experimental results showed that high feed temperatures, low downstream pressures, high initial feed concentrations and high pH values enhanced ammonia removal efficiency ($1-C_p/C_f$) [35]. Among all studied parameters, pH of the feed aqueous solution was found to dominate the membrane performance. The increase of the pH value favoured the presence of ammonia resulting in an enhancement of ammonia removal efficiency. It was concluded that ammonia removal efficiencies greater than 90% with separation factors (Eq. (12.25)) higher than eight could be achieved in VMD when a PTFE membrane with a mean pore size of 0.23 μm and effective porosity of 916.2 m^{-1} was used. The increase of ammonia concentration in the feed aqueous solution resulted in an increase of the total permeate flux but both the ammonia separation factor and the overall mass transfer coefficient (Eqs. (12.36) and (12.37)) were not strongly influenced by the initial ammonia concentration [35]. However, the increase of the downstream pressure reduced both the total VMD permeate flux and the overall mass transfer coefficient and increased the separation factor, whereas the increase of the feed temperature increased both the total permeate flux and the overall mass transfer coefficient but the separation factor was slightly decreased. The mass transfer resistance was found to switch from being predominately located at the feed membrane side for low feed temperatures, low feed flow rates and high downstream pressures, to being partially dominated by the membrane resistance for higher temperatures, higher flow rates and lower downstream pressures.

Another VMD application is the concentration of Chinese ginseng (plant of genus Panax) extracts aqueous solutions [32,33]. Ginseng is believed to have medicinal properties. A decrease of the permeate flux was observed during VMD operation and the initial permeate flux was high at high feed temperatures but a serious concentration polarization effect and membrane fouling were observed claiming the existence of a critical vapour pressure difference above which membrane fouling can be prevented [32]. In general, the selected values of the operating parameters (feed temperature, initial concentration, temperature and downstream pressure) affected considerably both the initial VMD permeate flux and its decline, confirming the existence of critical fouling operating conditions in VMD process [33].

A potential application of VMD to concentrate sucrose aqueous solutions was tested by Al-Asheh et al. [29] investigating the effects of different parameters, including feed temperature (40–70 °C) and feed flow rate (i.e. Reynolds number varying from 2000 to 7500) as well as the initial sucrose concentration (up to 20 wt%), on the permeate flux and separation factor. A significant increase of the VMD permeate flux with the feed flow rate was observed in laminar flow regime (up to 5000 Reynolds number), but this increase was marginal for higher Reynolds numbers. Moreover, the increase of the initial sucrose concentration resulted in a slight decrease in the VMD permeate flux (about 5.5% when sucrose concentration was increased from 0 to 17 wt%)

without affecting the separation factor. In all experiments the obtained permeate was pure water.

It is important to mention here that VMD process has been miniaturized [58]. A multilayered microfluidic chip or microdistillation chip was designed in a horizontal position, fabricated and tested in VMD using ethanol aqueous solution. A PTFE membrane of 0.2 μm mean pore size and 74% porosity was used. The chip was heated by an aluminum hotplate and the feed solution was delivered into the chip by means of a syringe pump (1 ml gastight syringe). A portable aspirator was employed to control the downstream pressure using a digital pressure sensor. The used downstream pressure was 75 ± 2 kPa. The retentate was collected in a micro-vial cooled with ice water, while the permeate was collected in another micro-vial cooled by dry ice and alcohol. The temperatures were measured by a digital radiation temperature sensor along the channel. The feed temperature was 80 °C, the feed flow rate was 5 μl/min and the methanol concentration in the aqueous feed solution was 0.28 M. The obtained concentration in the permeate was 0.54 M, more than nine times the concentration of the retentate, which was 0.06 M. Permeate flux was not reported.

It is worth noting that until now, attention has not been paid to MD microfluidic channels or miniaturized MD in order to guide the liquids and vapour currents for avoiding the problems of conventional MD and to be applied for micro-scale rectification in order for improving MD efficiency.

To conclude this section, it must be emphasized that in all the previously cited application areas, VMD process seems to be a promising technique; however, further studies and economical analysis together with detailed comparative works within the different MD configurations and with the pressure-driven separation processes as well as other separation techniques are needed to be done.

COMPARISON TO PERVAPORATION

The two membrane processes, PV and VMD, are similar and often confusion occurs between these processes. In both techniques, vacuum or low pressure is applied on the permeate side in order to establish the driving force. The fundamental difference between them is the role that the membrane plays in the separation. As it was reported earlier, VMD employs porous and hydrophobic membranes and the mechanism of transport is based on flow through pores as mentioned in the section 'Theoretical models'; whereas PV requires dense and selective membranes and the separation is based on the relative solubility and diffusivity of each component in the membrane material as it is schematically presented in Fig. 12.12.

The mass transport in PV is generally described by the solution-diffusion mechanism. The model consists of sorption of the feed liquid molecules at the upstream side of the membrane, diffusion through the membrane and desorption into vapour phase at the downstream side of the membrane. Therefore, the PV selectivity and permeability are governed by solubility and diffusivity of each component of the feed mixture. It is to be noted here that in PV, the permeated molecules vaporize somewhere between the upstream and the downstream side of the membrane; therefore, the permeate is obtained in vapour phase but is collected in liquid phase in nitrogen filled cold traps. As a consequence, PV can be used not only for the separation of aqueous mixtures but also for the separation of organic/organic mixtures. Depending on the permeating component two main application areas can be identified: (i) hydrophilic PV and (ii) organophilic PV. In the first case, the target compound water is preferentially permeated through the membrane and it is separated from an aqueous/organic mixture, while in the second

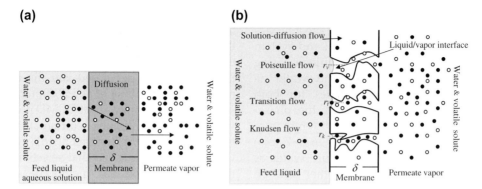

FIGURE 12.12 Mechanisms of mass transfer in PV (a) and VMD (b) processes.

case, the target organic compound is separated from an aqueous organic mixture or from an organic/organic mixture by being preferentially permeated through the membrane.

Based on the solution-diffusion mechanism, the PV permeate flux of a single component (i) through a dense membrane can be described by Fick's first law [3].

$$J_i^d = -D_i \frac{dC_i}{dx} \quad (12.38)$$

where D_i is the diffusion coefficient and C_i is the concentration of the component (i) in the membrane.

Assuming a uniform concentration gradient, the PV permeate flux can be written as follows:

$$J_i^d = D_i \frac{C_{i,m} - C_{i,p}}{\delta} = D_i \frac{S_i p_{i,m} - S_i p_{i,p}}{\delta} \quad (12.39)$$

where C_i is the concentration of the component i inside the membrane, p_i is the vapour pressure of the component i and S_i is the solubility coefficient applicable for the component i. Subscripts m and p are upstream and downstream side of the membrane, respectively. In fact, $p_{i,m}$ is the vapour pressure of the component i in equilibrium with the liquid that is in contact with the upstream side of the membrane.

Finally, the PV permeate flux can be expressed as:

$$J_i^d = D_i S_i \frac{p_{i,m} - p_{i,p}}{\delta} = K_i \frac{p_{i,m} - p_{i,p}}{\delta} \quad (12.40)$$

where $K_i = D_i S_i$ is the membrane permeability.

Studies of parallelism and differences of PV and VMD processes for the removal of VOCs from aqueous streams were performed [3,20]. Urtiaga et al. made the comparison on the basis of the simulation of the chloroform separation with time using commercial polydimethylsiloxane hollow fibres for PV experiments and microporous PP hollow fibres for VMD experiments. However, different fluid dynamic conditions were used for the two processes, which may affect both concentration and temperature polarization on the feed side. As well, different downstream pressures were used. It was found that the rate of chloroform removal was the same in both PV and VMD systems since the kinetics of the chloroform removal are limited by the transport of chloroform in the aqueous phase. Increase in Reynolds number had only a slight effect on the PV performance. Of course, rigorous comparison of PV and VMD processes can only be made by using the same system, the same membrane material and the same operating conditions.

Khayet and Matsuura [3] compared the performance of both PV and VMD techniques using the same membrane material and system operated under the same conditions using aqueous chloroform solutions and water as feed. PVDF flat sheet membranes for both processes have been fabricated by the phase-inversion method. Membranes of different pore sizes (0.02−0.7 µm) and porosities (26.8−79.6%) were prepared for VMD using pure water as a non-solvent additive in the polymer solution containing DMAC. On the other hand, dense membranes were fabricated for PV experiments by casting solutions of high polymer concentrations without non-solvent additive. It was found that the VMD permeate flux was higher than that of PV, whereas the chloroform selectivity was higher for the PV membranes. The chloroform selectivity by VMD decreased with the increase of the pore size and porosity of the PVDF membranes. This result may be partly due to the fact that in VMD the selectivity is mainly determined by the VLE at the feed/membrane interface although the diffusion across the porous membrane may impart some favour to the fluxes of lighter molecules.

Most of MD authors believe that the membrane acts only as a barrier to hold the liquid/vapour interfaces and it is not involved in the transport phenomena. The membrane is not necessary to be selective as required in PV because of the risk of membrane pore wetting. Moreover, in MD the transport of adsorbed molecules or atoms on membrane solid surfaces (i.e. surface diffusion) is neglected due to the fact that the diffusion area of the membrane matrix is small compared with the pore area. As well, the transport by solution-diffusion through the non-porous portion of the membrane is not considered in MD in spite of the possible existence of high 'affinity' between the species to be separated and the membrane material. In 1992, Fujii et al. [59,60] reported that surface diffusion may affect DCMD performance when using membranes with small pore sizes (<0.02 µm). In 2002, Rivier et al. [61] and García-Payo et al. [62] indicated that selectivity in thermostatic sweeping gas membrane distillation (TSGMD) depends on the difference between the involved components not only in volatility but also in the diffusion rate across the membrane as well as in the gas-gap. Therefore, the separation performance in MD is not only based on VLE of the involved components in the feed mixture but also on the thermodynamics (operating temperature) and kinetics.

It is worth quoting that for hydrophobic membranes, which are commonly employed in MD, the 'affinity' between water and the membrane material is very low and it may be allowed to neglect the contribution of transport through the membrane polymer matrix, especially for porous membranes with large pore sizes and high porosities. However, when those compounds that have strong affinity to the membrane material are present in the feed, the solution-diffusion mechanism must be included especially for membranes with low porosity and small pore size. In this case, the total permeate flux calculated previously by Eq. (12.22) should be modified to the sum of permeate flux through the membrane pores and the non-porous portion of the membrane:

$$J_i = \frac{N}{\tau\delta}\left(\sum_{j=1}^{m(r=0.05\lambda_i)} G_i^k f_j r_j^3 \Delta p_i + \sum_{j=m(r=0.05\lambda_i)}^{p(r=50\lambda_i)} (G_i^k f_j r_j^3 \Delta p_i + G_i^v f_j r_j^4 \overline{p}_i \Delta p) + \sum_{j=p(r=50\lambda_i)}^{n(r=r_{max})} G_i^v f_j r_j^4 \overline{p}_i \Delta p \right)$$

$$+ \frac{1}{\delta}\left(1 - N\sum_{j=1}^{n} f_j \pi r_j^2\right) K_i \Delta p_i \qquad (12.41)$$

This is a general theoretical model taking into account mass transport through the membrane matrix by solution-diffusion mechanism as well as Knudsen flow term, Poiseuille flow term and the combined Knudsen/Poiseuille flow term through the corresponding membrane pores. The model can be applied to both VMD and PV processes. Both experimental and theoretical studies have been performed considering PV and VMD processes. Simulation of the VMD performance of water and dilute binary chloroform/water mixtures was carried out. It was found that the model could reconstruct the trend observed in the experimental VMD results. The simulation proved fairly accurate in predicting VMD fluxes for pure water but resulted in an over-prediction of the permeate concentration for dilute aqueous chloroform solutions. This result was attributed to the experimental errors because of the high chloroform volatility.

It was found that the contribution to the total flux of the solution-diffusion flux through the non-porous portion of the MD membranes decreased from 30.4 to 0.3% for water and from 40.1 to 0.7% for chloroform as the pore size and porosity of the membranes were increased. This indicates that the solution-diffusion contribution can be neglected when using the commercial membranes commonly used in MD having large pore size ($>0.1\ \mu m$) and porosity ($>53\%$).

Systematic studies using different MD configurations are needed to clarify the contribution of the membrane material in the selectivity and permeability of the membrane.

References

[1] Bodell BR. Silicone rubber vapor diffusion in saline water distillation, United States Patent Serial No. 285,032 (1963).

[2] Bodell BR. Distillation of saline water using silicone rubber membrane, United States Patent Serial No. 3,361,645 (1968).

[3] M. Khayet, T. Matsuura, Pervaporation and vacuum membrane distillation processes: Modeling and experiments, AIChE J. 50 (2004) 1697−1712.

[4] M. Khayet, T. Matsuura, Preparation and characterization of polyvinylidene fluoride membranes for membrane distillation, Ind. Eng. Chem. Res. 40 (2001) 5710−5718.

[5] M. Khayet, K.C. Khulbe, T. Matsuura, Characterization of membranes for membrane distillation by atomic force microscopy and estimation of their water vapor transfer coefficients in vacuum membrane distillation process, J. Membr. Sci. 238 (2004) 199−211.

[6] J. Li, Z. Xu, Z. Liu, W. Yuan, H. Xiang, S. Wang, Y. Xu, Microporous polypropylene and polyethylene hollow fiber membranes: Part 3. Experimental studies on membrane distillation for desalination, Desalination 155 (2003) 153−156.

[7] B. Li, K.K. Sirkar, Novel membrane and device for vacuum membrane distillation-based desalination process, J. Membr. Sci. 257 (2005) 60−75.

[8] B. Wu, X. Tan, K. Li, W.K. Teo, Removal of 1,1,1-trichloroethane from water using a poly(vinylidene fluoride) hollow fiber membrane module: Vacuum membrane distillation operation, Sep. Purf. Tech. 52 (2006) 301−309.

[9] B. Wu, K. Li, W.K. Teo, Preparation and characterization of poly(vinylidene fluoride) hollow fiber membranes for vacuum membrane distillation, J. Appl. Polymer Sci. 106 (2007) 1482−1495.

[10] Z. Jin, D.L. Yang, S.H. Zhang, X.G. Jian, Hydrophobic modification of poly(phthalazinone ether sulfone ketone) hollow fiber membrane for vacuum membrane distillation, J. Membr. Sci. 310 (2008) 20−27.

[11] Z. Jin, D.L. Yang, S.H. Zhang, X.G. Jian, Removal of 2,4-dichlorophenol from wastewater by vacuum membrane distillation using hydrophobic PPESK hollow fiber membrane, Chin. Chem. Lett. 18 (2007) 1543−1547.

[12] Z. Jin, D.L. Yang, S.H. Zhang, X.G. Jian, Hydrophobic modification of poly (phthalazinone ether sulfone ketone) hollow fiber membrane for vacuum membrane distillation, Chin. Chem. Lett. 19 (2008) 367−370.

[13] S. Cerneaux, I. Stuzynska, W.M. Kujawski, M. Persin, A. Larbot, Comparison of various membrane distillation methods for desalination using hydrophobic ceramic membranes, J. Membr. Sci. 337 (2009) 55−60.

[14] D.E. Suk, T. Matsuura, H.B. Park, Y.M. Lee, Development of novel surface modified phase inversion membranes having hydrophobic surface-modifying macromolecule (nSMM) for vacuum membrane distillation, Desalination 261 (2010) 300−312.

[15] N. Tang, Q. Jia, H. Zhang, H. Li, S. Cao, Preparation and morphological characterization of narrow pore size distributed polypropylene hydrophobic membranes for vacuum membrane distillation via thermally induced phase separation, Desalination 256 (2010) 27–36.

[16] K.W. Lawson, D.R. Lloyd, Membrane Distillation. I. Module design and performance evaluation using vacuum membrane distillation, J. Membr. Sci. 120 (1996) 111–121.

[17] M. Khayet, M.P. Godino, J.I. Mengual, Possibility of nuclear desalination through various membrane distillation configurations: A comparative study, Int. J. Nuclear Desalination 1 (2003) 30–47.

[18] G.C. Sarti, C. Gostoli, S. Bandini, Extraction of organic components from aqueous streams by vacuum membrane distillation, J. Membr. Sci. 80 (1993) 21–33.

[19] N. Couffin, C. Cabassud, V. Lahoussine-Turcaud, A new process to remove halogenated VOCs for drinking water production: Vacuum membrane distillation, Desalination 117 (1998) 233–245.

[20] A.M. Urtiaga, E.D. Gorri, G. Ruiz, I. Ortiz, Parallelism and differences of pervaporation and vacuum membrane distillation in the removal of VOCs from aqueous streams, Sep. Purf. Tech. 22–23 (2001) 327–337.

[21] A.M. Urtiaga, G. Ruiz, I. Ortiz, Kinetic analysis of the vacuum membrane distillation of chloroform form aqueous solutions, J. Membr. Sci. 165 (2000) 99–110.

[22] S. Bandini, C. Gostoli, G.C. Sarti, Separation efficiency in vacuum membrane distillation, J. Membr. Sci. 73 (1992) 217–229.

[23] M.A. Izquierdo-Gil, G. Jonsson, Factors affecting flux and ethanol separation performance in vacuum membrane distillation (VMD), J. Membr. Sci. 214 (2003) 113–130.

[24] Y. Xu, B. Zhu, Y. Xu, Pilot test of vacuum membrane distillation for seawater desalination on a ship, Desalination 189 (2006) 165–169.

[25] D. Wirth, C. Cabassud, Water desalination using membrane distillation: Comparison between inside/out and outside/in permeation, Desalination 147 (2002) 139–145.

[26] X. Wang, L. Zhang, H. Yang, H. Chen, Feasibility research of potable water production via solar-heated hollow fiber membrane distillation system, Desalination 247 (2009) 403–411.

[27] T. Mohammadi, M.A. Safavi, Application of Taguchi method in optimization of desalination by vacuum membrane distillation, Desalination 249 (2009) 83–89.

[28] J. Mericq, S. Laborie, C. Cabassud, Vacuum membrane distillation of seawater reverse osmosis brines, Water Res. 44 (2010) 5260–5273.

[29] S. Al-Asheh, F. Banat, M. Qtaishat, M. Al-Khateeb, Concentration of sucrose solutions via vacuum membrane distillation, Desalination 195 (2006) 60–68.

[30] F. Banat, S. Al-Asheh, M. Qtaishat, Treatment of waters colored with methylene blue dye by vacuum membrane distillation, Desalination 174 (2005) 87–96.

[31] A. Criscuoli, J. Zhong, A. Figoli, M.C. Carnevale, R. Huang, E. Drioli, Treatment of dye solutions by vacuum membrane distillation, Water Res. 42 (2008) 5031–5037.

[32] Z. Zhao, F. Ma, W. Liu, D. Liu, Concentration of ginseng extracts aqueous solution by vacuum membrane distillation. 1. Effects of operating conditions, Desalination 234 (2008) 152–157.

[33] Z. Zhao, F. Ma, W. Liu, D. Liu, Concentration of ginseng extracts aqueous solution by vacuum membrane distillation. 2. Theory analysis of critical operating conditions and experimental confirmation, Desalination 267 (2011) 147–153.

[34] T. Mohammadi, M. Akbarabadi, Separation of ethylene glycol solution by vacuum membrane distillation (VMD), Desalination 181 (2005) 35–41.

[35] M.S. EL-Bourawi, M. Khayet, R. Ma, Z. Ding, Z. Li, X. Zhang, Application of vacuum membrane distillation for ammonia removal, J. Membr. Sci. 301 (2007) 200–209.

[36] S. Bandini, G.C. Sarti, Concentration of must through vacuum membrane distillation, Desalination 149 (2002) 253–259.

[37] R. Bagger-Jorgensen, A.S. Meyer, C. Varming, G. Jonsson, Recovery of volatile aroma compounds from black currant juice by vacuum membrane distillation, J. Food Eng. 64 (2004) 23–31.

[38] V. Soni, J. Abildskov, G. Jonsson, R. Gani, Modelling and analysis of vacuum membrane distillation for the recovery of volatile aroma compounds from black currant juice, J. Membr. Sci. 320 (2008) 442–455.

[39] N. Diban, O. Cristina, A. Urtiaga, I. Ortiz, Vacuum membrane distillation of the main pear aroma compound: Experimental study and mass transfer modelling, J. Membr. Sci. 326 (2009) 64–75.

[40] J. Mericq, S. Laborie, C. Cabassud, Evaluation of Systems coupling vacuum membrane distillation and solar energy for seawater desalination, Chem. Eng. J. 166 (2011) 596–606.

[41] S. Bandini, A. Saavedra, G.C. Sarti, Vacuum membrane distillation: Experiments and modeling, AIChE J. 43 (2) (1997) 398–408.

[42] J.I. Mengual, M. Khayet, M.P. Godino, Heat and mass transfer in vacuum membrane distillation, Int. J. Heat Mass Transfer 47 (2004) 865–875.

[43] R.W. Schofield, A.G. Fane, C.J.D. Fell, Heat and mass transfer in membrane distillation, J. Membr. Sci. 33 (1987) 299–313.

[44] K.W. Lawson, D.R. Lloyd, Review: Membrane distillation, J. Membr. Sci. 124 (1997) 1–25.

[45] A.O. Imdakm, M. Khayet, T. Matsuura, A Monte Carlo simulation model for vacuum membrane distillation process, J. Membr. Sci. 306 (2007) 341–348.

[46] C.R. Wilke, P. Chang, Correlation of diffusion coefficients in dilute solutions, AIChE J. 1 (2) (1955) 264–270.

[47] F. Banat, F.A. Al-Rub, K. Bani-Melhem, Desalination by vacuum membrane distillation: Sensitivity analysis, Sep. Purf. Tech. 33 (2003) 75–87.

[48] A. Criscuoli, M.C. Carnevale, E. Drioli, Evaluation of energy requirements in membrane distillation, Chem. Eng. Process. 47 (2008) 1098–1105.

[49] F.A. Banat, J. Simandl, Removal of benzene traces from contaminated water by vacuum membrane distillation, Chem. Eng. Sci. 51 (1996) 1257–1265.

[50] S. Bandini, G.C. Sarti, Heat and mass transport resistances in vacuum membrane distillation per drop, AIChE J. 45 (1999) 1422–1433.

[51] E. Hoffman, D.M. Pfenning, E. Phillippsen, P. Schwahn, M. Sieber, R. When, D. Woermann, Evaporation of alcohol/water mixtures through hydrophobic porous membranes, J. Membr. Sci. 34 (1987) 199–206.

[52] C. Cabassud, D. Wirth, Membrane distillation for water desalination: How to choose an appropriate membrane? Desalination 157 (2003) 307–314.

[53] J. Walton, H. Lu, C. Turner, S. Solis, H. Hein, Solar and waste heat desalination by membrane distillation, Desalination and water purification research and development program report no. 81, University of Texas at El Paso, El Paso, TX, April 2004.

[54] F. Suarez, S.W. Tyler, A.E. Childress, A theoretical study of a direct contact membrane distillation system coupled to a salt-gradient solar pond for terminal lakes reclamation, Water Res. 44 (2010) 4601–4615.

[55] Z. Wang, Z. Gu, S. Feng, Y. Li, Application of vacuum membrane distillation to lithium bromide adsorption refrigeration system, Int. J. Refrigeration 32 (2009) 1587–1596.

[56] W. Zanshe, G. Zhaolin, F. Shiyu, Li. Yun, Applications of membrane distillation technology in energy transformation process-basis and prospect, Chin. Sci. Bull. 54 (2009) 2766–2780.

[57] M. Sudoh, K. Takuwa, H. Iizuka, K. Nagamatsuya, Effects of thermal and concentration boundary layers on vapor permeation in membrane distillation of aqueous lithium bromide solution, J. Membr. Sci. 131 (1997) 1–7.

[58] Y. Zhang, S. Kato, T. Anazawa, Vacuum membrane distillation on a microfluidic chip, Commun. Chem. Commun. (2009) 2750–2752.

[59] Y. Fujii, S. Kigoshi, H. Iwatani, M. Aoyama, Selectivity and characteristics of direct contact membrane distillation type experiment: I. Permeability and selectivity through dried hydrophobic fine porous membranes, J. Membr. Sci. 72 (1992) 53–72.

[60] Y. Fujii, S. Kigoshi, H. Iwatani, M. Aoyama, Y. Fusaoka, Selectivity and characteristics of direct contact membrane distillation type experiment: II. Membrane treatment and selectivity increase, J. Membr. Sci. 72 (1992) 73–89.

[61] C.A. Rivier, M.C. García-Payo, I.W. Marison, U. von Stockar, Separation of binary mixtures by thermostatic sweeping gas membrane distillation: I. Theory and simulations, J. Membr. Sci. 201 (2002) 1–16.

[62] M.C. García-Payo, C.A. Rivier, I.W. Marison, U. von Stockar, Separation of binary mixtures by thermostatic sweeping gas membrane distillation: II. Experimental results with aqueous formic acid solutions, J. Membr. Sci. 198 (2002) 197–210.

CHAPTER 13

Air Gap Membrane Distillation

OUTLINE

Introduction	361	Air Gap Width	381
Theoretical Models	367	Effects of Some Membrane Parameters	382
		Other Effects	384
Experimental: Effects of Process Conditions	374	AGMD Applications	384
Effects of Feed Temperature	374	AGMD Technology in Desalination and Solar Units	384
Effects of Solute Concentration in the Feed Aqueous Solution	376	AGMD Technology in Food Processing	390
Effects of Feed Flow Rate	377	Treatment of Aqueous Alcohol Solutions	391
Coolant Temperature	378	Break of Azeotropic Mixtures	392
Cold Side Flow Rate	380	Extraction of Volatile Organic Compounds	394
Non-Condensable Gases	380	Other Applications of AGMD Technology	394

INTRODUCTION

In this MD configuration, called air gap membrane distillation (AGMD), the membrane module contains a stagnant air gap interposed between the membrane and a condensation surface placed inside the membrane module. The temperature difference between the feed aqueous solution and the cold surface is the driving force for evaporation of water and volatile compounds at the hot liquid/vapour interfaces formed at the feed membrane surface as can be seen in Fig. 13.1. Mass transfer occurs according to the following four steps:

(i) Movement of the transferring species from the bulk liquid feed towards the membrane surface.
(ii) Evaporation at the liquid/vapour interface formed at the membrane pores.
(iii) Transport of the evaporated species through the membrane pores and diffusion through the stagnant gas gap.
(iv) Condensation over the cold surface.

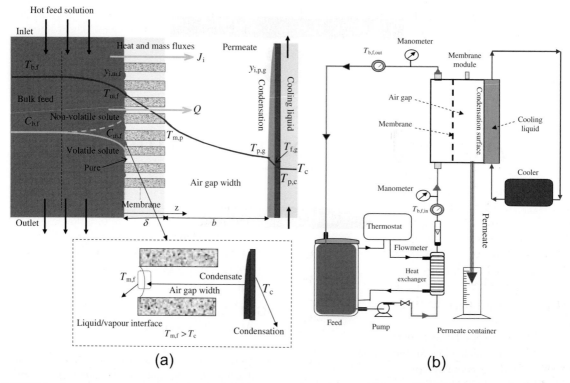

FIGURE 13.1 Schemas of heat and mass transfer through a flat sheet porous and hydrophobic membrane used in AGMD process (a) and a typical experimental AGMD laboratory system (b).

The temperature of the cold surface must be lower than the temperature at the feed/membrane interface. In this case, the evaporated volatile molecules cross both the membrane pores and the air gap and finally condense over the cold surface inside the membrane module. The vapour is transported from the permeate surface of the membrane to the condensation surface by natural convection in the air gap. The natural convection takes place because of the temperature difference in the air space.

The air gap is placed inside the membrane module in order to solve the problem of heat loss by conduction through the membrane, which leads to relatively low efficiency of the MD process. However, the permeate flux has to take place across the air barrier and therefore it is drastically reduced depending on the effective air gap width. On the other hand, because permeate is condensed on a cold surface without direct contact with the membrane surface, AGMD can be applied in the fields where direct contact membrane distillation (DCMD) applications are rather limited, such as for the removal of volatile organic compounds (VOCs) from aqueous solutions.

Similar to other MD configurations, simultaneous heat and mass transfer through the membrane is also taking place in AGMD. Figure 13.1 presents schemas of heat and mass transfer in AGMD process through a flat sheet porous and hydrophobic membrane and a schema of a typical experimental AGMD laboratory system.

It is to be noted that both flat sheet and hollow fibre membranes have been applied in AGMD configuration by means of different membrane modules. The flat sheet membranes were installed not only in plate and frame membrane modules but also in spiral wound modules.

It is worth quoting that in the beginning of the 1980s two different types of AGMD modules were developed indicating that the reduction of the conductive heat loss through the membrane was considered very important. Gore & Associated Co. proposed the use of a spiral wound module for AGMD configuration and termed the process 'Gore-Tex Membrane Distillation' [1], while the Swedish National Development Co. (Svenska Utvecklings AB (SU)) presented a plate and frame AGMD membrane module called 'SU Membrane Distillation system' [2–4]. The spiral wound AGMD module was fabricated using Gore-Tex membranes made of polytetrafluoroethylene (PTFE) with thicknesses as low as 25 µm, porosities up to 80% and pore sizes from 0.2 to 0.45 µm. The plate and frame AGMD module was fabricated using MD cassettes stacked together by injection-moulded plastic frames and assembled with the help of two end plates. More details may be found in Chapter 9. It was claimed that the SU Membrane Distillation system can be combined in a series of modules to meet the need for large capacities (i.e. 5 m^3 per 24 h and module). The reported advantages offered by this type of modules were that the system can utilize natural temperature gradients of waste energy, has low energy costs (i.e. 2.25 kWh/m^3 fresh water). Moreover, the energy requirement did not vary with the salinity, the problem of corrosion was avoided and the capital costs were reduced by using plastic pipes.

Since the 1980s, papers related to the use of AGMD in various applications increased [5–31] in the literature. Recently, the Swedish companies XZero AB and Scarab Development AB presented a plate and frame AGMD module (63 cm long, 17.5 cm wide and 73 cm high) for desalination [32,33]. This module is similar to 'SU Membrane Distillation system' [2–4]. Moreover, during last 8 years, a spiral wound AGMD module with an integrated heat recovery for the design of solar-powered desalination system was presented in various studies [20,27–30]. This module is also similar to 'Gore-Tex Membrane Distillation' [1]. Another alternative, Memstill® technology, was developed during last 10 years by the research and technology organization TNO (Netherlands Organization for Applied Scientific Research) for desalination of seawater by AGMD [25]. In their system the cold saline water flows through a condenser with non-permeable walls, increasing its temperature due to the condensing permeate, and then passes through a heat exchanger where additional heat is added before entering in direct contact with the membrane [35]. Details of this type of AGMD modules may be found in Chapter 9.

As it was reported previously, one of the advantages of the AGMD configuration is the low conductive heat loss through the membrane due to the presence of air in the permeate side of the membrane. However, the presence of air in the permeate side between the membrane and the condensing surface induces an increased mass transfer resistance leading to low permeate flux. Moreover, the design of AGMD modules is not simple compared with the other MD configurations because the condensing surface must be placed inside the membrane module. As a consequence, within the published papers in International Journals only 15.5% of the laboratory studies dealt with AGMD configuration (see Fig. 1.3). Nevertheless, AGMD is considered the most versatile configuration showing a great perspective for the MD future. Actually AGMD technology has been considered for scaling-up MD and construction of various pilot plants that are under evaluation although the production per square meter of membrane area is lower compared with DCMD [20,25,27–31].

In general, 50.4% of the MD publications dealt with theoretical models. In the case of AGMD configuration, 48.4% of the published papers up to December 2010 involved theoretical models. However, very few authors have focused their researches on the fabrication of membranes and modules for AGMD applications [34–39]. Compared to the other MD configurations, most of the fabricated MD membranes were tested in DCMD.

A polyvinylidene fluoride (PVDF) nano-fibrous membrane was fabricated by the electro-spinning method for desalination by AGMD process [34]. Similar trends to those of commercial and other types of MD membranes were observed in AGMD experiments. Permeate fluxes of 11.1 kg/m^2·h and 10.2 kg/m^2·h with NaCl rejection factors higher than 99% were obtained when using a cooling temperature of 22 °C, a feed temperature of 82 °C, a feed flow rate of 350 ml/min and NaCl feed concentrations of 3.5 and 6 wt%, respectively. It was also found that the nano-fibre membrane was maintained intact and unplugged after 25 days of desalination by AGMD operation. Studies on this new promising generation of MD membranes using other polymers, solvents, additives and different electro-spinning parameters are carried out in order to improve the MD performance of nano-fibrous membranes not only for desalination and rejection of non-volatile solutes but also in other MD applications such as the removal of volatile solutes from water.

Grafting of different commercial ceramic membranes such as alumina, titania and zirconia and alumino-silicate have been carried out using fluorosilanes (1H,1H,2H,2H-perfluorodecyltriethoxysilane) for desalination by AGMD process [35–38]. The grafted membranes exhibited good hydrophobic character (i.e. contact angles up to 200°). It is to be mentioned that commercial ceramic membranes are commonly prepared from metal oxides such as alumina, zirconia and/or titania. These membranes are originally hydrophilic due to the presence of the surface hydroxyl (-OH) groups, which are easily bound to water molecules. High salt NaCl rejection factors ranging from 95 to 100% were obtained depending on the pore size of the used ceramic membrane. For example, hydrophobic zirconia (50 nm pore size) and titania (5 nm pore size) tubular ceramic membranes were fabricated for different MD configurations (vacuum membrane distillation, VMD; DCMD and AGMD) [38]. The internal surface of the tubular membranes was chemically modified by grafting perfluoroalkylsilane molecule $C_8F_{17}(CH_2)_2Si(OC_2H_5)_3$ achieving coating layers of 10 μm for zirconia and 5 μm for titania. Salt rejection factors higher than 99% have been obtained for all tested MD configurations when using salt (NaCl) aqueous solutions of concentrations 0.5 and 1 M. The highest permeate fluxes obtained using the modified zirconia membrane are 180 l/m^2·day, 95 l/m^2·day and 113 l/m^2·day in desalination by VMD, DCMD and AGMD of a salt NaCl aqueous solution of 0.5 M, respectively. The corresponding permeate fluxes of titania modified membrane are 146 l/m^2·day for VMD, 20 l/m^2·day for DCMD and AGMD.

Superhydrophobic glass membranes with integrated and ordered arrays of nano-spiked microchannels with pore sizes of about 3.4 μm (inter-pore spacing 2 μm, porosity 26%, thickness 500 μm, water contact angles of about 165°) have been modified by means of differential chemical etching for desalination by AGMD [39]. Different salt (NaCl) concentrations in water ranging from 2.5 to 20 wt% were used. The obtained permeate fluxes were as high as 11.3 kg/m^2·h. It must be pointed out that the measured liquid entry pressure (*LEP*) of water of these membranes was found to be below 5 kPa, which is very low for MD purposes and the produced permeate water contained salt because of pore wetting.

Table 13.1 reviews the highest permeate fluxes observed in AGMD for some commercial and laboratory fabricated membranes.

TABLE 13.1 Reported Permeate Flux (J) of Different Types of Commercial, Fabricated and Modified Membranes for AGMD Applications

Membrane type	J (10^{-3} kg/m^{2}s)	Observation	Reference
TF200	1.3	$T_{b,f} = 70\,°C$, $T_p = 30\,°C$, distilled water	[7]
	1.1	$T_{b,f} = 67.5\,°C$, $T_p = 32.5\,°C$, wt% ethonal in water (selectivity 2.05)	
	3.19	$T_{b,f} = 55\,°C$, $T_p = 35\,°C$, $b = 10^{-3}$ m, $\phi_1 = 2\,l/min$, 120 g/l isopropanol in water	[17]
	2.69	$T_{b,f} = 42.5\,°C$, $T_p = 27.5\,°C$, $\phi_1 = 2\,l/min$, 120 g/l isopropanol in water	
	3.06	≈240 g/l isopropanol in water	
	1.28	≈420 g/l ethanol in water	
		≈640 g/l methanol	
TF450		$T_{b,f} = 42.5\,°C$, $T_p = 27.5\,°C$, 650 g/l methanol in water	[17]
	0.261	methanol permeate flux	
	0.081	water permeate flux	
PTFE (Gore, 0.45 μm pore size)	4.72	$T_{b,f} = 55\,°C$, $T_p = 35\,°C$, $b = 10^{-3}$ m, $\phi_1 = 2\,l/min$, 120 g/l isopropanol in water	[17]
FALP	3.89	$T_{b,f} = 62\,°C$, $T_p = 27.5\,°C$, $b = 4$ mm, $v_l = 0.063$ m/s 1 wt% NaCl	[13]
	3.83	5 wt% NaCl	
FHLP	2.78	$T_{b,f} = 55\,°C$, $T_p = 7\,°C$, tape water ($\psi_f = 297$ μs/cm, $\psi_p = 7$ μs/cm)	[10]
GVHP	2.28	$T_{b,f} = 50\,°C$, $T_p = 20\,°C$, $\phi_1 = 70\,l/h$, $b = 18$ mm, distilled water	[14]
	2.19	90 g/l sucrose in water	
	1.39	$T_{b,f} = 42.5\,°C$, $T_p = 27.5\,°C$, $\phi_1 = 2\,l/min$, ≈85 g/l isopropanol in water	[17]
	1.33	≈145 g/l ethanol in water	
	1.28	200 g/l methanol	
	2.78	$T_{b,f} = 70\,°C$, $T_p = 20\,°C$, $b = 0.19$ cm, seawater model solution ($\psi_p = 4 \pm 1$μS/cm)	[11]
HVHP	7.3	$T_{b,f} = 82\,°C$, $T_p = 7\,°C$, $b = 0.8$ mm, tape water ($\psi_f = 297$ μS/cm)	[10]
	1.94	$T_{b,f} = 52\,°C$; $T_p = 7\,°C$, tape water ($\psi_f = 297$ μS/cm, 99% salt rejection)	
	1.67	$T_{b,f} = 52\,°C$; $T_p = 7\,°C$, seawater model solution ($\psi_f = 37.6$ mS/cm, $\psi_p = 1100$ μS/cm)	
	0.87	$T_{b,f} = 50\,°C$, $T_p = 20\,°C$, $\phi_1 = 5\,l/min$, $b \approx 0.35$ cm, 10.2 wt% ethanol in water (selectivity 2.5)	[12]
	0.55	0.83 wt% ethanol in water (selectivity 3.1)	

(Continued)

TABLE 13.1 Reported Permeate Flux (J) of Different Types of Commercial, Fabricated and Modified Membranes for AGMD Applications—Cont'd

Membrane type	J (10^{-3} kg/m²s)	Observation	Reference
HVHP	2.4 2	$T_{b,f}$ = 70 °C, T_p = 20 °C, ϕ_l = 5 l/min, $b \approx 0.35$ cm, 9.7 wt% ethanol in water (selectivity ≈ 3) 1.55 wt% ethanol in water (selectivity ≈ 3.1)	[12]
PVDF nano-fibre	3.22 3.08 2.83	$T_{b,f}$ = 82 °C, T_p = 22 °C, ϕ_l = 21 l/h, $b \approx 2$ mm, 1 wt% NaCl 3.5 wt% NaCl 6 wt% NaCl; (>98.7% salt rejection)	[34]
Modified nanospiked glass[a]	3.14 2.68	$T_{b,f}$ = 95 °C, T_p = 22 °C, 2.5 wt% NaCl 20 wt% NaCl	[39]
Modified ZrO$_2$ (M1)[b]	1.88 1.67	$T_{b,f}$ = 95 °C, T_p = 5 °C, ϕ_l = 198–240 l/h, $b \approx 10$ mm, distilled water ≈2 M NaCl ≈4.6 M NaCl (≈ 100% salt rejection)	[36]
Modified ZrO$_2$ (M3)[b]	1.81 1.51	$T_{b,f}$ = 95 °C, T_p = 5 °C, ϕ_l = 198–240 l/h, $b \approx 10$ mm, (0.001–0.01) M NaCl ≈1 M NaCl (≈ 100% salt rejection)	[36]
Modified ZrO$_2$ (Zr50)[b]	1.45 1.27 1.1 1.1	$T_{b,f}$ = 95 °C, T_p = 5 °C, ϕ_l = 198–240 l/h, 0.5 M (Mediterranean seawater) 1 M NaCl 2 M NaCl 3 M NaCl (95–100% salt rejection)	[37]
Modified Al$_2$O$_3$ (Al200)[b]	1.16	$T_{b,f}$ = 95 °C, T_p = 5 °C, ϕ_l = 198–240 l/h, 2 M NaCl (≈100% salt retention)	[37]
Modified Al$_2$O$_3$ (Al800)[b]	0.69	$T_{b,f}$ = 95 °C, T_p = 5 °C, ϕ_l = 198–240 l/h, 2 M NaCl (≈94% salt retention)	[37]
Modified aluminosilicate (AlSi400)[b]	0.69	$T_{b,f}$ = 95 °C, T_p = 5 °C, ϕ_l = 198–240 l/h, 2 M NaCl (≈96% salt retention)	[37]
Modified TiO$_2$ (Ti5)[b]	0.231	$T_{b,f}$ = 95 °C, T_p = 5 °C, 0.5 M NaCl (99.1% salt rejection)	[38]
Modified ZrO$_2$ (Zr50)[b]	1.308	$T_{b,f}$ = 95 °C, T_p = 5 °C, 0.5 M NaCl (99.8% salt rejection)	[38]

[a] Fabricated superhydrophobic glass membrane with ordered arrays of nano-spiked microchannels modified by differential chemical etching.
[b] grafted ceramic tubular membranes by 1H,1H,2H,2H-perfluorodecyltriethoxysilane (Zirconia, ZrO$_2$: d_p = 50 nm); (Zirconia, ZrO$_2$: d_p = 200 nm); (Titania, TiO$_2$: d_p = 5 nm); (Alumina, Al$_2$O$_3$: d_p = 200 nm); aluminosilicate (AlSi400) [36–38].

$T_{b,f}$: bulk feed temperature; T_p: cooling temperature; ϕ_l: liquid flow rate; v_l: liquid circulation velocity; b: gap width; ψ_p: electrical conductivity of the permeate; ψ_f: electrical conductivity of the feed.

One of the potential applications of AGMD is desalination for the production of high-purity water and concentration of non-volatile solutes such as sucrose and isotopic compounds (i.e. ^{18}O) [5,6,8,10,11,14,18–20,27–31,34–39]. Moreover, AGMD process was used successfully in food processing such concentration of orange juice and milk [5]. The possibility of AGMD to concentrate hydriodic acid (HI) and sulphuric acid (H_2SO_4) aqueous solutions was demonstrated permitting MD to be applied for hydrogen generation [26]. AGMD process exhibits potential interest in breaking azeotropic mixtures (i.e. break the azeotrope into two solutions, one hypoazeotropic and the other hyperazeotropic) such as hydrochloric acid (HCl)/water, propionic acid (CH_3CH_2COOH)/water, formic acid (HCOOH)/water and HI/water mixtures [9,15,26]. Ethanol recovery from fermentation broth and treatment of alcohol aqueous solutions such as methanol and isopropanol aqueous solutions are other applications of AGMD technology [7,12,17]. Furthermore, AGMD was used for selective removal and extraction from water of other volatile compounds such as propanone, acetic acid (CH_3COOH) and nitric acid (HNO_3) [5,16]. Additionally, the possibility of using solar energy enables AGMD technology to be the MD configuration with high perspectives for construction of industrial plants. Preliminary experiments of AGMD were conducted for water purification in thermal cogeneration plants that require purified or treated water for a number of processes such as boiler/district heat make-up water systems and flue gas condensate treatment [31]. More investigations are needed to prove the viability and efficiency of the geothermal and solar pilot plants as well as thermal cogeneration plants.

THEORETICAL MODELS

The vapour transport across the membrane in AGMD mode is generally described by the theory of molecular diffusion including the air and non-condensable gases inside the pores of the membrane and in the air gap between the membrane and the condensation surface on the permeate side [5,7,9,11,12,40,41]. Moreover, separation of feed aqueous mixtures is not only governed by the vapour/liquid equilibrium (VLE) of the system under consideration but also by the diffusion rates of the feed components in the inert gas, which fills the membrane pores and the air gap that separates the membrane from the condensation surface.

The low solubility of air in water and the negligible air flux through the membrane permit to establish a pressure gradient opposite to the water vapour flux. Therefore, the air can be treated as a stagnant film and the steady-state diffusion of component i through a stationary air film is written as (see ordinary molecular diffusion in Chapter 11):

$$J_i = \frac{1}{RT} \frac{\varepsilon}{\tau \delta} \frac{PD_i}{p_a} \Delta p_i \qquad (13.1)$$

where J_i is the permeate flux of component i, p_a is the mean air pressure, D_i is the binary diffusion coefficient, P is the total pressure inside the pore, ε is the porosity of the membrane, δ is the thickness of the membrane, τ is the pore tortuosity and Δp_i is the transmembrane partial pressure difference of component i.

Stefan diffusion (i.e. Fick's equation of molecular diffusion) was used to describe multicomponent mass transfer in AGMD systems [11,12]. The permeate flux of a specie i across the membrane and the air gap is given by the following equation, which concerns a mixture of air and a specie i in vapour phase [42].

$$J_i = -cD_{ia}\frac{dy_i}{dz} + y_i(J_i + J_a) \qquad (13.2)$$

where the first term in the right side of this equation is the mass transfer by molecular diffusion, the second one the convective flux, y_i, is the mole fraction of the component i, z is the direction perpendicular to the membrane surface (See Fig. 13.1), D_{ia} is the binary diffusion coefficient (species i in air) and c is the molar density. If the ideal gas law is assumed c is written as:

$$c = \frac{P}{RT} \quad (13.3)$$

If the solubility of air in water is considered negligible, no air permeate flux will occur inside the pores ($J_a = 0$) and Eq. (13.2) becomes [11,12]:

$$J_i = -\frac{cD_{ia}}{1-y_i} \frac{dy_i}{dz} \quad (13.4)$$

It is to be noted that in the above derivation Fickian approach assumes that the rate of diffusion of species depends only on its concentration gradient and coupling interactions between the diffusing species, which may occur, were neglected.

Under steady-state conditions, the mass flux through the air phase from the membrane interface to the condensing surface is constant:

$$-\frac{dJ_i}{dz} = 0 \quad (13.5)$$

Combining Eqs. (13.4) and (13.5):

$$\frac{d}{dz}\left(-\frac{cD_{ia}}{1-y_i}\frac{dy_i}{dz}\right) = 0 \quad (13.6)$$

Eq. (13.6) can be integrated with the boundary conditions:

$$z = 0, \; y_i = y_{i,m,f} \text{ and } z = \delta\tau + b = b',$$
$$y_i = y_{i,p,g} \text{ (see Fig. 13.1a)} \quad (13.7)$$

where b is the air gap thickness.

The following expression is then obtained:

$$\frac{1}{1-y_i}\frac{dy_i}{dz} = C_0 \quad (13.8)$$

where C_0 is the integration constant.

Integrating Eq. (13.8):

$$\int_{y_{i,m,f}}^{y_{i,p,g}} \frac{dy_i}{1-y_i} = C_0 \int_0^{b'} dz \quad (13.9)$$

Obtaining C_0 from Eq. (13.9) and then using Eqs. (13.4) and (13.8), the AGMD permeate flux becomes:

$$J_i = -\frac{cD_{ia}}{b'} \ln\left(\frac{1-y_{i,m,f}}{1-y_{i,p,g}}\right) \quad (13.10)$$

To account for the effective membrane surface area, the AGMD permeate flux can be expressed as [11,12]:

$$J_i = -\frac{\varepsilon \, cD_{ia}}{b'} \ln\left(\frac{1-y_{i,m,f}}{1-y_{i,p,g}}\right) \quad (13.11)$$

For a binary system (species i and air) Eq. (13.11) can be given as:

$$J_i = \frac{\varepsilon \, cD_{ia}}{b' y_{aln}} (y_{i,m,f} - y_{i,p,g}) \quad (13.12)$$

where y_{aln} is the log mean of the air mole fractions at $z = 0$ ($y_{a,m,f}$) and at $z = b'$ ($y_{a,p,g}$):

$$y_{aln} = \frac{y_{a,m,f} - y_{a,p,g}}{\ln \frac{y_{a,m,f}}{y_{a,p,g}}} \quad (13.13)$$

In terms of pressure, the mass flux can be written in the following form considering Eq. (13.3):

$$J_i = \frac{\varepsilon \, D_{ia} P}{RT \, b' |p_a|_{ln}} (p_{i,m,f} - p_{i,p,g}) \quad (13.14)$$

where p is the partial pressure and $|p_a|_{ln}$ is the log mean of the air pressures at $z = 0$ ($p_{a,m,f}$) and at $z = b'$ ($p_{a,p,g}$):

$$|p_a|_{ln} = \frac{p_{a,m,f} - p_{a,p,g}}{\ln \frac{p_{a,m,f}}{p_{a,p,g}}} \quad (13.15)$$

It is to be noted that various authors assumed negligible natural convection in the air gap region when modelling the AGMD process. This natural convection takes place because of the temperature difference between the membrane surface and the condensing surface. Its direction and intensity depend on the orientation of the air gap relative to the gravity vector and its relative importance is proportional to the Rayleigh number (Ra). In general, the transport of a gas across an air gap is assumed to occur by diffusion alone if Ra defined by Eq. (13.16) is less than a critical value, usually on the order of 103 [43,44].

$$Ra = \frac{g\beta\Delta T_{m,p} b^3}{\nu_a \alpha_a} \quad (13.16)$$

where g is the gravitational acceleration, $\Delta T_{m,p}$ is the temperature difference in the air gap between the membrane and the cold surface, β is the thermal expansion coefficient, ν_a is the kinematic viscosity of the air and α_a is the thermal diffusivity of the air. For example, for $\Delta T_{m,p} = 40\,°C$ and $b = 3$ mm, which is typical for the AGMD system, Ra = 85. Therefore, natural convection is allowed to be ignored compared with mass transfer by diffusion.

Stefan–Maxwell equations were used by various authors to model AGMD process in multicomponent systems [7,15,45–49]. Stefan–Maxwell formalism takes into consideration all diffusional interactions between the diffusing species. The predictions of the Stefan–Maxwell-based models were found to be closer to the experimental data than the Fickian mass transfer-based mathematical model.

For one-dimensional steady-state transfer in n-component ideal gas mixture, Stefan–Maxwell equations are written as [7,45,46]:

$$\frac{dy_i}{d\omega} = RT \sum_{j=1, j\neq i}^{n} \frac{y_i J_j - y_j J_i}{P \frac{D_{ij}}{\delta\tau + b}} \quad i = 1,\ldots,n \quad (13.17)$$

where $\omega = \dfrac{z}{\delta\tau + b}$ and D_{ij} is the diffusivity of the binary gas i-j in the vapour phase.

In this case the following boundary conditions were considered [46]:

At $z = 0$ and $\omega = 0, y_i = y_{i,m,f}$ and at
$z = \delta\tau + b = b'$ and $\omega = 1, y_i = y_{i,p,g}$
$\quad (13.18)$

Analytical solutions of the Stefan–Maxwell equations for a system of n components may be found elsewhere [50–55]. Stefan–Maxwell equations provide $(n-1)$ relations, which form a differential system written in a matrix form as follows [51,53]:

$$\widehat{J} = c_t [\beta][\kappa][\xi]\Delta\widehat{y} \quad (13.19)$$

where $\widehat{J}$ is the vector permeate flux composed by all individual fluxes, c_t is the gas molar density, $[\beta]$ is the 'bootstrap' matrix, $[\kappa]$ is the matrix of multicomponent mass transfer coefficients, $[\xi]$ is the matrix of correction factors and $\Delta\widehat{y}$ is the vector driving force composed by all individual driving forces.

The matrix of correction factors is defined as:

$$[\xi] = [\phi]\{\exp[\phi] - [I]\}^{-1} \quad (13.20)$$

where the coefficients ϕ_{ii} and ϕ_{ij} are written as:

$$\phi_{ii} = \frac{RT}{P}\left(\frac{J_i}{k_{in}} + \sum_{k=1,k\neq i}^{n} \frac{J_k}{k_{ik}}\right) \quad (13.21)$$

$$\phi_{ij} = -\frac{J_i RT}{P}\left(\frac{1}{k_{ij}} - \frac{1}{k_{in}}\right) \quad (13.22)$$

where the subscripts i, j, k and n are indexes denoting component number.

The matrix $[\kappa]$ is defined as:

$$[\kappa] = [R]^{-1} \quad (13.23)$$

where the coefficients R_{ii} and R_{ij} are expressed as:

$$R_{ii} = \frac{y_i}{k_{in}} + \sum_{k=1,k\neq i}^{n} \frac{y_k}{k_{ik}} \quad (13.24)$$

$$R_{ij} = -y_i\left(\frac{1}{k_{ij}} - \frac{1}{k_{in}}\right) \quad (13.25)$$

The bootstrap coefficients of the matrix $[\beta]$, Stefan diffusion coefficients in this case, are defined by:

$$\beta_{ik} = \delta_{ik} + \frac{y_i}{y_n} (J_a = 0) \quad (13.26)$$

where δ_{ik} is the Kronecker delta.

Banat et al. [45,46] made corrections taking into account the effective membrane surface area and reported the following expression:

$$\widehat{J} = \frac{\varepsilon P}{RT} [\beta][\kappa][\xi] \Delta \widehat{y} \quad (13.27)$$

The tortuosity and the membrane thickness were taken into consideration in the calculation of the diffusion path length [45,46].

For a binary mixture a volatile solute, s, in water, w, the matrix $[\beta]$ was defined as [48]:

$$[\beta] = \begin{bmatrix} 1 + \dfrac{y_w}{y_a} & \dfrac{y_w}{y_a} \\ \dfrac{y_s}{y_a} & 1 + \dfrac{y_s}{y_a} \end{bmatrix} \quad (13.28)$$

and the matrix of mass transfer coefficients was written as:

$$[\kappa] = \begin{bmatrix} \dfrac{k_{wa}(y_w k_{sa} + (1-y_w)k_{ws})}{y_w k_{sa} + y_s k_{wa} + y_a k_{ws}} & \dfrac{y_w k_{sa}(k_{wa} - k_{ws})}{y_w k_{sa} + y_s k_{wa} + y_a k_{ws}} \\ \dfrac{y_s k_{wa}(k_{sa} - k_{ws})}{y_w k_{sa} + y_s k_{wa} + y_a k_{ws}} & \dfrac{k_{sa}(y_s k_{wa} + (1-y_s)k_{ws})}{y_w k_{sa} + y_s k_{wa} + y_a k_{ws}} \end{bmatrix} \quad (13.29)$$

where y and k are the gaseous molar fraction and mass transfer coefficient for a binary mixture, respectively.

Wesselingh and Krishna [56] proposed the use of the following approximated expression, which represents a set of linear algebraic equations that can be solved numerically to obtain the molar permeate fluxes:

$$\widehat{J} = [H]^{-1} \Delta \widehat{y} \quad (13.30)$$

where the coefficients of the matrix $[H]$ are defined by:

$$H_{ii} = \frac{RT}{P} \sum_{j=1, j \neq i}^{n} \frac{\overline{y}_j}{k_{ij}} \quad (13.31)$$

$$H_{ij} = \frac{RT}{P} \frac{\overline{y}_i}{k_{ij}} \quad (13.32)$$

where $\overline{y}$ is the average mole fraction.

As can be observed, complex mathematical equations are involved in the Stefan—Maxwell-based model. This has made many researchers to use more simple approaches such as the Fickian one assuming that the rate of diffusion of species i depends only on its concentration gradient neglecting the coupling interactions between the diffusing species (Eq. (13.10)).

It was observed that the predictions of the Stefan—Maxwell-based model were closer to the experimental data than the Fickian mass transfer-based mathematical model. This may be due to the effects of coupling interactions between diffusing species neglected in the Fickian approach [7,45—47]. Figure 13.2 shows, as an example, the predicted AGMD permeate flux and ethanol selectivity by the exact and approximate solution of the Stefan—Maxwell equations as well as the Fickian-based model. The values obtained by the exact and approximate Stefan—Maxwell model are higher than those of the Fickian model and the increase of temperature enhances the deviation between the Stefan—Maxwell and the Fickian model. This is attributed to the effect of ethanol/water interactions that depends on the temperature (Eqs. (20)—(25)). In this case this effect should not be neglected especially because the mutual diffusivity of ethanol/water is higher

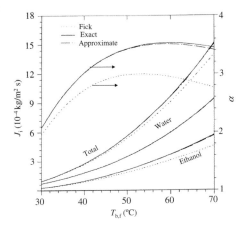

FIGURE 13.2 Predicted total and partial AGMD permeate fluxes (J_i) and selectivity (α, Eq. (12.25) in Chapter 12) as a function of the feed temperature ($T_{b,f}$) considering HVHP membrane (see Table 2.1 in Chapter 2), a module length of 1 m, a module width of 0.5 m, a stainless steel cooling plate of 2 mm thickness, a cooling temperature of 20 °C, ethanol in water concentration of 15 wt%, Reynolds numbers of the feed and cooling chambers of 1800 and an air gap thickness of 0.35 cm. *Source: Reprinted from [46]. Copyright 1999, with kind permission from Elsevier.*

than the mutual diffusivity of ethanol/air system.

As it can be observed in all these equations, the pore size is not considered although experimental studies proved the dependence of the AGMD flux on this parameter [5,14]. In fact, systematic studies on the effect of pore size and pore size distribution on the AGMD process performance are of a great interest. This fact will be discussed later on in this chapter.

Dusty gas model that takes into account all membrane parameters including the pore size of the membrane to describe the simultaneous Knudsen diffusion, molecular (ordinary) diffusion and viscous type of flows was used also in AGMD process to predict its performance [21,23]. The complete equations of dusty gas model were reported in Chapter 10 (Eqs. 28–32). Surface diffusion is commonly assumed negligible unless the pore area is much less than the membrane surface area (i.e. high porosity). The dusty gas model described correctly the dependence of the AGMD permeate flux on feed temperature, feed flow rate, temperature difference, pressure in the air gap and membrane type [23].

Since the solubility of air in water is low, the air is assumed to be a stagnant film and the viscous flow is neglected [57]. The AGMD permeate flux can then be modelled as a combined Knudsen and a molecular diffusion model as it is schematized in Fig. 13.3.

It was reported that in the case of AGMD carried out under atmospheric pressure, the dominant resistance was the molecular diffusion term (Fig. 13.3) because of the stagnant non-condensable gases trapped within the membrane pores and in the gap between the membrane and the condensing surface [57]. Because the gap width is generally 10–100 times the thickness of the membrane, the effects of the membrane characteristics were neglected in various studies.

It is worth quoting that the dusty gas model equations were reduced to the following expression that is somewhat similar to Eq. (13.14) [42,57]:

$$J_i = D^0_{iae} \frac{T^{(\gamma-1)/\gamma}}{R\delta|p_a|_{lm}} \Delta p_i \quad (13.33)$$

where γ is a constant, which for water is equal to 2.334, Δp_i is the transmembrane partial pressure difference and

$$D^0_{iae} = \frac{\varepsilon}{\tau} P D_{ia} \quad (13.34)$$

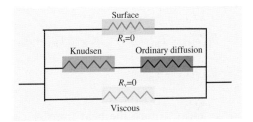

FIGURE 13.3 Mass transport mechanisms through a membrane used in AGMD process with a surface and viscous flow resistances assumed negligible.

It is worth quoting that the molecular diffusion model has been applied successfully in AGMD process although the pore size of the used membranes is as low as 0.2 μm [5,7,8,40,41].

It is to be noted that in all AGMD theoretical models the mass transport through the membrane material was not considered. As it is explained in Chapter 12 for VMD, when some feed components have strong interaction or 'affinity' with the membrane material, the transport mechanism through the material of the membrane may have a significant effect and must be included in the theoretical models. The separation performance in MD is not only based on VLE of the involved components in the feed mixture but also on the thermodynamics (i.e. operating temperature) and kinetic effects (i.e. diffusion of the components through the membrane and the gas-gap) [48,49].

In AGMD process, the heat transport consists of the following four steps:

(i) heat transport through the feed boundary layer
(ii) heat transport from the liquid/membrane interface, through the membrane and the gas gap, to the vapour/condensate liquid interface
(iii) condensation at the cold surface and heat transport through the condensate liquid boundary layer and
(iv) heat transfer to the cooling water.

At steady-state conditions, the heat flux from the bulk feed solution to the membrane surface is written as [5,11]:

$$Q_f = h_f(T_{b,f} - T_{m,f}) + \sum_i J_i C_{p,l,i}(T_{b,f} - T_{m,f})$$
$$= h_{hf}(T_{b,f} - T_{m,f})$$
(13.35)

where h_f and $C_{p,l,i}$ are the heat transfer coefficient and the specific heat of the liquid feed, respectively. The temperatures $T_{b,f}$ and $T_{m,f}$ are indicated in Fig. 13.1. The heat transfer coefficient h_f can be estimated from the empirical heat transfer correlations indicated previously in Chapter 10 (Eq. (13.54)) using dimensionless numbers.

The heat transfer from the liquid/membrane interface to the vapour/condensate liquid interface takes place via sensible heat flux, Q_s, and the heat associated to mass transfer:

$$Q_{p1} = Q_s + \sum_i J_i \Delta H_{v,i}$$
$$= h^*(T_{m,f} - T_{p,g}) + \sum_i J_i \Delta H_{v,i} \quad (13.36)$$

where $\Delta H_{v,i}$ is the evaporation enthalpy of specie i at the absolute temperature T and the heat transfer coefficient h^* is expressed as [11]:

$$h^* = h_y \left(\frac{\theta}{1 - e^{-\theta}}\right) \quad (13.37)$$

with

$$\theta = \frac{\sum_i J_i C_{p,g,i}}{h_y} \quad (13.38)$$

$C_{p,g,i}$ is the specific heat in the gas phase and θ is a dimensionless heat transfer rate factor in the form of Peclet number, which is the ratio of heat transfer by convection to conduction and h_y is the heat transfer coefficient in the gaseous phase written as follows:

$$h_y = \frac{k}{b} \quad (13.39)$$

where k is the gas phase thermal conductivity.

In Eq. (13.37), the Akerman correction factor $(\theta/(1-e^{-\theta}))$ takes into account the sensible heat transferred by the diffusing vapours. It gives the effect of finite mass transfer rates on the heat transfer coefficient h_y. If evaporation/condensation does not occur, this factor is unity but increases positively as the rate of evaporation increases.

The heat transport from the condensation layer interface to the bulk cold liquid is written as:

$$Q_{p2} = h_d(T_{p,g} - T_{f,g}) = \frac{k_c}{l}(T_{f,g} - T_{p,c})$$
$$= h_c(T_{p,c} - T_c) = h_p(T_{p,g} - T_c) \quad (13.40)$$

where h_d is the condensate heat transfer coefficient, k_c is the thermal conductivity of the condensing plate, l is its thickness, h_c is the coolant film heat transfer coefficient and h_p is the total heat transfer coefficient from vapour/condensate liquid interface to cooling water, which is expressed in the following form:

$$h_p = \left(\frac{1}{h_d} + \frac{l}{k_c} + \frac{1}{h_c}\right)^{-1} \quad (13.41)$$

For vertical surfaces h_d was calculated using the following equation shown as an example, although other equations can also be used [5,7,11,12,58]:

$$h_d = 0.943 \left(\frac{g\rho_p^2 \Delta H_v k_p^3}{L\mu_p(T_{p,g} - T_{f,g})}\right)^{1/4} \quad (13.42)$$

where ρ_p, k_p and μ_p are, respectively, the fluid density, thermal conductivity and dynamic viscosity at the condensate film temperature, L is the height of air gap and g is the gravitational acceleration.

In the absence of a special treatment of the condensation surface, the condensate forms a liquid film of a thickness δ_p giving rise to a heat transfer resistance that is defined as:

$$R_d = \frac{\delta_p}{k_p} \quad (13.43)$$

By using Eqs. (13.35), (13.36) and (13.40), the interfacial temperatures can be determined from the following equations:

$$T_{m,f} = T_{b,f} - \frac{H}{h_{h,f}}\left((T_{b,f} - T_c) + \frac{\sum_i J_i \Delta H_{v,i}}{h^*}\right) \quad (13.44)$$

$$T_{p,g} = T_c + \frac{H}{h_p}\left((T_{b,f} - T_c) + \frac{\sum_i J_i \Delta H_{v,i}}{h^*}\right) \quad (13.45)$$

where

$$H = \left(\frac{1}{h_{h,f}} + \frac{1}{h^*} + \frac{1}{h_p}\right)^{-1} \quad (13.46)$$

At steady-state conditions, the heat transfer flux, Q, is expressed as:

$$Q = Q_f = Q_{p1} = Q_{p2}$$
$$= H\left((T_{b,f} - T_c) + \frac{\sum_i J_i \Delta H_{v,i}}{h^*}\right) \quad (13.47)$$

It should be pointed out that in AGMD process the heat transfer coefficient on the permeate side, h_p, is much smaller than that of the feed side and therefore dominates the overall heat transfer coefficient, H, given in Eq. (13.46).

From Eqs. (13.44) and (13.45) the temperature polarization effects can be evaluated by using one of the following equations defined in [14,17]:

$$\theta = \frac{T_{m,f} - T_{p,g}}{T_{b,f} - T_{f,g}} \quad (13.48)$$

$$\theta = \frac{T_{m,f} - T_{p,g}}{T_{b,f} - T_c} \quad (13.49)$$

Due to water vapour transport though the membrane pores, the concentration of the non-volatile solutes at the membrane surface becomes higher than the concentration in the bulk feed aqueous solution, whereas the concentration of the volatile compounds at the membrane surface becomes smaller than the corresponding concentration in the bulk liquid feed. Both temperature and concentration polarization are schematically shown in Fig. 13.1. The equations derived for the other

MD configurations to estimate the concentration polarization coefficient were shown in Chapters 10–12 and they can be employed also in AGMD process.

It should also be noted that in a AGMD module with feed liquid flowing tangentially temperature drop occurs along the length of the module. This means that the inlet feed temperature is different from the outlet feed temperature depending among other parameters on the applied feed flow rate and the type of the membrane used. Therefore, a local permeate flux and local temperature and concentration polarization coefficients must be considered.

A two-dimensional theoretical model in which a simultaneous numerical solution of the momentum, energy and diffusion equations of the feed and permeate channels has been proposed to determine temperature, velocity and concentration profiles in the membrane module [58]. The effects of various AGMD operating parameters on the AGMD permeate flux as well as on the process thermal efficiency have been analyzed. The thermal efficiency (η) is defined as the ratio between the heat that contributes to evaporation and the total heat transferred from the feed to the permeate. Stefan diffusion has been considered to determine the permeability of the membrane. The predicted AGMD fluxes were validated in comparison with available experimental results.

Another numerical study was developed for AGMD process using elliptic procedure and a control-volume method and full vorticity transport equation together with the stream function, mass and energy equations [59,60]. The combined Knudsen and Stefan–Maxwell equations were used to determine the permeability of the membrane. Numerical solutions to the established Navier–Stokes mass and energy equations for laminar natural convection have been obtained for small Ra numbers. The effects of various AGMD operating parameters on the AGMD permeate flux were simulated. Fields of temperatures in the air gap space were built. Good agreements between the obtained results and experimental ones were observed.

From MD literature, it can be observed that the reported experimental studies are carried out varying one of the independent parameters maintaining the others fixed. Following this classical or conventional method of experimentation many experimental runs are necessary and interaction effects between parameters are ignored. Response surface methodology (RSM) that involves statistical design of experiments in which all factors are varied simultaneously is a possible method permitting to study the interaction effects between parameters and optimization of the MD process. A quadratic RSM model was developed for desalination by AGMD modules using Fortran code in Aspen Plus® platform [61]. The considered response was the produced water per unit of feed liquid flow rate and auxiliary heat input, whereas the considered variables were only two, the feed temperature and the feed flow rate. An optimum separation efficiency (i.e. ratio of produced water to the feed) of 5.8% was predicted.

Special attention should be devoted to optimization of different AGMD systems in order to study rigorously the interaction effects between parameters, increase the AGMD performance and decrease energy consumption.

EXPERIMENTAL: EFFECTS OF PROCESS CONDITIONS

Effects of Feed Temperature

The feed temperature effect on the AGMD performance has been widely investigated [10,11,14,40,46,58]. In general, the rate of evaporation is strongly affected by the feed temperature. Exponential trends between the AGMD permeate flux and the feed temperature were observed. By increasing the feed temperature from 40 to 80 °C the permeate flux can be enhanced 9-fold. As explained in the previous chapters this behaviour is due to the exponential increase of the vapour pressure of the feed aqueous solution with temperature (i.e. Antoine Equation), which enhances the driving force (i.e. vapour pressure difference) for both water and the volatile solutes present in the feed solution.

Taking into consideration the relationship between the vapour pressure and temperature a linear trend was found between the AGMD permeate flux and the vapour pressure difference [8,10,11]. The proportional factor is a function of the characteristics of the membrane, the permeating compounds, the feed flow rate including turbulence conditions and the membrane module configuration.

It is to be mentioned here that for a constant temperature difference between the bulk feed and the cooling surface ($\Delta T = T_{b,f} - T_c$), an exponential increase of the AGMD permeate flux with the feed temperature was found theoretically [40]. Moreover, an exponential increase of the AGMD permeate flux was experimentally confirmed as a function of the mean temperature (T_m) following the Arrhenius-type equation ($J_i \alpha \exp(-B/T_m)$) maintaining the bulk feed and permeate temperature difference, ΔT, constant. On the other hand, a linear relation was found between the permeate flux and ΔT, when T_m was kept constant at different levels [14,17]. These behaviours were observed not only when distilled water and aqueous sucrose solutions were used as feed but also for solutions containing volatile solutes such as isopropanol.

It should be pointed out that the heat loss by conduction through the membrane decreased with the rise of the feed temperature [11,40]. This also may explain the increase of the AGMD permeate flux with the increase of the feed temperature. Moreover, the thermal efficiency was improved by increasing the feed temperature [11,58]. For example, an increase of the feed temperature from 40 to 80 °C could enhance the thermal efficiency by 12% [58]. At high feed temperatures the quantity of heat loss by conduction becomes negligible compared with the heat transferred by the diffusing compounds. Therefore, it is better to operate AGMD at high temperatures than at low temperatures.

It is worth quoting that the temperature polarization coefficient (θ) decreases with the increase of the feed temperature and for a constant ΔT value θ also decreases with the increase of the mean temperature (T_m). However, for the same T_m the calculated temperature polarization coefficient did not change very much for different ΔT values [14,17].

No apparent effect of feed temperature was observed on the salt rejection factor (i.e. salt concentration in the permeate was less than 5 ppm, 4 ± 1 µS/cm) [11]. However, as shown in Fig. 13.2, when feed solutions containing volatile solutes are treated by AGMD process, the selectivity, α, of the volatile solute (Eq. (12.25) in Chapter 12) first increases with the feed temperature reaching a maximum and then declines [12,46]. This is attributed to the combined effects of feed concentration, vapour pressure, activity coefficient on the driving forces for the different components (i.e. water and volatile solutes). As a consequence, at higher feed temperatures, higher total and partial permeate fluxes and lower selectivity values are obtained. These results were predicted based on the mathematical equations presented in the previous section [46].

The selectivity of the volatile solute defined by the following equation (Eq. (13.50)) exhibited only a slight change of about 0.64 when the feed (17.7 wt% aqueous propionic acid solution) temperature was increased from 40 to 80 °C and the cooling water temperature was maintained at 30 °C [9,45] The predicted selectivity based on the Stefan–Maxwell approach agreed very well with the experimental one. The dependence with the feed temperature was attributed to the temperature polarization effect that became higher at higher feed temperatures.

$$\alpha = \frac{x_{b,p}}{x_{b,f}} \quad (13.50)$$

where $x_{b,f}$ and $x_{b,p}$ are the mole fraction of the volatile solute in the feed liquid and permeate liquid, respectively.

In the case of formic acid/water mixture of 77.5 wt%, the selectivity defined by Eq. (13.50) was below unity and changed from 0.94 to 0.88 with an increase of the feed temperature from 40 to 80 °C [47].

Effects of Solute Concentration in the Feed Aqueous Solution

It is known that the increase of the non-volatile solute concentration in the aqueous feed solution results in a reduction of the AGMD permeate flux due to the decrease of the water vapour pressure (i.e. driving force) [5,10,34,58]. Moreover, there is also the contribution of the concentration polarization effect. However, the latter contribution is very small compared with the temperature polarization effect. Furthermore, distilled water was obtained in the permeate indicating very high rejection factors for the non-volatile solutes, such as salts and sucrose, present in the feed aqueous solutions [5,10,11,14,34]. The following figure shows, as an example, the AGMD permeate flux as a function of the salt NaCl concentration in the feed aqueous solution (Fig. 13.4). A slight decline of the AGMD production, by about 6%, was observed when the NaCl concentration was increased from 0 to 10 wt%. This result indicates that AGMD process can be applied for the treatment of highly concentrated aqueous solutions without suffering from a considerable decrease in solute rejection, which are often observed in other membrane processes such as reverse osmosis (RO). Model simulations also indicated a very

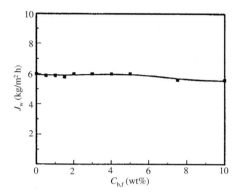

FIGURE 13.4 Effect of the feed salt (NaCl) concentration ($C_{b,f}$) on the AGMD permeate flux (J_w) of HVHP membrane (see Table 2.1) for a feed temperature of 55 °C and a cooling temperature of 7 °C. *Source: Reprinted from [10]. Copyright 1994, with kind permission from Elsevier.*

slight decrease of the thermal efficiency with increasing the salt feed concentration [58]. For example, a decrease of only 2% was predicted with the increase of NaCl concentration from 20,000 to 50,000 ppm.

As reported for DMCD process applied for the treatment of aqueous solutions of very high salt concentrations, scale formation or crystallization fouling is expected to occur also in AGMD. This phenomenon has not been rigorously studied yet for AGMD configuration.

When volatile solutes are present in the feed aqueous solution the trends observed for the total and partial AGMD permeate flux with the feed concentration are different than those observed for non-volatile solutes [5,12,46]. Figure 13.5 presents an example for dilute ethanol/water mixtures. The total permeate flux as well as the partial permeate flux of the volatile solute increased with the increase of the volatile solute concentration in the feed aqueous solution, whereas the partial permeate flux of water decreased slightly. These results are due to the effect of the concentration on the partial pressures of the different species and on their activity coefficients. Moreover, as shown in Fig. 13.5 for ethanol/water mixtures,

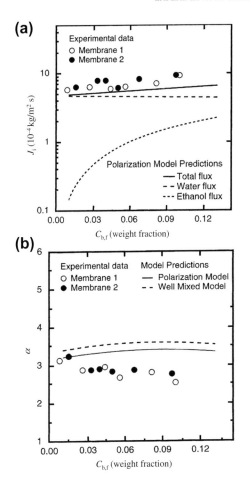

FIGURE 13.5 Effects of ethanol concentration in feed ($C_{b,f}$) on the AGMD partial permeate fluxes of water and ethanol (J_i) as well as on the total AGMD permeate flux (a) and on ethanol selectivity (α, Eq. (12.25)) (b) of HVHP membrane (see Table 2.1, two membrane samples 1 and 2) at a feed temperature of 50 °C, a feed flow rate of 5 l/min and a cooling water temperature of 20 °C. *Source: Reprinted from [12]. Copyright 1999, with kind permission from Elsevier.*

1–30 wt% [9]. However, for HCl/water mixtures and HNO_3/water mixtures in the concentration range of 10^{-3} to 10 mol/l, the permeate concentration increased with an increase in the feed concentration. As for the rejection factor it was constant for lower solute concentrations but declined for higher solute concentrations [5]. From the theoretical models two types of results were obtained: (i) increase in selectivity of the volatile solute by increasing its concentration up to an optimum value and then decrease for higher concentrations in the feed aqueous solution (i.e. ethanol/water mixtures with concentrations up to 30 wt%) [46], (ii) a relatively constant selectivity (i.e. ethanol/water mixture with concentrations up to 12 wt%, see Fig. 13.5b) [12]. It was reported that these variations may be associated to the air gap width and the convection effect.

It is to be noted that when using feed aqueous solutions containing volatile solutes, the upper limit of the concentration is dictated by the risk of membrane pore wetting. When this occurs high permeate fluxes are obtained but the selectivity values are decreased.

It was noticed that when salt is added to a feed aqueous solution containing volatile solutes, the selectivity of the volatile solute was improved. This is attributed to the reduction of the water vapour pressure while the concentration of the volatile solute is increased in the permeate solution. For example, an addition of 1 wt% of NaCl in ethanol/water solution of 9.7 wt% reduced the total permeate flux slightly but the selectivity enhancement was significant [12].

a continuous decreasing trend of the selectivity with the feed concentration may be obtained depending on the range of the feed concentration and also on the property of the volatile solute [9,12]. A similar trend was observed for propionic acid (CH_3CH_2COOH)/water mixtures with concentrations in the range

Effects of Feed Flow Rate

One possible way to reduce the temperature and concentration polarization effects in AGMD process is to increase the feed flow rate (i.e. feed circulation velocity) in order to establish adequate hydrodynamic conditions and work under turbulent flow regime. The consequence is that the heat transfer coefficient in the feed boundary layer, h_f in Eq. (13.35), and the temperature as well as the concentration at the membrane surface approach the bulk ones. This means a reduction of the boundary layer thickness when Reynolds number increases resulting in an increase of the driving force and higher AGMD permeate flux [5,11,12,17,46]. Within the laminar region, the AGMD permeate flux is expected to increase by increasing the feed flow rate prior to reaching an asymptotic value.

Figure 13.6 shows an example of the effect of feed circulation velocity on the AGMD permeate flux of distilled water and aqueous isopropanol solutions [17]. An asymptotic value at high feed flow rates is obtained indicating that the turbulent flow regime is reached.

It was found that the thermal efficiency (η) is affected slightly by varying the feed flow rate because the AGMD permeate flux as well as the heat transferred by conduction through the membrane and the heat associated to mass transfer all increase by increasing the feed flow rate [58].

When treating aqueous solutions containing non-volatile solutes, no effect of the feed flow rate on the permeate concentration was detected [11]. However, for feed solutions containing volatile solutes a slight increase of the selectivity was observed at higher feed flow rates [12]. For example, ethanol selectivity was increased from 3.25 to about 3.5 when the feed flow rate was increased from 2 to 5 l/min and the concentration of ethanol in water was 1.55 wt%. This was attributed mainly to the reduced effect of the concentration polarization.

Another way to increase AGMD permeate flux is to introduce an adequate spacer in the feed channel that guarantees an enhancement of the permeate flux with the lowest possible pressure drop. This study was carried out using different types of spacers with distinct geometries (round, twisted tapes, non-woven, woven, different flow angles, different angles between filaments, etc.) [22]. Figure 13.7 shows the convention followed to describe spacers. It was found that the highest AGMD permeate flux is for two types of spacers, those having round rods as filaments with flow attack angle of 45° and angle between filaments of 90°, and those made of twisted tapes with flow attack angle of 30° and angle between filaments of 120°.

Coolant Temperature

In general, in AGMD studies the permeate temperature varies between 7 and 30 °C. A slight decrease of the permeate flux with an increase of this temperature was observed due to the decrease of the partial pressure gradient, which is the driving force. Compared to the feed temperature effect, the AGMD permeate flux is not very sensitive to the coolant temperature. For example, the AGMD permeate flux

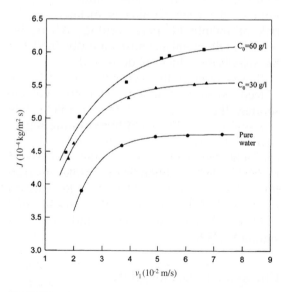

FIGURE 13.6 Effect of feed velocity (v_l) on the total AGMD permeate flux (J) of GVHP membrane (see Table 2.1) when using distilled water as feed and different isopropanol aqueous solutions at a mean temperature (T_m) of 27.5 °C and a temperature difference (ΔT) of 5 °C. *Source: Reprinted from [17]. Copyright 2000, with kind permission from Elsevier.*

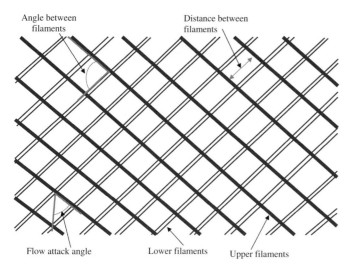

FIGURE 13.7 General description of spacers.

of the membrane HVHP (see Table 2.1 in Chapter 2) decreased from 4 to 3 kg/m^2·h when the cooling temperature was increased from 7 to 30 °C for a feed temperature of 60 °C, a feed flow rate of 5.5 l/min, an air gap distance of 0.35 cm and a saline solution with concentrations of salts similar to seawater [11]. For the same membrane, a decrease of the AGMD permeate flux from about 17.5 to 15 kg/m^2·h was observed experimentally for ethanol/water mixture of concentration 10.4 wt% when the coolant temperature was increased from 7 to 20 °C. The feed temperature was the same (60 °C) and the feed flow rate was 5 l/min [12]. Ethanol selectivity was decreased from 3.5 to 3. The small effect of the coolant temperature is due to the heat transfer coefficient in the air gap that dominates the overall heat transfer coefficient and to the low sensitivity of water vapour pressure at low temperatures. Compared to DCMD configuration the increase of the permeate flux is greater when the permeate temperature is increased.

It is worth quoting that, based on theoretical calculations, the calculated thermal efficiency decreased only by 2.5% when decreasing the cooling temperature from 45 to 5 °C. However, by increasing the feed temperature from 40 to 80 °C, 12% increase was obtained for the thermal efficiency. Therefore, little benefit would be derived from decreasing the cooling temperature. If the cooling temperature is reduced the AGMD permeate flux is increased slightly but also the thermal efficiency is decreased slightly. On the other hand, by increasing the feed temperature not only the AGMD permeate flux is increased, but also the thermal efficiency is improved. Therefore, it is more convenient to increase the feed side temperature rather than to decrease the coolant temperature. Ambient temperature can be applied for cooling.

When simulation was made for dilute ethanol/water mixtures using the Fickian and the Stefan–Maxwell mathematical model, a decrease in both ethanol and water partial permeate fluxes as well as ethanol selectivity was observed with an increase in the cooling temperature [46]. The predicted parameters using the Fickian model were found to be close to those obtained from the Stefan–Maxwell model at high cooling temperatures. This was associated partly to the reduced ethanol/water diffusion interactions at high cooling temperatures. However, when using a formic acid/water solution around its azeotropic point the calculated selectivity using the Stefan–Maxwell model was maintained practically the same over the whole range of cooling temperature, 10–30 °C [47].

Cold Side Flow Rate

Very few studies have been carried out on the effect of the cold side flow rate (i.e. cooling channel) on the AGMD performance [11,12, 46,58]. The objective of increasing the cold side flow rate is to decrease the air condensate interfacial temperature ($T_{f,g}$ in Fig. 13.1). However, as it is reported previously the coolant temperature effect is marginal. Therefore, a negligible effect of the cold side flow rate on the AGMD permeate flux is expected. This may be the reason of the few research studies conducted on this subject. When using aqueous ethanol solutions, very small changes were observed in the AGMD partial permeate fluxes and in the ethanol selectivity [12,46]. Similar result was obtained when using aqueous salt solutions [11]. This is also due partly to the heat transfer coefficient in the air gap, which is much smaller than the heat transfer coefficients in the feed and in the cold side.

Simulations indicated that the AGMD permeate flux may increase by 3% with the increase of the coolant velocity compared with 11% increase with the feed circulation velocity, and practically no change was detected in the thermal efficiency with the variation of the coolant velocity for a membrane length of 0.2 m [58].

It is to be noted here that the effects of the coolant flow rate on the AGMD performance may depend on the module height as the temperature of the cooling surface changes with the module height.

Non-Condensable Gases

Non-condensable gases refer to air in general and the dissolved gases in the feed aqueous solutions such as carbon dioxide produced from the thermal decomposition of bicarbonates. These gases are trapped inside the membrane pores and also are present in the permeate membrane side. The presence of these gases increases the resistance to mass transfer, reduces the rate of condensation and decreases the permeate flux. By deaerating the feed aqueous solution before entering the module and/or deaerating the permeate membrane side (i.e. reduction of the pressure in the gap between the membrane and the condensing surface), the resistance to mass transfer decreases improving the permeate flux. It was observed that a reduction of the air gap pressure down to a pressure equal to the water vapour pressure of the feed aqueous solution increased the AGMD permeate flux by a factor of up to 2.5–3 compared with the obtained AGMD permeate flux at atmospheric pressure [23]. In this case, the thermal efficiency was increased from about 78 to 95%. This result can be attributed to the decrease of the heat transfer loss by conduction through the membrane by reducing the air gap pressure.

The effects of the gas type on the AGMD performance have been studied using three different inert gases ranging from light to heavy (helium, air and sulphur hexafluoride) [15]. The air present in the gap space between the membrane and the condensing surface was replaced by the inert gas under study. Remember that separation in MD occurs also due to the diffusion of vapour through the gas-filled membrane pores and the gas gap width. Therefore, it is expected that changing the inert gas affects the diffusion step and the selectivity in AGMD process. In other words, since diffusion is molecular transport, the size and collision frequency of the molecules affect significantly their diffusion rate. Based on the Stefan–Maxwell mathematical model, a reduction of the AGMD permeate flux was noticed when using heavier gases than air such as sulphur hexafluoride, and an enhancement was noticed when using lighter gases such as helium. However, the selectivity was better when using the heaviest gas at the expense of the AGMD permeate flux reduction.

Air Gap Width

The length of the diffusion path in the air gap is another significant factor affecting the AGMD performance. The diffusion path is made up of the membrane thickness (δ) plus the length of the air gap perpendicular to the membrane surface (b in Fig. 13.1). AGMD membrane modules are normally designed to allow width variations of its thickness using different gaskets. In this way, it is possible to vary the length of the diffusion path over which the vapour molecules pass prior to condensation over the cooling surface. This is one of the advantages of AGMD technology.

It is to be expected when the gap width is below a critical value the heat and mass transfer rates in the condensing channel will be controlled by diffusion, and when it is larger than this critical value they will be controlled by free convection.

It was noticed that the air gap width and the feed temperature are the most important factors affecting AGMD performance. By reducing the air gap width the temperature gradient within the vapour compartment increases and consequently an increase of the water permeate flux in the case of feed solutions containing nonvolatile solutes occurs. As well, the total and partial permeate fluxes increase in the case of feed solutions containing volatile solutes. This increase is more pronounced at higher feed temperatures. In other words, the AGMD permeate flux is inversely proportional to the air gap width [5,11,12,17]. By reducing the air gap length from 5 to 1 mm, the experimental permeate flux increased 2–3.5-fold [40,58]. Furthermore, it was found theoretically that the effect of the air gap became much more significant for gap widths thinner than 1 mm [40]. It was also found that the heat transfer by conduction increased 3.4-fold when the air gap width decreased from 5 to 1 mm, whereas only a slight decrease was found for the thermal efficiency [40,58]. This is attributed to the large increase of the AGMD permeate flux as can be seen in Fig. 13.8, which is shown as an example [40]. Similarly, when the feed temperature is increased the AGMD permeate flux is also increased but the heat loss is reduced.

Remember that the purpose of using air gap in the permeate membrane modules is to reduce the heat loss by conduction in MD. The longer the gap width, the longer is the diffusion path and the higher is the mass transfer resistance, and therefore, the AGMD permeate flux is lower.

It is important to mention here that changing the membrane thickness (δ) will not affect the rate of evaporation as long as the air gap width (b) is larger [40]. Moreover, the degree of the natural convection depends on the geometry of the air gap compartment. If the height of the permeate side is kept constant whereas the gap width is changed, the geometric ratio, which is

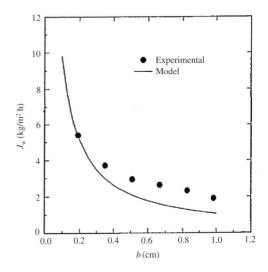

FIGURE 13.8 Effect of the air gap width (b) on the AGMD permeate flux (J_w) of HVHP membrane (see Table 2.1) at a feed temperature of 60 °C, a cooling temperature of 20 °C, a feed flow rate of 5.5 l/min and a saline solution with concentrations of salts similar to seawater. *Source: Reprinted from [11]. Copyright 1998, with kind permission from Taylor & Francis.*

the ratio between vertical and horizontal dimension of the permeate side, is changed, thereby the mechanism of heat transfer may be shifted from pure conduction and diffusion to natural convection. When the geometric ratio approaches unity, the contribution of the convective heat transfer approaches a maximum. However, when the geometric ratio approaches zero or infinity, the contribution of the convective heat transfer becomes negligible (i.e. Nusselt number (Nu) approaches 1). Natural convection is expected to contribute at greater air gap widths.

As far as the separation factor is concerned, it was observed that increasing the air gap width from 1 to 10 mm did not significantly affect the propionic acid selectivity (both theoretical Stefan–Maxwell and experimental) [9,45] nor the formic acid selectivity [15,47]. However, in the case of dilute ethanol/water mixtures it was predicted by the Fickian mathematical model, taking into consideration the temperature and concentration polarization effects, that the selectivity increases with the air gap width reaching an asymptotic value [12]. On the other hand, the selectivity predicted by a well mixed model (i.e. without including the temperature and polarization effects) did not change over the whole range of air gap width (1–10 mm). These discrepancies may be attributed partly to the significance of interactions between the volatile solute and water, which may decrease with increasing the air gap width. Moreover, the selectivity increase with the air gap width may be due to the reduced effect of the permeate flux on temperature and concentration polarization. Therefore, it is better to operate at as small as possible gap widths because the membrane module is more compact, less expansive and high production is obtained without sacrificing the selectivity.

Effects of Some Membrane Parameters

In all membrane processes including MD, the permeate flux is inversely proportional to the membrane thickness. This membrane parameter plays a significant role on the heat and mass transfer and must be optimized in order to obtain a high MD permeability with low conductive heat transfer. In general, the MD permeate flux is inversely proportional to the membrane thickness. This is a general trend in all MD configurations except for AGMD process because, as it is mentioned previously, the effect of membrane thickness seems negligible due to the predominant mass transfer resistance of the stagnant air gap. In other words, the thickness does not affect the AGMD permeate flux as long as the gap width is much larger than the thickness. This seems obvious theoretically [40]. Nevertheless, careful experimental studies are necessary using membranes with different thicknesses while maintaining all other membrane characteristics the same. Moreover, in MD, no systematic studies have been performed yet on the effects of pore tortuosity on the MD permeate flux.

It was found that the AGMD permeate flux was reduced 2- and 1.63-fold with the increase of the thermal conductivity of the membrane material from 0.05 to 0.3 W/m·K when the membrane porosity was fixed at 74 and 84%, respectively. This was a theoretical prediction and was attributed to the reduction of the thermal resistance and the subsequent increase of the heat transfer by conduction by 35 and 33% for the membranes porosity 74 and 84%, respectively [58]. As a consequence, the thermal efficiency of the AGMD process was found to decrease with the increase of the thermal conductivity of the membrane material by 9 and 5% for the membrane porosity 74 and 84%, respectively. It is worth to mention that the thermal conductivity of the membrane material has more significant effect on the overall mass transfer resistance in AGMD than in DCMD, suggesting its stronger impact on the performance of AGMD than DCMD [70].

As stated throughout this book, the MD membrane should be porous and at least one of the layers of the membrane should be made of a hydrophobic material. The pore size range may be from several nanometers to few micrometers and the pore size distribution should be as narrow as possible.

It is generally agreed upon that the MD permeate flux increases with an increase in the membrane pore size under some restrictions, which means that the pore size should allow a sufficiently high LEP in order to avoid wettability of membrane pores. This means that the pore size must be small enough, which leads to a conflict with the requirement of higher MD permeability. In modelling AGMD process, both Fickian and Stefan–Maxwell-based models do not take the membrane pore size into consideration. As can be seen in Fig. 13.9, when the pore size is greater than 1 μm there is some dependence of the AGMD permeate flux on the pore size. But the permeate concentration, given by the electrical conductivity, does not depend on the pore size. From Fig. 13.9, one may think that the AGMD performance is independent of the pore size if the pore size is smaller than 1 μm, which is the largest value for the commonly used membranes for MD [5]. However, when using different types of membranes having different porosities, pore sizes and materials (GVHP, HVHP, TF200, PTFE membranes from Gore, see Table 2.1 in Chapter 2) under the same AGMD operating conditions, it was found that different AGMD permeate fluxes could be obtained when distilled water and isopropanol/water mixtures with concentrations ranging from 45 to 225 g/l were used as feed [17]. For the same type of membranes, for example GVHP and HVHP, the permeate flux was higher for the membrane having greater pore size. It was concluded that the AGMD permeate flux increased as the membrane pore size increased. Moreover, when using aqueous sucrose solutions, different AGMD permeate fluxes were obtained when different membranes were used under the same operating conditions [14]. Recently, the influence of the pore size on the AGMD desalination performance was studied using chemically modified ceramic membranes by 1H,1H,2H,2H-perfluorodecyltriethoxysilane [37]. The pore sizes ranged from 50 to 800 nm. When the pore size was below 200 nm, a decrease of the AGMD permeate flux was observed with an increase of the pore size. However, for larger pore sizes practically no significant change was detected for the permeate flux. The salt (NaCl) rejection factor decreased with the increase of the pore size and a rejection factor as low as 89.2% was obtained confirming pore wetting. Therefore, it is necessary to determine an optimum pore size taking into consideration the wetting of membrane pores for each feed solution. This will limit the maximum permitted pore size of the membrane. More studies on the effect of

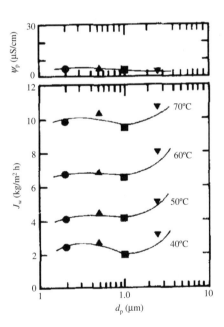

FIGURE 13.9 Effect of membrane pore size (d_p) on the AGMD permeate flux (J_w) of PTFE membranes and electrical conductivity of the permeate (Ψ_p) at different feed temperatures, a cooling temperature of 2 °C and a feed aqueous salt (NaCl) solution of concentration 3.8%. Source: Reprinted from [5]. Copyright 1987, with kind permission from Elsevier.

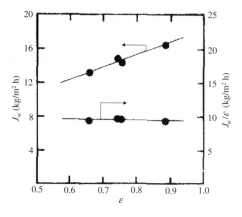

FIGURE 13.10 Effect of membrane porosity (ε) on the AGMD permeate flux (J_w) of PTFE membranes at a feed temperature of 60 °C, a cooling temperature of 20 °C and pure water used as feed. *Source: Reprinted from [5]. Copyright 1987, with kind permission from Elsevier.*

pore size and pore size distribution on the AGMD performance should be carried out in order to confirm the independence of the AGMD permeate flux from the membrane pore size.

The membrane porosity is another important membrane parameter affecting the MD process in general. In the case of AGMD process it was observed that the permeate flux increased linearly with the membrane porosity as shown in Fig. 13.10 [5]. Theoretically, the porosity should not affect the permeate flux as far as the gap width is large [40]. This result should be confirmed experimentally. The increase of membrane porosity reduces the effect of the thermal conductivity of the membrane material and increases the thermal efficiency of the AGMD process [58]. Therefore, the thermal conductivity of the membrane material should be small.

Other Effects

In order to enhance the AGMD permeate flux ultrasonic stimulation of resonance frequency 20 kHz and irradiation power up to 90 W was applied to a plate and frame AGMD membrane module [13]. A piezoelectric type of acoustic horn was adhered to the stainless steel shell in the feed side of the membrane module and the acoustic power was supplied from a power generator with a maximum power output of 100 W. An increase of production up to 25% was observed compared to without ultrasonic irradiation when FALP membrane (see Table 2.1 in Chapter 2) was used with a temperature difference of 55 °C, a gap width of 4 mm and feed aqueous salt (NaCl) solutions with concentrations up to 5 wt%. This enhancement of AGMD performance was attributed to the reduced effects of the temperature polarization as well as to the fouling reduction by particulate removal due to ultrasonic vibration, acoustic microstreaming and ultrasonic cavitation. Attempts were made to identify the most significant mechanism following distinct tests such as intermittent or ultrasonic stimulations. A developed theoretical model indicated that an increase of the AGMD permeate flux up to 200% is possible by applying an ultrasonic intensity up to 5 W/cm^2 [62].

AGMD APPLICATIONS

AGMD Technology in Desalination and Solar Units

Desalination is one of the potential applications of AGMD technology for production of high-purity water [5,6,8,10,11,18,20,27–30, 34–39]. As can be seen in Table 13.1, the AGMD permeate fluxes are as high as the DCMD permeate fluxes. Recently, a comparison between DCMD and AGMD configurations has been made using the same membrane, modified tubular zirconia of pore size 50 nm (Zr50), which was applied for desalination of 0.5 M NaCl solution at different feed temperatures [38]. For both MD modes very high rejection factors ranging from 99 to 100% were obtained but the AGMD production was higher than that of DCMD. For example, for a feed temperature of 95 °C the obtained AGMD permeate flux was about 1.2 times higher than that of DCMD permeate flux. For VMD configuration the obtained permeate flux was even higher (i.e. about 1.6 times) than that of AGMD but the NaCl rejection factor of VMD process was lower (i.e. 96.1%). Similar membranes were prepared by the same research group and tested for desalination [36]. It was found that permeate fluxes ranged from 0.7 to 7 $l/m^2 \cdot h$ with salt NaCl rejection factors ranging from 86 to 100% for the feed salt concentrations varying from 0.001 to 1 M. Lower salt rejection factors were obtained for dilute aqueous solutions and low feed temperatures. The highest AGMD permeate flux of 7 $l/m^2 \cdot h$ was obtained with a feed temperature of 95 °C and a cooling temperature of 5 °C when distilled water was used as feed. For 1 M NaCl feed aqueous solution, the highest AGMD permeate flux of 5.4 $l/m^2 \cdot h$ was obtained for a feed temperature of 99 °C and a cooling temperature of 5 °C with a separation factor of 100%. Other hydrophobic ceramic membranes were prepared for desalination by AGMD process and the results reported [35,37].

Kimura and Nakao [5] carried out AGMD tests of 3.8% NaCl aqueous solution using PTFE membranes of different pore sizes manufactured by Nitto Electric Industries Co. Ltd (Japan) and observed that the permeate concentration was independent of the pore size with an electrical conductivity less than 5 µS/cm (i.e. salt rejection factors higher than 99.9%). Moreover, the AGMD permeate flux was maintained practically the same for the membranes with pore sizes less than 1 µm but increased for membranes with pore sizes larger than 1 µm. It also increased linearly with the porosity. However, when the surface active agent sodium dodecylbenzene sulfonate (SDS) was added to the feed aqueous salt solution, both the AGMD permeate flux and the salt concentration in the permeate were enhanced indicating wetting of the membranes pores. This explains the reason for using membranes with small pore sizes in AGMD.

When aqueous solutions containing sulphuric acid (H_2SO_4) and sodium hydroxide (NaOH) were used to change the pH value, no changes were detected in the permeate flux, permeate pH and permeate electrical conductivity. Those values were comparable to those obtained when aqueous NaCl solutions were used for feed [5].

Experiments on seawater desalination by AGMD were conducted by Kubota et al. [6] using two test plate and frame membrane modules; one containing porous PTFE membrane (0.1 mm thickness, 75% porosity and 1.92 m^2 effective area) and the other silicone/polysulphone compositae hydrophobic dense membrane (0.25 mm thickness and 2.93 m^2 effective area). Suppliers and other specifications of the membranes were not reported. The obtained AGMD permeate fluxes of the PTFE membrane (i.e. maximum flux 10 kg/h) were higher than those of the dense membrane (i.e. maximum flux 8 kg/h) and for both membrane modules, the product water quality was very good with an electric conductivity of about 10 µS/cm when the produced AGMD permeate flux was high. However the quality of the produced water was not good when the AGMD permeate flux was low.

Kurokawa et al. [8] performed both theoretical and experimental studies using a PTFE membrane with a pore size of 0.2 µm, a porosity of 76% and an effective membrane area of $1.42 \cdot 10^{-2}$ m^2. A decrease of the AGMD permeate flux was observed with increasing concentration of lithium bromide (LiBr) and H_2SO_4 with rejection factors higher than 99.9% (i.e. electrical conductivities below 10 µS/cm). The maximum concentration of LiBr in the feed aqueous solutions was 56% while that of H_2SO_4 was 83%. The highest AGMD permeate fluxes obtained for both aqueous solutions were around 10^{-3} kg/m^2·s for air gap width of 3 and 5 mm. Good agreements were observed between the experimental and theoretical results.

Banat and Simandl [11] obtained permeates with electrical conductivities of 4 ± 1 µS/cm when treating by AGMD process a saline solution with a composition similar to seawater (Cl: 18,600 ppm, Na: 10,400 ppm, Mg: 1290 ppm, Ca: 410 ppm, K: 380 ppm, Br: 62 ppm, B: 4.9 ppm, F: 1.9 ppm) using HVHP membrane (see Table 2.1 in Chapter 2). The highest AGMD permeate flux was about 10 kg/m^2·h at a feed temperature of 70 °C, a cooling temperature of 20 °C and a gap width of 0.19 cm. A 10 days AGMD experimental test was carried out observing a constant electrical conductivity of the permeate ranging between 3 and 5 µS/cm with an increase of the permeate flux during the first 25 h from 1.2 to 1.6 kg/m^2·h and then it was maintained around this value for a feed temperature of 53 °C, a cooling temperature of 7 °C and an air gap distance of 0.99 cm. In another study, Banat and Simandl [10] showed that the AGMD permeate flux of GVHP and FHLP membranes (see Table 2.1 in Chapter 2) was maintained steady over more than 6 weeks AGMD run using tap water with an electrical conductivity of 297 µS/cm and seawater with an electrical conductivity of 37.6 mS/cm. It was also observed that the AGMD permeate flux was affected only slightly by the increase of the salt concentration. On the other hand, the electrical conductivity of the permeate of the GVHP membrane was steady at around 3 µS/cm (i.e. 99% salt rejection factor) when using tap water as feed but it increased up to 1600 µS/cm when using seawater (i.e. 95% salt rejection factor). This was attributed to the formation of salt crystals on the membrane surface leading to partial wetting of the membrane. However, it was reported that fouling and scaling problem were not encountered. After seawater experiments, tap water was used again as feed obtaining a permeate with an electrical conductivity of 18 µS/cm indicating that the initial value of 3 µS/cm for the distillate was never reproduced. In the case of FHLP membrane, the highest AGMD permeate flux was similar to that of GVHP membrane (i.e. about 10 kg/m^2·h), the steady-state flux was about 7 kg/m^2·h and the electrical conductivity of the permeate was 7 µS/cm when using tap water as feed.

Bouguecha and Dhahbi [18] looked at coupling fluidized bed crystallizer and AGMD process as possible solution to geothermal waste desalination. The purpose of the crystallizer is to reduce an important portion of feed hardness (i.e. production of calcium carbonate $CaCO_3$) without significant loss of temperature. Model solutions containing NaCl with concentrations varying from 3 to 35 g/l were treated obtaining a permeate with an electrical conductivity of 6 µS/cm. Permeate fluxes up to 8 kg/m^2·h were reported but the used membrane type was not indicated. Preliminary experiments were conducted and more investigations are needed to prove the viability and efficiency of the combined system.

The combined use of AGMD and solar energy was investigated experimentally by various authors [20,27–30,63]. Fraunhofer ISE (Germany) developed solar-driven desalination plants based on AGMD process with capacities up to 10 m^3/day [20,27–30]. Some plants integrating both solar thermal and photovoltaic panels (PV) were intended for autonomous operation in arid and semi-arid remote regions with a lack of electricity and drinkable water but with high solar irradiation. It is claimed that that technical simplicity, long maintenance-free operation periods, high quality produced water, no need for external electricity, no additional water for cooling and no impact on the environment are the important aims of their systems with heat sources resistant to seawater and corrosion free heat exchangers. By using a similar AGMD module to that employed previously by Bier and Plantikow

[63], Koschikowski et al. [20] discussed first the design of a semi-pilot plant using spiral wound PTFE membrane module working at 60–85 °C feed temperature. The used membrane has a pore size of 0.2 µm, a thickness of 35 µm, a porosity of 80% and an effective area of 8 m². The height of the membrane module is 700 mm and its diameter is 460 mm. The maximum distillate reported was 15 kg/h (i.e. 81 l/day) with a feed flow rate of about 225 l/h and the maximum evaporator inlet temperature was 90 °C. The maximum AGMD permeate flux obtained during the test period of summer 2002 was about 130 l/d under meteorological conditions of Freiburg (Germany). The gained out put ratio (GOR) defined by Eq. (13.51) was 5.5 at 350 l/h flow rate and 75 °C evaporator inlet temperature. It is worth quoting that surprisingly the authors did not report on the water quality.

$$GOR = \frac{J_w \Delta H_{v,w}}{Q_{in}} \quad (13.51)$$

where J_w is the AGMD permeate flux, $\Delta H_{v,w}$ is the enthalpy of evaporation of water and Q_{in} is the input energy supplied to the system.

Simulations were also carried out for three different locations (Eilat in Israel; Muscat in Oman and Palma de Mallorca in Spain) and the results during summer indicated that the plant is able to distillate 120–160 l/day with a solar collector area less than 6 m² and without heat storage [20].

Years later seven solar AGMD pilot plants fabricated using spiral wound AGMD PTFE membrane modules (Fraunhofer ISE) were installed in five countries: two in Jordan [27,28], one in Egypt [29], one in Morocco (Kelaa Sraghna) and three in Spain (Islas Canarias) [30]. Each unit consists of flat plate solar collectors, PV panels, spiral AGMD module(s) and a data acquisition system as can be seen in Figs. 13.11 and 13.12. The height of the used membrane modules was 450–800 mm, the diameter was 300–400 mm, the effective membrane area was 7–12 m², the specific thermal energy consumption was 100–200 kWh/m³ and the GOR of about 3–6.

In Jordan, a compact pilot plant with 10 m² membrane area in the spiral wound module was installed in the northern part of Jordan (Jordan University of Science and Technology, Irbid) and has been operated with brackish

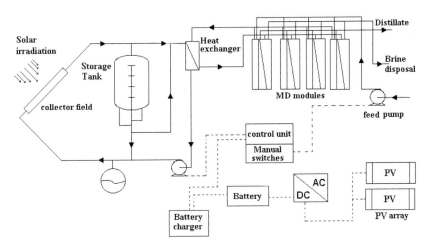

FIGURE 13.11 Schematics of the solar-driven AGMD plant installed in Aqaba (Jordan). *Source: Reprinted from [28]. Copyright 2007, with kind permission from Elsevier.*

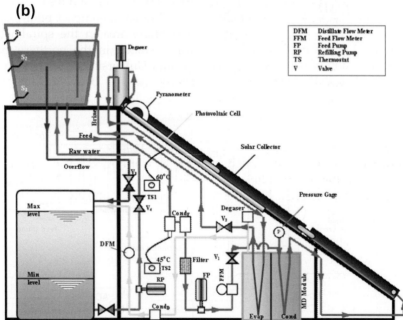

FIGURE 13.12 Photo (a) and side view (b) of the AGMD compact plant. *Source: Reprinted from [29]. Copyright 2008, with kind permission from Elsevier.*

water since August 2005 [27]. The area of the used flat solar collector (corrosion-free) is 5.73 m^2 while the PV module of 1000 W/m^2 has an area of (1310 × 654) mm^2. The plant has been operated continuously producing AGMD permeate fluxes as high as 120 l/day with an electrical conductivity of about 5 μS/cm and thermal energy requirement ranging between 200 and 300 kWh/m^3. However, in this case the GOR was found to be lower than unity (i.e. 0.3–0.9). A second solar AGMD pilot plant with four spiral AGMD modules assembled in parallel (Fig. 13.11) was installed in the south of Jordan (Aqaba port) and has been operated with untreated (i.e. without chemical treatment usually used in RO) seawater from red sea since February 2006 [28]. The electrical conductivity of the seawater is 55 mS/cm. Solar heat storage of 3 m^3 together with a battery bank was used in this system to store thermal and electrical energy. Moreover, the collector loop was separated from the seawater loop of the spiral AGMD membrane modules having an effective area of 10 m^2 per module. 72 m^2 flat solar collector and 12 PV modules, each having 120 Wp, (Watts peak) were used. The effect of the feed flow rate was studied observing a linear increase of the AGMD permeate flux from 1.45 to 2 l/m^2·h with a slight decrease of its quality (i.e. increase of the electrical conductivity from 50 to 100 μS/cm). This result was attributed to the reduction of both the temperature and concentration polarization effects. Long-term experiment was conducted for 90 days and it was found that production varied from 5 to 27 l/m^2·h with an electrical conductivity varying between 20 and 250 μS/cm (i.e. average salt rejection factor of 98%). The GOR values were found to be in the range 0.4–0.7. It was concluded that the range of the water production was from 2 to 11 l/day per unit area of solar collector with a specific energy consumption of 200–300 kWh/m^3. It is to be noted that typical production of traditional solar stills is between 2 and 6 l/m^2·day.

Another solar AGMD pilot plant similar to that installed in Ibrid (Jordan) was installed in Alexandria (Alexandria University, Egypt) with one spiral AGMD module with a membrane area of 10 m^2 and was operated since June 2005 [29]. Figure 13.12 shows a photo of the AGMD compact system. It was observed that the production was directly proportional to the daily total radiation. For example, for clear days the production was 64 l/day (i.e. 11.2 l/m^2·day) with an electrical conductivity of the produced water of 3 μS/cm, whereas that of the feed was 526 μS/cm and the accumulated solar energy was 41.6 kWh/day (i.e. 7.25 kWh/m^2·day). For cloudy days the production was 23.6 l/day (i.e. 4.11 l/m^2·day) with an electrical conductivity of 2 μS/cm (i.e. 99.5% salt rejection) and the accumulated solar energy was 29.5 kWh/day (i.e. 5.15 kWh/m^2·day).

In Gran Canaria, another solar AGMD pilot plant similar to that installed in Aqaba (Jordan) was installed but in this system five membrane modules were used with a maximum design capacity of 1600 l/day instead of 900 l/day, a solar collector area 90 m^2 instead of 72 m^2 and a PV area of 1.92 kWp instead of 1.44 kWp [30]. This plant has been operated since January 2005 [30]. As an example, a feed flow rate of about 415 l/h and a maximum inlet temperature of the evaporator 85 °C were observed. The maximum distillate was about 15 l/h (i.e. 75 l/day).

Recently economic analysis of these solar thermal AGMD units was reported [64]. Based on their calculations the estimated cost of the produced water is 15 and 18$/m^3 by the compact unit and large unit, respectively. The authors pointed out that membrane lifetime and plant lifetime were the key factors in determining the water production cost. The cost decreases with increasing membrane and/or plant lifetime. More details about cost analysis are reported in Chapter 15. It is to be mentioned here that modelling and optimization of a solar-driven AGMD desalination system was carried out at the Aspen Custom Modelerr® (ACM) platform in which a spiral wound membrane module was used. Optimization study suggests that simple thermal storage tank and low feed flow rates should be used for high water production.

The desalination performance, operation and maintenance procedures of an AGMD system were investigated by Walton et al. [65]. Low grade thermal energy (13–75 °C) supplied by a salt gradient solar pond was used. The system containing an AGMD module with an effective membrane area of 2.94 m^2 plus the controlling pumps and heaters was built by Scarab Development AB (Sweden) as reported previously. Hot brine was pumped from the bottom of the solar pond and passed through a heat exchanger to supply heat. Cold water from the solar pond surface was passed through a heat exchanger to provide cooling. Fluctuations of the AGMD permeate flux were observed reaching a maximum of 6 l/m^2·h.

AGMD Technology in Food Processing

Few attempts have been made to apply AGMD process in food processing. The number of studies carried out for this purpose is much smaller compared with DCMD and VMD processes (See Chapters 10 and 12).

AGMD was examined for the concentration of fruit juices. Kimura and Nakao [5] tested for the first time the use of AGMD for the concentration of mandarin juice. Figure 13.13 shows the permeate fluxes of AGMD and RO processes for different feed concentration ratios. The AGMD tests were carried out using a feed temperature of 48 °C and a cooling temperature of 10 °C, whereas RO process was carried out at a transmembrane pressure of 50 bar. It can be seen in Fig. 13.13 that for high values of feed concentration ratio (>1.5), the AGMD permeate fluxes were higher than those of RO, whereas at low feed concentrations the RO performance was better (i.e. permeate flux about 1.7 times higher than that of AGMD).

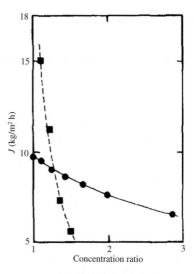

FIGURE 13.13 Comparison of AGMD performance and RO performance for concentration of mandarin orange juice. *Source: Reprinted from [5]. Copyright 1987, with kind permission from Elsevier.*

The AGMD was applied for the concentration of milk but a sharp decrease of the AGMD permeate flux with the increase of the feed concentration ratio and an increase of the electrical conductivity of the product water were observed. Moreover, it was found that the membranes tended to be fouled by the adhesion of fat to the membrane surface resulting in very low AGMD permeate fluxes [5]. Sugar and gelatine solutions were also treated by AGMD process [5,14]. In both cases, a decrease of the permeate flux was observed with the increase of the feed concentration. This was attributed partly to the increase of the viscosity of the feed solution and heat transfer resistance. When using 150 g/l aqueous sucrose solution, the produced permeate during 30 h was distilled water and its flux was maintained at the same level after 30 days of AGMD operation [14].

Taking into consideration that AGMD can be applied successfully in food processing, AGMD studies should be carried out in this promising application area.

Treatment of Aqueous Alcohol Solutions

As it was stated previously, alcohols are preferentially vapourised from aqueous feed solutions and are concentrated in the permeate. The permeate is enriched in alcohol content compared with the feed. Care must be taken in order to avoid membrane wetting. The risk of membrane pore wetting is high especially at high alcohol concentration in water, because of the low surface tension of alcohols. Figure 13.14 shows, as an example, the decrease of the minimum *LEP* of ethanol/water mixtures (*LEP* defined in Chapter 8) for TF200 membrane (see Table 2.1 in Chapter 2) with the increase in ethanol concentration. Therefore, dilute aqueous solutions of alcohols were treated in most of the AGMD studies.

The potential advantage for ethanol recovery from fermentation broth was discussed by Gostoli and Sarti [7] using TF200 membrane (see Table 2.1 in Chapter 2) for ethanol/water separation by AGMD. It was found that the separation factor was increased with the increase of the temperature difference across the membrane. When testing with aqueous ethanol solutions of concentrations 3–7 wt%, the measured total AGMD flux was practically the same as that of pure water used as feed (up to $1.3 \; 10^{-3}$ kg/m$^2\cdot$s), and the distillate was more concentrated in ethanol, the separation factor being up to 2 that is far below the values observed for a single distillation. Additionally, when 23.8 wt% ethanol solution was treated, the separation factor was extremely low, close to unity. In other words, the process was ethanol selective for low ethanol content in the feed, while it became water selective for high ethanol content. Stefan–Maxwell-based model for ethanol/water/air system was developed without incorporating the concentration polarization effect. It was found that the separation factor was highly sensitive to the feed composition at constant average temperature and temperature difference.

Banat and Simandl [12] investigated the removal of ethanol from dilute aqueous solutions by AGMD using HVHP membrane (see Table 2.1 in Chapter 2). The feed ethanol concentration tested was varied from 0.83 to 10 wt% within the feed temperature range of 40–70 °C. The effects of various AGMD operating parameters on permeate flux and ethanol selectivity were investigated. Ethanol selectivities ranging from 2 to 3.5 were achieved with permeate fluxes lower than 9 kg/m$^2\cdot$h. It was found that the Fickian-based mathematical model, which neglects the concentration and polarization effects, could not adequately predict the experimental results. Banat and Simandl [12] also examined the effect of salt on the AGMD process performance during the concentration of aqueous ethanol solutions and found better ethanol selectivity with only a slight decrease in total permeate flux. The exact Stefan–Maxwell and the approximate Stefan–Maxwell models were analysed in another study [46]. It was noticed that the results of these models were consistent between them, whereas some differences existed

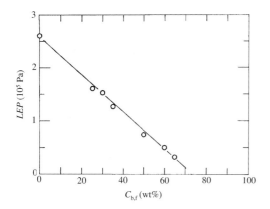

FIGURE 13.14 Minimum liquid entry pressure of ethanol/water mixture (*LEP*) in TF200 membrane for different ethanol concentrations in water ($C_{b,f}$). *Source: Reprinted from [7]. Copyright 1989, with kind permission from Elsevier.*

between them and the Fickian-based model. Moreover, it was concluded that the temperature and concentration variation along the membrane module length had to be considered at low air gap widths and high feed temperatures.

García-Payo et al. [17] also studied the removal of alcohol (methanol, ethanol, isopropanol) from their binary mixtures using five membranes (PTFE, PVDF) of different parameters. The effects of the relevant AGMD parameters were investigated. For membranes prepared with the same material, the AGMD fluxes were found to be higher when the membrane pore size is larger and the porosity is higher. It was also noticed that generally the AGMD process performance is better when using PTFE membranes. Moreover, for the PTFE membranes having the same pore size, the AGMD flux was lower when the membrane is on a support. AGMD permeate fluxes as high as 2.7 $kg/m^2 \cdot h$ were obtained at an isopropanol concentration of 225 g/l using the unsupported PTFE membrane with pore size 0.45 μm (Gore, Table 2.1 in Chapter 2). Unfortunately, the authors did not provide detailed alcohol selectivity values nor the permeate partial fluxes.

Break of Azeotropic Mixtures

Azeotropic mixtures are impossible to separate by simple fractional distillation that depends on VLE characteristics of the liquid mixture. Solutions with maximum or minimum azeotropic point reach a state in which boiling does not change the liquid composition. Therefore, other methods have been applied such as pervaporation, adsorptive distillation, capillary distillation and diffusion distillation [66–69]. Table 13.2 shows examples of water azeotropic systems. Udriot et al. [9] has shown that AGMD is of potential interest in breaking azeotropic mixtures. Until now the azeotropic aqueous mixtures tested in AGMD are HCl/water, propionic acid/water, formic acid/water mixtures [9,15,45,47,48]. The objective is to eliminate the azeotropic point that exists if the system is only governed by the VLE relationship.

Udriot et al. [9] conducted experiments, for the first time, in a plate and frame membrane module with azeotropic mixtures of HCl/water and propionic acid/water using supported PTFE membranes on polyester with pore sizes 0.2 and 0.45 μm (Schleicher & Schuell, Richen, Switzerland). Acid selectivity values (Eq. (13.50)) between 0.6 and 0.8 were achieved instead of unity given by the VLE. Additionally, it was seen that the HCl/water azeotrope was shifted to higher acid concentration, whereas the azeotropic point of the propionic acid/water system disappeared. This phenomenon was attributed to the differences in the acid/air and water/air diffusion rates across the membrane and the air gap of the different components of the azeotropic mixtures. The gas gap shifted the selectivity away from unity. For example, as the diffusivity of water in air is higher than the diffusivity of propionic acid in air, water will be enriched in the permeate more than what it is expected from the VLE. Therefore, the azeotropic point is completely eliminated. Furthermore, it was reported that the selectivity can be reduced by about 12% for the propionic acid/water system when going from DCMD to AGMD with a 7 mm air gap width.

Banat et al. [45] applied Stefan–Maxwell-based mathematical model to predict AGMD performance in separating the azeotropic mixture of propionic acid/water. The available experimental data obtained by Udriot et al. [9] were used to validate the model that accounts for the coupling interactions between the diffusing species as well as the temperature and the concentration polarization effects (see previous theoretical section). Good agreement between the predicted and experimental data was found reproducing reasonably well the effects on the propionic acid selectivity of the air gap width, feed composition and feed temperature [45].

Further theoretical investigations were carried out also by Banat et al. [15,47] using

TABLE 13.2 Aqueous Azeotropic Binary Systems

Compound	Azeotropic point (wt% of compound in water)
Benzyl alcohol ($C_6H_5CH_2HO$)	9.0
Cyclohexanol (($CH_2)_5$)HOHC	20
Furfuryl alcohol ($C_5H_6O_2$)	20
Allyl alcohol (CH_2CHCH_2HO)	72.9
Nitric acid (HNO_3)	68.5
Hydrochloric acid (HCl)	20.2
Butyric acid ($CH_3CH_2CH_2COOH$)	18.4
Formic acid (HCOOH)	77.5
Propionic acid (CH_3CH_2HOOC)	17.7
Hydriodic acid (HI)	57
Perchloric acid ($HClO_4$)	71.6
Diaceton alcohol ($CH_3C(O)CH_2C(OH)(CH_3)_2$)	13.0
Phenol (C_6H_5HO)	9.2

Fickian-based model and Stefan–Maxwell-based model to predict AGMD performance in separating the azeotrope mixture of formic acid/water, which is water poor azeotrope (i.e. 77.5 wt% formic acid). Therefore, to obtain selectivities far away from unity water should be removed from the feed and enriched in the permeate, while formic acid concentration in the feed is enhanced above its azeotropic composition. The effect of the inert gases, helium, air and sulphur hexafluoride, in breaking the formic acid/water azeotropic mixture was also examined. The formic acid selectivity was found to be larger when using helium (around 0.96), followed by air (about 0.9) and then sulphur hexafluoride (0.85—0.86). Similarly, the AGMD permeate fluxes were higher for helium, followed by air and sulphur hexafluoride. It was found that both the AGMD permeate flux and selectivity were governed by the VLE relations and the different diffusivities in the inert gas showing that the heavy inert gases such as sulphur hexafluoride help more in eliminating totally the azeotropic point than the lighter ones such as air and helium. Furthermore, it was found again that the predicted results by the Stefan–Maxwell model which incorporated temperature and concentration polarization effects were closer to previously reported experimental data than the Fickian-based model.

It should be noted that the tests carried out by Udriot et al. [9] showed the feasibility of breaking azeotropic mixtures. Kimura and Nakao [5], on the other hand, looked at the separation of the systems HNO_3/water, HCl/water and formic acid/water at different concentrations without mentioning the possibility of AGMD to eliminate the azeotropic point of these systems.

It is worth to indicate here that very few experimental studies have been carried out in this interesting and useful research area. Testing other aqueous azeotropic mixtures will be of great interest.

Extraction of Volatile Organic Compounds

The potential application of AGMD for the extraction of VOCs from dilute aqueous solutions would be of great interest [5,16].

Kimura and Nakao [5] examined the separation of volatile solutes such as HNO_3 and HCl at different concentrations in water and noticed similar trends for both components, which were different from the commonly observed trends when treating aqueous solutions containing non-volatile solutes. Moreover, the acid concentration in the permeate was found to increase with an increase in acid concentration in the feed. When testing CH_3COOH aqueous solutions (0.02–0.85 mol/l) and formic acid (HCOOH) aqueous solutions (0.14–1.1 mol/l) at 50 °C feed temperature and 18 °C cooling temperature, the measured AGMD fluxes were 9.5–10 $kg/m^2 \cdot h$ with rejection factors 40–44% for CH_3COOH/water solutions, while higher AGMD performance for formic acid/water solutions (12.8–13.0 $kg/m^2 \cdot h$ permeate fluxes and 57–59% rejection factors) [5].

Banat and Simandl [16] also used AGMD for the selective removal of propanone (CH_3COCH_3) from aqueous solutions using PVDF membrane (HVHP in Table 2.1 in Chapter 2), within feed temperature range of 40–70 °C and propanone concentration up to 6 wt%. The effects of feed concentration, feed flow rate, feed temperature and the temperature of the condensing surface on the AGMD permeate flux and propanone selectivity were investigated. It was found that the selectivity ranged from 2 to 6.

Studies are required to test AGMD technology for the treatment of wastewater containing other VOCs as well as non-volatile solutes.

Other Applications of AGMD Technology

One of the possible applications of AGMD is isotopes separation in aqueous solutions. Kim et al. [19] applied AGMD for the ^{18}O isotopic water separation using FGLP membrane (see Table 2.1 in Chapter 2). For example, ^{18}O-enriched water (>90%) is used as a target in the cyclotron for the production of β-emitting radioisotope ^{18}F, which is essential for positron emission tomography. It was observed that the AGMD flux and the degree of ^{18}O separation were higher for large temperature gradient across the membrane. More should be done in this AGMD application field. Not only for the separation of isotopic compounds but also for the concentration of waste aqueous solutions containing isotopes and a simultaneous production of pure water.

AGMD was applied successfully for concentration of HI and H_2SO_4 aqueous solutions using a flat sheet PTFE membrane with 0.2 μm pore size, 400 μm thickness and 90% porosity (Pall 'Emflon sheet') [26]. These solutions are of great interest for hydrogen production following sulphur–iodine thermochemical water-splitting cycle. Different HI and H_2SO_4 concentrations in water were tested in AGMD. The highest obtained retentate concentration for HI was about 8 mol/l. This concentration is higher than the azeotropic point of the mixture, which is 7.57 mol/l indicating that the azeotropic mixture was broken. It is to be mentioned here that distillation followed by gaseous HI thermal decomposition is difficult requiring high pressures because of the presence of the azeotrope HI/water system at about 57 wt%. In AGMD experiments, the lowest used feed concentration of HI in water was 0.3 mol/l. For example, it was observed that, when the initial feed concentration was 7.2 mol/l, the concentration of HI in the permeate was 5.2 mol/l and the average AGMD permeate flux was 7.8 $l/m^2 \cdot h$ for a feed temperature of 80 °C, a cooling temperature of 15 °C and a gap width of 10 mm. DCMD process was also applied for the same feed HI aqueous solutions, and it was found that this configuration was not useful when the azeotropic concentration was needed to be passed to obtain pure gaseous HI for decomposition [26]. In the case of H_2SO_4 aqueous mixture, this was concentrated in the feed up to 10.1 mol/l without presence of H_2SO_4 in the permeate. The lowest used feed concentration of H_2SO_4 in water was 0.9 mol/l. For example, when the initial concentration of H_2SO_4 in feed was 8.5 mol/l the average AGMD permeate flux was 0.7 l/m²·h for a feed temperature of 80 °C, a cooling temperature of 15 °C and a gap width of 10 mm. It was concluded that the obtained results demonstrated that AGMD could be applied for hydrogen generation. In fact, hydrogen production from water and alternative non-fossil sources is intended to be indispensable in order to satisfy the continuously increasing demand of hydrogen. Thermochemical cycles using heat from solar or nuclear sources represent one of the promising methods for hydrogen production at a large scale. Nowadays, the most extensively studied thermochemical water-splitting cycle is sulphur–iodine process consisting of the following three reactions:

(i) Bunsen reaction, exothermic at 20–120 °C, to produce the aqueous solution of two acids:

$$2H_2O + I_2 + SO_2 \rightarrow H_2SO_{4(aq)} + 2HI_{(aq)}$$

(ii) Sulphuric acid decomposition, endothermic at 800–900 °C:

$$H_2SO_{4(gas)} \rightarrow H_2O_{(gas)} + SO_{2(gas)} + 1/2 O_{2(gas)}$$

(iii) HI decomposition, endothermic at 300–600 °C:

$$2HI_{(gas)} \rightarrow I_{2(gas)} + H_{2(gas)}$$

Before vapourisation and subsequent high temperature decomposition, the sulphuric acid must be concentrated up to at least 90 wt% (i.e. 16.7 mol/l), while hydrogen iodide must be separated from the HI_x solution ($HI/H_2O/I_2$ mixture) from the Bunsen reaction prior to decomposition for hydrogen generation.

The first attempt has been made to apply AGMD process in thermal cogeneration plants that require use of pure water [31]. Both municipal water and flue gas condensate were employed as feed using plate and frame membrane modules (Scarab Development AB). Each module had a membrane area of 2.3 m², 10 cassettes, 9 feed channels, 9 cooling channels, 73 cm height, 63 cm width, 17.5 cm stack thickness, a gap width of 1 mm and a PTFE membrane with pore size 0.2 μm, 80% porosity, thickness 0.2 mm. District heating supply line was used for heating whereas municipal water was used for cooling. When employing as feed municipal water with an electrical conductivity lower than 467 μS/cm and two modules connected in series, a linear increase of the permeate flux (11–26 l/h) with the applied flow rate was observed at a feed temperature of 70 °C and a cooling temperature of 15 °C. Electrical conductivity values of the permeate in the range 1–3 μS/cm were reported. The AGMD performance was stable during 370 h after that the permeate flux decreased due to scale formation. When using flue gas condensate as feed with an electrical conductivity of 5.25 μS/cm, that of the permeate was 233 μS/cm. The high electrical conductivity of the permeate was attributed to the volatile ammonia that passed through the membrane and to the presence of hydrogen carbonate (HCO_3^-). Pretreatments reducing pH values of the feed solution below 6 as well as the use of membrane contactors to remove carbon dioxide, cause of alkalinity, from the feed solution may be of great interest in this case.

References

[1] D.W. Gore, Gore-Tex membrane distillation, Proceedings of the 10th Annual Convention of the Water Supply Improvement Association. Honolulu, USA, July 25–29, 1982.

[2] L. Carlsson, The new generation in sea water desalination: SU membrane distillation system, Desalination 45 (1983) 221–222.

[3] S.I. Andersson, N. Kjellander, B. Rodesjo, Design and field tests of a new membrane distillation desalination process, Desalination 56 (1985) 345–354.

[4] N. Khellander, Design and field tests of a membrane distillation system for seawater desalination, Desalination 61 (1987) 237–243.

[5] S. Kimura, S. Nakao, Transport phenomena in membrane distillation, J. Membr. Sci. 33 (1987) 285–298.

[6] S. Kubota, K. Ohta, I. Hayano, M. Hirai, K. Kikuchi, Y. Murayama, Experiments on seawater desalination by membrane distillation, Desalination 69 (1988) 19–26.

[7] C. Gostoli, G.C. Sarti, Separation of liquid mixtures by membrane distillation, J. Membr. Sci. 41 (1989) 211–224.

[8] H. Kurokawa, K. Ebara, O. Kuroda, S. Takahashi, Vapor permeate characteristics of membrane distillation, Sep. Sci. Technol. 25 (1990) 1349–1359.

[9] H. Udriot, A. Araque, U. von Stockar, Azeotropic mixtures may be broken by membrane distillation, Chem. Eng. J. 54 (1994) 87–93.

[10] F.A. Banat, J. Simandl, Theoretical and experimental study in membrane distillation, Desalination 95 (1994) 39–52.

[11] F.A. Banat, J. Simandl, Desalination by membrane distillation: A parametric study, Sep. Sci. Tech. 33 (1998) 201–226.

[12] F.A. Banat, J. Simandl, Membrane distillation for dilute ethanol: Separation from aqueous streams, J. Membr. Sci. 163 (1999) 333–348.

[13] C. Zhu, G.L. Liu, C.S. Cheung, C.W. Leung, Z.C. Zhu, Ultrasonic stimulation on enhancement of air gap membrane distillation, J. Membr. Sci. 161 (1999) 85–93.

[14] M.A. Izquierdo-Gil, M.C. García-Payo, C. Fernández-Pineda, Air gap membrane distillation of sucrose aqueous solutions, J. Membr. Sci. 155 (1999) 291–307.

[15] F.A. Banat, F. Abu Al-Rub, R. Jumah, M. Shannag, On the effect of inert gases in breaking the formic acid-water azeotrope by gas-gap membrane distillation, Chem. Eng. J. 73 (1999) 37–42.

[16] F.A. Banat, J. Simandl, Membrane distillation for propanone removal from aqueous streams, J. Chem. Tech. Biotech. 75 (2000) 168–178.

[17] M.C. García-Payo, M.A. Izquierdo-Gil, C. Fernández-Pineda, Air gap membrane distillation of aqueous alcohol solutions, J. Membr. Sci. 169 (2000) 61–80.

[18] S. Bouguecha, M. Dhahbi, Fluidised bed crystalliser and air gap membrane distillation as a solution to geothermal water desalination, Desalination 152 (2002) 237–244.

[19] J. Kim, S.E. Park, T.S. Kim, D.Y. Jeong, K.H. Ko, Isotopic water separation using AGMD and VEMD, Nukleonika 49 (2004) 137–142.

[20] J. Koschikowski, M. Wieghaus, M. Rommel, Solar Thermal-driven desalination plants based on membrane distillation, Desalination 156 (2003) 295–304.

[21] C.M. Guijt, G.W. Meindersma, T. Reith, A.B. Haan, Air gap membrane distillation: 1. Modeling and mass transport properties for hollow fiber membranes, Sep. Pur. Tech. 43 (2005) 233–244.

[22] M.N. Chernyshov, G.W. Meindersma, A.B. Haan, Comparison of spacers for temperature polarization reduction in air gap membrane distillation, Desalination 183 (2005) 363–374.

[23] C.M. Guijt, G.W. Meindersma, T. Reith, A.B. Haan, Air gap membrane distillation: 2. Model validation and hollow fiber module performance analysis, Sep. Pur. Tech. 43 (2005) 245–255.

[24] G.W. Meindersma, C.M. Guijt, A.B. Haan, Desalination and water recycling by air gap membrane distillation, Desalination 187 (2006) 291–301.

[25] J.H. Hanemaaijer, J.V. Medevoort, A.E. Jansen, C. Dotremont, E.V. Sonsbeek, T. Yuan, L.D. Ryck, Memstill membrane distillation – a future desalination technology, Desalination 199 (2006) 175–176.

[26] G. Caputo, C. Felici, P. Tarquini, A. Giaconia, S. Sau, Membrane distillation of HI/H_2O and H_2SO_4/H_2O mixtures for the sulphur-iodine thermochemical process, Int. J. Hydrogen Energy 32 (2007) 4736–4743.

[27] F. Banat, N. Jwaied, M. Rommel, J. Koschikowski, M. Weighaus, Desalination by a 'compact SMADES' autonomous solar-powered membrane distillation unit, Desalination 217 (2007) 29–37.

[28] F. Banat, N. Jwaied, M. Rommel, J. Koschikowski, M. Weighaus, Performance evaluation of the 'large SMADES' autonomous desalination solar-driven membrane distillation plant in Aqaba, Jordan, Desalination 217 (2007) 17–28.

[29] H.E.S. Fath, S.M. Elsherbiny, A.A. Hassan, M. Rommel, M. Wieghaus, J. Koschikowski, M. Vatansever, PV and thermally driven small-scale stand-alone solar desalination systems with very low maintenance needs, Desalination 225 (2008) 58–69.

[30] J. Koschikowski, M. Wieghaus, M. Rommel, V.S. Ortin, B.P. Suarez, J.R.B. Rodríguez, Experimental investigations on solar driven stand-alone membrane distillation systems for remote areas, Desalination 248 (2009) 125–131.

[31] A. Kullab, A. Martín, Membrane distillation and applications for water purification in thermal cogeneration plants, Sep. Pur. Tech. 76 (2011) 231–237.

[32] www.scarab.se

[33] www.xzero.se

[34] C. Feng, K.C. Khulbe, T. Matsuura, R. Gopal, S. Kaur, S. Ramakrishna, M. Khayet, Production of drinking

water from saline water by air-gap membrane distillation using polyvinylidene fluoride nanofiber membrane, J. Membr. Sci. 311 (2008) 1–6.
[35] A. Larbot, L. Gazagnes, S. Krajewski, M. Bukowska, W. Kujawski, Water desalination using ceramic membrane distillation, Desalination 168 (2004) 367–372.
[36] S.R. Krajewski, W. Kujawski, M. Bukowska, C. Picard, A. Larbot, Application of fluoroalkylsilanes (FAS) grafted ceramic membranes in membrane distillation process of NaCl solutions, J. Membr. Sci. 281 (2006) 253–259.
[37] L. Gazagnes, S. Cerneaux, M. Persin, E. Prouzet, A. Larbot, Desalination of sodium chloride solutions and seawater with hydrophobic ceramic membranes, Desalination 217 (2007) 260–266.
[38] S. Cerneaux, I. Stuzynska, W.M. Kujawski, M. Persin, A. Larbot, Comparison of various membrane distillation methods for desalination using hydrophobic ceramic membranes, J. Membr. Sci. 337 (2009) 55–60.
[39] Z. Ma, Y. Hong, L. Ma, M. Su, Superhydrophobic membranes with ordered arrays of nanospiked microchannels for water desalination, Langmuir Lett. 25 (2009) 5446–5450.
[40] A.S. Jonsson, R. Wimmerstedt, A.C. Harrysson, Membrane distillation: A theoretical study of evaporation through microporous membranes, Desalination 56 (1985) 237–249.
[41] C. Gostoli, G.C. Sarti, S. Matulli, Low temperature distillation through hydrophobic membranes, Sep. Sci. Technol. 22 (1987) 855–872.
[42] R.B. Bird, W.E. Stewart, E.N. Lightfoot, Transport Phenomena, second ed., Wiley, New York, 2002.
[43] R.K. MacGregor, A.P. Emery, Free convection through vertical plane layers: Moderate and high Prandtl number fluids, J. Heat Transfer 91 (1969) 391–403.
[44] A.F. Mills, Basic Heat and Mass Transfer, second ed., Prentice Hall, New Jersey, 1999.
[45] F.A. Banat, F. Abu Al-Rub, R. Jumah, M. Al-Shannag, Application of Stefan-Maxwell approach to azeotropic separation by membrane distillation, Chem. Eng. J. 73 (1999) 71–75.
[46] F.A. Banat, F. Abu Al-Rub, M. Shannag, Modeling of dilute ethanol-water mixture separation by membrane distillation, Sep. Pur. Tech. 16 (1999) 119–131.
[47] F.A. Banat, F. Abu Al-Rub, R. Jumah, M. Shannag, Theoretical investigation of membrane distillation role in breaking the formic acid water azeotropic point: Comparison between Fickian and Stefan-Maxwell-Based models, Int. Comm. Heat Mass Transfer 26 (1999) 879–888.

[48] C.A. Rivier, M.C. García-Payo, I.W. Marison, U. von Stockar, Separation of binary mixtures by thermostatic sweeping gas membrane distillation: I. Theory and simulations, J. Membr. Sci. 201 (2002) 1–16.
[49] M.C. García-Payo, C.A. Rivier, I.W. Marison, U. von Stockar, Separation of binary mixtures by thermostatic sweeping gas membrane distillation: II. Experimental results with aqueous formic acid solutions, J. Membr. Sci. 198 (2002) 197–210.
[50] R. Taylor, R. Krishna, Multicomponent mass transfer, Wiley, New York, 1993.
[51] R. Krishna, G.L. Standart, A multicomponent film model incorporating an exact matrix method of solution to the Maxwell–Stefan equations, AIChE J. 22 (1976) 383–389.
[52] R. Krishna, C.B. Panchal, Condensation of a binary vapor mixture in the presence of an inert gas, Chem. Eng. Sci. 32 (1977) 741–745.
[53] R. Krishna, G.L. Standart, Mass and energy transfer in multicomponent systems, Chem. Eng. Commun. 3 (1979) 201–275.
[54] R. Taylor, On exact solutions of the Maxwell–Stefan equations for the multicomponent film model, Chem. Eng. Commun. 10 (1981) 61–76.
[55] R. Taylor, More on exact solutions of the Maxwell–Stefan equations for the multicomponent film model, Chem. Eng. Commun. 14 (1982) 361–362.
[56] J.A. Wesselingh, R. Krishna, Mass transfer, Ellis Horwood, Chichester, 1990.
[57] K.W. Lawson, D.R. Lloyd, Review: Membrane distillation, J. Membr. Sci. 124 (1997) 1–25.
[58] A.M. Alklaibi, N. Lior, Transport analysis of air gap membrane distillation, J. Membr. Sci. 255 (2005) 239–253.
[59] S. Bouguecha, R. Chouikh, M. Dhahbi, Numerical study of the coupled heat and mass transfer in membrane distillation, Desalination 152 (2002) 245–252.
[60] R. Chouikh, S. Bouguecha, M. Dhahbi, Modeling of a modified air gap distillation membrane for the desalination of seawater, Desalination 181 (2005) 257–265.
[61] H. Chang, J.S. Liau, C.D. Ho, W.H. Wang, Simulation of membrane distillation modules for desalination by developing user's model on Aspen Plus platform, Desalination 249 (2009) 380–387.
[62] C. Zhu, G.L. Liu, Modeling of ultrasonic enhancement on membrane distillation, J. Membr. Sci. 176 (2000) 31–41.
[63] Bier C, Plantikow U, Solar powered desalination by membrane distillation, IDA World Congress on Desalination and Water Science, Abu Dhabi, 1995, pp. 397–410.

[64] F. Banat, N. Jwaied, Economic evaluation of desalination by small-scale autonomous solar-powered membrane distillation units, Desalination 220 (2008) 566–573.

[65] J. Walton, H. Lu, C. Turner, S. Solis, H. Hein, Solar and waste heat desalination by membrane distillation, Desalination and water purification research and development program Report No. 81, College of Engineering, University of Texas at El Paso, 2004.

[66] H. Gooding, F.J. Bahouth, Membrane-aided distillation of azeotropic solutions, Chem. Eng. Commun. 35 (1985) 267–279.

[67] F. Abu Al-Rub, J. Akili, R. Datta, Distillation of binary mixtures with capillary porous plates, Sep. Sci. Technol. 33 (1998) 10–15.

[68] G.C. Yeh, B.V. Yeh, B.J. Ratigan, Separation of liquid mixtures by capillary distillation, Desalination 81 (1991) 129–160.

[69] D. Fullarton, E.U. Schlunder, Diffusion distillation — a new separation process for azeotropic mixtures, Chem. Eng. Process 20 (1986) 255–263.

[70] A.M. Alklaibi, N. Lior, Heat and mass transfer resistance analysis of membrane distillation, J. Membr. Sci. 282 (2006) 362–369.

CHAPTER

14

Membrane Distillation Hybrid Systems

OUTLINE

Introduction	399
Hybrid Systems with Membrane Processes	401
Ultrafiltration and Membrane Distillation	401
Nanofiltration, Reverse Osmosis and Membrane Distillation	402
Direct Osmosis and Membrane Distillation	405
Pervaporation, Microfiltration and Membrane Distillation	406
Hybrid Systems with other Processes	407
Precipitation and Membrane Distillation	407
Fermentation and Membrane Distillation	411
Catalysis and Membrane Distillation	412
Traditional Distillation and Membrane Distillation	413
Nuclear-Powered MD	414
Cooling Tower and Membrane Distillation	415
Diesel Waste Heat and Membrane Distillation	416
Alternative Energy Sources and MD Applications	418
Solar Thermal Collectors, Photovoltaic Panels and Membrane Distillation	418
Membrane Modules Integrating Solar Absorbers	421
Salt-Gradient Solar Ponds and Membrane Distillation	423
Geothermal Energy and Membrane Distillation	425

INTRODUCTION

In general, a hybrid process refers to a combination of different unit operations that are interlinked with each other and optimized in order to achieve a predefined task. The operation units can perform the same or a different function. In this case, one of the operation units is membrane distillation (MD) whereas the others may be a membrane process (reverse osmosis, RO; nanofiltration, NF; ultrafiltration, UF; microfiltration, MF; bioreactor, etc.) or a conventional separation process (distillation, precipitation, crystallization, photocatalysis, etc.) or a combination between them. Moreover, MD-integrated systems, alternative energy sources combined with MD, multistaged unit operations composed of cascades of different MD configuration units and heat recovery systems can be included here. The MD-based systems are named hereafter MD hybrid. The main objectives are to make the whole innovative installation to have superior efficiency, higher productivity and quality, lower energy consumption, moderate cost to performance ratio, lower environmental emissions, compact and modular in design, more stable and safer than each individual operation unit working alone. Until now the MD hybrid applications are still under optimization at a theoretical or laboratory levels. Special care should be taken because hybrid processes might not be economical at small and laboratory scale.

It is interesting to point out that membrane technology plays an important role in hybrid processes because of the possible flexibility and modular design. For example, operating conditions can be adjusted to compensate possible changes of the membrane with time or possible changes in the feed aqueous solutions in order to maintain high efficiency of the whole system. The advantages of hybrid membrane processes were recognized in the beginning of the 1990s [1]. Various pervaporation (PV) hybrid systems have been designed and applied not only at laboratory scale but also at industrial scale [2—5]. In comparison, very few have been carried out including MD technology benefiting from the low temperature and transmembrane hydrostatic pressure required to perform MD operations, together with the high quality of produced water especially when treating aqueous solutions containing non-volatile solutes [6—20,22,23]. Moreover, the possibility of using industrial waste heat such as nuclear-powered based systems and/or alternative energy sources, such as solar and geothermal based energy, enables MD technique to possible cooperation with other processes in integrated systems making it a more promising separation technique for an industrial scale.

The development of hybrid MD systems may be considered one of the most important advances of MD applications for water treatment in general and desalination in particular. The possibility to combine MD with other systems for water treatment offers various advantages such as the increase of water quality and efficiency of the whole installation.

Optimization of the MD process consists of maximizing the permeate flux together with the salts rejection factor and minimizing the heat losses within the MD system. Several strategies for energy saving may be made in MD systems, based on the use of well-designed MD membranes and modules with low-temperature polarization effects, minimum heat losses through the membrane to the environment, multiple effect operation and heat recovery in heat exchangers. Use of heat recovery systems is an essential part of any efficient MD plant. This means an increase of capital investment costs but benefiting the operating costs.

MD has a potential to be integrated in pressure-driven separation plants as schematized in Fig. 14.1. The idea is to use MD to further concentrate the discharged brines, which may allow a high recovery integrated process leading to an enhancement in the process productivity by increasing the overall water recovery and, at the same time, reducing the environmental impact due to the brine disposal. In general, the retentate volumes of the pressure-driven separation processes are significant and thus represent a loss of valuable water resources and a major disposal challenge.

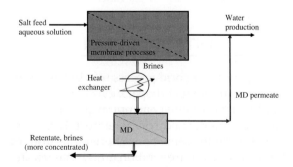

FIGURE 14.1 Example of pressure-driven (RO, NF, UF, MF) MD-integrated systems.

It is worth quoting that special care must be made when coupling MD process with pressure-driven membrane processes using their brines especially because water recoveries are limited by precipitation of solutes such as inorganic salts on the membrane surface leading to fouling and/or scaling and a decline of MD performance as well as wetting of the membrane pores. Cleaning strategies should be taken into consideration in order to restore the initial MD performance. Furthermore, complex wastewaters that cannot be treated by one process only could be treated effectively using a well-designed MD hybrid system.

A number of interesting possibilities on MD hybrid systems have been suggested, for which the technical and economical feasibility seem promising.

HYBRID SYSTEMS WITH MEMBRANE PROCESSES

Ultrafiltration and Membrane Distillation

Oily wastewater generated by various industries and subsequently discharged into the natural environment creates a major ecological problem through the world. The traditional methods for the separation of oil emulsion can be classified as chemical, mechanical and thermal. The chemical methods based on the neutralization of detergents need further purification in order to meet today's effluent standard for the sewage systems. The mechanical methods based on the phenomenon of gravitational emulsion breaking require heating step that may increase the size of oil droplet significantly. The thermal process requires a large amount of energy, and therefore this process is not cost effective. One of the recently utilized solutions in the treatment of oily wastewater is the biological method. However, this method has disadvantages such as low efficiency, operational difficulties and high operational costs.

A hybrid system combining UF and MD processes was proposed by Gryta et al. [6] for the treatment of oily wastewater. Figure 14.2 is a schematic diagram of the hybrid UF–MD process, in which direct contact membrane distillation (DCMD) is applied as a final purification method. Wastewater was treated first by UF and the resulting UF permeate was further purified by means of DCMD. The UF permeate is heated in a heat exchanger prior to entering into the MD modules connected in parallel mode. The distillate is collected outside the MD modules and the oil concentrate is recycled to the UF modules as feed. A tubular UF module equipped with polyvinylidene fluoride (PVDF, type FP100, 1.25 cm inner diameter, 1.2 m length and 0.9 m^2 membrane area, 100 kDa dextran molecular weight cut-off, PCI Membrane System) and a shell and tube MD module with capillary polypropylene (PP) membranes (1.8 mm inner diameter, 2.6 mm outer diameter, 0.2 μm nominal pore size, 73% porosity and 220 cm^2 effective membrane area) were tested using a typical bilge water (oil content 124–360 ppm) collected from the Szczecin harbour (Poland) without pretreatment. The permeate obtained from the UF process generally contains less than 5 ppm of oil. A further purification of the UF permeate by MD results

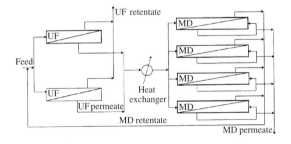

FIGURE 14.2 A schematic diagram of a UF/MD hybrid process for treatment of oily wastewater. *Source: Reprinted from [6]. Copyright 2001, with kind permission from Elsevier.*

in a complete removal of oil from wastewater with a very high reduction of the total organic carbon (99.5%) and total dissolved solids (TDSs) (99.9%).

The performance of the DCMD unit was tested by using 1 wt% NaCl solution or the UF permeate (oil content, 3–10 ppm) as feed. In the MD module, the feed circulated through the lumen side of the capillary while the permeate circulated through the inter-tubular space in a countercurrent mode. The feed inlet temperature was changed in a range of 40–80 °C and the permeate inlet temperature was controlled to a fixed temperature of 20 °C. The total pressure of the distillate was maintained at 10^3 Pa higher than that of the feed.

The UF study was carried out at two levels of oil contents, 124 and 360 ppm. For 124 ppm feed the initial flux was 1300 kg/m²·day, which decreased to 875 kg/m²·day after 50 h of operation. For 360 ppm feed the decrease in permeate flux was from 768 to 528 kg/m²·day. The quality of the obtained UF permeate is shown in Table 14.1.

An UF permeate with 3.2 ppm of oil was subjected to MD experiments. The MD permeate flux data at 20 °C was collected. As well, DCMD experiments were carried out for feed NaCl solution. The MD permeate flux increased with an increase in flow rate because the temperature and concentration polarization effects were reduced. The MD permeate flux for the oil feed was slightly lower than the feed aqueous NaCl solution. The oil rejected in DCMD process was accumulated in the vicinity of the membrane surface, which impeded the water evaporation and the decrease of the permeate flux to become severer as the oil concentration was increased. The quality of the produced water by the UF/MD hybrid system is also presented in Table 14.1. The integration of MD with UF permitted the complete removal of oil pollutants as well as the remaining soluble compounds in bilge water collected from the Szczecin harbour (Poland). Based on these experimental data, the flow sheet illustrated in Fig. 14.2 was proposed by Gryta et al. [6].

Nanofiltration, Reverse Osmosis and Membrane Distillation

Drioli et al. [7] integrated MD with RO process for water desalination following the procedure reported in Fig. 14.1. MD was proposed to treat RO brine with a concentration of 75 g/l at a temperature of 35 °C in order to enhance both efficiency and water recovery factor in seawater desalination (45 g/l at 25 °C). More fresh water can be produced because MD is much less sensitive to salt concentration than RO and the RO brine volume can be further reduced leading to lower environmental impact. For RO brine of 75 g/l, a feed temperature of 35 °C and a permeate

TABLE 14.1 Quality of UF permeate and UF/MD permeate

Water sample	Concentration (mg/l)			Rejection (%)		
	Oil	TOC[a]	TDS[b]	Oil	TOC	TDS
Bilge water	360	400	3790	-	-	-
UF permeate	4.9	8.6	3700	98.64	97.85	2.4
UF/MD permeate	0	1.8	1.4	100	99.55	99.96

[a] Total organic carbon.
[b] Total dissolved solids.
Source: Reprinted from [6]. Copyright 2001, with kind permission from Elsevier.

temperature of 25 °C, the permeate flux of the MD system was found to decrease from 2.4 to 1.4 kg/m$^2 \cdot$h corresponding to the saturation feed solution (5.33 M). The MD recovery factor was 77% and that of the RO process was 40% whereas for the RO/MD hybrid process the recovery factor was 87.6%. Moreover, costs analysis was performed taking into consideration $116/m^2 of the MD membrane. It was found that the RO/MD hybrid process produced more than twice as much water as the RO process alone at the same water cost, and the MD process alone produced as much water as the RO/MD hybrid process but at a water cost about 5% higher (i.e. $1.25/m^3 for RO alone with a production 0.391 time the feed flow rate, $1.32/m^3 for MD alone with a production 0.856 times the feed flow rate and $1.25/m^3 for RO/MD with a production 0.856 times the feed flow rate).

Karakulski et al. [8] proposed the application of the MD process in combination with NF and RO as well as UF in order to increase the quality of treated tap water (Szczecin, Poland), which was subjected to seasonal fluctuations. The content of solutes in this tap water was 415–430 ppm TDSs, in which 25 to 30 ppm of inorganic carbon and 5 to 10 ppm of organic compounds were detected by TOC analysis (i.e. total organic carbon). The concentrations of major elements were 28.8, 60.5, 17.1, 6.2 and 1.7 ppm for sodium (Na), calcium (Ca), magnesium (Mg), potassium (K) and silicon (Si), respectively. Its electrical conductivity was ranging from 600 and 700 µS/cm.

Different combinations have been considered on pilot plants equipped with UF, NF, RO and MD modules. The UF and NF modules were tubular (B1, PCI, equipped with FP100 membrane for UF and AFC30 membrane for NF), whereas RO modules were a spiral wound module (BW3040, Film-Tec) and a tubular module (B1, equipped with AFC99 membrane). The feed pressures were in the range of 5×10^5 Pa to 25×10^5 Pa. Shell and tube membrane modules (MD020CP2 N, Table 2.2) were used for DCMD experiments. The inlet temperatures of the feed and distillate were 77 and 20 °C, respectively.

The UF process was found to be efficient only in the removal of suspended solid and colloids, which allowed to reduce Silt Density Index from 7 to 10 of tap water to 2. Hence, the product water from UF unit could be directly used in the RO process. The quality of the produced water from UF/RO-integrated membrane system (UF/RO hybrid process) corresponds to the quality of distilled water.

The NF process was found to be very efficient in the rejection of organic matter from water (0.8–2.1 ppm TOC). The integration of UF with NF process (UF/NF hybrid process) increased the effectiveness of removal of organic compounds.

Demineralized water could be produced by DCMD applied directly to tap water. However, it caused precipitation of calcium carbonate ($CaCO_3$) onto the membrane surface, resulting in rapid decrease of the DCMD permeate flux. The rinsing of the module with 2% hydrochloric acid (HCl) allowed to clean the membrane and to restore its initial permeability. The preliminary softening of tap water with the NF process allowed to reduce fouling but still the permeate flux decline continued during the first 80 h of DCMD operation (see Fig. 14.3). This is because of the scale formation by carbonate that was permeated through NF membrane. Acidification of NF permeate with HCl to pH = 5 removed the carbonate deposit on the membrane surface and the permeate flux increased. When the RO process was used for the pretreatment of the tap water, fouling due to carbonate deposition was avoided. Therefore, the combination of RO and MD system produced the best results.

MD has a potential to be integrated in large seawater RO desalination plants to be operated in the brine stream as shown in Fig. 14.1. The

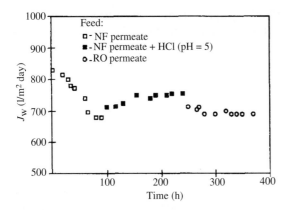

FIGURE 14.3 Variation of MD permeate flux (J_w) as a function of time for different feed solutions (NF permeate, NF permeate with HCl at pH 5 and RO permeate). *Source: Reprinted from [8]. Copyright 2002, with kind permission from Elsevier.*

idea is to use MD to further concentrate RO brines, which allow a high recovery integrated process for seawater desalination.

Recently, Mericq et al. [9] reported on the integration of vacuum membrane distillation (VMD) to RO technology for seawater desalination (see Fig. 14.1). RO brines from a plant installed in the Mediterranean sea (total concentration of 50 g/l, electrical conductivity of 47.4–51.4 mS/cm and 1.3–1.6 ppm TOC) was used as feed for the VMD unit containing a plate and frame membrane module of polytetrafluoroethylene (PTFE) membrane (0.22 μm nominal pore size, 0.175 mm thickness, 40% porosity, 4.2×10^5 Pa liquid entry pressure, 3.5 cm width, 16.5 cm length, 5.78×10^{-3} m² effective membrane area, Fluoropore supplied by Millipore). A 5-day test was conducted at a feed temperature of 51.5 °C with a Reynolds number (Re) of 3900 and a downstream pressure of 9 kPa for the first 2 days and 7.5 kPa for the other days. The obtained initial VMD permeate flux was 13.7 l/m²·h with an electrical conductivity of 40 μS/cm. After the first two days of operation, a decline of the VMD permeate flux was observed and then it was maintained constant for the rest of the days. It was reported that no scaling, organic fouling or biofouling effects were detected. Only very few crystal deposits (calcium sulphate, $CaSO_4$) were observed on the membrane surface but did not cover the membrane pores [9].

Simulations on the coupling of VMD with a 40,000 m³/day RO plant (40% recovery and 38.9 g/l seawater feed) were performed for optimization of the operating conditions. The considered VMD unit further concentrated the RO brine of 64.8 g/l up to 300 g/l under a feed temperature of 50 °C, Re number of 4000 and a downstream pressure of 6 kPa. High recovery factors up to 89% were obtained with a brine volume reduction by a factor of 5.5, an increase in the production by a factor more than 2 and a concentration of brine 7.6 times higher than that of seawater.

DCMD configuration was applied for the concentration of the primary RO retentate by Qu et al. [10] in order to enhance water recovery. PVDF hollow fibre membranes (i.e. 0.6 mm inner diameter, 1 mm inner diameter, 0.3 μm average pore size, 80% porosity and 94.2 cm² effective membrane area) assembled in a polyester shell and tube module (i.e. 20 mm outer diameter, 15 mm inner diameter and 100 mm effective module length) were applied in the DCMD process. The used RO retentate had an electrical conductivity varying from 980 to 1200 μS/cm, an average pH of 7.9, a TOC smaller than 1 ppm and the average concentrations of the major ions were 376 ppm Ca^{2+}, 203 ppm Mg^{2+}, 56.75 ppm Cl^-, 11.22 SiO_3^-, 9.25 mmol/l HCO_3^- and 0.35 mmol/l CO_3^{2-}. The feed temperature was 50 °C, the permeate temperature was 20 °C and the feed flow rate was 0.6 m/s. An initial permeate flux of about 11.4 kg/m²·h was measured and then reduced to about 7.5 kg/m²·h after 50 h DCMD operation with a decrease of the electrical conductivity from 5 to 2.5 μS/cm. The decrease of the permeate flux was associated to $CaCO_3$ deposition and by rinsing the membrane module with 2% HCl the permeate flux was restored. Subsequently, after acidification and adjustment of pH value to 6, the observed decline of the permeate flux was only 20% after 200 h of DCMD operation with an electrical conductivity of the produced water around 3.5 µS/cm. The RO retentate was concentrated 40 times with an enhancement of the water recovery factor up to 98.8% compared with 50% observed for the RO process alone using tap water as feed. $CaSO_4$ crystallization was observed because Ca^{2+} and SO_4^{2-} exceeded the solubility limit of $CaSO_4$ for the achieved high levels of water recovery. The scaling problem was eliminated after adjusting the pH to 4.

A hybrid MD system composed of NF as a pretreatment step, RO and MD has been suggested for water desalination by El-Zanati and El-Khatib [11]. In the proposed flow diagram, NF and RO units were combined together to form the feed solution of the MD unit, which is VMD configuration; while the permeate of both RO and MD processes were mixed giving the last product. Only simulations have been carried out and it was found that the water recovery would increase from 30–35% to 76.2% and the water production cost would be about $0.92/m^3·day, which is competitive to potable water produced by seawater RO systems.

MD was also investigated for water recovery enhancement in desalination of brackish water by Martinetti et al. [12] using concentrated RO brines (i.e. Brine 1: concentrate of primary RO process with 7500 mg/l TDS. Brine 2: secondary RO concentrate, which is the primary RO brine treated by the following different steps: first by sodium hydroxide then clarification and filtration, and finally the filtrate was treated by sulphuric acid and by the scale inhibitor Vitec 3000, Avista Technologies, Inc., with 17500 mg/l TDS). To avoid precipitation of sparingly soluble salts in the primary RO systems, water recovery was limited to 70%. Vacuum-enhanced DCMD configuration was considered using PTFE membrane supported on PP mesh (0.2 µm nominal pore size, GE Osmonics, Minneonka, MN, USA) and the pressure in the permeate membrane side was controlled using a needle valve at the inlet of the permeate membrane module. The lowest pressure applied on the permeate side was 360 mmHg, whereas the highest feed temperature was 60 °C with a permeate temperature of 20 °C. When using the RO Brine 1, initial permeate fluxes as high as $40 l/m^2·h$ were obtained and were maintained relatively constant up to a concentration factor (i.e. ratio between the concentration of the feed solution at any time and the initial feed concentration) of approximately 1.75, after which a rapid decline of the permeate fluxes was observed due to the precipitation of silica (SiO_2) and calcium sulphate ($CaSO_4$). Salt rejection factors were greater than 99.9%. When using RO Brine 2, much higher concentration factors and water recoveries were achieved than when using RO Brine 1. This was attributed to the residual scale inhibitor, which was used to reduce the formation of $CaSO_4$ and SiO_2 during the secondary RO treatment (RO Brine 2). The use of scale inhibitor increased the water recovery. This was greater than 89% for Brine 1 and greater than 98% for Brine 2 compared with those obtained for the RO process alone, which were 70% for brine 1 and 89% for Brine 2 [12].

Direct Osmosis and Membrane Distillation

Direct osmosis (DO) is an isothermal process that refers to the diffusion of water through a semipermeable membrane (i.e. dense and water selective membrane, hydrophilic membrane such as cellulose triacetate) from a higher water concentration side to the osmotic agent side. In contrast, osmotic distillation (OD) process is also an isothermal process but a porous and hydrophobic membrane is used and the evaporated molecules at the liquid/vapour interface at each membrane pore entrance are transported in vapour phase through the membrane to the liquid osmotic agent on the permeate membrane side (more details on OD may be found in Chapter 10).

In both DO and OD, the driving force is the concentration difference across the membrane. This is established using an osmotic agent. On the other hand, membrane osmotic distillation (MOD) is a combination of DCMD and OD in which the driving force is both the concentration and the temperature difference across the membrane. In MOD process an osmotic agent is also used on the permeate side.

MD and MOD were combined with the DO process for the pretreatment of wastewater (hygiene wastewater from NASA, Ecolab Inc., mixed with humidity condensate and urine) before treatment by the RO process. The objective is to increase the permeate flux and the solute rejection factor of the combined system [13]. It was reported that the permeate fluxes of the MD hybrid systems, both DO/MD and DO/MOD, were improved compared with those of the DO/OD combination and complete urea removal was achieved. The permeate flux of the DO/MD process was found to be 4–20 times greater than that of the DO/OD process and no fouling was detected over a period of 15 days operation. Better permeate flux enhancements were obtained for the DO/MOD process, 8–25 times higher than that of the DO/OD process because of the additional driving force induced by the osmotic agent.

Pervaporation, Microfiltration and Membrane Distillation

Johnson et al. [14] investigated a three-stage hybrid PV/MF/MOD process for the concentration of ethanol/water extracts of the *Echinacea* plant. Both the osmotic pressure of the strip solution and the temperature difference between the feed and the strip solution in the permeate are the driving force for water permeation through the membrane. Figure 14.4 shows a schematic diagram of the PV/MF/MOD hybrid process. PV retentate was subjected to MF and MF permeate was subjected to MOD.

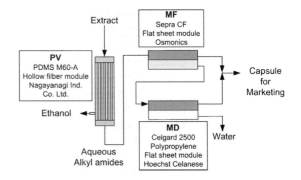

FIGURE 14.4 A schematic diagram of PV-MF-MOD hybrid process. *Source: Reprinted from [4]. Copyright 2006, with kind permission from Taylor & Francis.*

It is worth to indicate here that the most commonly used form of *Echinacea* herbal medicine is the tincture, which is a liquid ethanol/water extract from *Echinacea purpurea*, the purple coneflower. This has been widely used as an herbal medicine because of its perceived properties as an immunostimulant when ingested. Ethanol should be totally or partially removed from the tincture to satisfy government regulations. In the case of tincture preparation, the extraction has typically been carried out using finely chopped plant and solvent (e.g. 45% ethanol–55% water by volume) over a period of several days at ambient temperature. Following extraction of the active components, ethanol removal has traditionally been carried out by thermal evaporation under reduced pressure. However, the latter step caused degradation of thermally labile compounds and evaporation of volatile compounds.

Preliminary experiments using MOD in a single-stage process resulted in severe membrane pore wetting by the compounds soluble in ethanol. In order to avoid this problem, the following three stages of the hybrid process were proposed:

(i) Stage 1: Ethanol removal from the neat extract was achieved by PV. This gave an ethanol-free aqueous product containing suspended alkyl amides that were suitable for marketing in tincture form.

(ii) Stage 2: The precipitated alkyl amides were removed from the Stage 1 product by MF.

(iii) Stage 3: The MF permeate was concentrated several-fold by MOD, followed by adding-back of the MF retentate containing the precipitated alkyl amides to the MOD retentate. This gave a highly concentrated product suitable for marketing in capsule form.

PV was carried out using the poly(dimethylsiloxane) M60-A (Nagayanagi Ind. Co. Ltd.) hollow fibre module with a membrane area of $0.34\,m^2$. Both MF and MOD were carried out using the flat sheet cell (Osmonics Sepra CF) with an effective membrane area of $0.0155\,m^2$.

A 1:2 E. purpurea extract was supplied by New Products Development Pty Ltd. The extract solvent was a mixture of 45% (by volume) ethanol in water mixture. The concentration of the extract was 20°Brix. The extract was continuously pumped through the shell side of the module (i.e. 14 cm in length with an inner diameter of 200 µm and a wall thickness of 60 µm) and recycled back to the feed reservoir. Dry compressed air at 24 °C passed through the lumen side of the hollow fibre to collect the permeate ethanol. The PV experiment was stopped when the solute concentration became 2.7 times as high as the initial solute concentration in feed.

Both feed and permeate channel of the MF flat sheet cell were 0.86 mm in height, 14 cm in length and 9.5 cm in width. The used membrane is a polysulfone membrane (DESAL EW 500) with a nominal pore size of 0.04 µm. The PV retentate was recirculated through the separation cell under a pressure of 150 kPa and at a temperature of 24 °C. The MF experiment was stopped when the permeate to retentate ratio became 9:1.

The used membrane in MOD was a PP flat sheet membrane (Hoechst Celanese Celgard 2500) with a nominal rectangular pore size of (0.05 µm × 0.19 µm), a thickness of 22.9–27.9 µm and a porosity of 0.37–0.48. Saturated calcium chloride ($CaCl_2$) solution (40 wt%) was used as the strip solution. The $CaCl_2$ concentration was maintained by a special devise despite possible dilution of the solution with permeate water. Both the feed and strip solution were pumped through their respective channel in the membrane module at a flow rate of 600 ml/min. The feed temperature was maintained at 40 °C, while the strip solution temperature was maintained at 24 °C.

After conducting PV experiments under different operating conditions, the optimum PV conditions were found to be a feed temperature of 40 °C, a sweep gas flow rate of 3.0 l/min and a feed flow rate of 800 ml/min. In the third stage PV/MF/MOD hybrid process shown in Fig. 14.4, the above PV conditions were applied in the first PV stage. A concentration factor of approximately 2.7 was readily achievable by PV. It was shown that all ethanol had been removed at a concentration factor of 2.02. After the second stage, MF permeate was concentrated by MOD in the third stage until a concentration factor of 4.2 was achieved. This corresponds to a concentration of 57°Brix. The process was carried out for 19 h, during which period no membrane pore wetting occurred. The MOD retentate was then combined with MF retentate to produce a concentrated *Echinacea* extract containing the precipitated material that was suitable for packaging in capsule form.

HYBRID SYSTEMS WITH OTHER PROCESSES

Precipitation and Membrane Distillation

Each year a large amount of RO concentrate discharges leading to a significant loss of water resource and disposal challenge. As a consequence, minimizing the amount of concentrate discharge has become essential for many countries to alleviate this environmental challenge. Qu et al. [15] have integrated DCMD process with accelerated precipitation softening (APS) for high-recovery desalting of primary reverse osmosis (PRO) concentrate. The integrated process involved mineral precipitation by sodium hydroxide, followed by solid–liquid separation, MF and DCMD desalination.

PRO concentrate in the experiments was obtained from the RO unit of the direct drinking water preparation system designed for the 29th Olympic Games in Beijing. The RO system was operated at 50% recovery from the Beijing tap water pretreated with ozone oxidation, catalytic oxidation and active carbon filtration. The PRO concentrate contained 470.7 ppm of total hardness as $CaCO_3$ and small amounts of other minerals. The laboratory scale APS/DCMD set-up is illustrated schematically in Fig. 14.5. The APS treatment was conducted in a 10 l crystallizer. The initial pH was adjusted to 10.1, the calcite dosage used to promote precipitation was 5 g/l and the agitating speed was 200 rpm. After 30 min of precipitation, the suspended solids were allowed to settle for 1 h and then the supernatant was filtered through a 2 μm Nylon cartridge filter before being stored in the feed reservoir of the DCMD system. PVDF hollow fibre membranes (i.e. 0.6 mm inner diameter, 1 mm inner diameter, 0.3 μm average pore size, 80% porosity, 250 kPa liquid entry pressure of water and 94.2 cm² effective membrane area) assembled in a polyester shell and tube module (i.e. 20 mm outer diameter, 15 mm inner diameter and 100 mm effective module length) were applied in the DCMD process. The feed solution and the permeate solution circulated through the MD module co-currently, feed through the lumen side and permeate through the shell side. The feed temperature was 50 °C and the permeate temperature was 20 °C.

In order to evaluate the optimal APS operating parameters, initial small scale calcium removal tests were carried out at 25 °C in a 1 l

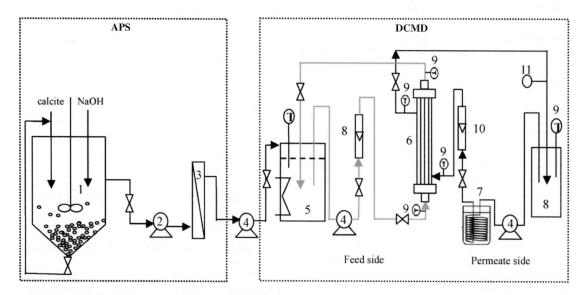

FIGURE 14.5 APS/DCMD process set-up. (1) crystallization reactor, (2) magnetic pump, (3) cartridge filter, (4) magnetic pump, (5) feed reservoir, (6) membrane module, (7) cooling coil, (8) permeate reservoir, (9) thermometer, (10) flow meter, (11) conductivity monitor. *Source: Reprinted from [15]. Copyright 2009, with kind permission from Elsevier.*

beaker with an agitator by following the calcium depletion rate. To aid rapid precipitation, a dose of 1 mol/l sodium hydroxide stock solution was added to the PRO concentrate to achieve the desired initial pH level. Calcite or quartz sand seeds were dispersed in the PRO concentrate prior to initiation of precipitation. Calcite seed load of 5 g/l and pH of 10.1 were found to be the best condition for the calcium removal. About 92% of calcium was removed at this condition within 15 min.

Figure 14.6 shows the results of DCMD experiments with PRO concentrate without (stage I) and with (stage II) APS treatment. In stage I the permeate flux decreases quickly from the initial value. The DCMD permeate flux however goes back to the above initial value after rinsing. When the module was subjected to stage II, DCMD experiment after rinsing, the permeate flux decline was much slower. Even after more than 50 h running period, permeate flux of 7.53 kg/m^2·h could be maintained. The rapid flux decline in the initial stage was attributed to be the $CaCO_3$ deposition at the hollow fibre lumen inlet. The permeate conductivity was maintained below 4 µS/cm during the first 300 h of stage II operation. In the last hours it went up to 6 µS/cm. During the first 300 h of operation, PRO retentate was concentrated effectively up to 40 times of the initial concentration and 98% of the mineral was recovered.

Liquid-phase precipitation (LPP) was combined with MD (LPP/MD) for the treatment of liquid waste stream containing small amounts of Transuranic (TRU) elements [16]. These are present in Idaho National Engineering and Environmental Laboratory (INEEL) tanks wastes. INEEL was established in the early 1950s to store and reprocess spent nuclear fuel for the recovery of uranium-235. The INEEL tanks waste contains predominantly nitric acid, nitrate and sodium, significant amounts of aluminum (or zirconium), potassium, appreciable amounts of sulphates, phosphates, chlorides, toxic metals and small amounts of TRU elements (plutonium, neptunium, americium and curium) and their fission products (plutonium, neptunium, americium and barium). The INEEL liquid waste was divided into high-activity waste and sodium-bearing liquid waste. The latter waste is still remaining in storage tanks. LPP/MD was proposed to potentially treat the INEEL sodium-bearing liquid waste shown in Table 14.2.

LPP experiments were first conducted by adding different amounts of isopropyl amine (IPA, 0.5, 1.0, 2.0, 4.0, 8.0 and 16.0 ml) to 20 ml of aqueous sodium-bearing liquid waste in order to precipitate minerals. The precipitate was then separated by filtration. The LPP experiments conducted on aluminum-nitrate system (aluminum, 16,998 mg/l and nitrate 117, 180 mg/l) showed that almost all aluminum was precipitated while only 45% of nitrate was removed when the IPA to water ratio was 0.8. Data were obtained also for other cations and anions.

Figure 14.7 shows the flow diagram of the proposed hybrid LPP/MD system. There are four main processing stages in the system. The first stage targets the removal of volatile nitric acid from the bulk of inorganic species by VMD. The second stage targets the separation of polyvalent cations (TRU elements, fission products, aluminum, chromium, phosphate

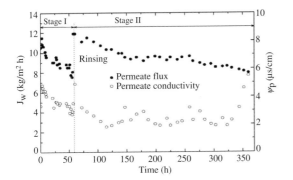

FIGURE 14.6 Change of the permeate flux (J_w) and permeate conductivity (ψ_p) with time. *Source: Reprinted from [15]. Copyright 2009, with kind permission from Elsevier.*

TABLE 14.2 Components of INEEL sodium-bearing liquid waste

Ions	Concentrations (mg/l)
Cations	
Na	41,841
K	7,429
Cs	3.84
Ca	1,844
Sr	1.18
Ba	7.76
Hg	320.9
Pb	414.4
Mn	714.2
Ni	117.4
Al	16,998
Cr	208.0
Fe	1,228.6
Eu	1,063.7
Zr	547.3
Anions	
F	1,273
Cl	1,205
NO_3	367,069
PO_4	1,519.5
SO_4	4,898.9
Salts	
$NaNO_3$	134,709.2
KNO_3	19,209
$CsNO_3$	146.2
$Ca(NO_3)_2\text{-}4H_2O$	10,862.9
$Sr(NO_3)_2$	241.3
$Ba(NO_3)_2$	190.8
$Mn(NO_3)_2\text{-}xH_2O$	2,326.4
$Ni(NO_3)_2\text{-}6H_2O$	581.6
$Al(NO_3)_3\text{-}9H_2O$	236,344.5
$Cr(NO_3)_3$	1,600.6
$Fe(NO_3)_3 9H_2O$	8,887.6
$Eu(NO_3)_3 5H_2O$	2,996.3
$Na_3PO_4 12H_2O$	4,054.4
Na_2SO_4	7,242
NaF	2,813.3
NaCl	1,987

Source: Reprinted from [16]. Copyright 2005, with kind permission from Elsevier.

and fluoride) from the predominant bulk of monovalent anions and cations (mainly sodium nitrate) by front-end LPP. Polyvalent cations (the precipitate) are subjected to vitrification. The filtrate is subjected to MD to concentrate the anions and cations, which are further precipitated by rear-end LPP. At this stage the MD configuration was not specified.

The author [16] reported various advantages of the LPP/MD hybrid system such as the use of small amount of additives and energy, treatment of multiple critical waste species, extraction of an economic value from some waste species, generation of minimal waste with suitable disposal paths and rapid deployment. Moreover, the author indicated that no effective or economic alternatives were available.

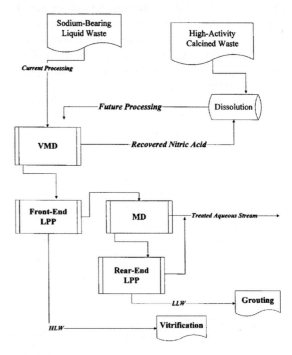

FIGURE 14.7 Flow sheet of LPP/MD hybrid system to treat the INNEL acidic waste. *Source: Reprinted from [16]. Copyright 2005, with kind permission from Elsevier.*

However, no data were presented from the operation of the hybrid system.

Fermentation and Membrane Distillation

Gryta [17] investigated ethanol production in a bioreactor integrated with DCMD system. The fermentation of sugar with *Saccharomyces cerevisiae* proceeds with the formation of by-products, which tend to inhibit the yeast productivity. The removal of by-products from the fermentation broth by MD process increased the efficiency and the rate of sugar conversion to ethanol. The separation of alcohol by MD enables achievement of a higher content of ethanol in the permeate than that in the broth. A beneficial effect of carbon dioxide (CO_2) on ethanol transport through the membrane was also observed. Schema of the experimental set-up of MD/fermentation hybrid process is shown in Fig. 14.8. The membrane module (MD020CP2 N, Table 2.2) was used. The bioreactor (1 l working volume) was packed with rings to a height of 80 cm. The fermentation broth was circulated through a heat exchanger and the DCMD module. The fermentation solution was prepared by dissolving 100 g of sugar in 1 l tap water into which 0.2 g of triammonium phosphate was added as a source of nitrogen and phosphorous for yeast. Subsequently, 5, 10 or 20 g of yeast was introduced to the solution. Commercially available dry distillery yeast was used as microorganism. Rehydration was run for 30 min, while the broth was periodically agitated. The fermentation process was carried out for several hours in continuous mode. The distillate cooled by a heat exchanger to 20 °C went into the lumen side of the membrane capillary. The fermentation broth coming out of the bioreactor at 34 °C was heated by another heat exchanger to 40 °C before being recycled to the bioreactor. The volume and the sugar concentration of the fermentation broth were kept constant at 1 l and 100 g/l. Bioreactor was also run without being connected to MD module.

The results of the fermentation experiments without and with integration of the DCMD process confirmed the advantages of the fermentation performed with continuous removal of fermented products by the membrane module.

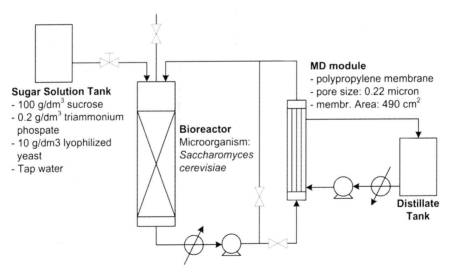

FIGURE 14.8 Schematics of the fermentation/MD hybrid experimental set-up. *Source: Reprinted from [4]. Copyright 2006, with kind permission from Taylor & Francis.*

Catalysis and Membrane Distillation

Coupling of photocatalytic membrane and DCMD for the degradation of organic contaminants in aqueous solutions has been proposed [18–20]. Titanium dioxide (TiO_2, Aeroxide® P25, Degussa, Germany) was used as a photocatalyst and different azo-dyes were employed as model contaminants in feed water (Acid Red 18, $C_{20}H_{11}N_2Na_3O_{10}S_3$; Acid Yellow 36, $C_{18}H_{14}N_3NaO_3S$; Direct Green 99, $C_{44}H_{28}N_{12}Na_4O_{14}S_4$). The photocatalyst and the feed aqueous solution were mixed in a feed tank illuminated with a mercury lamp emitting UVA light at a maximum wave length of 365 nm and an illumination intensity of 80 W/m² [20]. Before photodegradation of azo-dye, the aqueous mixture was stirred for 30 min in dark in order to allow adsorption of the dye molecules on the TiO_2 surface. Subsequently, the photocatalytic process was carried out for 5 h. Then, irradiation was carried out for a predetermined time and the feed solution was filtered through a 0.45 μm membrane and analysed. The membrane module (MD020CP2 N, Table 2.2) was used in DCMD and different process parameters were investigated. The tested feed temperature was increased up to 70 °C whereas the permeate temperature was maintained at 20 °C.

It is worth to mention that photodegradation rate depends on the initial dye concentration, photocatalyst loading and reaction temperature. Moreover, amongst the by-products of the photodegradation of azo-dyes both aromatic and aliphatic compounds are present, and the aromatic intermediates include either amines or phenolic compounds; while the main aliphatic species are formic acid and acetic acid. Other organic acids such as oxalic, glycolic, glyoxylic and malonic acids were also detected.

The photocotalysis/MD hybrid process was carried out in a laboratory scale installation for 5 h for each feed solution. It was found that all dyes were effectively degraded at the highest reaction temperature of 60 °C, the initial dye concentration had a significant effect on the composition of the permeate and the DCMD permeate flux was independent on the concentration of the photocatalyst. For example, with a feed temperature of 70 °C, a permeate temperature of 20 °C and 0.5 g/l of TiO_2 in water, the obtained DCMD permeate flux was about 325 l/m².day. Moreover, the model dyes were removed completely producing a permeate with an electrical conductivity lower than 3 μS/cm and pH values above 5.2, regardless of the applied process parameters. Almost complete retention of TOC, TDS and inorganic ions ($N-NO_3^-, N-NO_2^-, N-NH_4^+, SO_4^{2-}$) was achieved. Only a small amount of by-products of azo-dyes photodegradation such as the volatile organic acids, formic and acetic acid, was transported through the membrane (i.e. TOC concentration in the permeate was very low in a range of 0.4–1 mg/l). Moreover, inorganic nitrogen was detected in the permeate mainly in the form of ammonia (gaseous NH_3) with a concentration lower than 0.23 mg/l. The product quality was found to be significantly better than the hybrid photocatalysis and pressure-driven membrane processes [18].

Various advantages of the hybrid photocatalysis/MD process over the hybrid photocatalysis/pressure-driven (MF, UF and NF) processes have been identified as:

(i) Photodecomposition rate is higher at a temperature of 60 °C than at a room temperature. This indicates that the application of MD is more beneficial than the pressure-driven processes.
(ii) Complete separation of the dye and other non-volatile compounds, thus the product is practically pure water.
(iii) A continuous degradation of the compounds present in the feed aqueous solution avoids the increase of the feed concentration although the obtained permeate is practically pure water.

(iv) Cooling of the feed solution due to UV source is not necessary and the heat emitted can be utilized for heating the feed solution.
(v) The permeate flux is independent on the concentration of the applied photocatalyst.
(vi) Absence of fouling.
(vii) During DCMD process, the reduction of the feed volume did not affect the photodegradation rate of the azo-dyes.
(viii) Degradation and separation could be performed simultaneously in the same place, which minimizes the size of the installation.

Despite all these advantages, the hybrid photocatalysis/MD process is still under development but the pressure-driven membrane processes MF, UF and NF are potentially applied than the DCMD process. It is worth quoting that combined photocatalysis with pressure-driven processes have been carried out but resulted in a significant decrease of the permeate flux especially for UF and MF as well as significant fouling effects were observed. Moreover, the membranes commonly used in UF and MF do not have the capacity to remove low molecular organic compounds present in wastewaters as well as the products and by-products of their photodegradation [18–20].

Traditional Distillation and Membrane Distillation

Traditional distillation methods include simple effect distillation (i.e. evaporation followed by condensation or solar still), multi-effect distillation (MED) and multistage flash (MSF). MED is the oldest technique applied for seawater desalination and it is based on heat transfer from the condensing steam to feed seawater or brine in a series of stages or effects [21]. The condensed steam is recycled for evaporation of water from preheated seawater in the first effect and evaporation of the brine in the following effects that are operated at slightly lower temperature and pressure than in the previous effect. The energy retrieved from the condensing steam can be used for further water evaporation from the brine of the preceding effect. The number of effects is normally between 8 and 16 and the maximum used temperature in the first effect is about 120 °C because of the risk of scaling. A minimal temperature difference of 5 °C is required in each effect. On the other hand, the MSF process came into practice in the early 1960s and is based on a series of flash chambers where the evaporation of water from saline aqueous solutions results from a pressure drop and not from heat exchange with condensing steam. In MSF steam is generated from saline feed aqueous solutions at a progressively reduced pressure. The exhaust brine is partly recirculated to enhance water recovery. Like in MED process, the top brine temperature is limited to about 110 °C by the risk of scaling. The most important disadvantage of MSF is its limited performance ratio at about 11 resulting in much higher energy consumption. However, MSF is more reliable and easier to operate than MED.

The benefits of MD process could be used in combination with MED or MSF plants for pure water production. A combined one-stage MED/MD hybrid system working at atmospheric pressure was proposed by Andrés et al. [22]. A scheme of the MD hybrid plant is shown in Fig. 14.9. The relatively hot brine (28.3–57.3 °C) and distilled water (14.9–31.4 °C) of the MED of 1 m^3 volume are reused as feed and permeate in the DCMD modules (MD020CP2 N, Table 2.2). The MED alone produced up to 16 kg/h of distilled water with a gained output ratio (GOR, Eq. (13.51)) of 3.7 at a temperature of 85 °C and a circulation feed solution of 170 kg/h. The MED/MD hybrid system permitted an increase of the production by about 7.5%, with an electrical

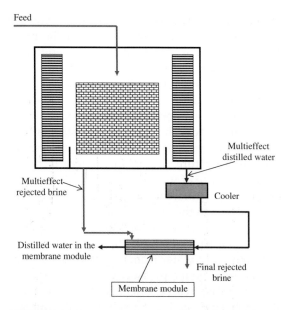

FIGURE 14.9 Schematics of one stage MED/MD hybrid plant. *Source: Reprinted from [22]. Copyright 1998, with kind permission from Elsevier.*

conductivity of about 12 μS/cm and an improvement of the energetic efficiency (GOR) by practically 10%.

Banat et al. [23] integrated DCMD process to a single-effect solar still for water desalination. The relatively hot brine in the solar still was used as a feed to the membrane module from its lumen side. The solar still was used for both brine heating and potable water production. A shell and tube membrane module (MD020TP2 N, Table 2.2 with an effective membrane area of 0.036 m^2) was used and the dimensions of the solar still are the following (base: 0.98 × 0.98 m; front wall: 0.1 m height; back wall: 0.47; inclination: 19°). A 14 mm diameter copper U-tube was employed for condensation fitted on the still wall at the lower end of the collection channel. The effects of solar irradiation, salt concentration (0.5–3.5 wt %), brine flow rate and temperature on the MD hybrid process performance was investigated realizing both indoor and outdoor experiments. It was observed that the brine flow rate affected significantly the permeate flux of the membrane module, whereas the effect of salt on the distillate of both the solar still and the membrane module was marginal. On the contrary, the fluxes of both the solar still and the membrane module increased exponentially with temperature. As it is logical, the production rates of the solar still and the DCMD module were a function of solar irradiation. As solar intensity increased the brine temperature increased and more potable water was produced. For example, the obtained permeate of the MD hybrid system during 7.5 h was about 9 kg/m^2·h for an average solar irradiation of 252 W/m^2. The distillate production of the solar still was found to be no more than 20% of the total production and reached maximum values of 2–3 h after the solar irradiation peak.

It is worth quoting that the integration of MD with MSF process or with other types of MED is not considered yet.

Nuclear-Powered MD

One of the advantages of MD is its ability to utilize waste heat from power sources such as nuclear reactors. Different MD configurations can be applied depending on each situation. Moreover, the possibility of nuclear desalination by MD was proposed by Khayet et al. [24,25]. It was confirmed that MD is feasible for processing low- and medium-level radioactive waste, giving high decontamination in only one stage. Processing radioactive liquid waste by MD in the place of their production is reasonable from an economic point of view. It simplifies all logistic activities and eliminates the costs of transportation and safeguarding. It also introduces the closed loop in liquid waste management, allowing not only the concentration of hazardous substances into a small volume, but also water recovery and recycling.

Nuclear desalination is one important option for solving the problem of fresh water resources in many parts of the world. The combination of nuclear reactor with MD plant is reasonable, as MD is an energy-intensive process that can utilize waste heat from the nuclear reactor and electricity produced by the nuclear power plants (NPP). Therefore, nuclear-powered MD desalination should be considered feasible and cost-effective. However, contradictory opinions exist about the sustainability of this method. The question of nuclear wastes remains a heavy burden on nuclear power desalination. The possible use of nuclear energy not only for production of electricity but also for production of fresh water by means of desalination has been explored by the International Atomic Energy Agency claiming that it is a cost competitive and feasible option.

When considering the possibility of implementing MD in the nuclear industry, namely at NPP for nuclear desalination, one cannot omit the problem of radioactive liquid waste treatment. The generated radioactive liquid waste in NPP and research centres needs processing for further safe storage and disposal. It is necessary to concentrate radionuclide wastes into small volumes and to separate them from the final effluent discharged to the environment to reach the levels permitted by sanitary standards. Different methods and multistage processes are employed for processing (i.e. chemical precipitation, sedimentation, ion exchange, thermal evaporation and pressure-driven membrane processes such as RO).

Khayet et al. [25] proposed the application of the DCMD process coupled with a nuclear reactor for water desalination and for low- and medium-level radioactive waste concentration. Both laboratory and pilot plant experiments were carried out. By using the heat and electricity generated in NPP, the DCMD units can produce fresh water for the local water distribution network. Simultaneously, the treatment of radioactive wastes takes place and the concentrate of radioactive substances obtained in DCMD installation can pass to the next required stages of processing (e.g. solidification by bitumization and cementation). The proposed approach of using DCMD for nuclear desalination, together with liquid radioactive waste treatment, creates an integrated system of water/wastewater management for NPP that can contribute to the idea of replacing conventional methods with clean technologies.

Cooling Tower and Membrane Distillation

In the MD process, researchers should not only look into different heating strategies of the feed aqueous solutions but also pay attention to the permeate side of the membrane modules that should be maintained at lower temperatures for condensation of the evaporated molecules. For example, in most industries recirculating cooling water (RCW) and fresh water production are completely separate systems. Recently, the DCMD process was integrated with traditional RCW for pure water production at laboratory level [26]. A schematic diagram of the RCW/DCMD hybrid system is shown in Fig. 14.10. The used membranes were PVDF hollow fibres fabricated by the dry/wet spinning method (0.18 μm nominal pore size, 76% porosity, 1 mm outer diameter, 0.7 mm inner diameter, and 150 kPa liquid entry pressure of water) and assembled in a polyester shell and tube module (50 cm effective membrane length, 70 cm length of the module, 96 mm outer diameter, 80 mm inner diameter, 1275 number of hollow fibre in a module and a total membrane area of about 2 m^2). The feed aqueous solution was circulated through the lumen side of the membrane module and its temperature was controlled by pumping through the electric heaters to simulate the heat gained in the condenser, whereas the permeate temperature was maintained by a condenser connected to a cooling tower. The cold temperature of the permeate at the inlet of the membrane

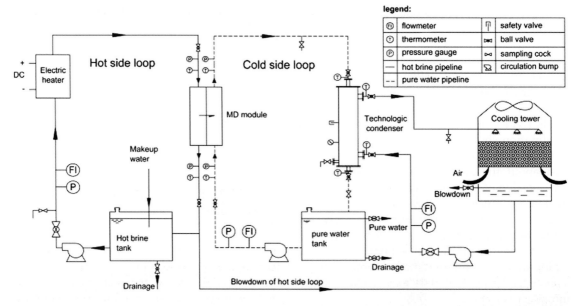

FIGURE 14.10 Schematics of the RCW/DCMD hybrid experimental system. *Source: Reprinted from [26]. Copyright 2008, with kind permission from Elsevier.*

module was 30–35 °C, whereas that of the feed was up to 70 °C. Permeate fluxes as high as 18 l/m²·h were reached with a temperature difference at the module inlets of 37 °C observing a linear increase of the production with the temperature difference. The effects of different operating parameters involved in the RCW/DCMD hybrid experimental system such as the flow rates were investigated and a comparison was made in terms of water savings and costs of a chemical plant in Shandong Province (China) that used the UF/RO hybrid process with a water recovery rate of 70% and 50 m³/h pure water production. Lower total investment and O&M (i.e. operation and maintenance) as well as lower produced water costs than in the UF/RO hybrid process were reported. It was concluded that this type of MD hybrid system was a promising technology and further efforts should be exerted for its development (i.e. long-term tests, fouling and scaling studies, cascade design of membrane modules).

Diesel Waste Heat and Membrane Distillation

Xu et al. [27] attempted to use the heat coming from the vessel engine for seawater desalination. A pilot VMD plant was constructed for seawater desalination on ships using polyvinyl chloride modules containing three bundles of PP hollow fibre membranes prepared by melt-stretching technique with an inner diameter of 327 μm and a thickness of 53 μm. Each bundle contained 5000 hollow fibres of 80 cm length with total membrane area of 12.3 m². The flow diagram and pictures of the plant are presented in Fig. 14.11. The seagoing vessel had two diesel engines, each having a power of 136 kW, and both were cooled by seawater. The temperature of the level-one cooling water was 65 °C, while that of the level-two cooling water was 35 °C. In MD operation, the level-one cooling water of 1800 kg/h flow rate was used as hot seawater that was fed to the VMD module. In the hot

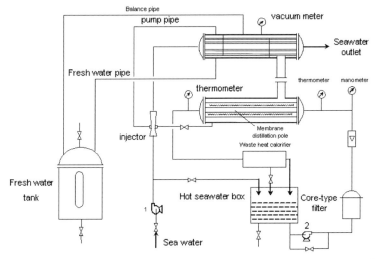

FIGURE 14.11 Flow sheet and pictures of the VMD used for seawater desalination. *Source: Reprinted from [27]. Copyright 2006 with kind permission from Elsevier.*

seawater line, the feed solution was pumped from the hot seawater box through the core-type filter into the lumen side of the hollow fibres. A part of water in the hot seawater evaporated by the aid of partial vacuum applied on the shell side of the hollow fibres and permeates through the hollow fibre wall. The retentate came out of the hollow fibre module and went through the waste heat calorifier to be heated by the diesel waste heat before returning to the hot water tank. The cold seawater circulated through an injector to generate low pressure in the permeate side. The vacuum can be broken by opening the valve attached to the fresh water tank to take out the condensed water. Different operating conditions have been investigated. It was found that the salt rejection factor of the pilot plant was up to 99.99% with a permeate flux of 5.4 kg/m$^2 \cdot$h at a feed temperature of 55 °C and a downstream pressure of 93 kPa. After 5 months, the high salt rejection factor and the permeate flux were maintained.

ALTERNATIVE ENERGY SOURCES AND MD APPLICATIONS

Different approaches can be applied for MD water treatment with low energy costs by optimizing and minimizing the energy consumption and/or the use of alternative energy sources. This should enable less-developed countries to have access to sufficient potable or desalted waters. Nowadays different MD systems including solar energy, salt gradient solar ponds (SGSP) and geothermal energy are among the most realistic options.

Solar Thermal Collectors, Photovoltaic Panels and Membrane Distillation

To benefit from the cheap solar energy supply several attempts have been made to develop solar-powered membrane distillation (SPMD) units as hybrid systems [23,28–33,35–39]. For instance, solar energy can be used either by producing the thermal energy required to generate the MD driving force for heat and mass transfer in the membrane modules or by producing electricity required to run the circulation or low-pressure pumps.

First attempt of solar-heated MD was made by Hogan et al. [28] as early as 1991 using DCMD process and heat recovery exchangers. The authors studied the feasibility of SPMD plant for the supply of domestic drinking water in the arid rural regions of Australia. The plant was designed and constructed applying a simulation, made previously, of the whole process. The flow diagram of the SPMD plant is given schematically in Fig. 14.12. The hot stream (shown as the solid line in Fig. 14.12), after heated at the solar collector (SC) and stored in storage tank, is circulated through the lumen side of the hollow fibre MD module, before entering the plate and frame heat exchanger I. The membrane module consists of 1100 fibres (17 cm length, 0.3 mm inner diameter and 0.6 mm outer diameter) with a pore size of 0.22 μm and 70% porosity. The cold stream (shown as the broken line in Fig. 14.12) flows in a countercurrent way through the shell side of the membrane module. The latent heat required to evaporate the hot water is recovered when the condensation occurs in the cold stream line. The heated cold stream is cooled at the heat exchanger I by the hot stream from the outlet of the MD module (even though it is called hot stream the temperature is lower than the cold stream in this heat exchanger). The cold stream is further cooled before entering the MD module.

By conducting simulation of the system and economic analysis, it was concluded that for the domestic sized plant of 50 kg/day the optimum configuration appeared to be a SC area of around 3 m^2, a membrane area of 1.8 m^2 and a total heat exchange area of 0.7 m^2. The capital cost for the unit was conservatively estimated at AUS$3500 in 1991.

In the activities of SMADES project financed by the European Commission, compact and large SPMD units based on the AGMD process were designed and tested in different locations (Germany, Jordan, Egypt, Morocco and Spain) [29–33]. Details of both the compact and the large SPMD plants together with some discussions may be found in Chapter 13 (Figs. 13.11 and 13.12). Recently among others, Banat's group at Jordan University of Science and Technology (JUST) has made by far the most important contribution to this subject [23,30,31,33]. Their report in desalination was made as early as 2002 and the report continued until the economic evaluation was made in 2008 [34].

In this chapter, an example of the compact SPMD plants is shown in Fig. 14.13. Figure 13.12

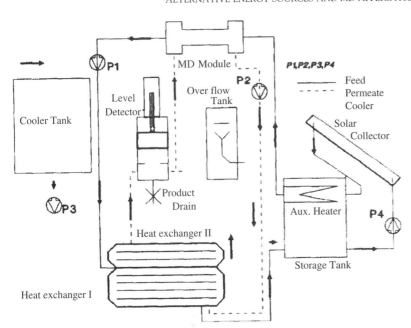

FIGURE 14.12 Schematic diagram of SPMD pilot plant. *Source: Reprinted from [28]. Copyright 1991 with kind permission from Elsevier.*

illustrates a picture and side view of a similar plant. The system consists of four major components: a spiral wound AGMD module (PTFE membrane with 0.2 μm nominal pore size, 35 μm thickness and 80% porosity; 10 m² effective membrane area in the module), a solar energy collector, a feed container and a photovoltaic panel (PV). Hot water coming from the solar energy collector is brought into the AGMD module and flows through the evaporator channel from its inlet (3) to outlet (4). The cold feed water is pumped from the feed container to the inlet of the condenser channel (1) of the AGMD module and flows to its outlet (2) in a countercurrent mode. Therefore, there is a heat flow from the hot (evaporator) channel to the cold (condenser) channel and the AGMD unit works as a heat exchanger. The temperature difference also causes the difference in partial vapour pressures of water, which will be the driving force for the flow of water vapour from the hot channel, through the membrane pore and the air gap to the metal surface that is cooled by the fluid in the condenser channel. The water vapour condenses on the metal surface and the condensate (distillate) is gained from the distillate outlet. The feed stream that is coming out from the outlet of the condenser channel (2) is further heated while passing through a 5.73 m² corrosion-free collector before being directed to the inlet of the evaporator channel (3). The retentate that comes out from the outlet of the evaporator channel (4) is recycled to the feed container.

As an example, the performance of the SPMD plant is shown in Fig. 14.14 for 1 day in August 2005 (Irbid, Jordan). The system started at 9:00 am when the solar irradiation was about 600 W/m². The feed flow rate depends on the current delivered by the PV and increases with increasing solar irradiation. The figure shows the parallel relationship between the feed flow rate and the solar irradiation. The permeate flux also followed the tendency in the irradiation intensity and reached its peak value (≈ 2.5 l/m²(membrane)·h) at about noon. The conductivity of the distillate went up to 40 μS/cm in the first few minutes of operation but then decreased to its normal value of 5 μS/cm.

Recently, an experimental solar MD system was used for arsenic removal from contaminated

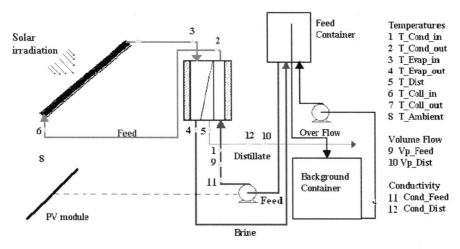

FIGURE 14.13 Flow diagram of the compact SPMD plant. *Source: Reprinted from [30]. Copyright 2007 with kind permission from Elsevier.*

groundwater [35,36]. DCMD configuration was considered using a plate and frame membrane module. Different membrane types were tested (PVDF, PTFE and PP, effective membrane area 0.0162 m^2). A SC storage tank was used together with two thermostats in order to control the temperatures of the feed and permeate. Therefore, the system was not fully solar powered. Almost 100% arsenic removal was achieved without any permeate flux decline and pore wetting was not detected. After 250 h DCMD operation, no arsenic was detected in the permeate. DCMD permeate flux up to 95 kg/m^2·h with almost 100% rejection factor was reported when using arsenic contaminated groundwater (396 ppb) with a feed temperature of 60 °C, a permeate temperature of 20–22 °C and a flat sheet PVDF membrane

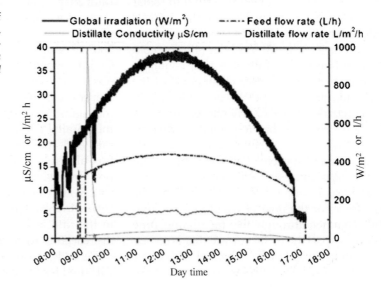

FIGURE 14.14 Daily change of global irradiation, feed flow rate, distillate conductivity and distillate flow rate. *Source: Reprinted from [30]. Copyright 2007 with kind permission from Elsevier.*

(Sepromembranes: 0.13 μm nominal pore size, 70–75% porosity and 150 μm thickness) [35]. No data were reported about the change of the permeate flux with solar irradiation intensity.

A solar VMD hybrid experimental system was proposed by Wang et al. [37] for potable water production from underground water and installed in Hangzhou (China). A 8 m^2 solar energy collector with 90% conversion efficiency was used to heat the feed solution having an electrical conductivity higher than 230 μS/cm. A PP hollow fibre membrane module (371 μm inner diameter, 35 μm thickness, 0.1 μm membrane pore size, 500 hollow fibres with a length of 0.14 m and a total membrane area of 0.09 m^2), an external condenser and a vacuum pump were employed. A water production of as high as 32.2 kg/m^2·h permeate was obtained with an electrical conductivity less than 4 μS/cm. The highest daily cumulated water production was 173.5 kg/m^2. It was claimed that this permeate flux could be more than 170 kg/m^2 under good weather conditions and more than 50 kg/m^2 in cloudy weather conditions. However, in this study, no energy consumption analysis was performed and no PVs were used to run the pumps.

Simulations and optimization of SPMD pilot plants have been also carried out in order to produce fresh water with reduced energy costs [38,39]. Ding et al. [38] considered DCMD configuration of both flat sheet and hollow fibre membrane modules and observed that the plant productivity could be improved by increasing its heat exchange capacity up to an optimum value, decreasing both the feed and the permeate flow rates and by optimizing the effective membrane surface area. It was concluded that for a certain SC area, heat recovery via an external heat exchanger from the permeate to the feed was effective to improve energy efficiency and it was an economical way to intensify the SPMD process. Chang et al. [39] considered spiral wound AGMD module for simulation and optimization of the SPMD pilot plant, which was similar to the one presented in Fig. 14.13. It was found that simple thermal storage tank and low flow rate through the AGMD module were advantageous for higher water production. Available experimental data showing the variation of the production rate with temperature and flow rate reported in [29,31] were simulated. Reasonably good agreements were found between the experimental and the simulated results.

It is noticed that only solar flat-plate thermal collectors have so far been considered. Other thermal collectors such as parabolic solar concentrators and spherical collectors incorporating heat storage tanks may be of great interest to increase the capacity of solar MD hybrid desalination systems.

Membrane Modules Integrating Solar Absorbers

A novel small system combining solar absorber and DCMD module was designed for water desalination at the temperature range 35–50 °C and for fluid heating [40]. A schematic diagram of the solar hybrid DCMD module is given in Fig. 14.15. The module was constructed with acrylic plate of 0.02 m thickness and insulated using Styrofoam of 0.05 m thickness. The effective length, width and height of the liquid flow channels were 0.21, 0.29 and 0.002 m, respectively. The module consisted of a blackened aluminum absorber plate 0.01 m in thickness with ditches, two spacers with 0.002 m thickness, an acrylic bottom plate with ditches and a hydrophobic porous membrane (PTFE supported by a PP net with 0.1 μm nominal pore size, 72% porosity, 130 μm thickness, 0.21 m length, 0.29 m width and 6.09×10^{-2} m^2 effective membrane area, Advantec®, Japan). The distance among the glass cover and the absorber plate was 0.01 m. A spacer size with 0.002 m thickness in both channels has been used. A few nylon fibres with 1.4×10^{-4} m diameter were used as supports of the membrane and were inserted between the spacer and membrane

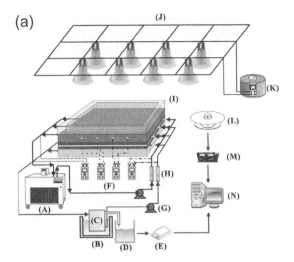

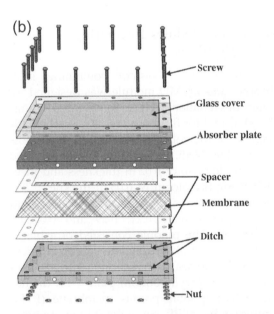

FIGURE 14.15 Schematic diagram of the solar hybrid DCMD module (a) and details of the module (b). (A) hot fluid thermostat, (B) cold fluid thermostat, (C) over flow tank, (D) beaker, (E) electronic balance, (F) temperature indicators, (G) pumps, (H) flow meters, (I) DCMD module, (J) artificial solar simulation, (K) transformer, (L) pyranometer, (M) data logger, (N) computer. *Source: Reprinted from [40]. Copyright 2010 with kind permission from Elsevier.*

surface. The solar absorber was inserted between the glass cover and the membrane in order to heat the aqueous saline feed solution that passed underneath the absorber plate. The incident solar radiation was measured and recorded using an Epply laboratory pyranometer and CR510 data logger (Epply Laboratory Inc.). A saline feed aqueous solution of 3.5 wt% and a permeate temperature of 25 °C were used.

As discussed in the previous chapters, an increase of the permeate flux with the feed flow rate (i.e. circulation velocity) and saline water temperature was observed. The saline feed temperature at the outlet of the membrane module increased with increasing the feed flow rate. Moreover, higher incident solar radiation ranging from 850 to 1100 W/m^2 induced higher temperature difference between the absorber plate and the cold liquid leading to greater permeate flux. Within 4 months of operation decline of the permeate flux was not observed. The maximum measured permeate flux was 4.1 kg/m$^2 \cdot$h at an inlet temperature of 50 °C, an incident solar radiation of 1100 W/m^2, a circulation velocity of 0.0259 m/s and with an electrical conductivity of 3 μS/cm. A theoretical model was developed to determine the water productivity and the temperature distribution in the solar hybrid DCMD module. A good agreement between the theoretical predictions and the experimental results was observed. The performance of the hybrid DCMD module was investigated with and without solar heating at different feed temperatures, circulation velocities and solar incident radiations. Enhancement of the permeate flux up to 16.56% was achieved for the highest solar radiation intensity and the highest feed circulation velocity, with a theoretical prediction of 85.1% for the collector efficiency.

This study opens up a new research area in MD. Long-term tests are necessary for sustainable development and improved design for this type of solar hybrid MD modules. Other configurations can be tested using more compact direct solar desalination modules and incorporating thinner spacers between the absorber plate and the membrane in order to reduce the thermal resistance. Not only flat-plate absorbers can be used, but also parabolic absorbers and thermosiphon systems are of great interest.

Although the majority of the solar MD hybrid systems concern the use of SCs to provide energy, other solar systems such as SGSP could be used.

Salt-Gradient Solar Ponds and Membrane Distillation

A SGSP is a large body of water with a depth between 2 and 5 m and a salinity gradient. As shown in Fig. 12.11 the top region of the pond is called the surface zone or upper convective zone (UCZ), the middle region is termed the main gradient zone or non-convective zone (NCZ) and the lower region is called the storage zone or lower convective zone (LCZ). Solar radiation can reach the bottom of the pond and is trapped in its lower region (LCZ) with higher salt concentration than the UCZ. The NCZ situated below the UCZ acts as a thermally insulating layer since natural convection currents are suppressed. In the LCZ the heat is stored and extracted. Thus, solar ponds combine both solar energy collection and long-term storage providing reliable thermal energy at temperatures raging from 50 to 90 °C and may be one of the cost-effective alternative energy systems for water desalination.

Both experimental and theoretical studies have been conducted on SGSP/MD hybrid systems for desalination using AGMD, DCMD and VMD configurations [41–43]. It is to be noted that the experimental SGSP/MD hybrid desalination system has been considered for the first time at the University of Texas, El Paso, in 1987, using the AGMD configuration [41,43]. Aqueous NaCl solutions of concentrations 35 and 269.6 g/l were used as feed solutions heated by low-grade thermal energy supplied from a salt-gradient solar pond through a heat exchanger. The obtained permeate fluxes were lower than 6.7 $l/m^2 \cdot h$. A water production of 0.158×10^{-3} $m^3/day/m^2$ of SGSP was achieved with a temperature difference across the membrane of 41 °C.

Simulations on coupled VMD and solar energy for seawater desalination have been carried out by Mericq et al. [42]. Both SGSP and SCs have been considered in the following different possibilities. The characteristics that were considered for the SGSP were a thickness of the UCZ of 0.15 m (from the top surface of the SGSP) with a salt concentration of 3 g/l and a temperature of 51 °C, a thickness of the LCZ of 0.45 m (from the bottom of the SGSP) with a salt concentration of 87.9 g/l and a temperature of 28 °C. The NCZ was situated between 0.15 m and 0.45 m with concentrations varying as, C (g/l) = 300.71 z (m)−50.1222, and temperatures varying as, t (°C) = 76.66 z (m) + 16.751, where z is the depth from the top of the SGSP. It was found that submerged membrane module in an SGSP-induced high concentration and temperature polarization effects reducing the VMD permeate flux considerably. The simulated permeate flux was found to be 3.7 and 70.6 $l/m^2 \cdot h$ when the membrane module was submerged in the LCZ without and with agitation, respectively. When the membrane module was fed from the LCZ the obtained permeate flux was found to be higher 71.2 $l/m^2 \cdot h$. Without agitation the effects of the temperature and concentration polarization were greater (i.e. 81% temperature polarization coefficient and 17% concentration polarization coefficient). The membrane characteristics and operating conditions used for the simulation were not reported. The energy consumption, technical feasibility, maintenance and costs were qualitatively analysed in order to select the best configuration. It was concluded by the simulation that the use of SC was the most interesting option producing a very high VMD permeate flux of 142 $l/m^2 \cdot h$ (i.e. 617 l/h) for a downstream pressure of 500 Pa and a Knudsen membrane permeability of 1.85×10^{-5} $s \cdot mol^{1/2} \cdot m^{-1} \cdot kg^{-1/2}$.

FIGURE 14.16 Possible SGSP/DCMD hybrid system. *Source: Reprinted from [43]. Copyright 2010 with kind permission from Elsevier.*

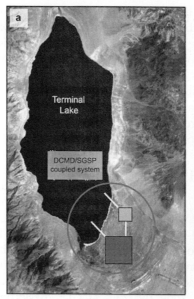
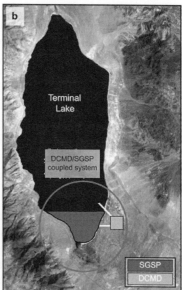
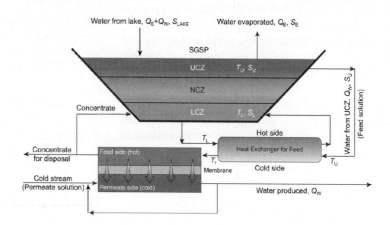

Recently, a theoretical study was performed by Suárez et al. [43] on a DCMD-integrated SGSP constructed inside and outside a terminal lake for sustainable fresh water production. A thermal model for the SGSP and a heat and mass transfer model in the membrane module were developed considering the meteorological data from Walker-Lake (Walker, NA, USA) although the models can be useful to any terminal lake. A PTFE membrane with 0.45 μm nominal pore size and 89% porosity was considered. The dimensions of the flow channels were 50 mm wide, 3 mm high and 200 mm long. The TDSs in the feed solution was typically 50–75 g/l. Figure 14.16 shows the possible coupling of the DCMD module to the SGSP. The heat extract from the LCZ is transferred to the feed solution using a heat exchanger, whereas the temperature of the permeate was maintained by a second heat exchanger connected to the lake not shown in Fig. 14.16. The temperature of the permeate was maintained equal to the temperature at a depth of 3 m in the lake, which was 13.9 °C. Different salinities in the LCZ were considered at different temperatures. It was observed that the highest permeate flux was at 53.9 °C with 10% salinity in the LCZ. The SGSP/DCMD hybrid system can produce up to 1.6×10^{-3} m^3/day/m^2 of SGSP with membrane areas ranging from 1 to 1.3×10^{-3} m^2. It was concluded that the hybrid system will be feasible only when the SGSP is constructed inside the terminal lake giving a net water production about 2.7×10^{-3} m^3/day/m^2 of SGSP.

Geothermal Energy and Membrane Distillation

In general, geothermal resources are recognized as geothermal water springs used for bathing and therapeutic treatments and not as geothermal energy sources. However, in some cases geothermal water is transferred in aqueducts to consumers. It is also used for desalination by RO or for heating greenhouses. It is to be noted that geothermal water has some inconveniences such as hardness that prevents its direct use for fresh water production by RO systems.

An MD system using sensible heat of geothermal water was reported by Bouguecha et al. [44]. A pretreatment by a conical fluidized bed crystallizer was used because of precipitation of $CaCO_3$ that took place due to the decrease of pressure and loss of CO_2 when water in the well moved close the surface. Three AGMD modules associated in series were considered. Each module consisted of two cells arranged in parallel configuration. As mentioned in the previous chapters, the permeate flux increased with the increase of the feed temperature, which was varied from 45 to 65 °C, and decreased with the increase of the feed salt (NaCl) concentration, which was varied from 1 to 36 g/l. The electrical conductivity of the permeate was constant at 6 μS/cm. The overall water conversion reached 25% (i.e. 10% in the first stage, 9% in the second stage and 6% in the third stage). Furthermore, the specific energy consumption was evaluated. It was found that the geothermal MD hybrid system consumed 111 kJ/kg of water and the cost of the produced water was $0.13/l compared with 82 kJ/kg and $0.08/l for RO system powered with PVs. Details of the calculations, the membrane type, some AGMD-operating conditions and geothermal water characteristics were not reported by the authors [44]. Therefore, more rigorous studies must be carried out in this line of research in order to confirm its feasibility both technically and experimentally.

As can be seen from the above examples, practically all tests for the MD hybrid systems were investigated at a laboratory level and pilot scale performance tests using actual power plants that are still operating for investigation. There is no doubt that the MD hybrid systems can be very attractive alternatives to individual processes. More multidisciplinary studies should be carried out in order to find an effective way for the construction and operation of industrial MD hybrid systems, while taking into account the environmental impact such as the discharge of the brine and the atmospheric emissions. Unfortunately, the environmental consideration has been rather neglected until now.

References

[1] R. Ray, R.W. Wytcherley, D. Newbold, S. Mccray, D. Friesen, D. Brose, Synergistic, membrane-based hybrid separation systems, J. Membr. Sci. 62 (1991) 347–369.

[2] F. Lipnizki, R.W. Field, P.K. Ten, Pervaporation-based hybrid process: a review of process design, applications and economics, J. Membr. Sci. 153 (1999) 183–210.

[3] F. Lipnizki, R.W. Field, Pervaporation-based hybrid processes in treating phenolic wastewater: technical aspects and cost engineering, Sep. Sci. Tech. 36 (2001) 3311–3335.

[4] D.E. Suk, T. Matsuura, Membrane-based hybrid processes: a review, Sep. Sci. Tech. 41 (2006) 595–626.

[5] B.V. Bruggen, E. Curcio, E. Drioli, Process intensification in the textile industry: the role of membrane technology, J. Env. Manag. 73 (2004) 267–274.

[6] M. Gryta, K. Karakulski, A.W. Morawski, Purification of oily wastewater by hybrid UF/MD, Water Res. 35 (2001) 3665–3669.

[7] E. Drioli, F. Lagana, A. Criscuoli, G. Barbieri, Integrated membrane operations in desalination process, Desalination 122 (1999) 141–145.

[8] K. Karakulski, M. Gryta, A. Morawski, Membrane processes used for potable quality improvement, Desalination 145 (2002) 315–319.

[9] J. Mericq, S. Laborie, C. Cabassud, Vacuum membrane distillation of seawater reverse osmosis brines, Water Res. 44 (2010) 5260–5273.

[10] D. Qu, J.W.B. Fan, Z. Luan, D. Hou, Study on concentrating primary reverse osmosis retentate by direct contact membrane distillation, Desalination 247 (2009) 540–550.

[11] E. El-Zanati, K.M. El-Khatib, Integrated membrane-based desalination system, Desalination 205 (2007) 15–25.

[12] C.R. Martinetti, A.E. Childress, T.Y. Cath, High recovery of concentrated RO brines using forward osmosis and membrane distillation, J. Membr. Sci. 331 (2009) 31–39.

[13] T.Y. Cath, D. Adamas, A.E. Childress, Membrane contactor processes for wastewater reclamation in space II. Combined direct osmosis, osmotic distillation, and membrane distillation for treatment of metabolic wastewater, J. Membr. Sci. 257 (2005) 111–119.

[14] R.A. Johnson, J.C. Sun, J. Sun, A pervaporation-microfiltration-osmotic distillation hybrid process for the concentration of ethanol-water extracts of the Echinacea plant, J. Membr. Sci 209 (2002) 221–232.

[15] D. Qu, J. Wang, L. Wang, D. Hou, Z. Luan, B. Wang, Integration of accelerated precipitation softening with membrane distillation for high-recovery desalination of primary reverse osmosis concentrate, Sep. Purif. Technol. 67 (2009) 21–25.

[16] M.S.H. Bader, A hybrid liquid-phase precipitation (LPP) process in conjunction with membrane distillation (MD) for the treatment of the INEEL sodium bearing liquid waste, J. Hazardous Mater. B121 (2005) 89–108.

[17] M. Gryta, The fermentation process integrated with membrane distillation, Sep. Purif. Technol. 24 (2001) 283–296.

[18] S. Mozia, A.W. Morawski, Hybridization of photocatalysis and membrane distillation for purification of wastewater, Catalysis today 118 (2006) 181–188.

[19] S. Mozia, A.W. Morawski, M. Toyoda, T. Tsumura, Effect of process parameters on photodegradation of Acid Yellow 36 in a hybrid photocatalysis-membrane distillation system, Chem. Eng. J. 150 (2009) 152–159.

[20] S. Mozia, A.W. Morawski, M. Toyoda, T. Tsumura, Integration of photocatalysis and membrane distillation for removal of mono- and poly-azo dyes from water, Desalination 250 (2010) 666–672.

[21] B.V. Bruggen, C. Vandecasteele, Distillation vs. membrane filtration: overview of process evolutions in seawater desalination, Desalination 143 (2002) 207–218.

[22] M.C. de Andrés, J. Doria, M. Khayet, L. Peña, J.I. Mengual, Coupling of a membrane distillation module to a multieffect distiller for pure water production, Desalination 115 (1998) 71–81.

[23] F. Banat, R. Jumah, M. Garaibeh, Exploitation of solar energy collected by solar stills for desalination by membrane distillation, Renewable Energy 25 (2002) 293–305.

[24] M. Khayet, M.P. Godino, J.I. Mengual, Possibility of nuclear desalination through various membrane distillation configurations: a comparative study, Int. J. Nuclear Desalination 1 (2003) 30–46.

[25] M. Khayet, J.I. Memgual, G. Zakrzewska-Trznadel, Direct contact membrane distillation for nuclear desalination, Part II: experiments with radioactive solutions, Int. J. Nuclear Desalination 2 (2006) 56–73.

[26] J. Wang, B. Fan, Z. Luan, D. Qu, X. Peng, D. Hou, Integration of direct contact membrane distillation and recirculating cooling water system for pure water production, J. Cleaner Production 16 (2008) 1847–1855.

[27] Y. Xu, B. Zhu, Y. Xu, Pilot test of vacuum membrane distillation for seawater desalination on a ship, Desalination 189 (2006) 165–169.

[28] P.A. Hogan, Fane, A.G. Sudjito, G.L. Morrison, Desalination by solar heated membrane distillation, Desalination 81 (1991) 81–90.

[29] J. Koschikowski, M. Wieghaus, M. Rommel, Solar Thermal-driven desalination plants based on membrane distillation, Desalination 156 (2003) 295–304.

[30] F. Banat, N. Jwaied, M. Rommel, J. Koschikowski, M. Weighaus, Desalination by a "compact SMADES" autonomous solar-powered membrane distillation unit, Desalination 217 (2007) 29–37.

[31] F. Banat, N. Jwaied, M. Rommel, J. Koschikowski, M. Weighaus, Performance evaluation of the "large SMADES" autonomous desalination solar-driven membrane distillation plant in Aqaba, Jordan, Desalination 217 (2007) 17–28.

[32] H.E.S. Fath, S.M. Elsherbiny, A.A. Hassan, M. Rommel, M. Wieghaus, J. Koschikowski, M. Vatansever, PV and thermally driven small-scale, stand-alone solar desalination systems with very low maintenance needs, Desalination 225 (2008) 58–69.

REFERENCES

[33] J. Koschikowski, M. Wieghaus, M. Rommel, V.S. Ortin, B.P. Suarez, J.R.B. Rodríguez, Experimental investigations on solar driven stand-alone membrane distillation systems for remote areas, Desalination 248 (2009) 125–131.

[34] F. Banat, N. Jwaied, Economic evaluation of desalination by small-scale autonomous solar-powered membrane distillation units, Desalination 220 (2008) 566–573.

[35] A.K. Manna, M. Sen, A.R. Martin, P. Pal, Removal of arsenic from contaminated groundwater by solar-driven membrane distillation, Env. Pollution 158 (2010) 805–811.

[36] P. Pal, A.K. Manna, Removal of arsenic from contaminated groundwater by solar-driven membrane distillation using three different commercial membranes, Water Res. 44 (2010) 5750–5760.

[37] X. Wang, L. Zhang, H. Yang, H. Chen, Feasibility research of potable water production via solar-heated hollow fiber membrane distillation system, Desalination 247 (2009) 403–411.

[38] Z. Ding, L. Liu, M.S. El-Bourawi, R. Ma, Analysis of a solar-powered membrane distillation system, Desalination 172 (2005) 27–40.

[39] H. Chang, G.B. Wang, Y.H. Chen, C.C. Li, C.L. Chang, Modeling and optimization of a solar driven membrane distillation, Renewable Energy 35 (2010) 2714–2722.

[40] T.C. Chen, C.D. Ho, Immediate assisted solar direct contact membrane distillation in saline water desalination, J. Membr. Sci. 358 (2010) 122–130.

[41] J. Walton, H. Lu, C. Turner, S. Solis, H. Hein, Solar and waste heat desalination by membrane distillation, Desalination and water purification research and development program report no. 81, University of Texas, El paso, TX, April 2004.

[42] J. Mericq, S. Laborie, C. Cabassud, Evaluation of Systems coupling vacuum membrane distillation and solar energy for seawater desalination, Chem. Eng. J. 166 (2011) 596–606.

[43] F. Suárez, S.W. Tyler, A.E. Childress, A theoretical study of a direct contact membrane distillation system coupled to a salt-gradient solar pond for terminal lakes reclamation, Water Res. 44 (2010) 4601–4615.

[44] S. Bouguecha, B. Hamrouni, M. Dhahbi, Small scale desalination pilots powered by renewable energy sources: case Studies, Desalination 183 (2005) 151–165.

CHAPTER 15

Economics, Energy Analysis and Costs Evaluation in MD

OUTLINE

Introduction	429	(ii) Annual Operating Costs	449
Energy Analysis in MD	431	Amortization or Fixed Charges (A_{fixed})	449
		O&M Costs ($A_{O\&M}$)	450
MD Costs Evaluations and Comparison to other Systems	441	Membrane Replacement Costs (A_{MR})	450
		Pretreatment Costs	450
Proposed Example of MD Costs Analysis	449	Plant Availability (f)	450
(i) Capital Cost (CC)	449	(iii) Other Annual Operating Costs	450

INTRODUCTION

There are only few studies reported in the literature on the economics, energy analysis and costs evaluations in membrane distillation (MD) process [1–18]. In most of the publications, only a small section is dedicated to this issue at their ends. Moreover, most of these studies are concerned with energy consumption and cost evaluation of water produced by MD process from seawater and brackish water. Comparisons with other processes such as reverse osmosis (RO) and conventional thermal distillation (i.e. simple effect distillation or solar still, multi-effect distillation, MED) and multistage flash (MSF) are also made. Usually, these conventional processes are well established industrially under strict optimization, whereas MD process is not. Therefore, optimized MD plants must be designed and developed first and then rigorous cost and economic analysis for comparison will become possible. More intensive and focused MD research efforts in this field are needed, not only theoretically but also experimentally in order to avoid scale-up issues.

As it was reported throughout this book, MD is an energy-intensive technology that exhibits the advantage of using renewable energy sources such as solar and geothermal energy or any low-grade and industrial waste heat.

Therefore, practically in all reports there is one common conclusion, i.e. the water production cost (WPC) can go down considerably if inexpensive heat source is available. As it will be shown in this chapter, economic analysis confirms that the water cost will be as low as $0.64/m^3$ if utilization of waste heat is considered. However, according to some other reports, the capital cost will increase considerably by installing solar systems.

It should be pointed out that various desalination plants are operating worldwide, converting about 27.5 million m^3/day of seawater and brackish water into fresh water. This quantity is only about 3% of the world's consumption of drinking and sanitation water. Therefore, desalination capacity is projected to grow with a decrease in costs. Actually, typical WPCs for large-scale units are below $1/m^3$, whereas for small-scale units they range from $1/m^3$ to $3/m^3$.

Solar energy-based plants are capital intensive, however market prices for renewable energy sources could gradually become lower in the future and would be competitive with conventional energy sources. If the selection of a desalination system (solar versus non-solar) would be based on the initial capital cost alone, the solar desalination system would rarely be selected. Non-solar desalination systems usually have relatively small initial costs and relatively large annual operating costs reflecting raw energy purchases. Solar desalination systems, however, are relatively expensive initially but have negligible non-solar energy cost during their lifetime.

It is worth noting that even though the WPC of small-scale solar powered MD (SPMD) systems looks considerably high, in remote isolated areas where the fuel transportation, the only energy alternative without having access to the electrical grid, is unreliable and expensive, the desalination with solar energy remains one of the most favourable processes.

Unlike RO, only little information is available on energy analysis and cost estimations for MD process. One of the reasons is that MD is not yet fully applied in commercial scale. Moreover, the costs of various principal components needed in an MD plant such as large-scale MD modules and membranes, and other cost-related information such as pre-treatments, optimum flow conditions, long-term MD performance, fouling and membrane life are not yet available at a satisfying level. Therefore, different and even conflicting conclusions are drawn depending on the laboratory system or the pilot plant from which the data for the analysis was produced. For example, this chapter will show that the WPC in MD varies, depending on the authors, from $0.3/m^3$ to $130/m^3$.

For MD systems more investigations on economics of the process should be conducted. Various factors should be considered in order to estimate adequately the WPC of an MD installation, including capital costs (system investment, auxiliary equipment investment, installation charges, feed pretreatment, security and control system, etc.), energy consumption both thermal and electrical, energy cost, capacity and feed water quality, membrane replacement and plant life, operating and maintenance (O&M), amortization and annual operating costs. The capital cost depends on the capacity of the MD system and its design. Annual operating cost should be calculated considering amortization or fixed charges, which account for the annual interest payments for borrowing the fund to cover the initial capital cost; O&M costs, which account among others for the annual payments for the O&M staff cost and the spare cost; cost required for membrane replacement and pretreatment costs as well as plant availability per year. Remember that nowadays the MD membranes and modules are expensive. There is no company selling modules for the different MD configurations. Furthermore, a commercially long life for the membrane is not yet guaranteed. Therefore, the logical first step is to produce working and reliable MD membrane modules for both laboratory and industrial scales. Furthermore, uniform economic analysis procedure should be followed for all MD systems to determine the WPC.

It is worth to mention that several strategies for energy savings can be used in MD such as the use of well-designed membranes, modules with low temperature and concentration polarization effects, minimum heat losses through the membrane and to the environment, multiple effect operation, heat recovery in heat exchangers, integration of other process such as RO. As shown in Chapter 14, MD hybrid systems lead to enhancement in the process productivity by increasing the overall water recovery and reduce the environmental impact due to the brine disposal.

ENERGY ANALYSIS IN MD

MD is an energy intensive technology and so energy economy is an important issue. Unlike the pressure-driven separation processes, energy consumption in MD systems includes both thermal energy necessary to heat the feed aqueous solution and to cool the permeate aqueous solution or condensation and the electrical energy required to run the circulation pumps, vacuum pumps or compressors. However, the heat energy requirement in MD can be by far more than 90% of the total energy requirement and increases drastically with the feed temperature. It is to note that electrical energy is more expensive than low-grade thermal energy.

The thermal efficiency (η) in MD was defined in Chapter 10 (Eq. (10.41)) and the effects of MD-operating conditions on (η) were discussed throughout this book. Generally, it is a ratio between the amount of heat brought into a MD system and the heat actually used for evaporation of the feed to produce fresh water or to concentrate the feed aqueous solution. It was observed that the thermal energy consumption in MD is very much sensitive to the feed temperature [1]. This is mainly due to the exponentially increased mass flux and to the decrease of the amount of heat lost by conduction through the membrane with the increase of the feed temperature. At high feed temperatures the heat transferred through the membrane by conduction will be negligible compared with the heat transferred due to the transmembrane mass flux. Therefore, the energy consumption per unit of distillate may be reduced appreciably at high operating feed temperatures.

The temperature polarization effect can be reduced by working under turbulent flow regime and with the aid of turbulence promoters, the heat loss by conduction through the membrane can be minimized by increasing pores size and membrane porosity as well as by employing optimum membrane thickness (i.e. well-designed MD membranes and modules), and adequate insulation materials for the membrane module(s), pipes and all plant accessories can reduce or even limit heat loss to environment.

To be rigorous, it is more adequate to use energy efficiency to characterize an MD system instead of thermal efficiency, since energy efficiency takes into consideration the global energy input, which includes both thermal energy (E_t) and electrical energy (E_e) instead of thermal energy alone:

$$\eta_E = \frac{\text{Effective heat for evaporation}}{\text{Energy input}}$$
$$= \frac{J_w A \Delta H_{v,w}}{E_t + E_e} \quad (15.1)$$

To improve energy efficiency in MD, a portion of the heat from the retentate should be recovered to preheat the feed aqueous solution using heat exchangers as shown in Fig. 15.1 so that the process heat requirement is reduced. The heat proportional to the temperature difference ($\Delta T_{b,f} = T_{b,f,out} - T_{b,f,in}$) is available for the

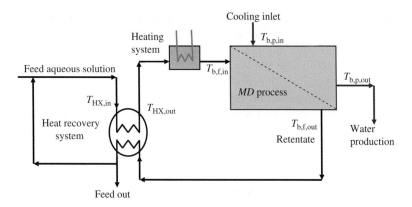

FIGURE 15.1 Schematics of MD process operating with heat recovery by heat exchanger.

evaporation process, while less thermal energy, which is proportional to ($\Delta T_{b,f} = T_{b,f,in} - T_{HX,out}$), is applied into the system. The MD process parameters should ensure that the temperature of the discharged retentate or concentrate is equal to the temperature of the solution to be treated. Multi-effect operation is also possible. The maximum heat recovery should be as high as possible. Instalment of heat recovery system increases the capital investment costs but at the same time brings benefit in the operating costs. The temperature and concentration polarization effects, thermal conduction through the membrane and heat loss to the surroundings should be minimized in order to increase energy efficiency. Among the MD configurations, VMD process may be the most energy efficient configuration, due to its low heat loss by conduction through the membrane and as result of its high permeate flux rates.

Few studies have been conducted on the effect of heat recovery on the process cost [1,2,4,8,12,13,19–22]. Schneider et al. [19] reported that heat recovery can reduce the WPC up to 4-fold; however, no cost analysis was reported.

The heat recovery factor refers to the number of energy uses and can be defined differently depending on the arrangements of the MD system. Fane et al. [1] defined it as:

$$H_R = \frac{\Delta T_{MD,module}}{\Delta T_{MD} + \Delta T_{HE}} \quad (15.2)$$

where $\Delta T_{MD,module}$ is the axial temperature drop along the MD module, ΔT_{MD} is the temperature difference across the membrane and ΔT_{HE} is the temperature difference across the heat exchanger(s).

Eq. (15.2) indicates that H_R is maximized by a high axial temperature drop along the MD module ($\Delta T_{MD,module}$) and by using small driving force (ΔT_{MD}). This indicates that energy recovery will be achieved at the expense of low permeate flux. For a feed temperature of 80 °C, H_R was found to be 3.6 with a heat loss by conduction through the membrane of 450 W/m²·K and an effective number of heat uses of 2.4 [1].

Hogan et al. [2] defined the heat recovery factor as the maximum heat recoverable in the main heat recovery exchanger divided by the heat transferred in the membrane module:

$$H_R = \frac{Q_{HE}}{Q_{MD}} \quad (15.3)$$

Higher H_R values were found when using greater membrane area and lower flow rates. This is due to the increased contact time of the fluids in the membrane module. However, lower permeate fluxes were obtained as can be seen in Fig. 15.2 not only for high membrane area but also for low flow rates in contrast to what it was shown in previous Chapters 10–12 when using MD systems without heat recovery. Therefore, the possibility of heat recovery and reuse is limited depending on the application under study.

It is to be noted that when small areas of solar thermal collectors are used or when less energy input is applied, greater heat recovery systems are required if the temperature of the feed aqueous solution is to be maintained.

The energy efficiency, termed gained output ratio (GOR), was defined in Chapter 13, Eq. (13.51), for solar systems with heat recovery. This is the ratio of the total latent heat of evaporation of the produced water to the input energy. In general, input thermal energy is considered [20–22]. Remember that only a part of the recovered energy is latent heat from the condensing process, while the other part is heat transferred by conduction through the membrane and the permeate channel. GOR value is less than unity for simple effect systems and higher than unity for multi-effect systems.

AGMD spiral wound membrane module with heat recovery was developed by Fraunhofer ISE (Germany) for solar-driven desalination plants with capacities up to 10 m^3/day [20–24] (see Fig. 9.11, Figs. 13.11, and 13.12 and Fig. 14.13). The feed aqueous solution to be treated passes first through the condenser channel and is gradually warmed by the latent heat of condensation. When it comes out of the condenser it is fed to the solar thermal collector for further heating before contacting the membrane. Among the reported characteristics of the membrane modules were the effective membrane area of 7–12 m^2, the specific thermal energy consumption of 100–200 kWh/m^3 and the GOR of about 3–6 [24]. Koschikowski et al. [20], by using a module with an effective membrane area of 7 m^2 in a compact SPMD plant (Fig. 14.13), reported a GOR value of 5.5 at 350 l/h flow rate and 75 °C evaporator inlet temperature with a specific energy consumption (i.e. ratio of supplied energy to the volume of produced fresh water) of about 117 kWh/m^3. GOR values between four and six and specific energy consumption ranging from 140 to 200 kWh/m^3 were claimed. The lower is the value of the specific energy consumption the more economical is the MD process.

When using a similar compact pilot plant with 10 m^2 membrane area operated with brackish water (Fig. 14.13), AGMD permeate fluxes as high as 120 l/day with an electrical conductivity of about 5 μS/cm and thermal energy ranging between 200 and 300 kWh/m^3 were obtained [20]. However, in this case the GOR values were found to be lower than unity (i.e. 0.3–0.9).

When using a large pilot plant with heat storage tank and four membrane modules with a total area of 40 m^2 operated with untreated (i.e. without chemical treatment usually used in RO) seawater from red sea

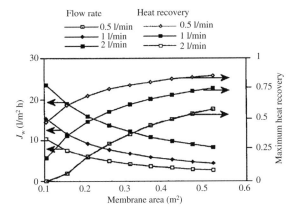

FIGURE 15.2 Simulated MD permeate flux and calculated maximum heat recovery factor at different flow rates and membrane area for the SPMD schematized in Fig. 14.12. Source: Reprinted from [2]. Copyright 1991 with kind permission from Elsevier.

(Fig. 13.11), fresh water production varying from 5 to 27 l/m²·h with electrical conductivity values varying between 20 and 250 μs/cm (i.e. average salt rejection factor of 98%) were produced [22]. In this case, the calculated GOR values were also found to be lower than unity, in the range 0.4–0.7. Specific energy consumption ranging from 200 to 300 kWh/m³ was reported for this pilot plant. More detailed discussions may be found in Chapter 13.

Another desalination alternative, Memstill® technology, also based on AGMD configuration with heat recovery was proposed by the research and technology organization TNO (Netherlands Organization for Applied Scientific Research) (see Fig. 9.10) [12,13]. A cold saline water flows through a condenser with non-permeable walls, increasing its temperature progressively due to the condensing permeate, and then passes through a heat exchanger where additional heat is added before entering in direct contact with the membrane in counter current flow configuration. The average energy consumption was claimed to be 73.75 MJ/m³. When the considered heat supply to the Memstill system was fuel fired or generated by cogeneration of heat and electricity the specific energy consumption was calculated to be 64.9 kWh/m³ (i.e. 231 MJ/m³ heat consumption plus 0.75 kWh/m³ electric energy supplied), whereas in the case of waste heat source the specific energy consumption was lower, 39.4 kWh/m³ (i.e. 139 MJ/m³ heat consumption plus 0.75 kWh/m³ electric energy supplied) [13].

Khayet et al. [8] performed various DCMD tests in a pilot plant designed with heat recovery systems as shown in Fig. 15.3. A spiral wound polytetrafluoroethylene (PTFE) membrane module (G-4.0-6-7, see Table 2.1) with an effective membrane area of 4 m² and three heat exchangers were used. For heat recovery a heat exchanger was employed (7 in Fig. 15.3).

In Fig. 15.3 the pump 5, through the heat exchanger 7 used for heat recovery, pumps the warm feed stream from reservoir 3. Then, it flows through a ceramic filter 9 and undergoes further warming up in the heat exchanger 10, which receives the heat from the hot stream delivered by boiler 6. The warm feed stream of temperature T1 enters the membrane module 1. After the module, the retentate of temperature T3 returns to the retentate reservoir. A cold permeate stream of temperature T6 is delivered from the permeate tank 2 by pump 4 to the heat exchanger 8, where it is cooled down by cooling tap water flowing in a counter current direction. The cold stream is introduced into the membrane module 1 at a temperature T2, and after passing the module 1, its temperature increases to T4. The distillate transfers the heat to the retentate stream in heat exchanger 7.

Experiments showed that the best conditions for running the process are at higher inlet feed temperatures (70–80 °C). As can be seen in Fig. 15.4, the specific energy consumption decreases with the increase of the inlet feed temperature. For this pilot plant one can obtain at a feed temperature ranging from 80 to 85 °C an energy consumption of about 600 kWh/m³. At high feed temperatures the permeate flux is higher and the specific energy consumption is lower partly due to the heat recovery.

Alklaibi [25] developed a mathematical model for a stand-alone MD process with heat recovery and tried to relate the energy required for heating and pumping to the heat and mass transfer in a MD unit. Different membrane permeability values were considered. A schematic diagram of the MD unit is shown in Fig. 15.5 together with the temperature profile in the MD module and in the heat exchanger employed for heat recovery. In this system, most of the heat required to increase the temperature of the feed aqueous solution to the inlet temperature of the hot channel of the MD module is recovered by the heat exchanger. The calculated values of the specific energy consumption for different membrane permeabilities and feed temperatures were found to be between 6 and 1500 kWh/m³. The obtained

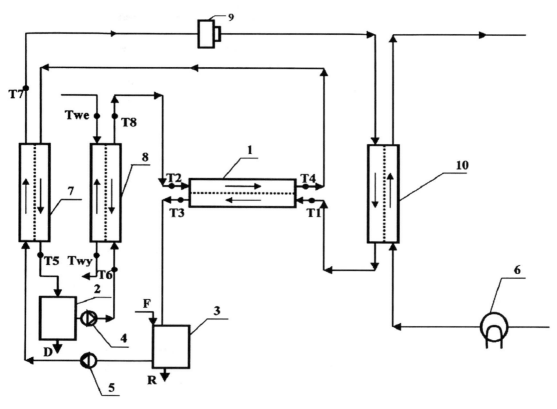

FIGURE 15.3 Schematics of a DCMD pilot plant with heat recovery used for water desalination and for treatment of low and medium radioactive liquid wastes (1: membrane module; 2: permeate tank; 3: feed tank; 4 and 5: circulation pumps; 6: heater; 7, 8 and 10: heat exchangers; 9: ceramic filter). *Source: Original figure was published in [8] (www.inderscience.com).*

specific energy consumption was reduced significantly by using a membrane with a high permeability and by discharging the feed at the outlet of the MD module at a high temperature. For example, for a feed temperature at the outlet of the membrane module of 25 °C, when the membrane permeability was increased 10 times, the calculated energy consumption was reduced more than three times; however, for a feed temperature at the outlet of the membrane module of 60 °C, the energy consumption was decreased more than five times for the same increase of the membrane permeability. Supposing that MD is economically acceptable if the specific energy consumption is lower than 50 kWh/m^3, he concluded that for an MD system to become competitive for seawater desalination, the feed solution should be discharged at a temperature over 50 °C with a high membrane permeability, and therefore a low-grade energy is needed.

In Table 15.1 the specific energy consumption reported for different MD systems and pilot plants are presented and compared with other separation processes used in desalination such as RO, solar still, MED and MSF.

Carlsson [26] stated that the specific energy consumption of a MD system can be as low as 1.25 kWh/m^3, the standard modules in large-scale MD systems would produce about 5 m^3/day per module and for large-scale plants the cost would be reduced. However, no

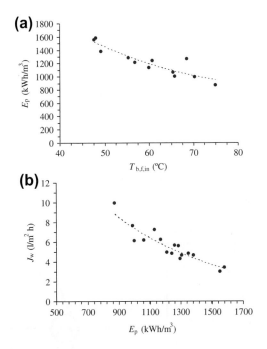

FIGURE 15.4 Specific energy consumption (E_p) of the DCMD pilot plant presented in Fig. 15.3 as a function of the applied feed inlet temperature (a) and the corresponding permeate flux (b). (feed flow rate 1380 l/h, permeate flow rate 1260 l/h and permeate temperature varying from 16.5 to 26.8 °C).

analytical details of the MD system and the operating parameters were reported.

Gazagnes et al. [5] indicated that AGMD process is able to produce a permeate flux of 5.2 l/m²·h with a specific energy consumption of about 1 kWh/m³.

By using sensible heat of geothermal water and AGMD system, Bouguecha et al. [3] obtained a specific energy consumption of 30.83 kWh/m³ (i.e. 111 kJ/kg). The theoretical value of the specific energy consumption was found to be lower 11.1–13.9 kWh/m³ (i.e. 40–50 kJ/kg). The difference between the experimental and theoretical values of the specific energy consumption was attributed by the authors to the exponential variation of the permeate flux with the temperature and also to the fact that the second and third stages of AGMD coupled to geothermal resources were operated at low temperatures leading to lower permeate fluxes and lower efficiencies. However, details of the calculations, membrane type, some of the operating AGMD conditions as well as geothermal water characteristics and real coupling were not reported [3].

Wang et al. [7] proposed a small solar VMD hybrid experimental system for potable water production from underground water. A PP hollow fibre membrane module (inner diameter 371 μm, thickness 35 μm, membrane pore size 0.1 μm, 500 hollow fibres with a length of 0.14 m, total membrane area 0.09 m²), an external condenser and a vacuum pump were employed. Figure 15.6 shows as an example the power consumption and the permeate flux. It can be observed in Fig. 15.6 that the power consumption of the vacuum pump (≈ 0.18 kW) and the feed circulation pump (≈ 0.37 kW) are much lower than the heat power consumption, especially for high permeate fluxes. A high water production of 29.75 kg/m²·h with a total power consumption of 21.69 kW, which is the same as a specific energy consumption of 8100.8 kWh/m³, was obtained. Higher specific energy consumption, 9079.5 kWh/m³ was calculated for a low water production of 3.07 kg/m²·h with a total power consumption of 3.62 kW. Compared to the values of the specific energy consumption reported in Table 15.1, the obtained values by Wang et al. are more than two orders of magnitude greater. One of the reasons of the observed high specific energy consumption is the small membrane module used without considering heat recovery systems. High specific energy consumption for laboratory MD systems is expected.

Evaluation of energy requirements in DCMD and VMD laboratories systems have been made by Criscuoli et al. [6] using plate-and-frame membrane modules of about 40 cm² effective membrane area. Longitudinal and transversal flow configurations were studied and different

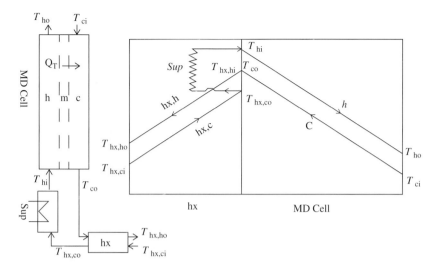

FIGURE 15.5 Schematics of a MD unit with a heat exchanger for heat recovery and profile of temperatures.
Source: Reprinted from [25]. Copyright 2008 with kind permission from Elsevier
Notes: H: hot feed; m: membrane' c: cold side; Sup: external heat source' subscripts; h: hot; c: cold; hx: heat exchanger; hi: inlet of hot channel; ho: outlet of hot channel; ci: inlet of the cold channel; co: outlet of cold channel.

MD-operating conditions were tested. The lowest obtained values of the specific energy consumption were 3546.3 kWh/m³ for longitudinal flow in DCMD and 1108.4 kWh/m³ for cross-flow VMD membrane module. For example, with respect to the VMD tests, when the feed temperature was increased from 53.9 to 59.3 °C in a cross-flow module with a membrane having 0.2 μm pore size and 91 μm thickness (Membrana, Germany), at a feed flow rate of 150 l/h and a downstream pressure of 10 mbar, the permeate flux increased from 29.7 to 51.5 kg/h·m², the energy consumption increased from 354.6 to 441.8 W but the specific energy consumption was decreased from 2984.8 to 2144.7 kW/m³ [6]. When the feed flow rate was increased from 150 to 235 l/h maintaining the feed inlet temperature at 59.1 °C and a downstream pressure at 60 mbar, the VMD permeate flux increased from 46.9 to 50.5 kg/m²·h but the energy consumption also increased from 199.5 to 223.9 W, whereas the specific energy consumption slightly increased from 1063.4 to 1108.4 kWh/m³. For DCMD process carried out on the longitudinal-flow membrane module, when the feed flow rate was increased from 100 to 300 l/h maintaining the feed temperature around 54 °C and the permeate temperature at 14.3 °C, the permeate flux increased from 15.6 to 19.1 kg/m²·h and the energy consumption from 227.7 to 274.2 W. This means that the specific energy consumption decreased from 3649 to 3589 kWh/m³. In terms of permeate flux, specific energy consumption and thermal efficiency VMD performance was found to be better than DCMD configuration.

Based on a developed computational analysis using a theoretical model describing the coupled heat and mass transfer, Cabassud and Wirth [9] studied energy consumption in two VMD systems, a continuous single-pass VMD system and a discontinuous VMD operation in which the retentate was returned back to the feed container. Theoretical model specifications were not reported. For the continuous

TABLE 15.1 Estimated specific energy consumption, E_p, of different MD systems and other separation processes used in desalination

E_p (kWh/m^3)	Observations	Reference	E_p (kWh/m^3)	Observations	Reference
MD systems			140–200	Production: 120 l/day	[20]
≈1	AGMD, Permeate flux: 5.2 l/m^2.h	[5]	200–300	SPMD compact plant	[21]
1.2	VMD discontinuous flow, Permeate flux 0.5 – 0.7 l/m^2.h	[9]	200–300	SPMD large plant, Permeate flux: 5–27 l/m^2.h	[22]
3.2	VMD single-pass flow, Permeate flux: 0.7 l/m^2.h		**Other desalination systems**		
1.25	Production: 5 m^3/day	[26]	1.3	RO of brackish water with photovoltaic panels & athermal collector, without energy recovery, Production: 0.2 m^3/day	[27]
20.5–66.7	Memstill® units, Production: 25–50 m^3/day per module	[13]			
2.05	MD hybrid system (MF/NF/MCr/RO/MD) without ERD[a,b]	[10]	2.5	RO min Production: 105,000 m^3/day	[13]
≈1.63	MD hybrid system (MF/NF/MCr/RO/MD) with ERD[a,b]		4.5	RO standard Production: 105,000 m^3/day	[13]
28.0	MD hybrid system (MF/NF/MCr/RO/MD) without ERD[a]		3	RO of brackish water	[22]
≈27.54	MD hybrid system (MF/NF/MCr/RO/MD) with ERD[a]		5–6	RO of seawater with energy recovery	
			17	RO of seawater, electric energy	
			22.8	RO powered with photovoltaic panels	[3]
30.8	AGMD, use of sensible heat of geothermal	[3]	26.4	MSF without ERD	[11]
55.6	Thermal and electrical energy	[2]	41	MSF	[13]
13	NF/RO/MD hybrid system	[11]	60–80	MSF, Average	[22]
15	RO/MD hybrid system		48–441	MSF	
2.25	RO/MD hybrid system[b]		640	Solar still, Production: 2–6 l/m^2.day	[21]
2.58	NF/RO/MD hybrid system[b]		416.7	Multi-effect solar still (MESS)	[3]
			≈30	MED thermal energy	[22]

[a] ERD: Energy recovery device.
[b] Thermal energy available in the plant or stream is already at the operating temperature of the MCr unit.

single-pass system the specific energy consumption was estimated to be 3.2 kWh/m^3, while for the discontinuous batch-wise system the energy consumption decreased to 1.2 kWh/m^3, which was practically the energy required by the vacuum pump. The authors further stated that for a membrane area 100 times larger than that of the experimental membranes and at low temperatures of about 25 °C, VMD can compete with RO on energy consumption (<2 kWh/m^3) with approximately the same as water production.

In VMD process, when the feed temperature is as low as the ambient temperature only the energy consumed by the vacuum pump and the feed circulation pump are necessary to be considered. At 25 °C, Cabassud and Wirth [9] found that the VMD permeate flux was

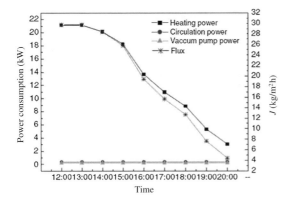

FIGURE 15.6 Power consumption and permeate flux of a small solar VMD pilot plant. *Source: Reprinted from [7]. Copyright 2009 with kind permission from Elsevier.*

independent of the Reynolds number (Re) and the minimum value of the specific energy consumption was 1.2 kWh/m^3 at a Re of 2000 when the feed salt concentration was 30 g/l and the downstream pressure was 100 Pa. Under these conditions, the VMD permeate flux was 13 l/m^2·h. On the other hand, when the feed concentration was 300 g/l, the minimum specific energy consumption was 1.3 kWh/m^3 at the Re of 500 with a VMD permeate flux of 8.8 l/m^2·h.

For higher feed temperatures than the ambient temperature, the main energy consumption will be for heating the aqueous feed solution. This heat energy is as high as 100 kWh/m^3; however, the vacuum energy consumption is as low as 1.3 kWh/m^3 and almost negligible as compared with the heating energy consumption. In contrast, the energy consumption for feed circulation increases with an increase of the flow rate, and consequently with an increase in Re. Cabassud and Wirth [9] indicated that the energy required for feed circulation was negligible compared with the vacuum pump energy until a salt concentration of 300 g/l and a Re of 7000 are reached. For example, for a aqueous salt solution of 30 g/l, a downstream pressure of 100 Pa, a membrane permeability coefficient of 10^{-5} s·mol$^{1/2}$/m·kg$^{1/2}$ and a single-pass VMD system, the permeate flux changes from 80 to 120 l/m^2·h as the Re increases from 1000 to 6000 and the specific energy consumption for the highest permeate flux was 1.5 kWh/m^3. It must be mentioned here that single-pass MD systems at high temperatures are not feasible because heating solutions of a large flow rate requires high investment costs. Another example is a discontinuous VMD system run at a feed salt concentration of 300 g/l, a downstream pressure of 100 Pa and a membrane permeability coefficient of 10^{-5} s·mol$^{1/2}$/m·kg$^{1/2}$. When the Re increased from 500 to 9000, the permeate flux was changed from 40 to 95 l/m^2·h. In this case, for a permeate flux of 85 l/m^2·h, the obtained specific energy consumption was 1.3 kWh/m^3. Cabassud and Wirth [9] concluded that VMD technology will be competitive from the point of view of energy consumption, only when free heating energy source, such as solar energy becomes available. Then, the energy consumption becomes nearly equal to 1.3 kWh/m^3 that comes from the vacuum pump energy and circulation energy consumption.

Criscuoli and Drioli [11] performed energetic and *exergetic* analysis of MD hybrid systems, RO/MD system (i.e. RO brine was treated by MD) and NF/RO/MD system (i.e. NF used as pretreatment step for the RO/MD system). The coupling of the MD with the RO unit introduced a thermal energy requirement, which increased the global energy of the coupling system with respect to RO alone, whereas NF applied for pretreatment increased the electrical energy requirements. The obtained specific energy consumption was 15 kWh/m^3 for the RO/MD hybrid system, whereas that of the NF/RO/MD hybrid system was found to be lower, 13 kWh/m^3. If thermal energy is already available, the specific energy consumption decreases considerably. In this case, values such as 2.25 and 2.58 kWh/m^3 were reported for the RO/MD hybrid system and the NF/RO/MD hybrid system, respectively. The integrated MD systems can be a very attractive alternative to RO if the thermal energy is available [11].

Macedonio et al. [10] made energetic and *exergetic* analysis as well as economic evaluation for seawater desalination by integrated membrane systems. One of the systems combined microfiltration (MF), nanofiltration (NF), membrane crystallization (MCr), RO and MD in one hybrid system. MCr was applied on NF retentate and MD was applied on RO retentate. Figure 15.7 shows a schematic diagram of this MD hybrid system. The specific energy consumption was found to be 28 kWh/m^3 without energy recovery. When the throttling valves on the brine stream were replaced by an energy recovery system such as a Pelton turbine or a pressure exchanger system, a slightly lower specific energy consumption, about 27.5 kWh/m^3, was obtained. However, when thermal energy was available in the plant or the stream was already at the operating temperature of the MCr unit, the specific energy consumption decreased by an order of magnitude, i.e. to 2.05 kWh/m^3 without energy recovery and to about 1.6 kWh/m^3 with heat recovery.

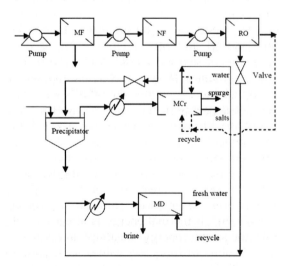

FIGURE 15.7 Schematic diagram of a MF/NF/MCr/RO/MD hybrid system. *Source: Reprinted from [10]. Copyright 2007 with kind permission from Elsevier.*

The word 'exergy' was introduced to express the quality of energy [4,10,11,28]. The total energy is divided into two parts, one *anergy* and the other *exergy* (or available work). The *anergy* is the part of energy that is forced to be given to the environment as heat in conditions of complete degradation. The *exergy* is the part of energy that is convertible into all other forms of energy. It represents the useful part of energy for a system in its environment. In other words, it can be defined as the maximum amount of work that the system can execute in its environment.

For a system in which the governing parameters are the temperature, pressure and composition, the *exergy* of a flow stream, E_X, was written as [4,28]

$$E_X = E_X^T + E_X^P + E_X^C \tag{15.4}$$

where E_X^T, E_X^P and E_X^C are the temperature, pressure and concentration *exergy* terms, respectively, that were defined as follows:

$$E_X^T = \dot{m} c_p \left[(T - T_0) - T_0 \ln\left(\frac{T}{T_0}\right) \right] \tag{15.5}$$

$$E_X^P = \dot{m} \left(\frac{P - P_0}{\rho} \right) \tag{15.6}$$

$$E_X^C = -\dot{m} c_p (n_s R T_0 \ln(x_s)) \tag{15.7}$$

where $\dot{m}$ is the mass flow rate, c_p is its specific heat, ρ is the density, n_s is the solvent concentration (mol/kg solution), x_s is the solvent mass fraction and the subscript 0 stands for reference state.

The indicator of how much exergy remains undestroyed during the operation of the system was termed *exergetic* efficiency and was defined as [4,10]:

$$\xi = \frac{Ex_{output}}{Ex_{input}} \times 100 \tag{15.8}$$

where Ex_{output} and Ex_{input} indicate *exergy* input and *exergy* output to the system, respectively.

Macedonio et al. [10] estimated the *exergetic* efficiency considering different hybrid systems with and without energy recovery device (ERD) from the high-pressure retentate of RO and as expected the *exergetic* efficiency improved with ERD. The efficiency was found to be higher for the hybrid systems without MCr and MD than for the hybrid systems including MCr and MD. However, for all considered cases, the efficiencies were far better than 1.12–10.4%, which was for MSF.

Al-Obaidani et al. [4] performed *exergy* calculations for a 24,000 m^3/day DCMD desalination plant operated with and without heat recovery system to reuse the heat from the brine to preheat the feed aqueous seawater solution with a heat recovery efficiency of 80%. The heat input was 45,036 and 39,690 kW for the plant without and with heat recovery system. The calculated *exergy* efficiency was 28.3 and 25.6% for the desalination plant with and without heat recovery, respectively; while the net *exergy* change between inlet and outlet was 71 and 353 kW with and without heat recovery, respectively. The specific heat consumption was 39.7 kWh/m^3 for the plant operated with heat recovery and 45 kWh/m^3 in case of the plant without heat recovery.

It is worth noting that the specific energy consumption is much higher for small laboratory MD systems compared with larger pilot plants with greater membrane areas. When heat recovery systems are used the specific energy consumption is reduced. Until now most of the studies have been only theoretical. Analysis based on realistic energy consumption is required in the future.

MD COSTS EVALUATIONS AND COMPARISON TO OTHER SYSTEMS

Few researchers have reported cost estimates for MD units [1,2–4,10,12–18]. The costs of various components required in MD units are not yet known because the technology is not fully implemented and applied in commercial scale.

More than 20 years ago, approximated cost analysis of a DCMD plant with heat recovery producing 5000 kg/h (44,000 t/year) fresh water was made by Fane et al. [1]. Hollow fibre membrane modules with total surface area of 800 m^2, three heat exchangers and two pumps were considered in their calculations. Scaled to 1987 $US, the calculated installed capital cost was $720,000, the operating cost was $3.3/t and the total cost was $4.9/t. According to their calculation, distilled WPCs decreased considerably with the increase of the applied feed temperature and at 90 °C feed temperature the production cost could be as low as $2/m^3. Moreover, the costs for a small SPMD pilot plant of production capacity of 50 kg/h (500 kg/day, daily production over 10 h) were estimated. They found that a capital cost would be in the range $10,000–$15,000 with a production cost of $10/m^3 to $15/m^3. These cost estimations were marginally higher than a RO desalination plant of the same capacity. However, it was concluded that for 5000 kg/h plant capacity the production costs could be similar to those of RO plants.

About 20 years ago Hogan et al. [2] have studied the feasibility of a SPMD pilot plant with heat recovery for the supply of domestic drinking water in the arid/rural regions of Australia (see Fig. 14.12). The SPMD pilot plant was designed and constructed based on the simulations they made earlier. It was found that the capital cost was very sensitive to the heat recovery factor and for a minimum capital cost the heat recovery factor should be between 60 and 80%. In other words, in order to reduce the capital cost of heat exchangers, their optimized design shifted towards higher heat recovery (in the range 60–80%) and smaller solar collector area. For a production capacity of 50 kg/day the optimum configuration was a solar collector area of around 3 m^2, a membrane area of 1.8 m^2 and a total heat exchanger area of 0.7 m^2 with a capital cost of $3500 (Australian in 1991). Hogan et al. [2] provided only capital itemized costs without providing production costs.

Sarti et al. [14] estimated the cost of benzene removal from wastewater containing 1000 ppm of benzene by VMD plant. The capital cost of the plant was $247,000 designed for 99% benzene removal with heat recovery of the retentate to preheat the liquid wastewater. Five stages operating at different downstream pressures were considered. In each stage the feed aqueous solution was the retentate of the previous stage. Capital depreciation, labour cost, module replacement and energy consumption were taken into account to estimate the treatment cost per unit volume of wastewater. The labour cost was 10% of the capital cost per year, the membrane cost was $450/m^2 in module, the assumed membrane life was 3 years, the depreciation was 15% of capital cost per year, the operation time was 7200 h/year, the pump efficiency was 0.8, the electricity cost was $0.085/kWh, the steam at low pressure was $0.013/kg and the cooling water cost was $15/m^3. For the above cost parameters the estimated production cost was $4.04/m^3; however, the VMD plant was not necessarily optimized.

In 1999, Drioli et al. [16] performed cost analysis of RO/MD hybrid system for water desalination. MD was proposed to treat RO brine with a concentration of 75 g/l at a temperature of 35 °C in order to enhance both efficiency and water recovery factor in seawater desalination (45 g/l at 25 °C). The cost of the MD membrane was considered to be $116/m^2. It was found that RO/MD hybrid process produced more than twice as much water as RO process alone at the same water cost, $1.25/m^3. The MD process alone produced as much water as the RO/MD hybrid process but at a water cost of $1.32/m^3 which is about 5% higher. Specific and detailed cost analysis was not indicated by Drioli et al. [16].

During last 5 years, some studies have been conducted on MD economics [3,4,10,13,15,18]. Bouguecha et al. [3] estimated the annual cost of 17 l/day AGMD process using sensible heat of geothermal water resource and compared it to the costs of RO powered by photovoltaic panels and multi-effect solar still of similar capacities. The total capital investment in the plants as well as the costs of O&M was taken into consideration. The O&M cost was estimated as 1/6 of annual cost. The obtained production cost was $130/m^3 for MD (i.e. $110/m^3 capital investment cost and $20/m^3 O&M cost). Again, detailed calculations were not reported by Bouguecha et al. [3].

Production costs of Memstill® technology (see Fig. 9.9) were reported by Meindersma et al. [13] and the results compared with those of RO for seawater desalination capacity of 105,000 m^3/day. The heat supply to the AGMD process was made by cogeneration of heat and electricity, fuel combustion or by a waste heat source. The results are summarized in Table 15.2. The specific energy consumption of Memstill® systems was discussed in the previous section. Since AGMD does not require high pressure pump, the hardware costs for AGMD are much lower than RO. However, the costs of AGMD modules are higher. Overall, the total fixed costs and O&M costs are lower for AGMD than for RO. When waste heat is available, the energy costs for Memstill® are comparable to or lower than RO. The unit water cost is the lowest (i.e. $0.26/m^3) for Memstill® if cheap waste heat of $0.1/GJ is available. Figure 15.8 shows the present and the future prospect of the unit WPC for different desalination processes including MD (Memstill® technology represents pressure normalized flux of 1.5×10^{-10} m^3/m^2·s·Pa; heat energy requirement in the range 80–240 MJ/m^3; production in the range 25–50 m^3/day·module and 50% recovery.). As can be seen in Fig. 15.8 the WPC of RO technology has decreased considerably since 1990s. This is attributed mainly to the decrease of the membrane module costs and to the development of energy efficient membrane

TABLE 15.2 Water production costs of Memstill® and RO process for seawater treatment of 105,000 m^3/day.

	Memstill® ($/m^3)				RO ($/m^3)	
	Fuel	Cogeneration	Waste heat		Minimal	Standard
Energy costs			Based on $0.5/GJ	Based on $0.1/GJ		
Heat costs	0.30	0.12	0.07	0.01	-	-
Electricity costs	0.03	0.03	0.03	0.03	0.1	0.18
Fixed costs						
Hardware costs	0.05	0.05	0.05	0.05	0.23	0.32
Module costs	0.11	0.11	0.12	0.12	0.02	0.03
Auxiliary costs						
O&M, chemicals, filters, etc.	0.05	0.05	0.05	0.05	0.10	0.10
Total water cost	0.54	0.35	0.31	0.26	0.45	0.63

Source: Reprinted from [13]. Copyrigt 2006 with kind permission from Elsevier.

modules as well as to more efficient ERDs. It seems for next coming years, the WPC of MD will be lower than that of RO. The water costs by MSF and MED will not be below $1/m^3. Therefore, only RO will be able to compete with MD in the coming future. However, Memstill® technology is still under test and should be proven on a larger scale before confirming the projected decrease of desalination costs by MD [12,13].

Wang et al. [15] integrated a recirculating cooling water (RCW) to a DCMD process for water production (see Fig. 14.10). Total investment and O&M costs were estimated for the MD hybrid system with a capacity of 50 m^3/h and compared with those of a hybrid unit, utltrafiltration UF/RO integrated to RCW (i.e. a chemical plant in Shandong, Province in China) with a water recovery rate of 70% and the same capacity. Lower total investment (US$280,000) and O&M costs ($0.139/m^3) were obtained for the MD hybrid plant compared with those of UF/RO hybrid plant (320,000 US$, $0.504/m^3). In their calculation the O&M costs included the cost of electricity ($0.067/kwh), chemicals and labours ($400/month·person). The lifetime of all facilities was 15 years whereas that of the membrane was 3 years. They did not report the total production costs.

El-Zanati and El-Khatib [17] proposed a hybrid system composed of NF, RO and VMD for water desalination. Simulations and cost analysis have been carried out for a water production of 76.2 m^3/day. It was found that a WPC would be about $0.92/m^3, which was claimed to be competitive to potable water produced by seawater RO systems. Again, in this case rigorous cost analysis was not reported.

Recently, much more detailed cost analysis studies in MD have been carried out by Macedonio et al. [10], Al-Obaidani et al. [4] and Banat and Jwaied [18]. Data and equations to be used for economic calculations as well as the assumptions involved were reported. The procedure of the economic analysis and the equations used are found at the end of this chapter.

Macedonio et al. [10] also did the economic evaluation of seven systems (RO alone and different RO hybrid systems) for seawater desalination. One of the proposed system integrated MF, NF, MCr, RO and MD in one hybrid system. MCr was applied on NF retentate and MD was applied on RO retentate (see Figure 15.7). Different possibilities have been considered in terms of heat recovery and

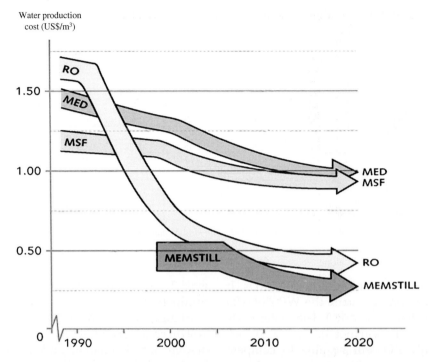

FIGURE 15.8 Expected cost development of large-scale desalination processes. *Source: Reprinted from [13]. Copyright 2006 with kind permission from Elsevier.*

available thermal energy in the plant. The obtained WPC ranged from 0.51 to $0.74/m^3$. An example is given in Fig. 15.9. The thermal energy demand in the hybrid system increases the WPC. This cost is decreased from $0.74/m^3$ to $0.55/m^3$ when the thermal energy is already available in the MD hybrid plant. The cost was lower with both heat recovery and available thermal energy in the plant, whereas higher cost was with neither energy recovery nor available thermal energy in the plant.

It is to be noted that for the MD hybrid system integrating MCr the WPC will be much lower if the salts sale is considered in the economic analysis. Another advantage is the elimination of the environmental problems caused by the brine disposal. In fact, one should not forget the cost derived from the brine transport by pipeline to a suitable discharge place. When the gain for the salts sale was taken into consideration, negative WPCs ($-0.36/m^3$ to $-0.13/m^3$) were obtained by Macedonio et al. [10]. The authors claimed that the gain for the salts sale covered more than entirely the cost of water production and therefore advised the integration of MCr in the desalination plant.

Al-Obaidani et al. [4] made studies on energy requirement and cost estimation for MD seawater desalination. The calculation was done for the DCMD plant of 24,000 m^3/day with and without heat recovery. A sensitivity test for different variables of DCMD on the process economics was also performed in order to identify the most sensitive parameters on the total water unit cost and to find the optimal operational conditions. One of such

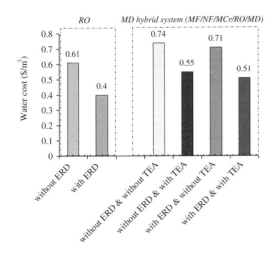

FIGURE 15.9 WPC of RO alone with and without energy recovery device (ERD, Pelton turbine) and MD hybrid system integrating MF, NF, MCr and RO with and without energy recovery device and with and without thermal energy available in the plant (TEA).

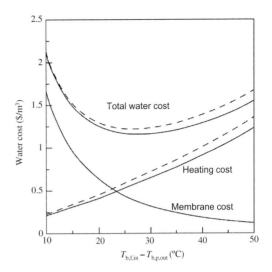

FIGURE 15.10 Effect of temperature difference ($T_{b,f,in} - T_{b,p,out}$) on WPC for DCMD without heat recovery system (broken line) and with heat recovery system (solid line). *Source: Reprinted from [4]. Copyright 2008 with kind permission from Elsevier.*

studies is the effect of the temperature difference between the feed input ($T_{b,f,in}$) and the permeate output ($T_{b,p,out}$) at the membrane module(s). It is well known that an increase in the temperature difference, which is the driving force in DCMD, enhances the permeate flux and therefore the membrane area may be reduced. Consequently, the capital cost is decreased. However, greater temperature difference requires a larger amount of heat input on the feed aqueous solution, which in turn increases the O&M cost and the total cost. Considering these two effects of the temperature difference, an optimum point is expected. Indeed, such an optimum value exists in the temperature difference as shown in Fig. 15.10. According to the figure, the minimum water cost obtained is $1.23/m³ for the feed inlet temperature of 55 °C (permeate outlet temperature of 25 °C) for DCMD without heat recovery. With heat recovery the minimum water cost is $1.17/m³ when the feed inlet temperature is 60 °C and the permeate outlet temperature is 30 °C.

In general, the WPC decreases with the enhancement of the water recovery factor, which is the ratio of the produced fresh water flow rate to the applied feed flow rate. When water recovery is as low as 50%, a considerable difference in water cost (i.e. 15%) was observed between the operation without and with heat recovery [4]. As water recovery increases the difference in water cost with and without heat recovery becomes smaller and at the water recovery of 90%, the difference becomes only 3%. This is because, the amount of the hot brine becomes small as the water recovery increases and only a limited amount of heat can be recovered by the heat exchanger. Therefore, for high water recovery factors the heat recovery does not affect the overall WPC. It is worth quoting that the overall water recovery of an MD hybrid system based on the pressure driven membrane processes (NF and RO) might reach 94% if MD is used to treat the concentrated brine.

The membrane cost contributed about 50% of the total capital cost and 30% of the O&M cost in the DCMD plant under consideration [4]. The effect of the membrane cost on the unit cost of water is given in Fig. 15.11. The figure shows that when the membrane cost (price of the membrane per unit area) is increased by 10%, the WPC increases 3.9% (with heat recovery) and 3.7% (without heat recovery).

From Fig. 15.11 the WPC is, as expected, quite sensitive to the steam cost. When the price of the steam per ton increases 10% the unit cost of water increases by 4.4 and 5%, respectively, for the MD operation with and without heat recovery.

Al-Obaidani et al. [4] concluded that the WPC becomes $1.23/m^3 and $1.17/m^3, respectively, for the MD operation without and with heat recovery at the optimum temperature difference. Since MD requires lower operating temperatures than the conventional distillation processes, it is possible to utilize low-grade energy sources from industrial processes. When waste heat is used, the WPC will go down even further to $0.64/m^3.

Banat and Jwaied [18] made economic evaluation of two SPMD plants based on AGMD spiral wound membrane module(s) with heat recovery (Fraunhofer ISE (Germany)). One is a compact unit installed in the northern part of Jordan (Irbid) and has been operated with brackish water since September 2005 (see Fig. 14.13) and the other is a large unit installed in the southern part of Jordan (Aqaba port) and has been operated with untreated seawater since February 2006 (see Fig. 13.11) [21,22]. Both units consist of flat plate collectors and photovoltaic (PV) panel and data acquisition system. Table 15.3 shows the technical characteristics of each SPMD plant together with the capital investment cost and annual operating costs. Specific energy consumption of this type of SPMD plants was discussed in Section 13.2. The obtained WPCs were $29.9/m^3 and $36/m^3 for the compact and large SPMD plants, respectively. It was reported that this cost was for very pure water and the production cost of drinkable water with a salt content

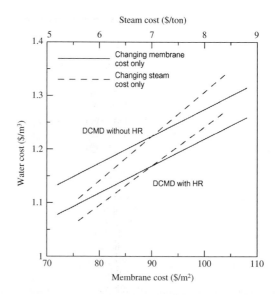

FIGURE 15.11 WPC versus membrane cost (steam cost constant at $7/ton) and steam cost (membrane cost constant at $90/m^2). *Source: Reprinted from [4]. Copyright 2008 with kind permission from Elsevier.*

of 500 ppm (dilution of brackish water using the produced pure water) will decrease the cost to one half, $15/m^3 for the compact unit and $18/m^3 for the large unit [18].

Banat and Jwaied [18] have also calculated the WPC when there is no interest payment required. Then, the cost of drinkable water became even lower; $10.2/m^3 and $11.9/m^3 for the compact and large system, respectively. Moreover, the influence of membrane and plant lifetime on WPC was simulated observing a significant reduction of WPC with increasing membrane lifetime as well as plant lifetime. For example, for 1 year membrane lifetime the WPC was found to be close to $30/m^3 for both compact and large SPMD plants; however for 7 year the cost was $14/m^3 and $17/m^3 for the compact and large plants, respectively. Therefore, increasing the reliability of the MD technology and MD membrane and plant lifetime could reduce the cost significantly.

TABLE 15.3 Summary of the values used for the cost estimation, capital investment cost and annual operating costs of the SPMD units.

	Compact unit	Large unit
Plant capacity (average, l/day)	100	500
Membrane area (m^2)	10	40
Solar collectors area (m^2)	5.73	72
Photovoltaic PV-module (kW$_p$)	0.106	1.44
Capital investment cost		
Membrane[a]	360	1,440
Membrane assembly	720	2,880
Solar collectors	600	7,200
Collectors racks	300	1,500
Photovoltaic PV-modules[b]	530	7,200
Heat exchanger	-	1,500
Piping and tanks	200	500
Thermal storage tank	-	700
Pumps	300	700
Battery	-	800
Monitoring equipment	2,000	2,500
Total equipment costs	5,310	36,360
Installation[c]	1,328	8,010
Instrumentation and control[d]	1,328	8,010
Total investment[e]	7,965	52,380
Annul cost		
Amortization or fixed charge	639	4203
O&M costs	128	841
Membrane replacement costs	216	864
Total costs	983	5908
Unit cost	$29.9/m^3	$36/m^3
Unit cost drinking water	15	18

[a] Membrane price; $36/m^2.
[b] PV-module price: $5/W$_p$ (Watts peak).
[c] Installation cost: 25% of the purchase cost.
[d] Instrumentation and control cost: 25% of the purchase cost.
[e] Land cost: zero.
Source: Reprinted from [18]. Copyright 2008 with kind permission from Elsevier.

Table 15.4 summarizes the WPC of MD and other separation processes applied using different energy sources. As can be seen distinct WPCs were claimed and reported. These vary from location to location depending on type and size of the plant, feed processed water, energy source, energy recovery systems, cost of energy, economic analysis procedure, etc.

Small SPMD plants are very attractive for remote areas although their WPCs are higher than those of other processes. According to Banat and Jwaied [18], the economic penalty comes mainly from the high initial capital investment. For potential commercial application of MD technology, much larger membrane modules having the required effective surface area should be investigated. One of the barriers for MD implementation is the high cost of commercial modules. Moreover, there is lack of commercially available MD units. However, some MD units claimed WPCs lower than that of the thermal desalination process.

TABLE 15.4 Estimated WPC of different MD systems and other separation processes used in desalination

WPC ($/m^3)	Observations	Year	Reference
MD systems			
10–15	SPMD plant, Production: 500 kg/day	1987	[1]
4.04	VMD plant for benzene removal	1993	[14]
1.25	MD hybrid system (RO&MD), $1.25/m^3 for only RO, $1.32/m^3 for only MD, installing costs ≈ $116/m^2 of MD membrane area.	1999	[16]
3.2	VMD single-pass flow, Permeate flux: 0.7 l/m^2.h	2003	[9]
1.2	VMD discontinuous flow, Permeate flux 0.5–0.7 l/m^2.h	2003	[9]
130	AGMD, use of sensible heat of geothermal, Production: 17 l/day	2005	[3]
0.26–0.54	Memstill® units, Production: 105,000 m^3/day, Heat supply generated by cogeneration of heat and electricity, fuel fired or by a waste heat source	2006	[13]
0.92	MD hybrid system (NF/RO/VMD), Production: 76.2 m^3/day	2007	[17]
0.74	MD hybrid system (MF/NF/MCr/RO/MD) without ERD[a,b]	2007	[10]
0.55	MD hybrid system (MF/NF/MCr/RO/MD) with ERD[a]	2007	[10]
0.71	MD hybrid system (MF/NF/MCr/RO/MD) without ERD[a,b]	2007	[10]
0.51	MD hybrid system (MF/NF/MCr/RO/MD) with ERD[a,b]	2007	[10]
≈1	AGM*D*, Permeate flux: 5.2 l/m^2.h	2007	[5]
1.23	MD with heat recovery, Production: 24,000 m^3/day	2008	[4]
1.17	MD without heat recovery, Production: 24,000 m^3/day	2008	[4]
0.64	MD with low-grade heat energy source	2008	[4]
15–29.9	SPMD compact, Production: 0.1 m^3/day	2008	[18]
18–36	SPMD large, Production: 0.5 m^3/day	2008	[18]
Other desalination systems			
12.05	RO, solar photovoltaic panels, Production: 1 m^3/day	1989	[29]
3.73	RO, solar photovoltaic panels, Production: 1 m^3/day	2002	[30]
2.7	RO, solar photovoltaic panels, Production: 500 m^3/day	2003	[31]
0.11	RO, electric, Production: 39,000 m^3/day	1994	[32]
80	RO, solar photovoltaic panels	----	[3]
0.45–0.63	RO, Production: 105,000 m^3/day	----	[13]
2.84	MSF, solar, Production: 1 m^3/day	1989	[29]
2.02	MSF, natural gas, Production: 20,000 m^3/day	1996	[33]
≈1.4	MSF	----	[4]
12	Solar still, Production: 1 m^3/day	1985	[34]

TABLE 15.4 Estimated WPC of different MD systems and other separation processes used in desalination—Cont'd

50	Multi-effect solarstill (MESS)	----	[3]
2	MED, solar, Production: 72 m³/day	1997	[35]
7–10	MED, solar, Production: 85 m³/day	1998	[36]
0.89	MED, solar, Production: 20,000 m³/day	1995	[37]
≈ 1	MED	----	[4]

^{a} ERD: Energy recovery device.
^{b} Thermal energy available in the plant or stream is already at the operating temperature of the MCr unit.

PROPOSED EXAMPLE OF MD COSTS ANALYSIS

The WPC depends, other than the process technology and investments, on plant capacity, type of feed aqueous solution to be treated (seawater, brackish water, etc.), pretreatment applied, energy consumption (type and cost), plant life and amortization. For instance, the major cost elements for desalination plants are the capital cost and the annual operating costs. These can be estimated as follows:

(i) Capital Cost (CC)

This covers purchasing cost of equipment including MD membrane module(s), auxiliary equipment, land, installation charges and pretreatment of the feed aqueous solution. The capital cost is a function of the process capacity and design features. For example, Table 15.3 shows the cost estimates of different components of two SPMD plants together with their total capital investment cost [18]. The MD membrane price was assumed as $36/m². The cost of MD membrane, $90/m², was assumed by Macedonio et al. [10] and Al-Obaidani et al. [4]. However, this price may change depending on the type of membrane (pore size, material, etc.), company and the quantity purchased. For example, the price of commercial polyvinylidene fluoride (PVDF) and PTFE membranes for small laboratory systems were calculated to be an order of magnitude greater than $36/m², about $300 m² for PVDF membrane and $800/m² for PTFE membrane (TF200 in Table 2.1), respectively.

The installation cost and the control instrumentation cost should also be considered. This is assumed to be 25% of the cost to purchase equipment. Furthermore, land or rental cost should be included. Most of authors considered zero land cost [18].

As shown in Table 2.3, the capital cost (CC) for the compact plant is $7,965 and the membrane module cost (MC = membrane cost + membrane assembly cost) is $1,080.

(ii) Annual Operating Costs

These costs are the total yearly costs of owing and operating a desalting plant. Among others, these costs include amortization or fixed charges, O&M costs and membrane replacement costs that can be calculated as follows:

Amortization or Fixed Charges (A_{fixed})

This accounts for the annual interest payments for borrowing the fund to cover the initial capital cost. It can be calculated by the amortization factor a which is given by:

$$a = \frac{i(1+i)^n}{(1+i)^n - 1} \qquad (15.9)$$

where i is the annual interest rate (%) and n is the year or life time of the plant. Banat and Jwaid [18] used in their calculations $i = 0.05$ (5%) and $n = 20$ and found an amortization factor of 0.080243/year.

The annual fixed charges is calculated as:

$$A_{fixed} = a \times CC \quad (15.10)$$

For the SPMD compact plant, the calculated A_{fixed} is \$639/year [18]. It is to be noted that direct capital costs (CC) are considered to calculate the fixed charges. The indirect capital cost is 10% of the total direct capital cost (CC) [10].

O&M Costs ($A_{O\&M}$)

This accounts for the annual payments for the operation and maintenance of the plant, staff cost, spare costs etc. The annual O&M costs ($A_{O\&M}$) were estimated to be 20% of the plant annual payment.

$$A_{O\&M} = 0.2 \times A_{fixed} \quad (15.11)$$

For the SPMD compact plant the calculated $A_{O\&M}$ is \$128/year.

Membrane Replacement Costs (A_{MR})

This accounts for the cost required for membrane replacement and it is a function of the produced water quality and this will vary depending on the type of membrane and module, MD process configuration, feed aqueous solutions, etc. Therefore, the membrane replacement time per year will vary. Membrane replacement cost can be estimated to be 20% of the membrane module cost (MC). For example, for the SPMD compact plant the calculated A_{MR} value is \$216/year.

Pretreatment Costs

This depends on the type of the feed aqueous solution to be treated. Banat and Jwaied [18] assumed zero pre-treatment cost.

Plant Availability (f)

This refers to working time of the plant and can be assumed to be 90% per year.

Finally, the total annual cost (A_{total}) and the WPC can be calculated as:

$$A_{total} = A_{fixed} + A_{O\&M} + A_{MR} \quad (15.12)$$

$$WPC = \frac{A_{total}}{f M 365} \quad (15.13)$$

where f is the plant availability and M the plant capacity.

The obtained total annual cost of the SPMD compact plant is \$983/year and the WPC is \$29.9/m^3 as summarized in Table 15.3.

(iii) Other Annual Operating Costs

If electrical energy is used, the annual electric power cost ($A_{electric}$) can be calculated using the following expression:

$$A_{electric} = c\,w\,M\,f\,365 \quad (15.14)$$

where c is the electric cost about \$0.09/kWh and w is the specific consumption of electric power (kWh/m^3).

The annual labour cost (A_{labour}) can be calculated as follows:

$$A_{labour} = g\,M\,f\,365 \quad (15.15)$$

where g is the specific cost of operating labour assumed as \$0.05/m^3 [10].

The annual brine disposal (A_{brine}) is expressed as:

$$A_{brine} = b\,M\,f\,365 \quad (15.16)$$

where b is the specific cost of brine disposal assumed as \$0.0015/m^3 [4].

References

[1] A.G. Fane, R.W. Schofield, C.J.D. Fell, The efficient use of energy in membrane distillation, Desalination 64 (1987) 231–243.

REFERENCES

[2] P.A. Hogan, A.G. Sudjito, G.L. Fane, Morrison, desalination by solar heated membrane distillation, Desalination 81 (1991) 81–90.

[3] S. Bouguecha, B. Hamrouni, M. Dhahbi, Small scale desalination pilots powered by renewable energy sources: Case studies, Desalination 183 (2005) 151–165.

[4] S. Al-Obaidani, E. Curcio, F. Macedonio, G.D. Profio, H. Al-Hinai, E. Drioli, Potential of membrane distillation in seawater desalination: thermal efficiency, sensitivity study and cost estimation, J. Membr. Sci. 323 (2008) 85–98.

[5] L. Gazagnes, S. Cerneaux, M. Persin, E. Prouzet, A. Larbot, Desalination of sodium chloride and seawater with hydrophobic ceramic membranes, Desalination 217 (2007) 260–266.

[6] A. Criscuoli, M.C. Carnevale, E. Drioli, Evaluation of energy requirements in membrane distillation, Chem. Eng. Proc. 47 (2008) 1098–1105.

[7] X. Wang, L. Zhang, H. Yang, H. Chen, Feasibility research of potable water production via solar-heated hollow fiber membrane distillation system, Desalination 247 (2009) 403–411.

[8] M. Khayet, J.I. Mengual, G. Zakrzewska-Trznadel, Direct contact membrane distillation for nuclear desalination. Part. II. Experiments with radioactive solutions, Int. J. Nuclear Desalination 2 (2006) 56–73.

[9] C. Cabassud, D. Wirth, Membrane distillation for water desalination: How to choose an appropriate membrane? Desalination 157 (2003) 307–314.

[10] F. Macedonio, E. Curcio, E. Drioli, Integrated membrane systems for seawater desalination: Energetic and exergetic analysis, economic evaluation, experimental study, Desalination 203 (2007) 260–276.

[11] A. Criscuoli, E. Drioli, Energetic and exergetic analysis of an integrated membrane desalination system, Desalination 124 (1999) 243–249.

[12] J.H. Hanemaaijer, J. van Medevoort, A.E. Jansen, C. Dotremont, E. van Sonsbeek, T. Yuan, L. De Ryck, Memstill membrane distillation – a future desalination technology, Desalination 199 (2006) 175–176.

[13] G.W. Meindersma, C.M. Guijt, A.B. de Haan, Desalination and water recycling by air gap membrane distillation, Desalination 187 (2006) 291–301.

[14] G.C. Sarti, C. Gostoli, S. Bandini, Extraction of organic compounds from aqueous streams by vacuum membrane distillation, J. Membr. Sci. 80 (1993) 21–23.

[15] J. Wang, B. Fan, Z. Luan, D. Qu, X. Peng, D. Hou, Integration of direct contact membrane distillation and recirculating cooling water system for pure water production, J. Cleaner Prod. 16 (2008) 1847–1855.

[16] E. Drioli, F. Laganá, A. Criscuoli, G. Barbieri, Integrated membrane operations in desalination process, Desalination 122 (1999) 141–145.

[17] E. El-Zanati, K.M. El-Khatib, Integrated membrane-based desalination system, Desalination 205 (2007) 15–25.

[18] F. Banat, N. Jwaied, Economic evaluation of desalination by small-scale autonomous solar-powered membrane distillation units, Desalination 220 (2008) 566–573.

[19] K. Schneider, W. Hölz, R. Wollbeck, Membranes and modules for transmembrane distillation, J. Membr. Sci. 39 (1988) 25–42.

[20] J. Koschikowski, M. Wieghaus, M. Rommel, Solar Thermal-driven desalination plants based on membrane distillation, Desalination 156 (2003) 295–304.

[21] F. Banat, N. Jwaied, M. Rommel, J. Koschikowski, M. Weighaus, Desalination by a 'compact SMADES' autonomous solar-powered membrane distillation unit, Desalination 217 (2007) 29–37.

[22] F. Banat, N. Jwaied, M. Rommel, J. Koschikowski, M. Weighaus, Performance evaluation of the 'large SMADES' autonomous desalination solar-driven membrane distillation plant in Aqaba, Jordan, Desalination 217 (2007) 17–28.

[23] H.E.S. Fath, S.M. Elsherbiny, A.A. Hassan, M. Rommel, M. Wieghaus, J. Koschikowski, M. Vatansever, PV and thermally driven small-scale, stand-alone solar desalination systems with very low maintenance needs, Desalination 225 (2008) 58–69.

[24] J. Koschikowski, M. Wieghaus, M. Rommel, V.S. Ortin, B.P. Suarez, J.R.B. Rodríguez, Experimental investigations on solar driven stand-alone membrane distillation systems for remote areas, Desalination 248 (2009) 125–131.

[25] A.M. Alklaibi, The potential of membrane distillation as a stand alone desalination process, Desalination 223 (2008) 375–385.

[26] L. Carlsson, The new generation in seawater desalination SU membrane distillation system, Desalination 45 (1983) 221–222.

[27] M. Khayet, M. Essalhi, C. Armenta-Déu, C. Cojocaru, N. Hilal, Optimization of solar-powered reverse osmosis desalination pilot plant using response surface methodology, Desalination 261 (2010) 284–292.

[28] H. Mehdizadeh, Membrane desalination plants from an energy-exergy point of view, Desalination 191 (2006) 200–209.

[29] R.K. Suri, A.M.R. Al-Marafie, A.A. Al-Homoud, G.P. Maheshwari, Cost-effectiveness of solar water production, Desalination 71 (1989) 165–175.

[30] G. Ahmad, J. Schmid, Feasibility study of brackish water desalination in the Egyptian deserts and rural regions using PV systems, Ener. Conver. Manag. 43 (2002) 2641–2649.

[31] G. Fiorenza, V.K. Sharma, G. Braccio, Techno-economic evaluation of a solar powered water desalination plant, Ener. Conserv. Manag. 44 (2003) 2217–2240.

[32] A. Khater, S. Dannish, M. Al-Ansari, Privatization as a financing alternative for desalination plants in Bahrain, Desalination 97 (1994) 281–290.

[33] J. Ayoub, R. Alward, Water requirements and remote arid areas: The need for small-scale desalination, Desalination 107 (1996) 131–147.

[34] E.E. Delyannis, A. Delyannis, Economics of solar stills, Desalination 52 (1985) 167–176.

[35] J. Kulhanek, T. Onishi, A. Nii, B. Milow, E. Zarza, Advanced MED solar desalination plants configurations, costs, future-seven years experience at the Plataforma Solar de Almeria (Spain), Desalination 108 (1997) 51–58.

[36] A. El-Nashar, M. Samad, The solar desalination plant in Abu Dhabi: 13 years of performance and operating history, Renewable Energy 14 (1998) 236–274.

[37] P. Glueckstern, Potential uses of solar energy for seawater desalination, Desalination 101 (1995) 11–20.

CHAPTER 16

Future Directions in Membrane Distillation

OUTLINE

Introduction	453	Less-Studied MD Research Areas	456
Well-Studied MD Research Areas	455	Perspectives	459

INTRODUCTION

As it is indicated throughout this book, membrane distillation (MD) is still under evaluation and far to fulfil all the expectations although it has been known for more than 40 years and used successfully in numerous application areas principally in desalination. Although the MD results seem very promising at the development stage of laboratory, several attempts of commercialization have failed due to difficulties in engineering aspects. Until now, practically all membrane modules were designed for academic purposes in the research laboratories rather than for industrial use. Moreover, different opinions exist concerning MD industrialization. MD economy has not yet received an adequate focus of research, and for potential commercial application of MD technology, scale-up membrane modules and MD plants should be investigated. Moreover, introduction of innovative MD units is needed.

The MD researchers discuss the high energy consumption of MD units, the difficulties with long-term operation with the simultaneous risk of membrane wetting and fouling, the lack of MD membranes and MD modules and the uncertain energetic and economic costs. Therefore, MD technology is taking more academic role than commercial, although it has some significant advantages over other processes, such as low sensitivity to feed concentration, operating at low temperatures, high resistance to fouling compared with other membrane separation processes, very high rejection factors of non-volatile solutes present in water, high system compactness and use of energy sources such as waste or solar heat.

As it is shown in this book, commercial membranes are often used although these membranes were fabricated for other purposes. Moreover, for the same membrane different operational parameters are reported and the permeate fluxes obtained by different groups seem to disagree sometimes by an order of magnitude. Accurate membrane characteristics such as pore size, porosity, liquid entry pressure, thermal conductivity, etc., should be supplied by the membrane manufacturers.

It is worth noting that the permeate fluxes of an order of magnitude difference were sometimes reported even when the experiments were performed under the same operating conditions using the same commercial membranes. This may be attributed to the different MD modules used. The magnitude of the permeate flux obtained in the MD process should be affected significantly by the module design, the MD configuration and its operating conditions.

Proposals have been made to improve the produced water quantity and quality and to reduce the energy consumption. They include MD hybrid systems in conjunction with other conventional processes such as distillation and pressure-driven membrane processes, use of alternative energy sources such as solar and geothermal energy, and waste heat recovery by installing the MD plant near the nuclear power plant.

During the last 5 years, the MD permeate flux has been increasing without scarifying the rejection factor especially when applied for desalination. Furthermore, the energy consumption has been decreasing, especially by using solar energy systems. As of today, the highest obtainable permeate flux reported for each MD configuration is

- direct contact membrane distillation (DCMD): 145.8 and 116.6 kg/m^2·h using distilled water and 1.3 mol% aqueous sodium chloride (NaCl) solution, respectively (feed temperature 74 °C, permeate temperature 20 °C, commercial membrane 3 ME supplied by 3 M Corporation (Table 2.1 in Chapter 2) [1]).
- sweeping gas membrane distillation (SGMD): 14.4 and 11.2 kg/m^2·h for distilled water used as feed with an electrical conductivity of 9 µS/cm and NaCl feed aqueous solution of 182 µS/cm, respectively (feed temperature 65 °C, air temperature 20 °C, PP Accurel® S6/2 MD020CP2 N membrane module (Table 2.2 in Chapter 2) [2]).
- vacuum membrane distillation (VMD): 372.6, 421.2 and 576.7 kg/m^2·h for the commercial membranes 3 MA, 3 MB and 3 MC, respectively (membranes supplied by 3 M Corporation (see Table 2.1 in Chapter 2), feed distilled water, feed temperature 74 °C, downstream pressure 3 kPa [3]).
- air gap membrane distillation (AGMD): 26.3 kg/m^2·h for the commercial membrane HVHP (see Table 2.1 in Chapter 2) tap water as feed with an electrical conductivity of 297 µS/cm, feed temperature 82 °C, cooling temperature 7 °C and air gap distance 0.8 mm [4].

As can be seen, there is more than one order of magnitude difference between the highest permeate fluxes of the various MD configurations. On the one hand, the future of MD looks bright considering the very high permeate fluxes of the 3 M membranes obtained in DCMD and VMD; while on the other hand, there remains a room for improvement for all other membranes and modules.

It is known that in MD the heat requirements represent a significant part of the process cost and the thermal energy consumption is very sensitive to the feed temperature. It is worth mentioning that some authors are too optimistic regarding the specific energy consumption of MD process and water production costs. For example, costs as low as \$0.26/m^3 were claimed for seawater desalination by MD units with production of 105,000 m^3/day (i.e. Memstill® technology) [5]. However, this technology is still under the pilot scale operation and the promised production cost is yet to be proven by a larger scale operation [5,6]. In most cases analytical details of the MD system, the operating parameters and the procedure followed to calculate the energy consumption are not provided. Most of the studies on economics and energy consumption have been made on a theoretical basis and evaluation of energy consumption based on more realistic system is required. Gazagnes et al. [7] indicated that AGMD process is able to produce a permeate flux of 5.2 $l/m^2 \cdot h$ with a specific energy consumption of about 1 kWh/m^3. Carlsson [8] stated that the specific energy consumption of a MD system can be as low as 1.25 kWh/m^3 for a MD system capable of producing about 5 m^3/day. Cabassud and Wirth [9] found a minimum value of specific energy consumption, when treating 30 g/l feed salt aqueous solution by VMD with a permeate flux of 13 $l/m^2 \cdot h$, to be 1.2 kWh/m^3. On the other hand, it was 1.3 kWh/m^3 when treating 300 g/l feed salt aqueous solution by VMD with a permeate flux of 8.8 l/m^2 h. Other authors reported more than three orders of magnitude higher specific energy consumption. The lowest obtained values of the specific energy consumption by Criscuoli et al. [10] were 3546.3 kWh/m^3 in DCMD and 1108.4 kWh/m^3 in VMD. Wang et al. [11] reported a specific energy consumption value as high as 9079.5 kWh/m^3 for a solar VMD. One of the reasons of the observed discrepancy in the values of the high specific energy consumption is the dimension of the MD system and the use of heat recovery devices.

Some MD studies have been carried out considering heat recovery. However, among them very few studied the effect of the heat recovery on the MD process cost. It was mentioned that heat recovery could reduce the MD process cost up to 4-fold but rigorous analysis has not been reported yet.

Practically all studies on MD hybrid and/or heat recovery systems were investigated at laboratory scale. Even the pilot scale performance was examined in few cases using power plants; they are currently still under investigation. More multidisciplinary studies should be carried out in order to find an effective way for industrial MD systems.

In the next section, we present some research areas where MD process is well studied and other research areas that would be of great research interest for the future direction of MD technology. The presented research areas are only recommendations in order to avoid repetitions of what is already investigated. Even though we encourage young researchers who are on the verge of launching their career in MD through the areas that we think are interesting and yet unexplored, there may be some other MD areas of which we are unaware.

WELL-STUDIED MD RESEARCH AREAS

A number of MD studies have been performed until now in a laboratory scale using small membrane modules. Most of the publications are concerned with experimental studies on the effects of the process operating conditions and theoretical models including heat and mass transfer mechanisms. Similarly, designed membrane modules in the form of either shell and tube or plate and frame were used. As a matter of fact, the effects of the operating parameters on the MD performance have been intensively studied and well understood. Nevertheless, some parameters are repeatedly studied and almost the same conclusions are obtained, especially for DCMD. On the other hand, configurations other than DCMD (e.g. SGMD and thermostatic SGMD) have rarely been investigated. More studies should be performed on the latter two configurations especially in terms of the effect of gas flow rate (i.e. circulation velocity) on the MD performance. It should also be noted that, while the effects of MD operating parameters on the MD performance have been intensively studied as thoroughly indicated in chapters 10–13, the effects of interactions between MD operating parameters are not well investigated.

The short-term performance in desalting solutions containing low concentration of salts (<30 g/l) have been thoroughly investigated by using different MD configurations, and the decrease of the MD permeate flux with the increase of the salt concentration is well understood. However, the long-term MD operation and the treatment of saline solutions with high salt concentrations up to 300 g/l are needed.

Some commercial membranes such as TF200 (Gelman), GVHP (Millipore) and PP Accurel® S6/2 (see Table 2.2 in Chapter 2) have been characterized and used extensively in various MD studies. However, other commercial membranes, which may exhibit even higher MD performance, have rarely been used or never tested in MD. Some of the examples are 3 MA, 3 MB, 3 MC, 3 MD and 3 ME supplied by 3 M Corporation (see Table 2.1 in Chapter 2).

During the last years, modelling of MD has been one of the main subjects of investigation. Kinetic theory of gases has been used extensively for the prediction of the MD permeate flux by means of the combined Knudsen/ordinary diffusion mechanism (Eq. (10.25) in Chapter 10) in DCMD and SGMD and Knudsen mechanism in VMD configuration. It is necessary to apply alternative theories.

It should be pointed out that most of the developed theoretical models involve at least one adjustment parameter, frequently pore tortuosity factor, to predict the MD permeate flux. In most cases, semi-empirical models were made applying the heat and mass transfer correlations originally developed for rigid nonporous heat exchangers, which has been questioned by a group of authors.

The MD permeate flux, the temperature polarization coefficient and the concentration polarization coefficient are commonly predicted taking into account the average (of feed and permeate) temperatures and concentrations, which are often calculated from the experimentally obtained temperatures and concentrations at the inlets and outlets of the membrane modules. It is known that both temperature and concentration change along the membrane module and consequently the local MD permeate flux, temperature and polarization coefficient also change. Therefore, adequate mathematical models should be developed to determine the temperature and concentration profiles along the membrane module to predict the overall MD permeate flux, temperature polarization coefficient and concentration polarization coefficient more precisely.

In general, a number of MD studies deal with desalination using NaCl solutions of different concentrations, but much less studies have been carried out using solutions containing organic compounds (esters, ethers, chlorinated hydrocarbons, aromatic compounds, etc.) and on the environmental impact of the brine wastes.

Most MD applications are concerned about the treatment of aqueous solutions for distilled water production and less attention is focused on the concentration and recovery of valuable components of solution, e.g. food processing.

Until now all MD applications have been focused on the treatment of aqueous solutions, in which water is the major component. The treatment of non-aqueous solutions by MD would be of great interest to test the applicability of the transport mechanisms through the dry pores of the membrane and the effects of operating conditions on the MD performance, which are already known for aqueous solutions.

LESS-STUDIED MD RESEARCH AREAS

As indicated throughout this book, a number of potentially interesting research areas within the MD field have been practically ignored, and others have been studied but only superficially. All of them may be very interesting subjects for further MD development. Some of the potential research activities are outlined below while more detailed discussions are made throughout this book.

- Design and fabrication of flat sheet and hollow fibre membranes specifically for MD applications are still lacking. Improved MD membranes with specific morphology and characteristics are highly demanded. Membranes with different pore sizes, porosities, thicknesses and materials as well as other novel features are required to carry out systematic MD studies for better understanding of mass transport in different MD configurations, thereby enabling to supply information necessary for the improvement of the MD performance and for MD industrialization. Examples of interesting membranes would be dual layer hydrophobic/hydrophilic or tri-layered hydrophobic/hydrophilic/hydrophobic membranes, nanostructured membranes using different materials, modified inorganic and metallic membranes (i.e. hydrophobic inorganic and metallic membranes).
- Despite the number of works carried out on the fabrication of membranes for MD applications, very few research groups have investigated the effects of membrane preparation parameters on MD performance. The effects on morphological and structural characteristics and MD performance have not been fully elucidated due to many interrelating variables involved in the membrane fabrication processes, especially for hollow fibre membranes. Even the effect of an individual parameter (e.g. air gap distance, shear rate or flow rate of the polymer solution in the dry/wet spinning process, coagulation temperature and solvent evaporation temperature of phase inversion membranes with surface modifying macromolecules, ambient parameters, etc.) on the membrane structure and MD performance has not been studied satisfactorily yet. Furthermore, membrane materials different from those already studied, including polyvinylidene fluoride, should be tested.
- The relationship between the membrane structure (physical and chemical) and the transport phenomena deserves more intensive investigation in MD research. Computational strategies would play an important role.
- Design and fabrication of modules with outstanding performance are necessary for different MD configurations. For example, thorough investigations on the fabrication of spiral wound modules for MD have not been carried out in the academic environment. Design, fabrication and testing of solar MD module are another new research area in MD. Theoretical studies which supplement the experimental studies are necessary to optimize membrane fabrication and MD module design parameters.
- Novel membrane characterization techniques should be utilized in MD. For example, optical techniques using high magnification camera or microscope combined with camera or video recorder can be employed to observe in situ microscopic views of possible MD membrane fouling or particle deposition near the membrane surface.
- The effects of membrane pore size and pore size distribution on the AGMD performance have not been deeply investigated. The temperature and concentration profile along the membrane module length must be considered especially when the air gap width is narrow and the feed temperature is high. The model based on the Stefan–Maxwell equations must be checked for membranes of different materials and characterization parameters (e.g. different pore sizes) not only for AGMD configuration but also for TSGMD. Few studies using hollow fibre membranes in AGMD were performed.

- To predict the permeate flux of SGMD configuration, the pore size distribution has not yet been considered. So was the dusty gas model, even though the latter model combines all available transport mechanisms.
- An extensive study on the concentration polarization in SGMD has not yet been performed. This should be done for aqueous solutions containing non-volatile solutes as well as organic volatile compounds.
- Like DCMD, crystallization may occur in SGMD when treating feed aqueous solutions above their saturation limit. However, SGMD crystallization, scale formation fouling or crystallization fouling have not yet been studied.
- Other applications of SGMD should be tested such as removal/concentration of organic compounds, e.g. esters, ethers, chlorinated hydrocarbons and aromatic compounds, from their dilute organic/water mixtures.
- In VMD configuration, heat transfer by conduction through the membrane should be thoroughly investigated to confirm that the internal heat loss by conduction through the membrane can be neglected.
- When treating solutions with components that have strong 'affinity' to the membrane material, the transport mechanism through the matrix of the membrane, namely solution-diffusion flow, may have a significant effect on the MD performance. Systematic studies are needed to clarify this issue and to know the effect of membrane material on the performance of different MD configurations.
- Long-term MD operations, membrane ageing and all types of fouling phenomena should be studied using different MD configurations.
- Optimization of MD systems and membrane modules should be made in order to study the interactions between operating parameters, to increase MD performance and to decrease energy consumption. This can be done using design of experiments, response surface methodology or artificial neural network.
- Adequate and detailed comparisons of different MD configurations in their performance (permeate flux, separation, energy consumption, costs, easy control and scale up, etc.) are the matters of further investigations.
- Detailed economical analysis of different MD systems and pilot plants are required.
- More attention should be paid to MD microfluidic channels or miniaturized MD technology to improve the MD efficiency.
- More and novel MD hybrid processes should be developed to make the process economically more feasible with more advanced waste-stream treatment processes not only for feed aqueous solutions containing non-volatile solutes but also those containing volatile solutes. For example, the integration of MD with multistage flash process or with different types of multi-effect distillers has not yet been attempted.
- Various energy recovery devices have to be used to decrease specific energy consumption and costs.
- MD is often coupled to solar energy systems. Other renewable energy sources such as wind energy or wave energy may also be of great interest for MD applications. Coupling renewable energy systems to SGMD process has not yet been attempted.
- Solar flat-plate thermal collectors have been considered in MD. Other thermal collectors such as parabolic solar concentrators and spherical collectors incorporating heat storage tanks should also be tested.
- Similar to other separation processes, multistage MD processes should be developed.
- A system combining two or more MD configurations and also other separation processes has not yet been considered. This should be a subject of investigation for

researchers working with aqueous solutions containing both volatile and non-volatile components.

Researchers must have a close look at the studies in already established membrane separation processes such as RO, pervaporation, nanofiltration, etc., and emerging novel technologies such as renewable energy systems, functional and molecular imprints materials, nanotechnology, etc., trying to incorporate innovative findings to MD technology.

PERSPECTIVES

In general, there are a plenty of opportunities for the MD separation process in all modern industrial sectors. Taking the advantages of the MD technology over the other water treatment processes into consideration, the perspectives for the future of MD are very promising especially in desalination and environmental applications. Recent developments of MD solar pilot plants permit to affirm that new wide perspectives in MD for a sustainable industrial growth are possible. However, much more intensive and continuous research efforts are needed in both basic and applied research operations, where the primary objectives should be MD process development methods, decrease of production costs with low energy consumption and less waste generation, and high flexibility and easy scaling up for the construction of competitive and efficient innovative pilot plants.

Despite the great potential of MD process, it is still far to fulfil all the expectations. To overcome the existing barriers, the authors of this book are convinced that the possible developments of MD are related with the different research lines indicated in the previous sections. One of the most interesting MD developments is related to the possibility of integrating various membrane operations in the same industrial cycle for a global efficiency improvement in terms of product quality, costs, energy consumption, etc. Authors should look for other multidisciplinary innovative and interesting research areas for MD applications taking into account the possible advantages and drawbacks of this technology.

The actual cost of membranes for MD is expensive to compete with the price of other membrane separation processes such as RO and the actual water production cost seems also higher than RO. Very few membrane modules are being commercialized by some companies. Therefore, there is an urge to start research focussing on the development of new, cheap, efficient and compact membrane modules with different dimensions for MD applications in order to promote industrial applications of MD.

One of the authors (T.M.) witnessed rapid commercialization of RO almost a half century ago, soon after the announcement of asymmetric cellulose acetate membrane for seawater desalination and the fundamental work that has followed. In contrast to RO, the fundamental research has preceded commercialization in case of MD. It is believed that time has matured for MD to be commercialized and there are a number of reasons that support the authors' confidence in this technology. Among others, the progress in the development of alternative energy sources in general and solar energy in particular makes us hope that MD will have access to energy of reasonable price in the near future. For example, a large demand in the solar panel shall inevitably bring down the solar energy costs considerably. However, it was only recent when RO desalination technology became financially profitable despite its long history. Similarly, it may be indeed a *challenging but very rewarding* task to make the MD process technically reliable and financially profitable.

References

[1] K.W. Lawson, D.R. Lloyd, Membrane distillation: II. Direct contact MD, J. Membr. Sci. 120 (1996) 123–133.

[2] M. Khayet, P. Godino, J.I. Mengual, Theoretical and experimental studies on desalination using sweeping gas membrane method, Desalination 157 (2003) 297–305.

[3] K.W. Lawson, D.R. Lloyd, Membrane Distillation. I. Module design and performance evaluation using vacuum membrane distillation, J. Membr. Sci. 120 (1996) 111–121.

[4] F.A. Banat, J. Simandl, Theoretical and experimental study in membrane distillation, Desalination 95 (1994) 39–52.

[5] G.W. Meindersma, C.M. Guijt, A.B. de Haan, Desalination and water recycling by air gap membrane distillation, Desalination 187 (2006) 291–301.

[6] J.H. Hanemaaijer, J. van Medevoort, A.E. Jansen, C. Dotremont, E. van Sonsbeek, T. Yuan, L. De Ryck, Memstill membrane distillation – a future desalination technology, Desalination 199 (2006) 175–176.

[7] L. Gazagnes, S. Cerneaux, M. Persin, E. Prouzet, A. Larbot, Desalination of sodium chloride and seawater with hydrophobic ceramic membranes, Desalination 217 (2007) 260–266.

[8] L. Carlsson, The new generation in seawater desalination SU membrane distillation system, Desalination 45 (1983) 221–222.

[9] C. Cabassud, D. Wirth, Membrane distillation for water desalination: how to choose an appropriate membrane? Desalination 157 (2003) 307–314.

[10] A. Criscuoli, M.C. Carnevale, E. Drioli, Evaluation of energy requirements in membrane distillation, Chem. Eng. Proc. 47 (2008) 1098–1105.

[11] X. Wang, L. Zhang, H. Yang, H. Chen, Feasibility research of potable water production via solar-heated hollow fiber membrane distillation system, Desalination 247 (2009) 403–411.

Index

Note: Page numbers followed by f indicate figures and t indicate tables.

A

ABB, 228–229
Accelerated precipitation softening (APS), 407–409, 408f, 409f
Accurel membranes, 237, 268f, 282, 278, 282–283, 329t–330t
  in fruit juice concentration, 285
  SGMD with, 318–319, 299t
  studies using, 456
  VMD withp0420, 343
Acetone, 330
ACM. *See* Aspen Custom Modeler
Adsorptive distillation, 392
AFM. *See* Atomic force microscopy
AGMD. *See* Air Gap Membrane Distillation
Air gap distance, 74, 75f, 76f, 381f, 381–382
Air Gap Membrane Distillation (AGMD), 2–4, 3f, 7–8, 295–297, 361–367
  advantages of, 363
  applications of, 363, 367, 384, 394–395
    aqueous alcohol solution treatment, 391–392, 391f
    azeotropic mixture breaking with, 367, 392–393, 393t
    ethanol recovery, 367, 391–392
    food processing, 367, 390
    geothermal waste desalination, 386
    HI and H_2SO_4 concentration, 394–395
    isotope separation, 394
    lithium bromide and, 385–386
    mandarin juice concentration, 390, 390f
    milk concentration, 390
    in solar units, 384–390, 387f, 388f
    in thermal cogeneration, 395
    VOC extraction, 394
    volatile compound separation, 367
  ceramic membranes for, 364, 365t–366t, 383–384
  cost evaluations of, 442, 446
  cylindrical membrane modules for, 231–232
  desalination with, 130, 384–390
    fluidized bed crystallizers in, 386
    geothermal waste and, 386
    heat recovery and, 434
    seawater, 385
    spiral wound membrane modules for, 228
  geothermal energy and, 425, 436
  heat recovery in, 433
  LEP in, 383–384
  mass transfer in, 361, 371f, 372
  membrane porosity and, 384
  in Memstill, 442–443
  permeate flux in, 365t–366t, 368, 371f
    feed temperature and, 374–375
    highest obtained, 454
    membrane thermal conductivity and, 382
    non-volatile solute concentration and, 376, 378
    pore size and, 385
    volatile solutes and, 376–377, 377f
  plate-and-frame membrane modules for, 230–232, 233f
  pore size in, 383–384, 383f
    predicting effect of, 456–459
  process condition effects, 374–376
    air gap width, 381–382, 381f
    cold side flow rte, 380
    coolant temperature, 378–380
    feed flow rate, 377–378
    feed temperature, 374–376
    membrane parameters, 382–384
    non-condensable gases, 380–381
    solute concentration, 376f, 376–377
    volatile solute selectivity and, 375–376
  process schematic, 362f
  salt rejection factor in, 375, 383–384
  SGSPs and, 423
  solar energy for, 367, 446
  spacer-filled channels in, 378, 379
  spiral wound modules for, 228, 241–242, 363, 387–389
  SPMD with, 418–419, 421
  superhydrophobic glass membranes for, 364
  temperature polarization in, 375
  theoretical models for, 367–374
    dusty gas model in, 371
    Fickian models in, 379–380
    heat transport in, 372–373, 382
    Knudsen diffusion in, 374
    mass flux in, 368–369
    Stefan diffusion in, 367–368, 370
    Stefan-Maxwell equations in, 369–371, 374, 379–380
    temperature drop along module length, 374
    vapour transport in, 367
  thermal efficiency of, 375, 379
  ultrasonic stimulation in, 384
  vapour transport in, 10
  VLE and, 367
Alternative energy sources, 418
  photovoltaic panels, 418–421
  salt-gradient solar ponds, 423–425
  solar absorbers integrated in membrane modules, 421–423
  solar thermal collectors, 418–421
Ambient permeate temperature, 270
Ammonia removal, 319, 320, 342, 353
Anergy, 440
Apple juice, 285

APS. *See* Accelerated precipitation softening
Aroma compound recovery, 348–349
Arsenic removal, 286–287
Aspen Custom Modeler (ACM), 389–390
Aspen Plus, 374
Asymmetric membranes, categories of, 122
Asymmetric porous membranes, 90
Atomic force microscopy (AFM), 27, 61, 76f, 148–149, 150f, 191
  membrane surface structure observation with, 204, 205f
  pore size determination with, 203–209
  pore size distribution from, 207f, 208f
  pore tortuosity evaluation with, 211
  sector analysis of membrane image from, 206f
  surface porosity determination with, 211
Azeotropic mixtures, 13
  AGMD for breaking, 367, 392–393, 393t
  DCMD for breaking, 287–288
  pervaporation and, 392
  TSGMD and, 312–313, 320
Azo-dyes, 412

B

Baffles, 239f
Beijing Institute of Plastic Research, 320
Benzene, 25, 346, 442
Biological solution concentration, 287
Bioreactors, 411
Biotechnology, 12–13, 287
Black currant juice, 348–349
Boron rejection, 287
Bubble point method, 193–196, 195f
Bureau for Reclamation, USA, 229

C

CA. *See* Cellulose acetate
Capillary distillation, 392
Capillary membranes, 238f
  baffles for, 239f
  permeate flux of, 275t, 280t
  in plate-and-frame modules, 227, 235–237
  in tubular modules, 227, 236–241

Carbon tetrachloride, 127
Casting solutions
  lithium chloride in, 23
  water in, 23–24
Catalysis, 412–413
Celgard LLC, 237
Cellulose acetate (CA), 18, 123
  plasma grafting to, 215
  in radiation graft polymerization, 127, 141–142
Cellulose nitrate (CT), 18
Cellulose triacetate (CTA), 18, 24–25
Ceramic membranes
  for AGMD, 364, 365t–366t, 383–384
  grafting, 123, 129–130, 130f
    process parameter effects in, 143–144, 143f
  tubular fibre, 31, 125
  for VMD, 327
CFD. *See* Computational fluid dynamics
Chemical stability, 220–221
Chloroform, 341–342, 342f, 346, 356
Coagulants
  bath temperature of, 55–56, 56f
  in dry/wet spinning, 64–65
  PVDF surface morphology and, 71–73, 72f
  spinning parameters and, 65–66
Co-extrusion spinning, 123–125, 141, 158–159
Combined liquid-liquid and solid-liquid phase separation, 100–102
Commercial membranes, 19–21, 20t, 21t
Composite membranes, 6–7
  bi-layered, 28–33
  desalination with, 123, 137
  dry/jet wet spinning of, 33
  hydrophobic/hydrophilic flat sheet, 29–30, 30f
  hydrophobic/hydrophilic hollow fibre, 33, 124–125, 127, 159
  in MD, 123
  methods for fabricating, 122
  multi-layered, 28–33, 122
  NF with, 122
  pore wetting and, 31–32, 125–126
  principles of formation, 126
  PVDF in, 124
  SMMs and, 122, 124
  thickness of, 221

UF with, 122
in VMD, 133
Computational fluid dynamics (CFD), 82–83, 243
Concentration polarization, 9
  in DCMD, 264, 265
  predicting, 456
  in SGMD, 307, 311–312
Contact angle measurements, 148–149, 215–216, 216f
Coolant liquid treatment, 352–353
Cooling tower and MD hybrid systems, 415–416
Copolymer membranes
  flat sheet, 27–28, 28f
  hollow fibre, 27–28, 28f, 63–64
Corrugations
  in hollow fibre membranes, 83–85, 84f
  spinnerets for, 83–85, 84f
Cost analysis, 449
  amortization or fixed charges, 449
  annual operating costs, 449
  capital cost, 449
  membrane replacement costs, 450
  O&M costs, 450
  plant availability, 450
  pretreatment costs, 450
Cost estimations
  for MD, 430
  of WPC for different MD systems, 448t–449t
Cost evaluations, 429–431
  of AGMD
    with geothermal, 442
    with solar power, 446
  of DCMD, 441, 443–446
  of desalination, 444–445, 444f
  factors for, 430–431
  heat recovery and, 431–432, 445–446
  of hybrid systems, 442–444
  MD compared to other systems, 441–449
  of Memstill, 442–443
  of RO, 442–443, 443t
  of SPMD, 441–442, 446–449, 447t
  of VMD, 442
Coulter Porometer II, 198
Coulter porosimetry, 198, 200f
CR510 data loggers, 421–422
Crystallization fouling
  in DCMD, 274
  in SGMD, 312

CT. *See* Cellulose nitrate
CTA. *See* Cellulose triacetate
Cylindrical membrane modules, 231f, 230
  for AGMD, 231–232
Cylindrical pores, 260–261

D

DAP. *See* Diamyl phthalate
DBP. *See* Dibutyl phthalate
DCMD. *See* Direct Contact Membrane Distillation
DEHP. *See* Di(2-ethylhexyl) phthalate
Desalination
  with AGMD, 130, 384–390
    fluidized bed crystallizers in, 386
    geothermal waste and, 386
    heat recovery and, 434
    seawater, 385
    spiral wound membrane modules for, 228
  composite membranes for, 123, 137
  concentrate waste stream from, 283
  cost development of, 444–445, 444f
  with DCMD, 123–125, 130, 137, 138, 250, 278–284, 384–385
    plate-and-frame modules for, 228
    radiation graft polymerization of membranes for, 142
  energy analysis of, 438t
  geothermal waste, 386
  heat recovery and, 433
  for high-purity water production, 12
  with MED, 413
  membrane crystallization and, 283
  with MSF, 413
  nuclear, 415
  RO and MD hybrid systems for, 402–403
  with SGMD, 6
  ship-based, 416–418, 417f
  solar-powered, 350–351, 351f, 363, 386–387, 387f, 433
  with VMD, 124–125, 144–145, 146f, 326–327, 341, 341, 349–351
Dialysis, 42
Diamyl phthalate (DAP), 102
Dibutyl phthalate (DBP), 99, 115
DIDP. *See* Diisodecyl phthalate
Diesel waste heat, 416–418
Di(2-ethylhexyl) phthalate (DEHP), 115

Differential scanning calorimeter (DSC), 96–97, 222
Diffusion distillation, 392
Diffusion-induced phase separation (DIPS), 90, 102, 117
Diisodecyl phthalate (DIDP), 113f
Dimethyl acetamide (DMAC), 22–23, 25
  LEP and, 213
  membrane morphology and, 54
  polymer concentration and, 51, 51f
  wet spinning with, 62
Dimethyl formamide (DMF), 22–23, 54, 175
Dimethyl sulfoxide (DMSO), 24–25
Dip coating. *See* Surface coating
Diphenyl ether (DPE), 96
DIPS. *See* Diffusion-induced phase separation
Direct Contact Membrane Distillation (DCMD), 2–5, 3f, 9, 249–254
  advantages of, 288
  air in membrane in, 10–11
  applications of, 251, 254, 278, 286–288
    ammonia separation, 320
    arsenic removal, 286–287
    azeotropic mixture breaking, 287–288
    biological solutions, 287
    biotechnology, 287
    boron rejection, 287
    desalination and crystallization, 278–284
    fruit juice concentration, 284–285
    pharmaceutical waste water treatment with, 286
    radioactive waste water treatment with, 286
    traditional Chinese medicine extracts, 286
    VOCs and, 287
    wastewater treatment, 285–286
  commercial membranes in, 252–253, 253t
  composite hydrophobic/hydrophilic flat sheet membranes for, 29–30, 30f, 137
  concentration polarization in, 264, 265
  conventional distillation and, 250
  copolymer membranes in, 27

cost analysis of, 441, 443–446
deaerated systems for, 259
demineralizing water with, 403
desalination with, 123–125, 130, 137, 138, 250, 278–284, 384–385
  plate-and-frame modules for, 228
  radiation graft polymerization of membranes for, 142
dual layer hydrophilic/hydrophobic hollow fibre membranes for, 63
energy analysis of, 436–437
ethanol production and, 411
exergetic efficiency of, 441
experimental setup for, 252f
feed solution temperature in, 250–251
flat sheet membranes for, 279t
heat and mass transfer in, 251f
heat recovery and, 434, 435f, 436f, 444–445
in hybrid systems, 401–402, 412, 413
LEP in, 250–252
long-term performance of, 282–283, 282f
mean free path in, 10
MED coupled with, 413–414
membrane materials for, 123
MSF coupled with, 413–414
nuclear-powered, 415
permeate flux in, 265–266, 267f, 268, 268f
  feed flow rates and, 271–272, 272f
  flat sheet membranes for, 279t
  highest obtained, 454
  hollow fibre membranes for, 280t–281t
  long-term performance and, 282, 282f
plate-and-frame membrane modules for, 228, 230–231, 231f, 234f, 235f, 236
process condition effects, 268
  feed and permeate flow rates, 271–272, 272f
  feed temperature, 268–269, 269f, 310f
  permeate temperature, 269–271, 269f, 270f, 271f
  solute concentration, 272–275, 272f
process of, 250

Direct Contact Membrane Distillation (DCMD) (*Continued*)
  radioactive waste decontamination with, 12
  RCW, recirculating cooling tower, 415–416, 416f, 443
  RO combined with, 404–405, 409, 409f
  SGSPs, solar gradient solar ponds, 423–425
  solar absorbers in modules for, 421–422, 422f
  solar stills coupled with, 414
  spacer-filled channels in modules for, 236
  spiral wound modules for, 241–242
  SPMD with, 418–421
  temperature polarization in, 263–265, 271, 273
  theoretical models of, 252–254
    assumptions for, 254
    dusty gas model, 260
    heat transfer in, 261–268, 267f, 268f
    heat transfer resistances in, 262, 263
    interconnected pores and, 260–261
    Kinetic Theory of Gases and, 256
    mass transfer in, 254–261
    permeate flux in, 265–266, 267f, 268f
    pore size in, 258–259, 266
    solute mass transfer coefficient in, 264
    temperature profile in, 262
    tortuosity factor in, 266
  thermal efficiency of, 268f, 269, 273
  transport mechanism in, 257, 257f
  vacuum-enhanced, 405
  VMD compared with, 328
  volatile solutes in, 251–252
Direct Osmosis (DO)
  in hybrid systems, 405–406
  wastewater treatment with, 406
District heating, 395
DMAC. *See* Dimethyl acetamide
DMF. *See* Dimethyl formamide
DMSO. *See* Dimethyl sulfoxide
DO. *See* Direct Osmosis
DPE. *See* Diphenyl ether
Dry/jet wet spinning, 26, 33
  co-extrusion, 124–125, 158–159, 158f
  electro-spinning compared with, 164
  PVDF in, 124–125, 127, 158–159
Dry/wet phase inversion spinning, 60
Dry/wet spinning, 24–25, 27, 60, 62–64, 78f
  coagulation in, 64–65
  LEP of membranes from, 214f
  PVDF in, 34f
  PVDF-HFP in, 27, 63–64
  shear induced orientation in, 80–81
DSC. *See* Differential scanning calorimeter
Dual layer hollow fibre membranes, 62–63
Dusty gas model, 10
  AGMD and, 371
  DCMD and, 260
  SGMD and, 304
  VMD and, 331
Dynamic mechanical thermoanalyzers, 219
Dynamometers, 219, 219f

E

Echinacea extract, 406, 407
Economics, 429–431
Effective porosity
  methods for measuring, 209–211
  modeling of, 210
Electrolux, 228–229
Electron microscopy, 200–203
Electro-spinning, 163–164
  bead and droplet formation in, 172, 173
  conventional spinning compared with, 164
  effects of system and process parameters, 170–185
  fibre crystallinity in, 179
  fibre packing in, 169
  fibre structure modifications in, 169
  field profile in, 168f
  phase separation and, 169
  polymers for, 164
  pore formation in, 175–176
  process of, 167–169
  process parameter effects in, 178–185
    ambient parameters, 184–185, 184f
    applied voltage, 178–179, 179f
    collector-needle tip gap distance, 180–182, 182f
    needle tip, 180
    polymer flow rate, 179–180, 181f
    post-treatment, 182
    temperature, 180
  system parameter effects, 172–178
    electrical conductivity of solution, 176–178
    molecular weight, 174–175
    polymer concentration, 172–174, 173f
    solvents, 175–176, 176f, 177f
  systems for, 166–169, 167f
  water contact angles of membranes from, 215
Electrostatic spinning. *See* Electro-spinning
Energy analysis, 429–431
  of DCMD, 436–437
  of desalination, 438t
  of MD, 429–431, 438t
    energy efficiency and, 431–432, 433
    heat recovery and, 431–434, 432f, 435f, 436, 436f
    temperature polarization and, 431
    thermal efficiency and, 431
  of MD hybrid systems, 439–440, 440f
  of MSF, 438t
  of RO, 438t
  of VMD, 436–439, 439f
Energy efficiency, 431–433
Energy recovery device (ERD), 441, 445f
Enka AG, 7–8, 228, 237
EPDM FDA. *See* Ethylene propylene diene monomer
ERD. *See* Energy recovery device
Ethanol, 72–73, 72f, 346, 347, 348
  AGMD for recovery of, 367, 391–392
  DCMD and production of, 411
  post-treatment with, 81
  PVDF selectivity for, 24–25
Ethyl acetate, 346
Ethylene glycol, 26
Ethylene propylene diene monomer (EPDM FDA), 229
Exergetic efficiency, 440, 441
Exergy, 440, 441
Extractants, 92

F

FALP membranes, 384
  AGMD with, 365t–366t
FAS. *See* Fluoroalkylsilanes
Fermentation, 411, 411f
FESEM. *See* Field emission scanning electron microscopy
FGLP membranes, 257–258
FHLP membranes
  AGMD with, 365t–366t, 386
  DCMD with, 257–258
Fibre membranes. *See* Hollow fibre membranes
Field emission scanning electron microscopy (FESEM), 191, 200–203, 203f
Film theory, 338
Flat sheet membranes
  commercial, 194t, 210t
  composite hydrophobic/hydrophilic flat sheet, 29–30, 30f
  copolymer, 27, 28f
  for DCMD, 279t
  for MF, 407
  for MOD, 407
  penetration pressure determination for, 212, 213
  in plate-and-frame modules, 227, 230–236
  porosity measurement of, 209, 210
  porous hydrophobic
    kinetics considerations in formation of, 47–48
    Loeb-Sourirajan method for, 43
    principles of formation of, 43–48
    thermodynamics considerations in formation of, 44–47
  porous hydrophobic TIPS
    diluents and structure of, 108–109
    drying and structure of, 111
    extractants and structure of, 110–111
    nucleating agents and structure of, 110
    polymer molecular weight and structure of, 108, 108f
    polymer type/concentration and structure of, 105–108, 106f, 107f
    polymer/diluent melting temperature and time effects on structure of, 109
    preparation of, 103–104, 103f
    process parameter effects on structure of, 105–112
    quenching conditions and structure of, 109–110, 110f
    stretching and structure of, 111–112
  process parameters and structure of, 48–58, 68–85
    coagulation bath temperature, 55–56, 56f
    drying effects, 56–58
    non-solvent additive concentration effects, 51–54, 51f, 52f, 53f, 54f
    non-solvent additive types and, 54, 55f
    polymer concentration, 50–51, 50f, 51f
    polymer type, 48, 49f
    solvent evaporation time, 55–56, 57f
    solvent or solvent mixture effects, 54
  SGMD with, 317
  single hydrophobic layer, 22–24
    copolymer, 27–28
    microporous, 24, 24t
  in spiral wound modules, 227, 241–242, 241f
  for VMD, 327
Flat-plate solar collectors, 421
Fluidized bed crystallizers, 386
Fluoroalkylsilanes (FAS), 125, 129–130
Fluorosilanes, 217, 217f
Food processing, 12, 367, 390
Fouling, 274
Fourier transform infrared spectroscopic analysis (FTIR), 222
Fraunhofer ISE, 241–242, 433
Fruit juice concentration
  with AGMD, 390, 390f
  with DCMD, 284–285
  with VMD, 348–349
FTIR. *See* Fourier transform infrared spectroscopic analysis

G

Gained output ratio (GOR), 433
Gas permeation, 42
  SMM characterization and, 148–149
Gas permeation test, 191–193, 192f
Gas separation, multi-layered composite membranes for, 122
Gas/liquid mass transfer, 237
Gel spinning, 164
Geothermal energy, 425, 436, 442
Geothermal waste desalination, 386
Ginseng extract concentration, 353
GOR. *See* Gained output ratio
Gore & Associated Co, 6, 228, 237, 241, 363
Gore-Tex, 6, 228
Gore-Tex Membrane Distillation, 363
GVHP membranes, 193, 194t, 198, 202, 203, 207, 207f, 208t, 210t
  AGMD with, 383–384, 365t–366t
  studies using, 456
  thermal conductivity of, 262
  VMD with, 329t–330t

H

Halogenated VOCs, 346
HDPE. *See* High-density polyethylene
Heat recovery, 431–432, 432f
  AGMD and, 433
  arrangements of system for, 432–433, 433f
  DCMD and, 434, 435f, 436f, 444–445
  geothermal energy and, 436
  MD cost evaluation and, 445–446, 446
  process cost and, 432, 455
  scale of studies, 455
  solar thermal collectors and, 433
  solar-powered desalination and, 433
  VMD and, 436
Heat transfer, 262, 263–264
  in AGMD, 372–373, 382
  in DCMD, 261–268, 267f, 268f
  in MD, 11
  in SGMD, 305–306, 307
  in TIPS, 91
  vapour, 262
  in VMD, 11–12, 335–336, 338, 456–459
Heat treatment, 124
  of hollow fiber membranes, 29, 81
  of nanofibre membranes, 182–183, 183f
  of pervaporation membranes, 81
  of PVDF, 29, 37–38
  of RO membranes, 81
Heavy metals, 286

Helium, 380–381
High-density polyethylene (HDPE), 114
Hollow fibre membranes
  baffles for, 239f
  ceramic, 125
  coating polymers in, 29
  commercially available, 60
  composite hydrophobic/hydrophilic, 29, 30f, 33, 124–125
  copolymer, 27, 28f, 63–64
  corrugations in, 83–85, 84f
  for DCMD, 280t
  development of, 59–64
  dry/jet wet spinning for, 26, 33
  dry/wet phase inversion spinning technique, 18
  dry/wet spinning technique for, 24–25, 27, 60, 62, 78f
    coagulation in, 64–65
    LEP of, 214f
    PVDF in, 34f
    PVDF-HFP in, 27, 63–64
    shear induced orientation in, 80–81
  dual porous hydrophobic/hydrophilic layers, 62
  future directions in, 85
  heat treatment of, 29, 81
  melt-extruded/cold-stretching for, 27
  module designs for, 238
  packing density of, 238–239, 243
  penetration pressure determination for, 212–213
  permeate flux of, 275t
  for pervaporation, 407
  phase inversion and preparation of, 60–61
  pore size in, 77
  porosity measurement of, 209, 210
  porous hydrophobic TIPS
    air gap distance and structure of, 116
    cold stretching and, 117
    diluents and structure of, 30f, 113–115, 114f
    HDPE in, 114, 114f
    polymer molecular weight, type, and concentration effects on structure of, 112–113
    post-treatment and structure of, 117
    preparation of, 104, 104f
    process parameter effects on structure of, 112
    spin draw ratio and, 116–117
    spinning temperature and structure of, 115–116
    take-up speed and, 116
    water bath temperature and, 116, 116f
  porous tubular supports in, 25–26
  PP in, 32, 60, 125, 326
  PPESK in, 326–327
  preparation of, 60
  principles of formation of, 64–68
  process parameters and structure of, 68–69
    additive content and, 69–71
    air gap distance, 74, 75f, 76f
    bore flow rate, 73–74
    coagulants, 71–73, 72f
    polymer concentration and, 69–71, 71f
    polymer solution flow rate, 80–81
    post-treatment effects on, 81
    shear stress and, 67–68
    spinneret design and, 81–85, 82f
    take-up speed, 78–79, 79f
  PVDF in, 24–26, 29, 60, 326
    shear rate and structure of, 80
    silicone-coated, 131, 132
  PVDF-HFP in, 27
  single hydrophobic layer, 24–27, 61–62
  spinnerets for, 80
    corrugations and, 83–85, 84f
    design and fabrication of, 81
    die swell and, 67–68, 68f
    dope flow angle in, 82, 83f
    dual layer, 65, 67f
    elongation rates in, 82–83
    membrane structure and, 81–85, 82f
    shear rates in, 82–83
    shear stress in, 67–68
    single layer, 65f
    smart, 83
  spinning parameters of, 64–65, 66t
    bore flow rate, 65–66, 73–74
    coagulation liquids, 65–66
    membrane structure and, 68–69
  spinning system for, 65f
  in tubular modules, 228, 236–241
  for VMD, 326, 347
  wet spinning technique for, 25, 60, 62, 73f, 78f
  shear induced orientation in, 80–81
HVHP membranes, 193, 194t, 198, 202–203, 207–208, 208f, 208t, 210t
  AGMD with, 365t–366t, 383–384, 386
  thermal conductivity of, 262
Hybrid systems. See Membrane distillation hybrid systems
Hydrogen carbonate, 395
Hydrogen production, 394–395
Hydrophilic polymers, in SMM process parameters
  concentration of, 149–151, 150f
  type of, 151–152
Hydrophilic surface coating, 123, 133–134, 145–146
Hydrophobic metallic membranes, 298
Hydrophobic porous membranes, 356
  measuring porosity of, 209–210
  preparation of, 190
Hydrophobic surface coating, 123, 131–133, 144–145
Hydrophobicity, 17–18

I

Idaho National Engineering and Environmental Laboratory (INEEL), 409, 410t
Immersion precipitation, 91
Industrial & Engineering Chemistry Process Design Development, 5
INEEL. See Idaho National Engineering and Environmental Laboratory
Institut Européen des Membranes, 298
Instron dynamometer, 219
International Atomic Energy Agency, 415
International Union of Pure and Applied Chemistry (IUPAC), 4
IPA. See Isopropyl alcohol; Isopropyl amine
iPP. See Isotactic polypropylene
iPP/DAP system, 102

iPP/DPE system, 96, 97f, 99f, 101f
iPP/TA system, 96
Isopropanol, 346
Isopropyl alcohol (IPA), 311, 312–313, 319, 347
Isopropyl amine (IPA), 409
Isostress model, 261–262
Isotactic polypropylene (iPP), 90–91, 96, 103f, 106f, 328
IUPAC. *See* International Union of Pure and Applied Chemistry

K

Kinetic Theory of Gases, 254, 256
Knudsen diffusion, 260, 302, 337, 374
Knudsen flow model, 10–11
  in AGMD models, 374
  in DCMD models, 256–259
  in SGMD models, 301, 302, 302f
  in VMD models, 334
Knudsen/Poiseuille transition region, 259

L

LBARS. *See* Lithium bromide adsorption refrigeration system
LEP. *See* Liquid entry pressure
Lewis cells, 234f, 235
Liqui-Cel Extra-Flow module, 237
Liquid entry pressure (*LEP*), 2
  in AGMD, 383–384
  in DCMD, 250, 251–252
  for hollow fibre membranes, 214f
  penetration pressure determination and, 211–216, 212f
  pore size and, 213
  in SGMD, 315
  SMM characterization and, 148–149
  in VMD, 325
Liquid-liquid phase separation, 92, 97–98, 109
Liquid-phase precipitation (LPP), 409–410, 410f
Lithium bromide, 385–386
Lithium bromide adsorption refrigeration system (LBARS), 352

Lithium chloride, 23, 25, 53, 53f, 54f
  LEP and, 192
  mechanical properties of PVDF membranes with, 219
  in wet spinning, 62
Lithium perchlorate trihydrate, 54
LIX54, 210
Loeb-Sourirajan method, 43
LPP. *See* Liquid-phase precipitation

M

Magnetic stirrers, 250
Mandarin juice, 390, 390f
MAS. *See* Membrane Air Stripping
MD. *See* Membrane Distillation
MDC. *See* Membrane distillation crystallization
ME. *See* Membrane evaporation
Mechanical stability, 218–220, 219f
MED. *See* Multi-effect distillation
Melt spinning, 164
Melt-extruded/cold-stretching, 27, 326
Membranes. *See also Specific types*
  AFM of
    pore size distribution from, 207f, 208f
    sector analysis, 206f
    surface structure, 204f, 205f
  asymmetric porous, 90
  categories of asymmetric, 122
  categorizing, 90
  ceramic tubular fibre, 31, 125
  characteristics of, 18, 189–191, 208t
  characterization methods for, 191, 221–222, 456–459
  chemical stability of, 220–221
  commercial, 19–21, 20t, 21t, 193, 210t
  composite, 6–7
    bi-layered, 28–35
    desalination with, 123, 137
    dry/jet wet spinning of, 33
    hydrophobic/hydrophilic flat sheet, 29–30, 30f
    hydrophobic/hydrophilic hollow fibre, 33, 124–125, 130f, 159
    in MD, 123
    methods for fabricating, 122
    multi-layered, 28–35, 122
    NF with, 122
    pore wetting and, 31–32, 125–126
    principles of formation, 126
    PVDF in, 124
    SMMs and, 122, 124
    thickness of, 221
    UF with, 122
    in VMD, 133
  copolymer
    flat sheet, 27–28, 28f
    hollow fibre, 27–28, 28f, 63–64
  cost of, 459
  deaeration of, 331
  defining, 41–43
  engineering for MD, 35–38
  flat sheet single hydrophobic layer, 22–24
    copolymer, 27
    microporous, 24, 24t
    heat-treated, 124
  hollow fibre single hydrophobic layer, 24–27
  hydrophobic metallic, 298
  hydrophobic porous, 190
  hydrophobicity of, 17–18
  Kinetic Theory of Gases and predicting performance of, 254
  laboratory-made, 21–35, 22f
  materials for, 18, 19, 35–38
    for DCMD, 123
  mechanical stability of, 218–220, 219f
  MF membranes, 18–19
  modified, 122–126, 141–159
    casting hydrophobic polymer over porous supports, 134–136, 135f, 136f, 146–148, 147f
    ceramic membrane grafting, 129–130, 130f, 143–144, 143f
    co-extrusion spinning, 141, 158–159
    hydrophilic surface coating, 145–146
    hydrophobic surface coating, 144–145
    methods for, 126–140
    plasma polymerization, 122, 123, 127–129, 128f, 142–143, 143f
    porous hydrophobic/hydrophilic composite membrane formation, 126
    process parameter effects on structure and performance of, 141–159
    radiation graft polymerization, 126–127, 141–142, 141f

Membranes (*Continued*)
    SMMs, 136—138, 148—149
        surface coating, 130—133,
            144—145
    nano-structured, 166
    optical techniques for inspection of,
        221—222
    penetration pressure determination
        for, 211—216, 212f, 214f
    performance parameters for,
        190—191
    phase inversion and preparation of,
        90—92
    plasma-modified CN, 28—29
    polymer-coated, 124
    pore size determination, 191—209
        AFM for, 203—209
        bubble point method, 95f,
            193—196
        electron microscopy for, 200—203,
            201f, 202f
        gas permeation test, 191—193
        mercury porosimetry, 199—200,
            201f
        wet/dry flow method, 193—198,
            197f, 199f
    pore tortuosity evaluation for, 211
    porosity and effective porosity
        in AGMD, 384
        methods for measuring, 209—211
        modeling, 210—211
    porous hydrophobic flat sheet
        kinetics considerations in
            formation of, 47—48
        principles of formation, 43—48
        thermodynamics considerations
            in formation of, 44—47
    preparation techniques for, 21—27,
        37—38
    requirements for, 17, 35—37, 190
    research in design and fabrication of,
        456—459
    roughness parameters for, 204—206
    surface pore density of, 204
    symmetric porous, 90—91
    synthetic, 42
    synthetic polymeric, 42, 43
    thermal conductivity of, 217, 218f
    thermal degradation measurements
        of, 217, 217f
    thermal stability tests of, 216—218
    track-etched, 34—35, 35f
    tubular, 235—236, 317

    types of, 41—43
    VLE and, 17—18
Membrane Air Stripping (MAS),
    239—240, 298, 320—321. *See
    also* Sweeping Gas Membrane
    Distillation
Membrane crystallization, 273—274,
    274f, 283—284
Membrane Distillation (MD). *See also
    specific configurations*
    advantages of, 1—2
    applications of, 12—13
        alternative energy sources and,
            418—425
        azeotropic mixture separation,
            13
        biotechnology, 12—13
        food processing, 12
        high-purity water production
            with, 12
        radioactive waste water
            processing, 414, 415
        VOC extraction, 13
        wastewater treatment with, 12
    categorizing membranes for, 90
    commercialization of, 459
    composite membranes in, 123
    concept of, 2—4
    configurations for, 2—4, 3f
    continuing research in, 453—455
    cost analysis for, 449—450
        amortization or fixed charges,
            449—450
        annual operating costs, 449
        capital cost, 449
        membrane replacement costs, 450
        O&M costs, 450
        plant availability, 450
        pretreatment costs, 450
    cost estimations for, 429, 448t—449t
    cost evaluations of, 429—431,
        441—449
    defining, 1
    diffusion area in, 11
    economics of, 429—431
    energy analysis in, 429—431, 438t
        energy efficiency and, 431—432,
            433
        heat recovery and, 431—434, 436,
            432f
        temperature polarization and, 431
        thermal efficiency and, 431
    feed temperatures for, 12

    fouling in, 274
    geothermal energy and, 425
    heat recovery in, 431—433, 432f
    heat requirements in, 454—455
    heat transfer in, 11
    historical survey of, 5—9
    increasing membrane permeability
        in, 331
    less-studied research areas in,
        456—459
    membrane engineering and material
        selection, 35—38
    microfluidic channels for, 354
    modeling of, 456
    module design for, 242—243
    multi-layered composite
        membranes for, 122
    nanofibre membranes in, 165—166
    nomenclature in, 4—5
    nuclear-powered, 414—415
    optimization of, 400
    photovoltaic panels and, 418—421
    pore wetting in, 31—32
    in pressure-driven separation
        plants, 400, 400f, 401
    research activity in, 8—9, 8f, 9f
    RO brine treatment with, 405
    salt-gradient solar ponds and,
        423—425
    scaling up, 453
    solar thermal collectors and,
        418—421
    transport mechanism in, 9—12
    well-studied research areas in,
        455—456
    wetting criteria for, 213
Membrane distillation crystallization
    (MDC), 283—284, 284f
Membrane distillation hybrid
    systems, 399—401, 407—418
    APS in, 407—408, 408f, 409f
    catalysis and, 412—413
    cooling towers and, 415—416
    cost evaluations of, 442, 443—444
    DCMD in, 401—402, 412, 413
    diesel waste heat and, 416—418
    energy analysis of, 439—440, 440f
    fermentation and, 411, 411f
    LPP in, 409—411, 410f
    MED in, 413—414, 414f
    with membrane processes, 401—407
        DO, 405—406
        NF, 402—405, 404f

RO, 402–405, 442
UF, 401, 401f, 402t
MF in, 406–407, 406f
MOD in, 406, 406f, 407
MSF in, 413–414
nuclear-powered, 414–415
pervaporation in, 400, 406–407, 406f
precipitation in, 407–418
SGSP in, 423
traditional distillation in, 413–414
VMD in, 404
wastewater treatment with, 401–402, 402t
Membrane evaporation (ME), 298
Membrane modules, 227–230
choice of, 229–230
cylindrical, 230, 231f
design of, 242–243
hollow fibre membranes and designs for, 238
for Memstill, 240f
permeate temperature changes in, 270
plate-and-frame, 227, 231f, 234f, 230–236
requirements for, 242–243
research on design and fabrication of, 456–459
for SGMD, 240
shell-and-tube, 227, 236–241, 236f
solar absorbers integrated with, 421–423
spiral wound, 227, 241–242, 241f, 387–389
temperature drop in, 374
for TSGMD, 239f
types of, 227
Membrane osmotic distillation (MOD), 405–406
flat sheet membranes for, 407
in MD hybrid systems, 406, 406f, 407
wastewater treatment with, 406
Membrane scaling, 283
Membrane transport processes, 249–250
Membrane wetting, water contact angles and, 214–215
Memstill, 229, 240–241, 240f, 363, 434, 442–443, 443t
Mercury porosimetry, 199, 201f
MESS. *See* Multi-effect solar stills
Methanol, 131
Methyl acetate, 346

Methyl tert-butyl ether (MTBE), 328, 344, 346
Methyltrimethoxysilane (MTMS), 132
MF. *See* Microfiltration
Microdistillation, 354
Microdyn Modulbau GmbH & Co. KG, 237
Microdyn-Nadir GmbH, 237
Microfiltration (MF), 1–2, 42, 62, 298
AFM of membranes for, 203
flat sheet membranes for, 407
in MD hybrid systems, 406–407, 406f
MD membranes and, 18–19
photocatalysis and, 413
Microfluidic channels, 354
Milk concentration, 390
Millipore membranes, 193
MOD. *See* Membrane osmotic distillation
Modified membranes, 122–126
methods for, 126–141
casting hydrophobic polymer over porous supports, 134–136, 135f, 136f
ceramic membrane grafting, 123, 129–130, 130f
co-extrusion spinning, 141
plasma polymerization, 122, 123, 125, 127–129, 128f, 143f
porous hydrophobic/hydrophilic composite membrane formation, 126
radiation graft polymerization, 126–141
SMMs, 136–138
surface coating, 130–131, 133–134
plasma-modified, 28–29
process parameter effects on structure and performance of, 141–159
casting hydrophobic polymer over porous supports, 146–148, 147f
ceramic membrane grafting, 143–144, 143f
co-extrusion spinning, 158–159
hydrophilic surface coating, 145–146
hydrophobic surface coating, 144–145
plasma polymerization, 142–143

radiation graft polymerization, 141–142, 141f
SMMs, 148–149
surface coating, 144–145
Molecular diffusion, 257, 258, 260, 302, 302f
MSF. *See* Multi-stage flash
MTBE, 328, 344, 346
MTMS. *See* Methyltrimethoxysilane
Multi-effect distillation, 429
Multi-effect distillation (MED), 240–241, 413, 414f
Multi-effect solar stills (MESS), 438t
Multilab ESCA 3000, 202
Multi-stage flash (MSF), 240–241, 413, 429, 438t
Must, 348

N

Nanofibre membranes, 33–34, 34f
benefits of, 164–165
electro-spinning process parameter effects on, 178–179
ambient parameters, 184–185, 184f
applied voltage, 178, 179
collector-needle tip gap distance, 180–182, 182f
needle tip, 180
polymer flow rate, 179–180, 181f
post-treatment, 182–184
temperature, 180
electro-spinning system parameter effects on, 172–174
electrical conductivity of solution, 176–178
molecular weight, 174–175
polymer concentration, 172–174, 173f
solvents, 175–176, 176f, 177f
heat treatment of, 182–183, 183f
in MD, 165–166
pore formation in, 175–176
principles of formation, 166–170
PVDF in, 170–171, 171f, 180–182
AGMD with, 364
solvent effects on structure of, 175
system and process parameters and structure of, 170–185
Nano-fibres, 163–166
applications of, 164, 165f
electro-spinning for, 163–164

Nano-fibres (*Continued*)
  surface area of, 164
Nanofiltration (NF), 1–2, 42
  AFM of membranes for, 203
  composite membranes for, 122
  demineralizing water with, 403
  in hybrid systems, 402–405, 404f
  multi-layered composite
    membranes for, 122
  photocatalysis and, 413
  pretreatment with, 405
  tubular modules for, 403
  UF combined with, 403
  VMD compared with, 328
Nano-scale materials, 163
Nanoscope III, 203–204
Nano-structured materials, 163, 166
Nanotechnology, 163
Netherlands Organization for
    Applied Scientific Research
    (TNO), 229, 240–241, 363, 434
New Products Development Pty Ltd,
    407
NF. *See* Nanofiltration
N-methyl-1-pyrrolidone (NMP), 26,
    62, 70f, 124–125
NMP. *See* N-methyl-1-pyrrolidone
NMR. *See* Nuclear magnetic
    resonance
NPP. *See* Nuclear power plants
Nuclear desalination, 415
Nuclear fuel reprocessing, 409
Nuclear magnetic resonance (NMR),
    222
Nuclear power plants (NPP), 415
Nuclear-powered MD, 414–415
Nucleating agents, 110

O

Octafluorocyclobutane (OFCB),
    28–29, 123–124, 128–129
OD. *See* Osmotic distillation
OFCB. *See* Octafluorocyclobutane
Olive mill wastewaters (OMW), 286
OMD. *See* Osmotic membrane
    distillation
OMW. *See* Olive mill wastewaters
Optical contact angle meter, 215–216
Optical techniques for membrane
    inspection, 221–222
Orange juice, 285
Ordinary molecular diffusion
    model, 10

Osmonics, 348–349
Osmotic distillation (OD), 31–32,
    237, 275–278, 405–406. *See
    also* Membrane osmotic
    distillation
  defining, 275–276
  mass transfer in, 277, 277f
  operating temperatures for, 276
  volatile compound concentration
    with, 276–277
Osmotic membrane distillation
    (OMD), 277

P

PAN. *See* Polyacrylonitrile
Parabolic solar concentrators, 421
PC. *See* Polycarbonates
PE. *See* Polyethylene
Pear aroma compound, 349
PEG. *See* Polyethylene glycol
PEI. *See* Polyetherimide
Penetration pressure determination,
    211–216, 214f, 276f
Permeate flux
  in AGMD, 368, 365t–366t, 371f
    feed temperature and, 374–375
    highest obtained, 454
    membrane thermal conductivity
      and, 382
    non-volatile solute concentration
      and, 376, 378
    pore size and, 385
    volatile solutes and, 376–377, 377f
  in capillary and hollow fibre
    membranes, 275t, 280t–281t
  in DCMD, 265–266, 268, 268f, 308
    flat sheet membranes for, 279t
    highest obtained, 454
    hollow fibre membranes for,
      280t–281t
    long-term performance and, 282,
      282–283, 282f
  feed concentration and, 273
  feed temperature and, 268–269
  in hollow fibre membranes, 275t
  membranes thickness and, 382
  permeate temperature and, 269–270
  in SGMD, 298–300, 299t
    highest obtained, 454
    organic volatile solute
      concentration and, 312–313
    pore size distribution and, 303
    predicting, 456–459

  sweeping gas flow rate and, 318
  theoretical, 300–303
  of TF200 membrane, 276f
  in VMD, 133, 328, 329t–330t
    feed temperature and, 325, 338,
      340f
    highest obtained, 454
Pervaporation (PV), 42. *See also*
    Sweeping gas pervaporation
  azeotropic mixtures and, 392
  feed temperature for, 407
  heat treatment of membranes for, 81
  hollow fibre membranes for, 407
  in hybrid systems, 400, 406–407,
    406f
  multi-layered composite
    membranes for, 122
  VMD compared with, 325, 334,
    354–357, 355f
  VOC extraction with, 355
PES. *See* Polyethersulfone
PET. *See* Polyethylene terephthalate
Pharmaceutical wastewater
    treatment, 286
Phase inversion. *See also* Thermally
    induced phase inversion
  defining, 90
  dry/wet spinning and, 60
  hollow fibre membrane preparation
    and, 60–61
  mechanical properties of
    membranes from, 219–220
  membrane preparation with, 90–92
  non-solvent additive concentration
    and, 51–54
  polymer concentration and, 50–51
  polymer types and, 48–50
  PVDF and PVDF-TFE in, 48–50, 49f
  thermally induced, 60
Phase separation. *See also* Diffusion-
    induced phase separation;
    Thermally induced phase
    separation
  combined liquid-liquid and solid-
    liquid, 100–102
  electro-spinning and, 169
  liquid-liquid, 92, 97–98, 109
  methods for, 90
  solid-liquid, 98–99, 100f, 109
Photocatalysis, 412–413
Photodegradation, 412
Photovoltaics (PV), 350, 386–387,
    418–421

Plasma deposition, 122
Plasma grafting, 215
Plasma polymerization, 122, 123, 127–129, 128f
  discharge time in, 142–143, 143f
  membrane properties and, 142–143
  pore size and, 142
Plasma-modified CN membrane, 28–29
Plate-and-frame membrane modules, 227, 230–236, 232f, 363
  for AGMD, 230, 231–232, 233f
  capillary membranes in, 227, 235–236, 237
  for DCMD, 230–231, 232f, 234f, 235f, 236
  for SGMD, 230, 232f
  spacer-filled channels in, 236
  for VMD, 230–231, 234f
PMMA. See Poly(methyl methacrylate)
PMMA-1,4-butanediol, 95. See also Poly(methyl methacrylate)
PMMA-sulfolane, 95
PMSP. See Poly(1-trimethylsilyl-t-propyne)
Poiseuille flow model, 10, 259, 260, 301–302, 301f, 357
Polyacrylonitrile (PAN), 24–25, 63, 124–125
Polyallylamine, 123
Polycarbonates (PC), 18
Poly(dimethylsiloxane), 407
Polyesters (PST), 18
Polyetherimide (PEI), 29–31
  LEP and, 213
  SMM-modified membranes, 124, 149, 151–152, 157, 157f
Polyethersulfone (PES), 63, 151–152
  surface-modified, 327
Polyethylene (PE), 5, 18, 60
  melt-extruded/cold-stretching membranes from, 326
Polyethylene glycol (PEG), 32, 63–64, 125–126
Polyethylene terephthalate (PET), 35f
Polyketone, 29, 124, 131
Polymer coating, 124
Polymer spinning solutions, 64–65, 69–71
  flow rate of, 80–81

Poly(methyl methacrylate) (PMMA), 95
Poly(phenylene oxide) (PPO), 24–25
Poly(phthalazinone ether sulfone ketone) (PPESK), 32–33, 132
  in hollow fibre membranes, 326–327
  surface coating and, 124–125, 144–145, 145f
  VMD with membranes from, 347
Polypropylene (PP), 5, 18, 19. See also Isotactic polypropylene; Syndiotactic polypropylene
  commercial membranes made from, 19, 21, 60
  in flat sheet membranes, 317, 407
  in hollow fibre membranes, 32, 60, 125, 326
  melt-extruded/cold-stretching membranes from, 326
  MOD with membranes from, 407
  SGMD membranes from, 317
  in SPMD systems, 419–421
  TIPS membranes from, 90–91, 99
  in tubular membranes, 317
Polystyrene (PS)
  electro-spinning of, 173–174
  plasma grafting of, 215
Polysulfone (PS), 24–25, 123, 407
Polytetrafluoroethylene (PTFE), 5, 6, 19, 125–126, 228
  AGMD with membranes from, 383–384, 365t–366t
  commercial membranes made from, 19, 21, 60
  in hollow fibre membranes, 60
  hydrophilic solution coating and, 133–134, 135f
  SGMD membranes from, 317
  in solar absorber-integrated modules, 421–422
  in spiral wound membrane modules, 241–242, 363
  in SPMD systems, 419–421
  thermal conductivity of membranes from, 262
  water contact angles and, 215
Polytrifluoropropylsiloxane, 124–125, 326–327
Poly(1-trimethylsilyl-t-propyne) (PMSP), 29, 131

Polyvinyl alcohol (PVA), 32
Polyvinyl chloride (PVC), 5
  ship-based desalination with membranes from, 416–418
  solvent effects on structure of nano-fibre membranes from, 175
Polyvinyl pyrrolidone, 327
Polyvinylacetate (PVAC), 178
Polyvinylidene fluoride (PVDF), 6, 11, 18, 19, 42
  chloroform separation and membranes from, 71–73
  coagulants and surface morphology of, 71, 72f
  co-extrusion dry/jet wet spinning of, 124–125, 158–159, 158f
  commercial membranes made from, 19, 60
  in composite membranes, 124
  dry/wet spinning of, 78f
  electro-spun, water contact angles of, 215–216
  ethanol post-treatment of, 81
  ethanol selectivity of, 24–25
  in flat sheet membranes, 22–23
  heat treating of, 29
  in hollow fibre membranes, 24–25, 26, 29, 60, 326
  shear rate and structure of, 80
  silicone-coated, 131, 132
  mechanical properties of membranes from, 219
  in nano-fibre membranes, 170–171, 171f, 180–182
  AGMD with, 364, 365t–366t
  solvent effects on structure of, 175
  non-solvent additives and, 51–53, 52f, 54f
  phase inversion and, 48–50, 49f
  plasma-modified CN membrane compared with, 28–29
  polymer coating of, 29
  polymer concentration and, 50–51, 51f
  spinning solution for, 61
  additives and morphology of, 70f
  concentration of, 69–71, 71f
  in SPMD systems, 419–421
  TCA separation and membranes from, 347

Polyvinylidene fluoride (PVDF) (*Continued*)
  thermal conductivity of membranes from, 262
  in TIPS membranes, 99, 100f
  UF membranes from, 61
  wet spinning of, 78f
  wet/dry flow method and, 196
Poly(vinylidene fluoride-co-tetrafluoroethylene) (PVDF-TFE), 19, 27
  mechanical properties of membranes from, 219–220
  non-solvent additives and, 54, 55f
  phase inversion and, 48–50, 49f, 50
  polymer concentration and, 50, 50f
Poly(vinylidene-hexalfuoropropylene) (PVDF-HFP), 19, 27, 28f
  dry/wet spinning hollow fibre membranes from, 27, 63–64
  LEP of hollow fibre membranes from, 214f
  spinning solution concentration and membrane structure of, 69–71
Poly(vinyl-pyrrolidone) (PVP), 26, 61
Pore size, 77, 456–459
  in AGMD, 383–384, 383f, 385, 456–459
  chloroform separation and, 346
  determining, 191–209
    AFM for, 203–209
    bubble point method, 193–196, 195f
    electron microscopy for, 200–203, 201f, 203f
    gas permeation test for, 191–209, 192f
    mercury porosimetry, 199–200
    wet/dry flow method, 193–198, 195f, 197f, 199f
  distribution of, 221
    AFM and determining, 207f, 208f
    DCMD and, 258
    SGMD and, 303
    VMD and, 332, 335f
  LEP and, 202
  mass transport mechanisms and, 258
  mean, 221
  plasma polymerization and, 142

  in SGMD, 301–302, 303
  theoretical modeling and, 258–259, 266, 301–302
  water contact angles and, 215
Pore space interconnectivity, 260–261
Pore tortuosity
  in DCMD theoretical models, 266
  evaluating, 35–38
  in SGMD theoretical models, 317
Pore wetting
  alcohol solutions and, 391
  composite membranes and, 31–32, 125–126
  membrane characteristics and, 189
  penetration pressure determination and, 211–212
  pressure-driven systems and, 401
POREFLON membranes, 311, 315, 299t
Porosity
  of AGMD membranes, 384
  methods for measuring, 209–211
  surface, 211
Porous hydrophobic flat sheet membranes
  kinetics considerations, 47–48
  Loeb-Sourirajan method for, 43
  principles of formation, 43–48
  thermodynamics considerations, 44–47
Porous hydrophobic TIPS membranes
  phase diagrams for, 97, 97f, 98f
  principles of formation, 92–104
    combined liquid-liquid and solid-liquid phase separation, 100–104
    flat sheet membrane preparations, 103–104, 103f
    hollow fibre membrane preparation, 103f, 104
    kinetic considerations, 102
    liquid-liquid phase separation, 97–98
    polymer/diluents systems for, 92–93, 94t, 95f, 100–101
    solid-liquid phase separation, 98–99, 100f
    thermodynamic considerations, 93–97, 95f, 96f
  process parameter effects on structure of, 105–111

    air gap distance, 116
    cold stretching, 117
    diluents, 108–109, 112–113, 113f, 114f
    drying and, 111
    extractants and, 110–111
    in flat sheet membranes, 105–108
    hollow fibre membranes, 112–117
    nucleating agents and, 110
    polymer composition, 112–113
    polymer molecular weight and, 108, 108f, 112–113
    polymer type and concentration, 105–108, 106f, 107f, 112–113
    polymer/diluent melting temperature and time, 109
    post-treatment, 117
    quenching conditions and, 109–110, 110f
    spin draw ratio and, 116–117
    spinning temperature and, 115–116
    stretching and, 111
    take-up speed and, 116
    water bath temperature and, 116, 116f
Positron emission tomography, 394
PP. *See* Polypropylene
PP Accurel. *See* Accurel membranes
PP Membrana, 329t–330t
PPESK. *See* Poly(phthalazinone ether sulfone ketone)
PPO. *See* Poly(phenylene oxide)
Precipitation. *See also* Accelerated precipitation softening; Thermally induced phase separation
  immersion, 91
  liquid-phase, 409–410, 410f
  in MD hybrid systems, 407–411
Pressure-driven separation plants, 400, 400f, 401
Primary reverse osmosis (PRO), 407–409
PRO. *See* Primary reverse osmosis
PS. *See* Polystyrene; Polysulfone
PST. *See* Polyesters
PTFE. *See* Polytetrafluoroethylene

PV. *See* Pervaporation; Photovoltaics
PVA. *See* Polyvinyl alcohol
PVAC. *See* Polyvinylacetate
PVC. *See* Polyvinyl chloride
PVDF. *See* Polyvinylidene fluoride
PVDF/DMP membranes, 110f
PVDF-HFP. *See* Poly(vinylidene-hexalfuoropropylene)
PVDF/NMP membranes, 124–125, 136
PVDF/PAN membranes, 124–125
PVDF-TFE. *See* Poly(vinylidene fluoride-co-tetrafluoroethylene)
PVP. *See* Poly(vinyl-pyrrolidone)
Pyd. *See* Pyridine
Pyranometers, 421–422
Pyridine (Pyd), 127, 141–142

R

Radiation graft polymerization, 123, 126–127
  CA in, 127, 141–142
  membrane properties and, 141–142, 141f
Radiation polystyrene grafting, 28–29
Radioactive waste water solutions, 286, 414, 415
RCW. *See* Recirculating cooling water
Rebound resilience, 219
Recirculating cooling water (RCW), 415–416, 416f, 443
Response surface methodology (RSM), 274–275, 374
Reverse osmosis (RO), 1–2, 6, 18, 42, 249–250, 429. *See also* Primary reverse osmosis
  AFM of membranes for, 203
  brine from, 405
  commercialization of, 459
  concentrate discharge from, 407–408
  cost evaluations, 442–443, 443t
  DCMD combined with, 404–405, 409, 409f
  desalination with, 402–403
  energy analysis of, 438t
  geothermal water and, 425
  heat treatment of membranes for, 81
  in hybrid systems, 402–405, 442

multi-layered composite membranes for, 122
  pretreatment with, 403
  solute concentration and, 376
  UF combined with, 403
  VMD coupled with, 328, 404
RO. *See* Reverse osmosis
Roughness parameters, 204–206, 209
RSM. *See* Response surface methodology

S

Saccharomyces cerevisiae, 411
Salinity gradient solar ponds (SGSP), 350–351, 351f
Salt gradient solar ponds (SGSP), 390, 418, 423, 424–425, 424f
Sandia National Laboratory, 228–229
SC. *See* Solar collectors
Scale formation fouling, 274
Scaling, 274, 283, 341
Scanning electron microscopy (SEM), 27, 61, 70f, 79, 79f, 148–149, 191, 200–203, 201f
Scarab Development AB, 228–229, 231–232, 233f, 363, 390, 395
Second World Congress on Desalination and Water Reuse, 228
SEM. *See* Scanning electron microscopy
SGMD. *See* Sweeping Gas Membrane Distillation
SGSP. *See* Salinity gradient solar ponds; Salt gradient solar ponds
SGSP/MD hybrid systems, 423
Shear induced orientation, 80–81
Shell-and-tube membrane modules, 227, 236–241, 236f, 318–319
Ship-based desalination, 416–418, 417f
Sigma 701 Tensiometer, 216
Silicone rubber, 29, 124, 326–327
Simple effect distillation, 413, 429
SMADES project, 418
Smart spinnerets, 83
SMM. *See* Surface-modifying macromolecules
Sodium alginate hydrogel, 133
Solar absorbers, 421–423, 422f

Solar collectors (SC), 351f, 350–351, 418, 419–421, 441–442
Solar distillation, 297
Solar hybrid DCMD modules, 421–422, 422f
Solar power
  AGMD and, 367, 387f, 388f, 384, 446
  capital requirements of, 430
  desalination with, 350–351, 351f, 363, 386–387, 387f, 433
  heat recovery and, 433
  VMD coupled with, 350–351, 351f, 423, 436
Solar stills, 414, 429, 438t
Solar thermal collectors, 418–421, 433
Solar thermal power, 386–390
Solar-powered membrane distillation (SPMD), 418, 419f, 420f
  cost evaluations of, 441–442, 446, 447, 447t
  small-scale, 430
  WPCs of, 430
Solid-liquid phase separation, 98–99, 109, 100f
Solute mass transfer coefficient, 264
Solution coating. *See* Surface coating
Solvent/non-solvent phase inversion, 91
Spacer-filled channels, 236, 378, 379f
Spherical solar collectors, 421
Spinnerets, 80
  corrugations and, 83–85, 84f
  design and fabrication of, 81
  die swell and, 67–68, 68f
  dope flow angle in, 82, 83f
  dual layer, 65, 167f
  elongation rates in, 82–83
  membrane structure and, 81–85, 82f
  shear rates in, 82–83
  shear stress in, 67–68
  single layer, 65f
  smart, 83
Spiral turbulent promoters, 236
Spiral wound membrane modules, 227, 241–242, 241f
  for AGMD, 228, 241–242, 363, 387–389
  for DCMD, 241–242
  flat sheet membranes in, 227, 241–242, 241f
  heat recovery and, 433

Spiral wound membrane modules (*Continued*)
  for RO, 403
SPMD. *See* Solar-powered membrane distillation
sPP. *See* Syndiotactic polypropylene
sPP/DPE system, 96, 97f
SPV. *See* Sweeping gas pervaporation
St. *See* Styrene
Stefan diffusion, 367–368, 370
Stefan-Maxwell-based models, 304–305, 320, 369, 370, 374
Styrene (St), 127
SU Membrane Distillation System, 363
Sucrose solution concentration, 353–354
Sugarcane juice, 285
Sulphur hexafluoride, 380–381
Superhydrophobic glass membranes, 364
Surface coating, 123, 130, 134f
  casting hydrophobic polymer over porous supports, 134–136, 135f, 136f
    process parameter effects, 146–148, 147f
  hydrophilic, 123, 133–134
    process parameter effects, 145–146
  hydrophobic, 123, 131–133
    process parameter effects, 144–145
    VMD desalination performance and, 144–145, 146f
  of PPESK, 124–125, 144–145, 145f
  process parameter effects, 144–145
Surface porosity, 211
Surface-modifying macromolecules (SMM), 19, 29–31, 63, 123
  characterization of, 138–140, 148–149, 140f
  chemical stability of membranes and, 220–221
  composite membranes and, 122, 124
  fluorine and, 220–221
  LEP and, 213
  membrane modification with, 136–138
  membrane preparation with, 140–141
  PEI membranes modified with, 129, 149, 151–152, 157, 157f

PES membranes with, 327
process parameter effects, 148
  evaporation time, 155, 155f, 156f, 157f
  hydrophilic polymer concentration, 149–151, 150f
  hydrophilic polymer type, 151–152
  SMM concentration, 127, 153f
  SMM stoichiometric ratio, 154–155
  SMM type, 152f, 127
  solvent effects, 154
synthesis of, 138, 139f
water contact angles and, 215
Swedish National Development Co, 7–8, 228, 363
Sweeping Gas Membrane Distillation (SGMD), 2–4, 3f, 5, 8, 295–300
  advantages of, 297
  applications of, 318–321
    ammonia removal, 319, 320
    IPA selectivity and, 319
    organic compound separation, 320–321
    research in, 456–459
    wastewater treatment, 319
  crystallization in, 312, 456–459
  defining, 295
  disadvantages of, 313, 320
  gas inlet temperatures in, 313–314
  gas temperature along membrane modules in, 295–297, 306
  LEP in, 315
  module design for, 240
  permeate flux in, 298–302, 299t
    highest obtained, 454
    organic volatile solute concentration and, 312–313
    pore size distribution and, 303
    predicting, 456–459
    sweeping gas flow rate and, 318
    theoretical, 300–303
  plate-and-frame membrane modules for, 230, 232f
  process condition effects, 309–318
    feed flow rate, 314, 316f
    feed temperature, 309, 310f
    gas flow rate, 315–318, 317f
    gas temperature, 313, 313f, 314f
    organic selectivity, 318

organic volatile solute concentration and permeate flux, 312–313
    permeate flux and, 318
    solutes in feed aqueous solution, 311–313, 312f
    temperature polarization in, 317
  process schematic, 296
  seawater desalination with, 6
  solar distillation and, 297
  theoretical models for, 300–309
    concentration polarization coefficients, 307, 311–312
    dusty gas model and, 304
    heat transport in, 305, 307
    Knudsen flow in, 301, 302, 302f
    mass transfer mechanism in, 302–303, 305
    mean free path of volatile molecules in, 301–302
    molecular diffusion in, 302, 302f
    permeate flux in, 300–303
    Poiseuille flow in, 301–302, 301f
    pore size in, 301–302, 303
    Stefan-Maxwell-based models, 304–305, 320
    temperature and concentration profiles, 306–307
    temperature polarization coefficients, 307–309, 311–312, 317
    thermal efficiency in, 305–306
    tortuosity factors in, 317
    transmembrane hydrostatic pressure in, 315
    VLE and, 303–304, 311
    VMD compared with, 328
Sweeping gas pervaporation (SPV), 298
Symmetric porous membranes, 90–91
Syndiotactic polypropylene (sPP), 96, 106f
Synthetic membranes, 42
Synthetic polymeric membranes, 42, 43

T

TA. *See* Tallowamine
Tallowamine (TA), 96
Taurine, 286

TCA. *See* Trichloroethane
TCM. *See* Traditional Chinese Medicine
Tear strength, 219
Teflon. *See* Polytetrafluoroethylene
TEM. *See* Track-etched membranes. *See also* Transmission electron microscopy
Temperature polarization, 9
   in AGMD, 375
   in DCMD, 263–264, 265, 271, 273
   energy analysis and, 431
   feed concentration and, 273, 311–312
   predicting, 456
   in SGMD, 307–309, 311–312, 317
   in VMD, 337, 339, 340f
Tensile stress-strain properties, 219, 220f
Tetrachloroethylene, 346
Tetrahydrofuran (THF), 173–174, 175
TF200 membranes, 198, 194t, 208t, 210t, 257–258
   AGMD with, 383–384, 365t–366t
   permeate flux of, 276f
   SGMD with, 299t, 302, 303, 315–317
   studies using, 456
   VMD with, 329t–330t
TF450 membranes, 193, 198, 194t, 208t, 210t
   AGMD with, 365t–366t
   SGMD with, 302, 303f, 315–317, 299t
TF1000 membranes, 193, 198, 194t, 208t, 210t, 257–258
TFA. *See* Trifluoroacetic acid
TGA. *See* Thermogravimetric analysis
Thermal cogeneration plants, 395
Thermal conductivity, 217–218, 218f
Thermal degradation measurements, 217
Thermal efficiency
   in AGMD, 375, 379
   of DCMD, 268f, 269, 273
   energy analysis and, 431
   feed concentration and, 273
   in MD, 431
   of SGMD, 305–306
   of VMD, 338–339
Thermal energy consumption, 454–455

Thermal membrane distillation, 6–7
Thermal precipitation. *See* Thermally induced phase separation
Thermal stability tests, 216–218
Thermally induced phase inversion (TIP), 60
Thermally induced phase separation (TIPS), 90–91, 117. *See also* Porous hydrophobic TIPS membranes
   advantages of, 92
   basic steps of, 91
   heat transfer in, 91
   liquid-liquid, 92, 97–98
   principles of membrane formation, 92–104
   solid-liquid, 98–99
   for VMD membranes, 328
Thermochemical water-splitting, 394–395
Thermodynamics
   porous hydrophobic flat sheet membranes and, 44–47
   porous hydrophobic TIPS membranes and, 93–97, 95f, 96f
Thermogravimetric analysis (TGA), 143, 217, 217f
Thermostatic Sweeping Gas Membrane Distillation (TSGMD), 2–4, 295–297, 3f, 297f, 298, 356
   azeotropic concentration and, 312–313, 320
   IPA and, 312–313
   modules for, 239f
   organic volatile solute concentration in, 312–313
THF. *See* Tetrahydrofuran
3MA membranes, 329t–330t
3MB membranes, 329t–330t
3MC membranes, 329t–330t
3ME membranes, 278–281
TIP. *See* Thermally induced phase inversion
TIPS. *See* Thermally induced phase separation
Titanium dioxide, 412
TNO. *See* Netherlands Organization for Applied Scientific Research
Toluene, 25

Track-etched membranes (TEM), 34–35, 35f
Traditional Chinese Medicine (TCM), 286–288
Traditional distillation, 413, 429
Transmembrane hydrostatic pressure, 315
Transmission electron microscopy (TEM), 200–201
Transuranic elements (TRU elements), 409–410
1,1,1-Trichloroethane (TCA), 25
Trichloroethane (TCA), 62, 326, 346, 347
Trifluoroacetic acid (TFA)
TRU elements. *See* Transuranic elements
TSGMD. *See* Thermostatic Sweeping Gas Membrane Distillation
Tubular fibre membranes. *See* Hollow fibre membranes
Tubular membranes, 236
   SGMD with, 317
Tubular modules
   capillary membranes in, 227, 236–241
   hollow fibre membranes in, 228, 236–241
   designs for, 238
   packing density of, 238–239, 243
   for NF, 403
   for SGMD, 240
   for TSGMD, 239f
   for UF, 403

U

Uddevalla Finmekanik AB (UFAB), 228–229
UF. *See* Ultrafiltration
UFAB. *See* Uddevalla Finmekanik AB
Ultrafiltration (UF), 1–2, 42
   AFM of membranes for, 203
   composite membranes for, 122
   in hybrid systems, 401–405, 401f, 402t
   multi-layered composite membranes for, 122
   NF combined with, 403
   photocatalysis and, 413
   PVDF membranes for, 61
   RO combined with, 403
   tubular modules for, 403
Ultrasonic stimulation, 384

V

Vacuum Membrane Distillation (VMD), 2–4, 3f, 297, 323–328
- advantages of, 325
- applications of, 328, 346–354
  - alcohol aqueous solution treatment, 347–349
  - ammonia separation, 320, 353
  - aroma compound recovery, 348–349
  - benzene removal, 442
  - coolant liquid treatment, 352–353
  - fruit juice concentration, 348–349
  - ginseng extract concentration, 353
  - lithium bromide adsorption refrigeration, 352–353
  - sucrose solution concentration, 353–354
  - textile wastewater treatment, 351–352
  - VOC extraction, 355, 346–347
- boundary layer resistance in, 11–12
- ceramic membranes for, 327
- composite membranes in, 133
- cost evaluations of, 442
- DCMD compared with, 328
- desalination with, 124–125, 144–145, 146f, 326–327, 341, 349, 351
- energy analysis of, 436–439, 439f
- energy requirements in, 339, 345–346
- flat sheet membranes for, 327
- heat recovery in, 436
- heat transfer in, 11–12, 335–336, 338, 456–459
- hollow fibre membranes for, 326, 347
- hybrid systems with, 404
- hydrophobic surface coating of membranes and, 144, 146f
- LEP in, 325
- mechanical stability of membranes in, 218–219
- membrane preparation for, 325–326, 326
- miniaturization of, 354
- other MD configurations compared with, 328
- permeate flux in, 133, 328, 329t–330t
  - feed temperature and, 339f, 340f, 338
- highest obtained, 454
- pervaporation compared with, 325, 334, 354–357, 355f
- plate-and-frame membrane modules for, 230–231, 234f
- PPESK membranes for, 347
- process condition effects, 338–346
  - downstream pressure, 344–346, 345f
  - feed flow rate, 342–344, 343f
  - feed temperature, 338–342
  - solute concentration, 340–342
  - volatile solutes, 341–342, 342f
- process description, 323–325
- process schematic, 324f
- publications about, 325
- RO coupled with, 328, 404
- scaling in, 341
- SGMD compared with, 328
- SGSPs and, 423
- ship-based systems for, 416–418, 417f
- solar energy coupling to, 350–351, 351f
- SPMD with, 421
- theoretical models for, 328–338
  - dusty gas model in, 331
  - film theory in, 338
  - heat transfer in, 335–336, 338
  - mass transport in, 331, 332, 335–336, 337
  - network model in, 335f, 334–335
  - pore assumptions in, 331
  - pore size distribution and, 332, 336f
  - temperature polarization in, 337, 339, 340f
- thermal efficiency of, 338–339
- TIPS membranes for, 328
- transport mechanisms in, 259
- VLE and, 344
- water as non-solvent additive in, 23–24

Vapor/liquid equilibrium (VLE), 4, 17–18
- AGMD and, 367
- azeotropic mixtures and, 392
- DCMD and, 287–288
- SGMD and, 303–304, 311
- VMD and, 344–345

Vapour heat transfer coefficients, 262

Vinyltrimethylsilicon/carbon tetrafluoride (VTMS/CF4), 28–29, 123–124, 128–129

Viscous flow model, 10
- in VMD models, 331

Vitec 3000, 405

VLE. See Vapor/liquid equilibrium

VMD. See Vacuum Membrane Distillation

VOCs. See Volatile organic compounds

Void volume fraction, 209–210, 209f, 210t

Volatile organic compounds (VOCs), 13, 320–321, 328
- AGMD extraction of, 394
- DCMD and, 287
- halogenated, 346
- pervaporation extraction of, 355
- VMD extraction of, 346–347, 355

VTMS/CF4. See Vinyltrimethylsilicon/carbon tetrafluoride

W

Wastewater treatment, 12
- DCMD for, 285–246
- hybrid systems for, 401–402, 402t
- MOD and DO for, 406
- pharmaceutical, 286
- SGMD for, 319
- textile, 351–352
- VMD for, 351–352

Water contact angles, 215
- measuring, 148–149, 215–216, 216f
- membrane supports and, 215
- membrane wetting and, 214–215
- pore size and, 215
- of PTFE foils, 215
- SMMs and, 215

Water production cost (WPC), 429–430, 446, 446f, 448t–449t
- with DCMD, 441
- of small-scale SPMD, 430
- water recovery factor and, 445

Water recovery factor, 445

Water-soluble carbodiimide (WSC), 125–126

Wet spinning, 25, 60, 62, 73f, 78f. See also Dry/jet wet spinning; Dry/wet spinning
- electro-spinning compared with, 164

shear induced orientation in, 80–81
of VMD membranes, 326
Wet/dry flow method, 193, 195f, 197f, 199f, 202
WL Gore & Associates. *See* Gore & Associated Co
WPC. *See* Water production cost
WSC. *See* Water-soluble carbodiimide

X

XPS. *See* X-ray photoelectron spectroscopy

X-ray diffraction (XRD), 222
X-ray photoelectron spectroscopy (XPS), 28–29, 137, 148–149, 220, 220f
XRD. *See* X-ray diffraction
XZero AB, 228–229, 231–235, 233f, 363